Smith's Elements of Soil Mechanics

Eighth Edition

Ian Smith
Napier University, Edinburgh

Blackwell Science Ltd, a Blackwell Publishing Company
Editorial Offices:
9600 Garsington Road, Oxford OX4 2DQ
Tel: +44 (0)1865 776868
Blackwell Science, Inc., 350 Main Street
Malden, MA 02148-5018, USA
Tel: +1 781 388 8250
Iowa State Press, a Blackwell Publishing Company, 2121 State Avenue, Ames, Iowa 50014-8300, USA
Tel: +1 515 292 0140
Blackwell Science Asia Pty, 54 University Street, Carlton, Victoria 3053, Australia
Tel: +61 (0)3 9347 0300
Blackwell Wissenschafts Verlag,
Kurfürstendamm 57, 10707 Berlin, Germany
Tel: +49 (0)30 32 79 060

First published by Crosby Lockwood & Son Ltd 1968
Second edition 1971
Third edition published by Crosby Lockwood Staples 1974
Reprinted 1975
Fourth edition 1978
Reprinted in paperback 1978
Reprinted by Granada Publishing 1979, 1981
Fifth edition 1982
Reprinted 1983, 1984
Reprinted by Collins Professional and Technical Books 1987
Sixth edition published by BSP Professional Books 1990
Reprinted by Blackwell Science 1990, 1992, 1993, 1994, 1995
Seventh edition published by Blackwell Science 1998
Reprinted 2000, 2003
Eighth edition published by Blackwell Publishing 2006

ISBN10: 1-4051-3370-8
ISBN13: 978-1-4051-3370-1

A catalogue record for this title is available from the British Library

Set in 10/12 pt Times
by Graphicraft Limited, Hong Kong
Printed and bound in Great Britain
by TJ International Ltd, Padstow, Cornwall

For further information on
Blackwell Publishing, visit our website:
www.blackwellpublishing.com

Contents

Preface

It took a whole quarter of a century to get there, but at last, in December 2004, the long-awaited *Eurocode 7: Geotechnical design – Part 1: General rules* was finally published. This European design standard is to be fully adopted by 2010 and its introduction and subsequent implementation mean a radical change to all aspects of geotechnical design across Europe. This affects practising engineers, university lecturers of geotechnical engineering and, of course, all students undertaking courses in civil engineering. The long-established, traditional approaches to geotechnical design must now be moved to one side to make way for the new limit state design approach advocated in Eurocode 7. This is a daunting thought for lecturers and students alike and so I have endeavoured to make the understanding of the new Code as simple and painless as possible by introducing it in this, the eighth edition of *Elements of Soil Mechanics*. Through several worked examples and clear explanatory text, the philosophy of Eurocode 7 and its design approaches are set out covering a whole range of topics including slope stability, retaining walls and shallow and deep foundations.

To help the reader follow many of the principles and worked examples in the book, I have produced a suite of spreadsheets and portable documents to accompany the book. The spreadsheets match up against many of the worked examples and these can be used by the reader to better understand the analysis being adopted in the worked example. This, I hope, will be particularly beneficial to understanding the Eurocode 7 design examples. In addition, I have produced the solutions to the exercises at the end of the chapters as a series of portable document format (pdf) files. All of these files can be freely downloaded from: http://sbe.napier.ac.uk/esm.

Whilst the introduction of Eurocode 7 has driven the bulk of the new material in this edition, I have also updated other aspects of the text throughout. This was done in recognition that some aspects of the book had become dated as a result of the introduction of new methods and standards. Furthermore, the format of the book has been improved to aid readability and thus help the reader in understanding the material. All in all, I believe I have produced a valuable and very up-to-date textbook on soil mechanics from which the learning of the subject should be made easier.

I must thank Dr Andrew Bond, Director of Geocentrix and UK delegate on the Eurocode 7 committee, for his feedback during the preparation of the material for the chapters dealing with Eurocode 7. Also, thanks must go to my colleague Dr John McDougall for his advice on the revisions I have made to the chapter on unsaturated soils.

G. N. Smith, 1927–2002

In April 2002 my father died. This edition of the book would not have been written had it not been for the popularity of the earlier editions that he wrote, and I am grateful that I had the opportunity to write this edition based on his previous accomplishment. This edition is as much his work as mine.

Ian Smith
October 2005

Notation Index

The following is a list of the more important symbols used in the text.

A	Area, pore pressure coefficient
A'	Effective foundation area
A_b	Area of base of pile
A_r	Area ratio
A_s	Area of surface of embedded length of pile shaft
B	Width, diameter, pore pressure coefficient
B'	Effective foundation width
C	Cohesive force, constant
C_C	Compression index, soil compressibility
C_r	Static cone resistance
C_s	Constant of compressibility
C_u	Uniformity coefficient
C_v	Void fluid compressibility
D	Diameter, depth factor, embedded length of pile
D_r	Relative density
D_{10}	Effective particle size
E	Modulus of elasticity, efficiency of pile group
E_d	Eurocode 7 design value of effect of actions
$E_{dst;d}$	Eurocode 7 design value of effect of destabilising actions
$E_{stb;d}$	Eurocode 7 design value of effect of stabilising actions
F	Factor of safety
F_b	Factor of safety on pile base resistance
$F_{c;d}$	Eurocode 7 design axial compression load on a pile
F_d	Eurocode 7 design value of an action
F_{rep}	Eurocode 7 representative value of an action
F_s	Factor of safety on pile shaft resistance
$G_{dst;d}$	Eurocode 7 design value of destabilising permanent vertical action (uplift)
G_s	Particle specific gravity
$G_{stb;d}$	Eurocode 7 design value of stabilising permanent vertical action (uplift)
$G'_{stb;d}$	Eurocode 7 design value of stabilising permanent vertical action (heave)
GWL	Groundwater level
H	Thickness, height, horizontal load
I	Index, moment of inertia
I_L	Liquidity index
I_P	Plasticity index
I_σ	Vertical stress influence factor

K	Factor, ratio of σ_3/σ_1
K_a	Coefficient of active earth pressure
K_0	Coefficient of earth pressure at rest
K_p	Coefficient of passive earth pressure
K_s	Pile constant
L	Length
L'	Effective foundation length
M	Moment, slope projection of critical state line, mass, mobilisation factor
M_s	Mass of solids
M_w	Mass of water
MCV	Moisture condition value
N	Number, stability number, specific volume for $\ln p' = 0$ (one-dimensional consolidation), uncorrected blow count in SPT
N'	Corrected blow count in SPT
N_c, N_q, N_γ	Bearing capacity coefficients
P	Force
P_a	Thrust due to active earth pressure
P_p	Thrust due to passive earth pressure
P_w	Thrust due to water or seepage forces
Q	Total quantity of flow in time t
Q_b	Ultimate soil strength at pile base
Q_s	Ultimate soil strength around pile shaft
Q_u	Ultimate load carrying capacity of pile
R	Radius, reaction, residual factor
$R_{b;cal}$	Eurocode 7 calculated value of pile base resistance
$R_{b;k}$	Eurocode 7 characteristic value of pile base resistance
R_c	Eurocode 7 compressive resistance of ground against a pile at ultimate limit state
$R_{c;cal}$	Eurocode 7 calculated value of R_c
$R_{c;d}$	Eurocode 7 design value of R_c
$R_{c;k}$	Eurocode 7 characteristic value of R_c
$R_{c;m}$	Eurocode 7 measured value of R_c
R_d	Eurocode 7 design resisting force
R_o	Overconsolidation ratio (one-dimensional)
R_p	Overconsolidation ratio (isotropic)
$R_{s;cal}$	Eurocode 7 calculated value of pile shaft resistance
$R_{s;k}$	Eurocode 7 characteristic value of pile shaft resistance
S	Vane shear strength
$S_{dst;d}$	Eurocode 7 design value of destabilising seepage force
S_r	Degree of saturation
S_t	Sensitivity
T	Time factor, tangential force, surface tension, torque
T_d	Eurocode 7 design value of total shearing resistance around structure
U	Average degree of consolidation
U_z	Degree of consolidation at a point at depth z
V	Volume, vertical load

V_a	Volume of air
$V_{dst;d}$	Eurocode 7 design value of destabilising vertical action on a structure
V_s	Volume of solids
V_v	Volume of voids
V_w	Volume of water
W	Weight
W_s	Weight of solids
W_w	Weight of water
X_d	Eurocode 7 design value of a material property
X_k	Eurocode 7 representative value of a material property
Z	Section modulus
a	Area, intercept of MCV calibration line with w axis
b	Width, slope of MCV calibration line
c	Unit cohesion with respect to total stresses
c'	Unit cohesion with respect to effective stresses
c_b	Undisturbed soil shear strength at pile base
c'_d	Eurocode 7 design value of effective cohesion
c_r	Residual value of cohesion
c_u	Undrained unit cohesion
$\bar{c}_u$	Average undrained shear strength of soil
$c_{u;d}$	Eurocode 7 design value of undrained shear strength
c_v	Coefficient of consolidation
c_w	Unit cohesion between wall and soil
d	Pile penetration, pile diameter
d_c, d_q, d_γ	Depth factors
e	Void ratio, eccentricity
f_s	Ultimate skin friction for piles
g	Gravitational acceleration
h	Hydrostatic head, height
h_c	Capillary rise, tension crack depth
h_e	Equivalent height of soil
h_w	Excess head
i	Hydraulic gradient
i_c	Critical hydraulic gradient
i_c, i_q, i_γ	Inclination factors
k	Coefficient of permeability
l	Length
m	Stability coefficient
m_B, m_L	Eurocode 7 load inclination factor parameters
m_v	Coefficient of volume compressibility
n	Porosity, stability coefficient
p	Pressure, mean pressure
p_a	Active earth pressure
p_c	Preconsolidation pressure (one-dimensional)
p'_e	Equivalent consolidation pressure (isotropic)

p_0	Earth pressure at rest
p'_m	Preconsolidation pressure (isotropic)
p'_o	Effective overburden pressure
p_p	Passive earth pressure
q	Unit quantity of flow, deviator stress, uniform surcharge
q_a	Safe bearing capacity
q_u	Ultimate bearing capacity
$q_{u\,net}$	Net ultimate bearing capacity
r	Radius, radial distance, finite difference constant
r_u	Pore pressure ratio
s	Suction value of soil, stress parameter
s_c, s_q, s_γ	Shape factors
s_w	Corrected drawdown in pumping well
t	Time, stress parameter
u, u_w	Pore water pressure
u_a	Pore air pressure, pore pressure due to σ_3 in a saturated soil
u_d	Pore pressure due to $(\sigma_1 - \sigma_3)$ in a saturated soil
$u_{dst;d}$	Eurocode 7 design value of destabilising total pore water pressure
u_i	Initial pore water pressure
v	Velocity, specific volume
w	Water, or moisture, content
w_L	Liquid limit
w_P	Plastic limit
w_s	Shrinkage limit
x	Horizontal distance
y	Vertical, or horizontal, distance
z	Vertical distance, depth
z_o	Depth of tension crack
α	Angle, pile adhesion factor
β	Slope angle
Γ	Eurocode 7 over-design factor, specific volume at ln $P' = 0$
γ	Unit weight (weight density)
γ'	Submerged, buoyant or effective unit weight (effective weight density)
$\gamma_{A;dst}$	Eurocode 7 partial factor: accidental action – unfavourable
γ_b	Bulk unit weight (bulk weight density), Eurocode 7 partial factor: pile base resistance
γ'_c	Eurocode 7 partial factor: effective cohesion
γ_{cu}	Eurocode 7 partial factor: undrained shear strength
γ_d	Dry unit weight (dry weight density)
γ_F	Eurocode 7 partial factor for an action
$\gamma_{G;dst}$	Eurocode 7 partial factor: permanent action – unfavourable
$\gamma_{G;stb}$	Eurocode 7 partial factor: permanent action – favourable
γ_M	Eurocode 7 partial factor for a soil parameter
$\gamma_{Q;dst}$	Eurocode 7 partial factor: variable action – unfavourable
γ_{qu}	Eurocode 7 partial factor: unconfined compressive strength

γ_R	Eurocode 7 partial factor for a resistance
γ_{Re}	Eurocode 7 partial factor: earth resistance
γ_{Rh}	Eurocode 7 partial factor: sliding resistance
γ_{Rv}	Eurocode 7 partial factor: bearing resistance
γ_s	Eurocode 7 partial factor: pile shaft resistance
γ_{sat}	Saturated unit weight (saturated weight density)
γ_t	Eurocode 7 partial factor: pile total resistance
γ_w	Unit weight of water (weight density of water)
γ_γ	Eurocode 7 partial factor: weight density
γ_{ϕ}'	Eurocode 7 partial factor: angle of shearing resistance
δ	Ground–structure interface friction angle
ε	Strain
θ	Angle subtended at centre of slip circle
κ	Slope of swelling line
λ	Slope of normal consolidation line
μ	Settlement coefficient, one micron, Poisson's ratio
ξ_1, ξ_2	Eurocode 7 correlation factors to evaluate results of static pile load tests
ξ_3, ξ_4	Eurocode 7 correlation factors to derive pile resistance from ground investigation results
ρ	Density, settlement
ρ'	Submerged, buoyant or effective density
ρ_b	Bulk density
ρ_c	Consolidation settlement
ρ_d	Dry density
ρ_i	Immediate settlement
ρ_{sat}	Saturated density
ρ_w	Density of water
σ	Total normal stress
σ'	Effective normal stress
σ_a, σ_a'	Total, effective axial stress
σ_e'	Equivalent consolidation pressure (one-dimensional)
σ_{oct}	Octahedral normal stress
σ_r, σ_r'	Total, effective radial stress
$\sigma_{stb;d}$	Eurocode 7 design value of stabilising total vertical stress
σ_v'	Effective overburden pressure
$\overline{\sigma_v'}$	Average effective overburden pressure
$\sigma_1, \sigma_2, \sigma_3$	Total major, intermediate and minor stress
$\sigma_1', \sigma_2', \sigma_3'$	Effective major, intermediate and minor stress
τ	Shear stress
τ_{oct}	Octahedral shear stress
ϕ	Angle of shearing resistance with respect to total stresses
ϕ'	Angle of shearing resistance with respect to effective stresses
$\phi_{cv;d}$	Design value of critical state angle of shearing resistance
ϕ_d'	Design value of ϕ'
χ	Saturation parameter
ψ	Angle of back of wall to horizontal

Chapter 1

Classification and Identification Properties of Soil

In the field of civil engineering, nearly all projects are built on to, or into, the ground. Whether the project is a structure, a roadway, a tunnel, or a bridge, the nature of the soil at that location is of great importance to the civil engineer. *Geotechnical engineering* is the term given to the branch of engineering which is concerned with aspects pertaining to the ground. Soil mechanics is the subject within this branch which looks at the behaviour of soils in civil engineering.

Geotechnical engineers are not the only professionals interested in the ground; soil physicists, agricutural engineers, farmers and gardeners all take an interest in the types of soil with which they are working. These workers, however, concern themselves mostly with the organic topsoils found at the soil surface. In contrast, geotechnical engineers are mainly interested in the engineering soils found beneath the topsoil. It is the engineering properties and behaviour of these soils which are their concern.

1.1 Agricultural and engineering soil

If an excavation is made through previously undisturbed ground the following materials are usually encountered (Fig. 1.1).

Topsoil

A layer of organic soil, usually not more than 500 mm thick, in which humus (highly organic partly decomposed vegetable matter) is often found.

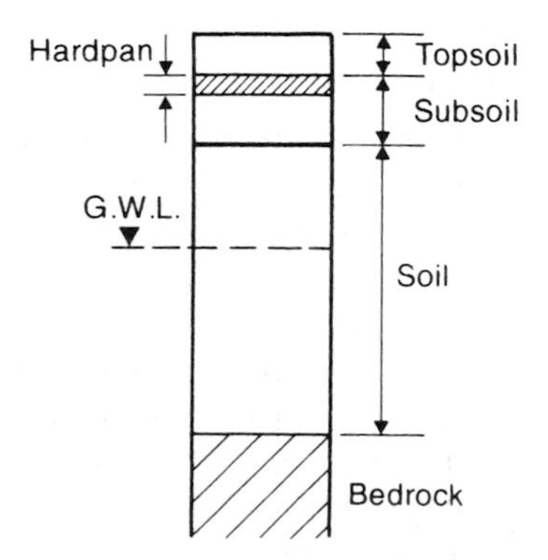

Fig. 1.1 Materials encountered during excavation.

Subsoil

The portion of the Earth's crust affected by current weathering, and lying between the topsoil and the unweathered soil below.

Hardpan

In humid climates humic acid can be formed by rain water causing decomposition of humus. This acid leaches out iron and alumina oxides down into the lower layers where they act as cementation agents to form a hard, rock-like material. Hardpan is difficult to excavate and, as it does not soften when wet, has a high resistance to normal soil drilling methods. A hardpan layer is sometimes found at the junction of the topsoil and the subsoil.

Soil

The soft geological deposits extending from the subsoil to bedrock. In some soils there is a certain amount of cementation between the grains which affects the physical properties of the soil. If this cementation is such that a rock-hard material has been produced then the material must be described as rock. A rough rule is that if the material can be excavated by hand or hand tools it is a soil.

Groundwater

A reservoir of underground water. The upper surface of this water may occur at any depth and is known as the water table or groundwater level (GWL).

1.2 Engineering definitions

Geologists class all items of the Earth's crust as rock, whether hard or soft deposits. Civil engineers consider rock and soil separately.

1.2.1 Rock

Rocks are made from various types of minerals. Minerals are substances of crystalline form made up from a particular chemical combination. The main minerals found in rocks include quartz, feldspar, calcite and mica. Geologists classify all rocks into three basic groups: *igneous*, *sedimentary* and *metamorphic*.

Igneous rocks

These rocks have become solid from a melted liquid state. *Extrusive* igneous rocks are those that arrived on the surface of the Earth as molten lava and cooled. *Intrusive* igneous rocks are formed from magma (molten rock) that forced itself through cracks into rock beds below the surface and solidified there.

Examples of igneous rocks: granite, basalt, gabbro.

Sedimentary rocks

Weathering reduces the rock mass to fragmented particles, which can be more easily transported by wind, water and ice. When dropped by the agents of weathering, they are termed sediments. These sediments are typically deposited in layers or beds called strata and when compacted and cemented together (lithification) they form sedimentary rocks.

Examples of sedimentary rocks: shale, sandstone, chalk.

Metamorphic rocks

Metamorphism through high temperatures and pressures acting on sedimentary or igneous rocks produces metamorphic rocks. The original rock undergoes both chemical and physical alterations.

Examples of metamorphic rocks: slate, quartzite, marble.

1.2.2 Soil

The actions of frost, temperature, gravity, wind, rain and chemical weathering are continually forming rock particles that eventually become soils. There are three types of soil when considering modes of formation.

Transported soil (gravels, sands, silts and clays)

Most soils have been transported by water. As a stream or river loses its velocity it tends to deposit some of the particles that it is carrying, dropping the larger, heavier particles first. Hence, on the higher reaches of a river, gravel and sand are found whilst on the lower or older parts, silts and clays predominate, especially where the river enters the sea or a lake and loses its velocity. Ice has been another important transportation agent, and large deposits of boulder clay and moraine are often encountered.

In arid parts of the world wind is continually forming sand deposits in the form of ridges. The sand particles in these ridges have been more or less rolled along and are invariably rounded and fairly uniform in size. Light brown, wind-blown deposits of silt-size particles, known as loess, are often encountered in thin layers, the particles having sometimes travelled considerable distances.

Residual soil (topsoil, laterites)

These soils are formed *in situ* by chemical weathering and may be found on level rock surfaces where the action of the elements has produced a soil with little tendency to move. Residual soils can also occur whenever the rate of break up of the rock exceeds the rate of removal. If the parent rock is igneous or metamorphic the resulting soil sizes range from silt to gravel.

Laterites are formed by chemical weathering under warm, humid tropical conditions when the rain water leaches out the soluble rock material leaving behind the insoluble hydroxides of iron and aluminium, giving them their characteristic red-brown colour.

Organic soil

These soils contain large amounts of decomposed animal and vegetable matter. They are usually dark in colour and give off a distinctive odour. Deposits of organic silts and clays have usually been created from river or lake sediments. Peat is a special form of organic soil and is a dark brown spongy material which almost entirely consists of lightly to fully decomposed vegetable matter. It exists in one of three forms:

- *Fibrous*: Non plastic with a firm structure only slightly altered by decay.
- *Pseudo-fibrous*: Peat in this form still has a fibrous appearance but is much softer and more plastic than fibrous peat. The change is due more to prolonged submergence in airless water than to decomposition.
- *Amorphous*: With this type of peat decomposition has destroyed the original fibrous vegetable structure so that it has virtually become an organic clay.

Peat deposits occur extensively throughout the world and can be extremely troublesome when encountered in civil engineering work.

1.2.3 Granular and cohesive soils

Geotechnical engineers classify soils as either *granular* or *cohesive*. Granular soils (sometimes referred to as *cohesionless* soils) are formed from loose particles without strong inter-particle forces, e.g. sands and gravels. Cohesive soils (e.g. clays, clayey silts) are made from particles bound together with clay minerals. The particles are flaky and sheet-like and retain a significant amount of adsorbed water on their surfaces. The ability of the sheet-like particles to slide relative to one another gives a cohesive soil the property known as *plasticity*.

1.3 Clays

It is generally believed that rock fragments can be reduced by mechanical means to a limiting size of about 0.002 mm so that a soil containing particles above this size has a mineral content similar to the parent rock from which it was created.

For the production of particles smaller than 0.002 mm some form of chemical action is generally necessary before breakdown can be achieved. Such particles, although having a chemical content similar to the parent rock, have a different crystalline structure and are known as clay particles. An exception is rock flour, rock grains smaller than 0.002 mm, produced by the glacial action of rocks grinding against each other.

1.3.1 Classes of clay minerals

The minerals constituting a clay are invariably the result of the chemical weathering of rock particles and are hydrates of aluminium, iron or magnesium silicate generally combined in such a manner as to create sheet-like structures only a few molecules thick. These sheets are built from two basic units, the tetrahedral unit of silica and the octahedral unit of the hydroxide of aluminium, iron or magnesium. The main dimension of a clay particle is usually less than 0.002 mm and the different types of minerals have been created from the manner in which these structures were stacked together.

The three main groups of clay minerals are as follows.

Kaolinite group

This mineral is the most dominant part of residual clay deposits and is made up from large stacks of alternating single tetrahedral sheets of silicate and octahedral sheets of aluminium. Kaolinites are very stable with a strong structure and absorb little water. They have low swelling and shrinkage responses to water content variation.

Illite group

Consists of a series of single octahedral sheets of aluminium sandwiched between two tetrahedral sheets of silicon. In the octahedral sheets some of the aluminium is replaced by iron and magnesium and in the tetrahedral sheets there is a partial replacement of silicon by aluminium. Illites tend to absorb more water than kaolinites and have higher swelling and shrinkage characteristics.

Montmorillonite group

This mineral has a similar structure to the illite group but, in the tetrahedral sheets, some of the silicon is replaced by iron, magnesium and aluminium. Montmorillonites exhibit extremely high water absorption, swelling and shrinkage characteristics. Bentonite is a member of this mineral group and is usually formed from weathered volcanic ash. Because of its large expansive properties when it is mixed with water it is much in demand as a general grout in the plugging of leaks in reservoirs and tunnels. It is also used as a drilling mud for soil borings.

Readers interested in this subject of clay mineralogy are referred to the publication by Grim (1968).

1.3.2 Structure of a clay deposit

Macrostructure

The visible features of a clay deposit collectively form its macrostructure and include such features as fissures, root holes, bedding patterns, silt and sand seams or lenses and other discontinuities.

A study of the macrostructure is important as it usually has an effect on the behaviour of the soil mass. For example the strength of an unfissured clay mass is much stronger than along a crack.

Microstructure

The structural arrangement of microscopic sized clay particles, or groups of particles, defines the microstructure of a clay deposit. Clay deposits have been laid down under water and were created by the settlement and deposition of clay particles out of suspension. Often during their deposition, the action of Van der Waals forces attracted clay particles together and created flocculant, or honeycombed, structures which, although still microscopic, are of considerably greater volume than single clay particles. Such groups of clay particles are referred to as *clay flocs*.

1.4 Soil classification

Soil classification enables the engineer to assign a soil to one of a limited number of groups, based on the material properties and characteristics of the soil. The classification groups are then used as a system of reference for soils.

Soils can be classified in the field or in the laboratory. Field techniques are usually based upon visual recognition. Laboratory techniques include several specialised tests.

1.4.1 *In the field*

Gravels, sands and peats are easily recognisable, but difficulty arises in deciding when a soil is a fine sand or a coarse silt or when it is a fine silt or a clay. The following rules may, however, help:

Fine sand	Silt	Clay
Individual particles visible	Some particles visible	No particles visible
Exhibits dilatancy	Exhibits dilatancy	No dilatancy
Easy to crumble and falls off hands when dry	Easy to crumble and can be dusted off hands when dry	Hard to crumble and sticks to hands when dry
Feels gritty	Feels rough	Feels smooth
No plasticity	Some plasticity	Plasticity

The dilatancy test consists in moulding a small amount of soil in the palm of the hand; if water is seen to recede when the soil is pressed, then it is either a sand or a silt.

Organic silts and clays are invariably dark grey to blue-black in colour and give off a characteristic odour, particularly with fresh samples.

The condition of a clay very much depends upon its degree of *consolidation*. At one extreme, a soft normally consolidated clay can be moulded by the fingers whereas, at the other extreme, a hard over consolidated clay cannot. Consolidation is described in Chapter 9.

1.4.2 *In the laboratory*

Granular soils – particle size distribution

A standardised sytem of classification helps to eliminate human error. The usual method is based on the determination of the particle size distribution by shaking a dried sample of the soil (usually after washing) through a set of sieves and recording the mass retained on each sieve. The classification system adopted by the British Standards Institution is the MIT (Massachusetts Institute of Technology) system. The boundaries defined by this system can be seen on the particle size distribution sheet in Fig. 1.2. The results of the sieve analysis are plotted with the particle sizes horizontal and the summation percentages vertical. As soil particles vary in size from molecular to boulder it is necessary to use a log scale for the horizontal plot so that the full range can be shown on the one sheet.

The smallest aperture generally used in soils work is that of the 0.063 mm size sieve. Below this size (i.e. silt sizes) the distribution curve must be obtained by sedimentation (pipette or hydrometer). Unless a centrifuge is used, it is not possible to determine the range of clay sizes in a soil, and all that can be done is to obtain the total percentage of clay sizes present. A full description of these tests is given in BS 1377: Part 2.

Examples of particle size distribution (or grading) curves for different soil types are shown in Fig. 1.8. From these grading curves it is possible to determine for each soil the total percentage of a particular size and the percentage of particle sizes larger or smaller than any particular particle size.

The effective size of a distribution, D_{10}

An important particle size within a soil distribution is the effective size which is the largest size of the smallest 10 per cent. It is given the symbol D_{10}. Other particle sizes, such as D_{60} and D_{85}, are defined in the same manner.

Grading of a distribution

For a granular soil the shape of its grading curve indicates the distribution of the soil particles within it.

If the shape of the curve is not too steep and is more or less constant over the full range of the soil's particle sizes then the particle size distribution extends evenly over the range of the particle sizes within the soil and there is no deficiency or excess of any particular particle size. Such a soil is said to be *well graded.*

If the soil has any other form of distribution curve then it is said to be *poorly graded.* According to their distribution curves there are two types of poorly graded soil:

- If the major part of the curve is steep then the soil has a particle size distribution extending over a limited range with most particles tending to be about the same size. The soil is said to be *closely graded* or, more commonly, *uniformly graded.*
- If a soil has large percentages of its bigger and smaller particles and only a small percentage of the intermediate sizes then its grading curve will exhibit a significantly flat section or plateau. Such a soil is said to be *gap graded.*

The uniformity coefficient C_u

The grading of a soil is best determined by direct observation of its particle size distribution curve. This can be difficult for those studying the subject for the first time but some guidance can be obtained by the use of a grading parameter, known as the uniformity coefficient.

$$C_u = \frac{D_{60}}{D_{10}}$$

If $C_u < 4.0$ then the soil is uniformly graded.

If $C_u > 4.0$ then the soil is either well graded or gap and a glance at the grading curve should be sufficient for the reader to decide which is the correct description.

Excel

Example 1.1

The results of a sieve analysis on a soil sample were:

Sieve size (mm)	Mass retained (g)
10	0.0
6.3	5.5
2	25.7
1	23.1
0.6	22.0
0.3	17.3
0.15	12.7
0.063	6.9

2.3 g passed through the 63 μm sieve.

Plot the particle size distribution curve and determine the uniformity coefficient of the soil.

Solution

The aim is to determine the percentage of soil (by mass) passing through each sieve. To do this the percentage retained on each sieve is determined and subtracted from the percentage passing through the previous sieve. This gives the percentage passing through the current sieve.

Calculations may be set out as follows:

Sieve size (mm)	Mass retained (g)	Percentage retained (%)	Percentage passing (%)
10	0.0	0	100
6.3	5.5	5	95
2	25.7	22	73
1	23.1	20	53
0.6	22.0	19	34
0.3	17.3	15	19
0.15	12.7	11	8
0.063	6.9	6	2
Pass 0.063	2.3	2	
Total mass	115.5		

e.g. sieve size 2 mm:

$$\text{Percentage retained} = \frac{25.7}{115.5} \times 100 = 22\%$$

$$\text{Percentage passing} = 95 - 22 = 73\%$$

The particle size distribution curve is shown in Fig. 1.2. The soil has approximate proportions of 30 per cent gravel and 70 per cent sand.

$$D_{10} = 0.17\text{ mm};\quad D_{60} = 1.5\text{ mm};\quad C_u = \frac{D_{60}}{D_{10}} = \frac{1.5}{0.17} = 8.8$$

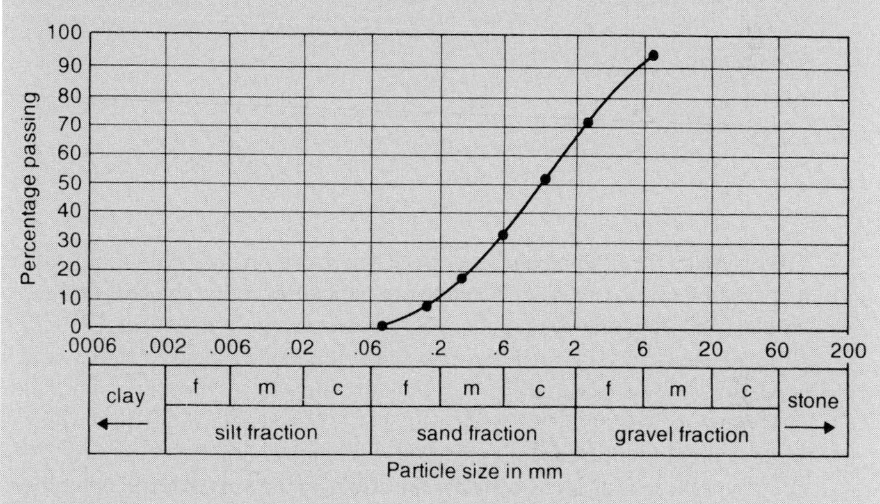

Fig. 1.2 Example 1.1.

Cohesive soils – consistency limit (or index) tests

Water content (w)

The most common way of expressing the amount of water present in a soil is the water content. The water content, also called the moisture content, is given the symbol w and is the ratio of the amount of water to the amount of dry soil.

$$w = \frac{\text{Weight of water}}{\text{Weight of solids}} = \frac{W_w}{W_s} \quad \text{or} \quad w = \frac{\text{Mass of water}}{\text{Mass of solids}} = \frac{M_w}{M_s}$$

w is usually expressed as a percentage and should be quoted to two significant figures.

Drying soils

Soils can be either oven or air dried. It has become standard practice to oven dry soils at a temperature of 105°C but it should be remembered that some soils can be damaged by such a temperature. Oven drying is necessary for water content and particle specific gravity (see Section 1.7.3) tests but air drying should be used whenever possible for other soil tests that also require the test sample to be dry.

Example 1.2

A sample of soil was placed in a moisture content tin of mass 19.52 g. The combined mass of the soil and the tin was 48.27 g. After oven drying the soil and the tin had a mass of 42.31 g.

Determine the water content of the soil.

Solution

$$w = \frac{M_w}{M_s} = \frac{48.27 - 42.31}{42.31 - 19.52} = \frac{5.96}{22.79} = 0.262 = 26\%$$

The results of the grading tests described above can only classify a soil with regard to its particle size distribution. They do not indicate whether the fine grained particles will exhibit the plasticity generally associated with fine grained soils. Hence, although a particle size analysis will completely define a gravel and a sand it is necessary to carry out plasticity tests in order to fully classify a clay or a fine silt.

These tests were evolved by Atterberg (1911) and determine the various values of water content at which changes in a soil's strength characteristics occur. As an introduction to these tests let us consider the effect on the strength and compressibility of a soil as the amount of water within it is varied. With a cohesionless soil, i.e. a gravel or a sand, both parameters are only slightly affected by a change in water content whereas a cohesive soil, i.e. a silt or a clay, tends to become considerably stronger and less compressible, i.e. less easy to mould, as it dries out.

Let us consider a cohesive soil with an extremely high water content, i.e. a suspension of soil particles in water. The soil behaves as a liquid and if an attempt is made to apply a shear stress there will be continual deformation with no sign of a failure stress value. If the soil is allowed to slowly dry out a point will be reached where the soil just begins to exhibit a small shear resistance. If the shear stress were removed it will be found that the soil has experienced a permanent deformation; it is now acting as a plastic solid and not as a liquid.

Liquid limit (w_L) and plastic limit (w_P)

The water content at which the soil stops acting as a liquid and starts acting as a plastic solid is known as the *liquid limit* (w_L or LL); see Fig. 1.3c.

As further moisture is driven from the soil it becomes possible for the soil to resist large shearing stresses. Eventually the soil exhibits no permanent deformation and simply fractures with no plastic deformation, i.e. it acts as a brittle solid. The limit at which plastic failure changes to brittle failure is known as the *plastic limit* (w_P or PL); see Fig. 1.3a.

Plasticity index (I_P)

The *plasticity index* is the range of water content within which a soil is plastic; the finer the soil the greater its plasticity index.

$$\text{Plasticity index} = \text{Liquid limit} - \text{Plastic limit}$$

$$I_P = w_L - w_P$$

or

$$PI = LL - PL$$

The shearing strength to deformation relationship within the plasticity range is illustrated in Fig. 1.3b.

Note The use of the symbols w_L, w_P and I_P follows the recommendations by the ISSMFE Lexicon (1985). However, the symbols LL, PL and PI are still used in many publications.

Liquidity index

The *liquidity index* enables one to compare a soil's plasticity with its natural moisture content (w).

$$I_L = \frac{w - w_P}{I_P}$$

If $I_L = 1.0$ the soil is at its liquid limit; if $I_L = 0$ the soil is at its plastic limit.

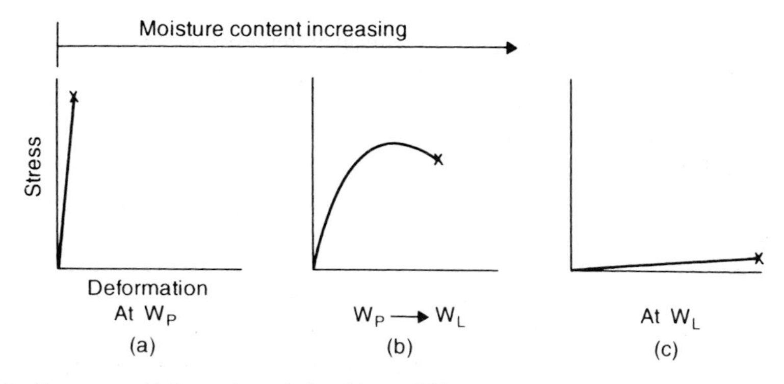

Fig. 1.3 Shear stress/deformation relationships at different water contents.

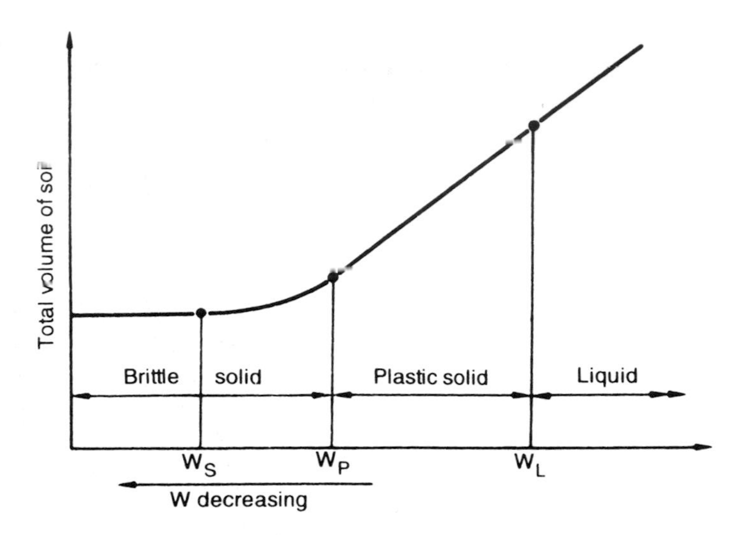

Fig. 1.4 Changes in total volume against moisture content.

Shrinkage limit

If the drying process is prolonged after the plastic limit has been reached the soil will continue to decrease in volume until a certain value of moisture content is reached. This value is known as the shrinkage limit and at values of moisture content below this level the soil is partially saturated. In other words, below the shrinkage limit the volume of the soil remains constant with further drying, but the weight of the soil decreases until the soil is fully dried.

In Fig. 1.4 the variation of the total volume of a soil with its moisture content is plotted, showing the positions of the liquid, plastic and shrinkage limits.

Determination of liquid and plastic limits

Liquid limit test

BS 1377: Part 2 specifies the following three methods for determining the liquid limit of soil.

(1) Cone penetrometer method (definitive method)
Details of the apparatus are shown in Fig. 1.5. The soil to be tested is air dried and thoroughly mixed. At least 200 g of the soil is sieved through a 425 μm sieve and placed on a glass plate. The soil is then mixed with distilled water into a paste.

A metal cup, approximately 55 mm in diameter and 40 mm deep, is filled with the paste and the surface struck off level. The cone, of mass 80 g, is next placed at the centre of the smoothed soil surface and level with it. The cone is released so that it penetrates into the soil and the amount of penetration, over a time period of 5 seconds, is measured.

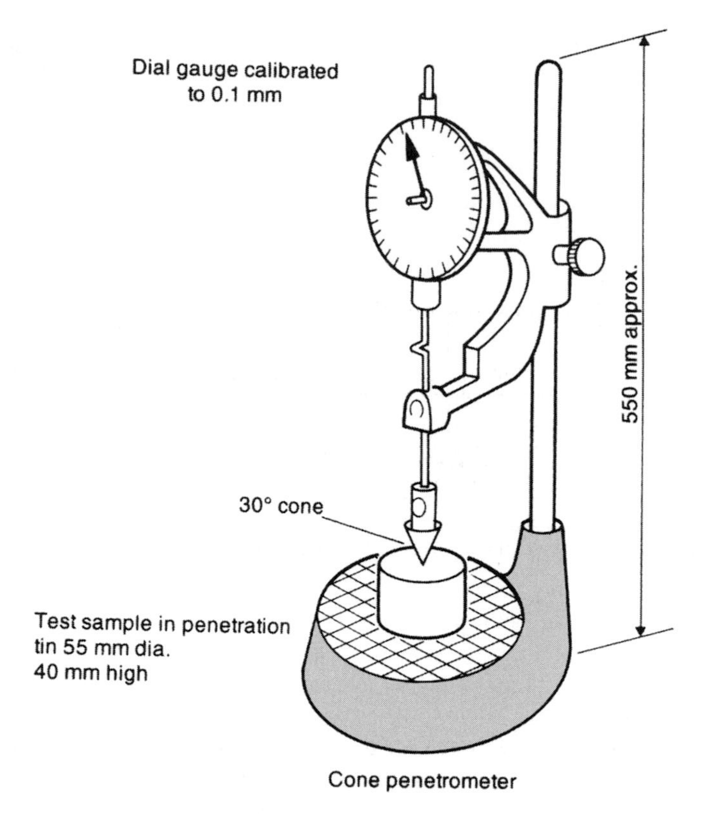

Fig. 1.5 Liquid limit apparatus.

The test is now repeated by lifting the cone clear, cleaning it and filling up the depression in the surface of the soil by adding a little more of the wet soil.

If the difference between the two measured penetrations is less than 0.5 mm then the tests are considered valid. The average penetration is noted and a moisture content determination is carried out on the soil tested.

The procedure is repeated at least four times with increasing water contents. The amount of water used throughout should be such that the penetrations obtained lie within a range of 15 to 25 mm.

To obtain the liquid limit the variation of cone penetration (plotted vertically) to moisture content (plotted horizontally) is drawn out (both scales being natural).

The best straight line is drawn through the experimental points and the liquid limit is taken to be the moisture content corresponding to a cone penetration of 20 mm (expressed as a whole number).

(2) One-point cone penetrometer method

In this test the procedure is similar to that described above, with the exception that only one point is required. The test is thus fairly rapid. Once the average penetration for the point is established, the moisture content of the soil is determined. The moisture content is then multiplied by a factor to give the liquid limit. The value of the factor is dependent on both the cone penetration and the range of moisture content within which the measured moisture content falls. The factors were determined through experimental work performed by Clayton and Jukes (1978).

(3) Method using the Casagrande apparatus

Until 1975 this was the only method for determining liquid that was recognised by the British Standards Institution. Although still used worldwide the test is now largely superseded by cone penetration techniques.

Plastic limit test

About 20 g of soil prepared as in the liquid limit test is used. The soil is mixed on the glass plate with just enough water to make it sufficiently plastic for rolling into a ball, which is then rolled out between the hand and the glass to form a thread. The soil is said to be at its plastic limit when it just begins to crumble at a thread diameter of 3 mm. At this stage a section of the thread is removed for moisture content determination. The test should be repeated at least once more.

It is interesting to note that, in Russia and P. R. China, the cone penetrometer is used to determine both w_L and w_P. The apparatus used consists of a 30° included angle cone with a total mass of 76 g. The test is the same as the liquid limit test in Britain, a penetration of 17 mm giving w_L and a penetration of 2 mm giving w_P.

Example 1.3

A BS cone penetrometer test was carried out on a sample of clay with the following results:

Cone penetration (mm)	16.1	17.6	19.3	21.3	22.6
Moisture content (%)	50.0	52.1	54.1	57.0	58.2

The results from the plastic limit test were:

Test no.	Mass of tin (g)	Mass of wet soil + tin (g)	Mass of dry soil + tin (g)
1	8.1	20.7	18.7
2	8.4	19.6	17.8

Determine the liquid limit, plastic limit and the plasticity index of the soil.

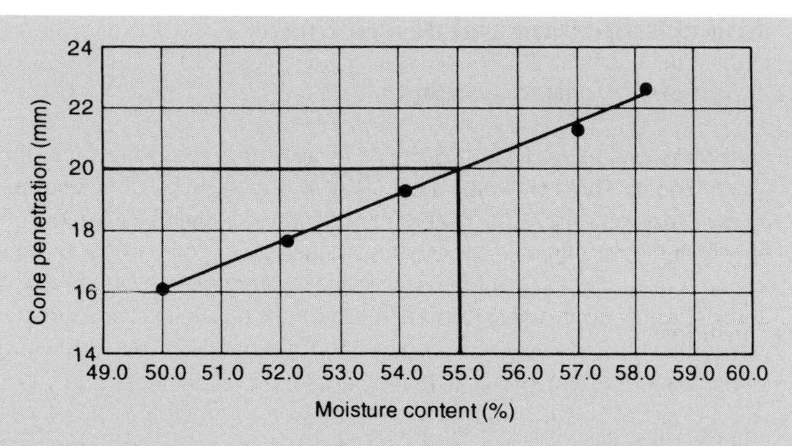

Fig. 1.6 Example 1.3.

Solution

The plot of cone penetration to moisture content is shown in Fig. 1.6. The liquid limit is the moisture content corresponding to 20 mm penetration, i.e. $w_L = 55\%$.

The plastic limit is determined thus:

$$w_P(1) = \frac{20.7 - 18.7}{18.7 - 8.1} \times 100 = 18.9$$

$$w_P(2) = \frac{19.6 - 17.8}{17.8 - 8.4} \times 100 = 19.1$$

Average $w_P = 19\%$

The plasticity index is the difference between w_L and w_P, i.e.

$$I_P = 55 - 19 = 36\%$$

1.5 Common types of soil

Soils are usually a mixture, e.g. silty clay, sandy silt, etc. Local names are often used for soil types that occur within a particular region. e.g. London clay, etc. Boulder clay is an unstratified and irregular mixture of boulders, cobbles, gravel, sand, silt and clay of glacial origin. In spite of its name boulder clay is not a pure clay. Moraines are gravel and sand deposits of glacial origin. Loam is a soft deposit consisting of a mixture of sand, silt and clay in approximately equal quantities.

'Fill' is soil excavated from a 'borrow' area which is used for filling hollows or for the construction of earthfill structures, such as dams or embankments.

1.6 Soil classification and description

1.6.1 Soil classification systems

Soil classification systems have been in use for a very long time, the first being created some 4000 years ago by a Chinese engineer. In 1896 a soil classification system was proposed by the Bureau of Soils, United States Department of Agriculture in which the various soil types were classified purely on particle size and it is interesting to note that the limiting sizes used are more or less the same as those in use today. Further improved systems allowed for the plasticity characteristics of soil and a modified form of the system proposed by Casagrande in 1947 is the basis of the soil classification system used in Britain.

The British Soil Classification System (BSCS)

The British Standard BS 5930 (1999), *Code of practice for site investigations*, gives a full description of the BSCS and the reader is advised to obtain sight of a copy.

The system divides soil into two main categories. If at least 35 per cent of a soil can pass through a 63 μm sieve then it is a *fine soil*. Conversely, if the amount of soil that can pass through the 63 μm sieve is less than 35 per cent then it is a *coarse soil*. Each category is divided into groups, depending upon the grading of the soil particles not passing the 63 μm sieve and upon the plasticity characteristics of the soil particles passing the 425 μm sieve.

A summary of the BSCS is shown in Table 1.1 and its associated plasticity chart in Fig. 1.7.

To use the plasticity chart it is necessary to plot a point whose coordinates are the liquid limit and the plasticity index of the soil to be identified. The soil is classified by observing the position of the point relative to the sloping straight line drawn across the diagram.

This line, known as the A-line, is an empirical boundary between inorganic clays, whose points lie above the line, and organic silts and clays whose points lie below. The A-line goes through the base line at $I_P = 0$, $W_L = 20$ per cent so that its equation is:

$$I_P = 0.73\,(w_L - 20\%)$$

or $$PI = 0.73\,(LL - 20\%)$$

The main soil types are designated by capital letters:

G	Gravel	M	Silt, M-soil
S	Sand	C	Clay
F	Fine soil, Fines	Pt	Peat

The classification 'F' is intended for use when there is difficulty in determining whether a soil is a silt or a clay.

Originally all soils that plotted below the A-line of the plasticity charts were classified as silts. The term 'M-soil' has been introduced to classify soils that plot below the A-line but have particle size distributions not wholly in the range of silt sizes.

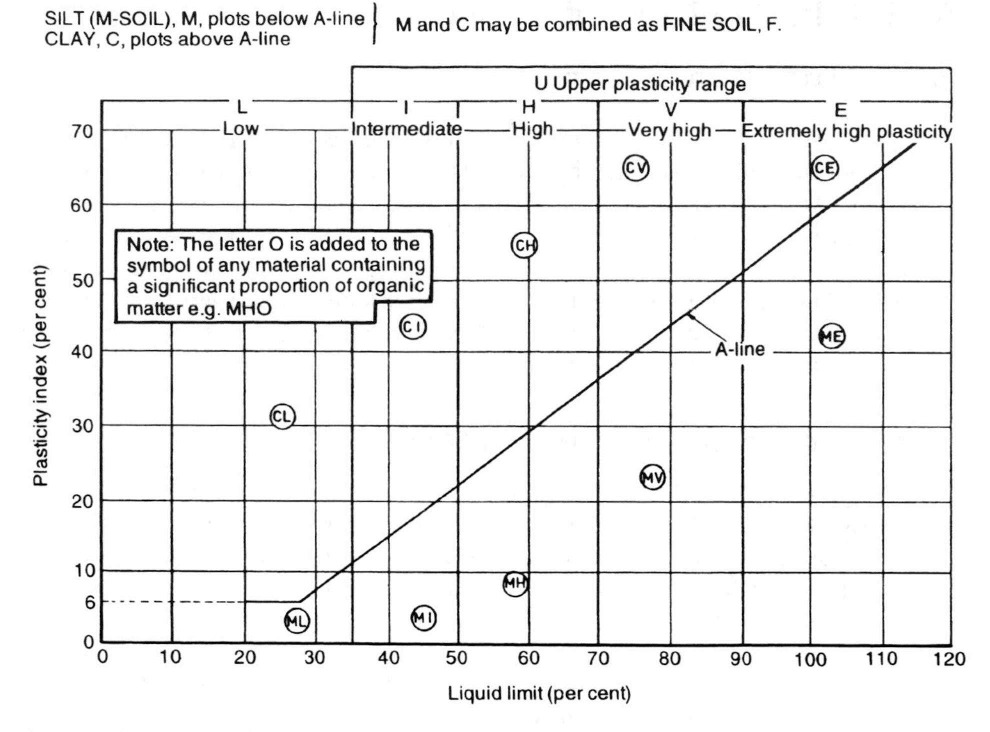

Fig. 1.7 Plasticity chart for the BSCS (after BS 5930: 1999).

Behind the letter designating the main soil type additional letters are added to further describe the soil and to denote its grading and plasticity. These letters are:

W	Well graded	L	Low plasticity	($w_L < 35\%$)
P	Poorly graded	I	Intermediate	($35 \leq w_L \leq 70$)
P_u	Uniform	H	High plasticity	($50 \leq w_L \leq 70$)
P_g	Gap graded	V	Very high	($70 \leq w_L \leq 90$)
O	Organic	E	Extremely high	($w_L > 90\%$)

The letter O is applied at the end of the group symbol for a soil, no matter what type, if the soil has a significant amount of organic matter within it.

Examples of the use of the symbols are set out below.

Soil description	Group symbol
Well graded silty SAND	SWM
Organic CLAY of high plasticity	CHO
Sandy CLAY of intermediate plasticity	CIS
Uniform clayey sand	SP_uF

Table 1.1 British Soil Classification System for Engineering Purposes (after BS 5930: 1999).

COARSE SOILS – less than 35% of the material is finer than 0.06 mm

Soil groups		Subgroups and laboratory identification				
GRAVEL and SAND may be qualified Sandy GRAVEL and Gravelly SAND, etc. where appropriate (See 41.3.2.2)		Group symbol	Subgroup symbol	Fines (% less than 0.06 mm)	Liquid limit %	Name
GRAVELS – More than 50% of coarse material is of gravel size (coarser than 2 mm)	Slightly silty or clayey GRAVEL	G: GW GP	GW GPu GPg	0 to 5		Well graded GRAVEL Poorly graded/Uniform/Gap graded GRAVEL
	Silty GRAVEL	G-F: G-M G-C	GWM GPM GWC GPC	5 to 15		Well graded/Poorly graded silty GRAVEL Well graded/Poorly graded clayey GRAVEL
	Very silty GRAVEL Very clayey GRAVEL	GF: GM GC	GML, etc GCL GCI GCH GCV GCE	15 to 35		Very silty GRAVEL; subdivide as for GC Very clayey GRAVEL (clay of low, intermediate, high, very high, extremely high plasticity)
SANDS – More than 50% of coarse material is of sand size (finer than 2 mm)	Slightly silty or clayey SAND	S: SW SP	SW SPu SPg	0 to 5		Well graded SAND Poorly graded/Uniform/Gap graded SAND
	Silty SAND Clayey SAND	S-F: S-M S-C	SWM SPM SWC SPC	5 to 15		Well graded/Poorly graded silty SAND Well graded/Poorly graded clayey SAND

FINE SOILS – more than 35% of the material is finer than 0.06 mm

	Very silty SAND Very clayey SAND	SF SM SC	SML, etc SCL SCI SCH SCV SCE	15 to 35		Very silty SAND; subdivided as for SC Very clayey SAND (clay of low, intermediate, high, very high, extremely high plasticity)
Gravelly or sandy SILTS and CLAYS 35% to 65% fines	Gravelly SILT Gravelly CLAY	FG MG CG	MLG, etc CLG CIG CHG CVG CEG		 <35 35 to 50 50 to 70 70 to 90 >90	Gravelly SILT; subdivide as for CG Gravelly CLAY of low plasticity of intermediate plasticity of high plasticity of very high plasticity of extremely high plasticity
SILT and CLAYS 65% to 100% fines	Sandy SILT Sandy CLAY	FS MS CS	MLS, etc CLS, etc			Sandy SILT; subdivide as for CG Sandy CLAY; subdivide as for CG
	SILT (M-SOIL) CLAY	F M C	ML, etc CL CI CH CV CE		 <35 35 to 50 50 to 70 70 to 90 >90	SILT, subdivide as for C CLAY of low plasticity of intermediate plasticity of high plasticity of very high plasticity of extremely high plasticity
ORGANIC SOILS	Descriptive letter 'O' suffixed to organic matter suspected to be a significant constituent. Example MHO: any group or sub-group symbol. Organic SILT of high plasticity.					
PEAT	Pt Peat soils consist predominantly of plant remains which may be fibrous or amorphous.					

When classification tests are carried out on a stony soil sample any particles nominally greater than 60 mm are removed by sieving (with a standard 63 mm sieve) and their percentage determined. The tests are then carried out on the remaining soil. The material removed is classed as cobbles, 66 to 200 mm in size, with symbol Cb, or boulders, greater than 200 mm in size, with the symbol B.

Fine and coarse soils that contain cobbles, or cobbles and boulders, are indicated in symbols by the use of the addition sign. For instance, a well graded SAND with gravel and cobbles would have the group symbol SWG + Cb.

Example 1.4

Classify the soil of Example 1.1 whose particle size distribution curve is shown in Fig. 1.2.

Solutions

The value for C_u already been found to be 8.8. From Fig. 1.2 it is seen that the grading curve has a regular slope and therefore contains roughly equal percentages of particle sizes. The soil is a well graded gravelly SAND with the group symbol SWG.

Note that when classifying the soil it is customary to indicate the main soil type in capital letters, i.e. SAND.

Excel

Example 1.5

A set of particle size distribution analyses on three soils, A, B and C, gave the following results:

Sieve size (mm)	Percentage passing		
	Soil A	Soil B	Soil C
20	90	–	–
10	56	–	–
6.3	47	–	–
2	44	–	–
0.6	40	95	–
0.425	–	80	–
0.300	29	10	–
0.212	–	3	–
0.150	–	–	100
0.063	5	1	91

Soil C: Since more than 10 per cent passed the 63 μm sieve, a pipette analysis (described in BS 1377: Part 2 and by Head (1992)) was performed. The results were:

Particle sizes (mm)	Percentage passing
	Soil C
0.04	78
0.02	61
0.006	47
0.002	40

Soil C was found to have a liquid limit of 48 per cent and a plastic limit of 21 per cent.

Plot the particle size distribution curves and classify each soil.

Solution

The particle size distribution curves for the three soils are shown in Fig. 1.8. The curves can be used to obtain the following particle sizes for soils A and B.

Soil	D_{10} (mm)	D_{30} (mm)	D_{60} (mm)
A	0.1	0.31	11.0
B	0.3	0.42	0.44

Soil A: From the grading curve it is seen that this soil consists of some 57 per cent gravel and 43 per cent sand and is therefore predominantly gravel. The curve has a horizontal portion indicating that the soil has only a small percentage of soil particles within this range. It is therefore gap graded.

The soil is a gap graded sandy GRAVEL. Group symbol GP_gS.

Soil B: From the grading curve it is immediately seen that this soil is a sand with most of its particles about the same size.

The soil is a uniformly graded SAND. Group symbol SP_u.

Soil C: It is interesting to note that, as the whole of soil C passed the 425 μm sieve, there would be no need to remove any of the soil before subjecting it to the consistency limit tests. From the grading curve, by considering particle sizes only, the soil is a mixture of some 10 per cent sand, 50 per cent silt and 40 per cent clay. The soil is undoubtedly fine and the group symbol could be F, although, as the silt particles are more dominant than the clay, it could be given the symbol MC. The liquid limit of the soil is 48 per cent which, according to BS 5930, indicates an intermediate plasticity. The group symbol of the soil could therefore be either FI or MCI.

However, for mixtures of fine soils BS 5930 suggests that classification is best carried out by the use of the plasticity chart shown in Fig. 1.7. The liquid limit of the soil = 48 per cent and the plasticity index, $(w_L - w_P) = 27$ per cent. Using Fig. 1.7 it is seen that the British system classifies the soil as an inorganic clay with the group symbol CI.

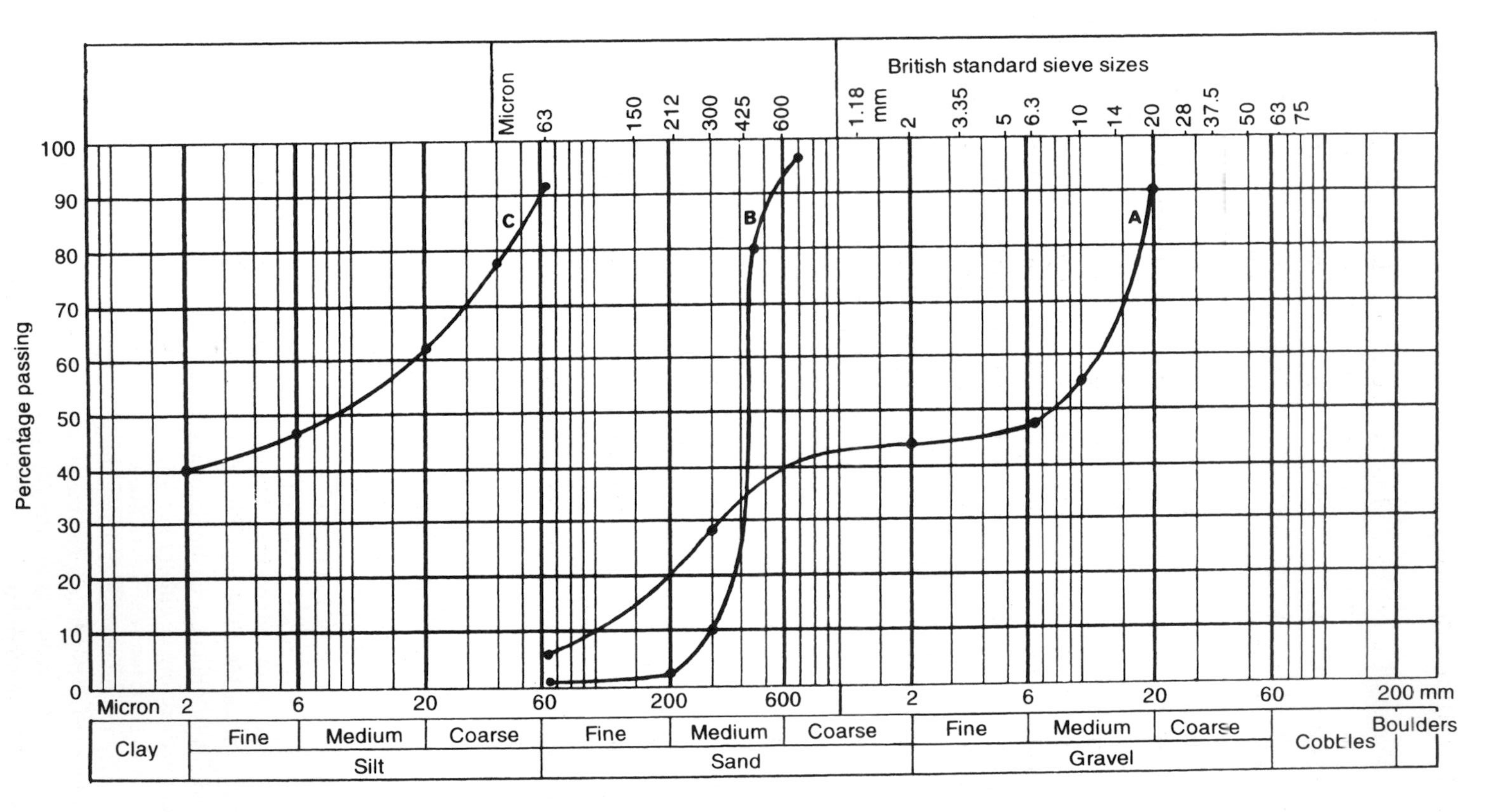

Fig. 1.8 Example 1.5.

1.6.2 Description of soils

Classifying and describing a soil are two operations which are not necessarily the same. An operator who has not even visited the site from which a soil came can classify the soil from the information obtained from grading and plasticity tests carried out on disturbed samples. Such tests are necessary if the soil is being considered as a possible construction material and the information obtained from them must be included in any description of the soil.

Further information regarding the colour of a soil, the texture of its particles, etc., can be obtained in the laboratory from disturbed soil samples but a full description of a soil must include its *in situ*, as well as its laboratory characteristics. Some of this latter information can be found in the laboratory from undisturbed samples of the soil collected for other purposes, such as strength or permeability tests, but usually not until after the tests have taken place and the samples can then be split open for proper examination. Other relevant information, such as bedding, geological details, etc., obtained from borehole data and site observations should also be included in the soil's description.

Further information is available in BS 5930 *Code of Practice for Site Investigations*, and Clayton *et al*. (1995).

1.7 Soil properties

From the foregoing it is seen that soil consists of a mass of solid particles separated by spaces or voids. A cross-section through a granular soil may have an appearance similar to that shown in Fig. 1.9a.

In order to study the properties of such a soil mass it is advantageous to adopt an idealised form of the diagram as shown in Fig. 1.9b. The soil mass has a total volume V and a volume of solid particles that summates to V_s. The volume of the voids, V_v, is obviously equal to $V - V_s$.

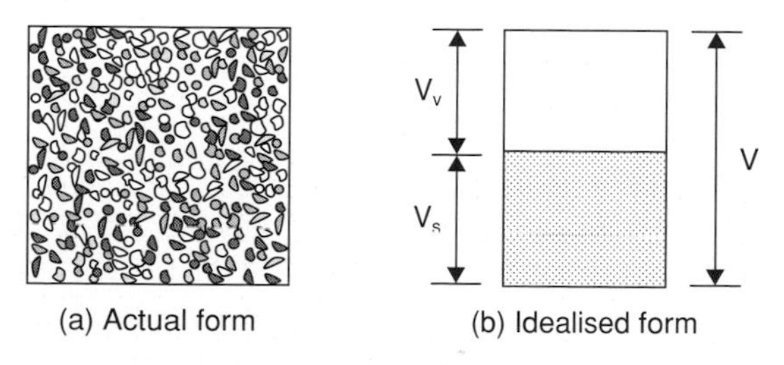

Fig. 1.9 Cross-section through a granular soil.

1.7.1 Void ratio and porosity

From a study of Fig. 1.9 the following may be defined:

Void ratio

$$e = \frac{\text{Volume of voids}}{\text{Volume of solids}} = \frac{V_v}{V_s}$$

Porosity

$$n = \frac{\text{Volume of voids}}{\text{Total volume}}$$

$$n = \frac{V_v}{V} = \frac{V_v}{V_v + V_s} = \frac{e}{1 + e}$$

1.7.2 Degree of saturation (S_r)

The voids of a soil may be filled with air or water or both. If only air is present the soil is dry, whereas if only water is present the soil is saturated. When both air and water are present the soil is said to be partially saturated. These three conditions are represented in Figs 1.10a, b and c.

The degree of saturation is simply:

$$S_r = \frac{\text{Volume of water}}{\text{Volume of voids}} = \frac{V_w}{V_v} \quad \text{(usually expressed as a percentage)}$$

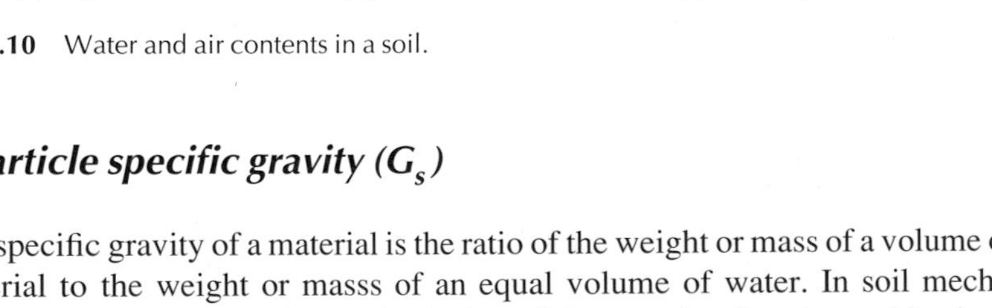

(a) Dry soil (b) Saturated soil (c) Partially saturated soil

Fig. 1.10 Water and air contents in a soil.

1.7.3 Particle specific gravity (G_s)

The specific gravity of a material is the ratio of the weight or mass of a volume of the material to the weight or masss of an equal volume of water. In soil mechanics the most important specific gravity is that of the actual soil grains and is given the symbol G_s.

From the above definition it is seen that, for a soil sample with volume of solids V_s and weight of solids W_s,

$$G_s = \frac{W_s}{V_s \gamma_w}$$

where γ_w = weight of water and, if the sample has a mass of solids M_s,

$$G_s = \frac{M_s}{V_s\rho_w}$$

where ρ_w = density of water (= 1.0 Mg/m^3 at 20°C) i.e.

$$G_s = \frac{M_s}{V_s\rho_w} = \frac{W_s}{V_s\gamma_w}$$

The density of the particles ρ_s is defined as:

$$\rho_s = \frac{M_s}{V_s}$$

therefore,

$$G_s = \frac{\rho_s}{\rho_w}$$

BS 1377: Part 2 specifies methods of test for determining the particle density. For fine, medium or coarse soils the Standard specifies the use of a one litre gas jar fitted with a rubber bung and a mechanical shaking apparatus which can rotate the gas jar, end over end, at some 50 rpm (Fig. 1.11).

The test consists briefly of placing oven dried soil (approximately 200 g for a fine soil and 400 g for a medium or coarse soil) into the gas jar along with some 500 ml of water at room temperature. The jar is sealed with the bung and shaken, first by hand and then in the machine, for some 20 to 30 minutes.

From various weighings that are made the specific gravity of the soil can be calculated. (See Example 1.6.)

If ρ_s is measured in units of Mg/m^3 and the water temperature is assumed to be 20°C, it follows that ρ_s and G_s are numerically equal. G_s is dimensionless.

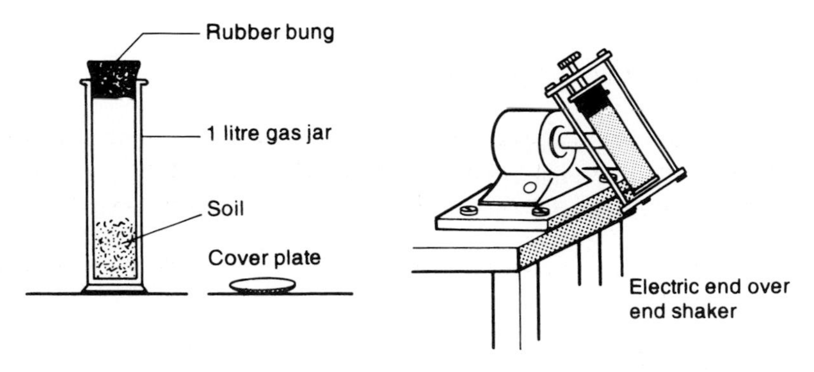

Fig. 1.11 Determination of particle density.

Soil contains particles of different minerals with consequently different specific gravities: G_s therefore represents an average value for the particles.

Generally sands have an average value of $G_s = 2.65$ and clays an average value of 2.75. The particle specific gravity of organic soils can vary considerably. An organic clay can have a G_s value of about 2.60 whereas a bog peat can have a value as low as 1.3.

For coal spoil heaps G_s can vary from about 2.0 for an unburnt shale with a high coal content, to about 2.7 for a burnt shale.

Excel

Example 1.6

The mass of an empty gas jar, together with its glass cover plate, was 478.0 g. When completely filled with water and the cover plate fitted the mass was 1508.2 g. An oven dried sample of soil was inserted in the dry gas jar and the total mass, including the cover plate, was 676.6 g. Water was added to the soil and, after a suitable period of shaking, was topped up until the gas jar was brim full. The cover plate was fitted and the total mass was found to be 1632.6 g.

Determine the particle specific gravity of the soil.

Solution

Mass of soil + water = 1632.6 − 478.0 = 1154.6 g
Mass of dry soil = 676.6 − 478.0 = 198.6 g
Mass of water present with soil = 1154.6 − 198.6 = 956.0 g
Mass of water when gas jar full = 1508.2 − 478.0 = 1030.2 g.
Therefore, mass of water of same volume as soil = 1030.2 − 956.0 = 74.2 g

$$G_s = \frac{\text{Mass of soil}}{\text{Mass of same volume of water}} = \frac{198.6}{74.2} = 2.68$$

Alternative solution
The specific gravity can be quickly found from a formula thus:

$$G_s = \frac{S}{[(J + W) - (J + W + S)] + S}$$

where

$$\begin{aligned} S &= \text{mass of dry soil (g)} \\ J + W &= \text{mass of jar + water (g)} \\ J + W + S &= \text{mass of jar + water + soil (g)} \end{aligned}$$

i.e.,

$$G_s = \frac{198.6}{(1508.2 - 1632.6) + 198.6} = 2.68$$

1.7.4 Density and unit weight

In a system of properly chosen units:

$$\text{Force} = \text{Mass} \times \text{Acceleration} \tag{1}$$

In the SI system of units the basic units are:

Mass	the kilogram	(kg)
Length	the metre	(m)
Time	the second	(s).

The derived force unit (the newton) is obtained by putting unit values into the right-hand side of Equation (1). It is given the symbol N.

$$1\ \text{N} = 1\ \text{kg} \times 1\ \text{m/s}^2 \tag{2}$$

From Equation (2) it is seen that the newton is that force which, when acting on a mass of one kilogram, produces an acceleration of one metre per second per second.

In soils we are mainly interested in the gravitational force exerted by masses (i.e. their weights). The acceleration term in Equation (1) therefore is g, the symbol used to denote the acceleration due to the Earth's gravitational field. The average value for g, at the Earth's surface, is 9.806 m/s^2.

The gravitational force which acts on a one-kilogram mass is therefore 1 × 9.806 = 9.806 N. The weight of a one-kilogram mass is 9.806 newtons.

We can therefore express the amount of material in a given volume, V, in two ways:

the amount of mass, M, in the volume, or
the amount of weight, W, in the volume.

If we consider unit volume, the two systems give the density and the unit weight of the material:

$$\text{Density,}\quad \rho = \frac{\text{Mass}}{\text{Volume}} = \frac{M}{V}$$

$$\text{Unit weight,}\ \gamma = \frac{\text{Weight}}{\text{Volume}} = \frac{W}{V}$$

As an example, consider water at 20°C:

$$\text{Density of water, } \rho_w = 1000\ \text{kg/m}^3 = 1.0\ \text{Mg/m}^3$$

The weight of a 1000 kg mass is 1000 × 9.806 N = 9806 N. Hence the unit weight of water, γ_w, is 9806 N/m^3 = 9.81 kN/m^3.

Soil densities are usually expressed in Mg/m^3 to the nearest 0.01. It should be noted that 1.0 Mg/m^3 is the same as 1.0 $tonne/m^3$ and 1.0 g/ml (1 tonne = 1000 kg).

Soil weights are usually expressed in kN/m^3.

In soils work it is generally more convenient to work in unit weights than in densities.

Note: Weight density

With the introduction of Eurocode 7 to geotechnical design (see Chapter 7) the term *weight density* is likely to eventually replace the term *unit weight*. The two terms are synonymous and since the term *unit weight* has been in use for many decades it will certainly remain in use for many years to come and for this reason it is used throughout this book.

Unit weight of soil

The unit weight of a material is its weight per unit volume. In soils work the most important unit weights are as follows.

Bulk unit weight (γ)

This is the natural *in situ* unit weight of the soil:

$$\gamma = \frac{\text{Total weight}}{\text{Total volume}} = \frac{W}{V} = \frac{W_s + W_w}{V_s + V_v}$$

$$= \frac{G_s V_s \gamma_w + V_v \gamma_w S_r}{V_s + V_v} = \gamma_w \frac{(G_s + eS_r)}{1 + e}$$

Saturated unit weight (γ_{sat})

$$\gamma_{sat} = \frac{\text{Saturated weight}}{\text{Total volume}}$$

When soil is saturated $S_r = 1$, therefore

$$\gamma_{sat} = \gamma_w \frac{G_s + e}{1 + e}$$

Dry unit weight (γ_d)

$$\gamma_d = \frac{\text{Dry weight}}{\text{Total volume}}$$

$$= \frac{\gamma_w G_s}{1 + e} \qquad (\text{as } S_r = 0)$$

Buoyant unit weight (γ')

When a soil is below the water table, part of its weight is balanced by the buoyant effect of the water. This upthrust equals the weight of the volume of the water displaced.

Hence, considering unit volume:

Buoyant unit weight = Saturated unit weight – Unit weight of water

$$= \gamma_w \frac{G_s + e}{1 + e} - \gamma_w = \gamma_w \frac{G_s - 1}{1 + e}$$

Buoyant unit weight is often referred to as the submerged unit weight or the effective unit weight.

Density of soil

Similar expressions can be obtained for densities:

Bulk density, $\rho_b = \rho_w \dfrac{(G_s + eS_r)}{1 + e}$

Saturated density, $\rho_{sat} = \rho_w \dfrac{(G_s + e)}{1 + e}$

Dry density, $\rho_d = \rho_w \dfrac{G_s}{1 + e}$

Buoyant density, $\rho' = \rho_w \dfrac{G_s - 1}{1 + e}$

Relationship between density and unit weight values

In the above expressions, G_s, e, S_r and the number 1 are all dimensionless.

Hence a particular unit weight = γ_w times a constant.
The corresponding density = ρ_w times the same constant.

Example 1.7

A soil has a dry density of 1.9 Mg/m^3. What is its dry unit weight?

Solution

$$\rho_d = 1.9 \text{ Mg/m}^3 = \rho_w \times 1.9$$

Now

$$\gamma_d = \gamma_w \times 1.9 = 18.64 \text{ kN/m}^3$$

Similarly, if

$$\gamma_d = 18.64 \text{ kN/m}^3$$

then

$$\rho_d = \frac{18.64}{9.81} = 1.90 \text{ Mg/m}^3$$

Quick approximation
If we assume the engineering approximation that $\gamma_w = 10\,\text{kN/m}^3$ (instead of 9.81), then in the previous example:

$$\rho_d = 1.9 \text{ Mg/m}^3 \quad \text{and} \quad \gamma_d = 1.9 \times 10 = 19 \text{ kN/m}^3$$

Hence, given a density in Mg/m^3, the unit weight (in kN/m^3) is found by multiplying by 10.

Example 1.8

Saturated density $= 2.4 \text{ Mg/m}^3$
Saturated unit weight $= 24 \text{ kN/m}^3$

(The more exact value for the unit weight is, of course, $2.4 \times 9.81 = 23.54 \text{ kN/m}^3$, but few engineers would hesitate to use the rounded-off figure of 24 kN/m^3, as soil is in any case so variable a material.)

Relationship between w, γ_d and γ

$$\gamma = \frac{W_w + W_s}{V} \tag{3}$$

$$\gamma_d = \frac{W_s}{V} \tag{4}$$

$$w = \frac{W_w}{W_s} \tag{5}$$

From Equation (5) $W_w = wW_s$ and, substituting in Equation (3),

$$\gamma = \frac{W_s}{V}(1 + w)$$

i.e.

$$\gamma_d = \frac{\gamma}{1 + w}$$

Thus to find the dry unit weight from the bulk unit weight, divide the latter by (1 + w) where w is the moisture content expressed as a decimal.

Relationship between e, w and G_s for a saturated soil

$$w = \frac{W_w}{W_s} = \frac{V_w \gamma_w}{V_s \gamma_w G_s} = \frac{V_v}{V_s G_s} = \frac{e}{G_s} \qquad (V_w = V_v \text{ if the soil is saturated})$$

i.e.

$$e = wG_s$$

Relationship between e, w and G_s for a partially saturated soil

$$w = \frac{W_w}{W_s} = \frac{V_w \gamma_w}{V_s \gamma_w G_s} = \frac{V_v S_r}{V_s G_s} = \frac{eS_r}{G_s}$$

i.e.

$$e = \frac{wG_s}{S_r}$$

Example 1.9

In a bulk density determination a sample of clay with a mass of 683 g was coated with paraffin wax. The combined mass of the clay and the wax was 690.6 g. The volume of the clay and the wax was found, by immersion in water, to be 350 ml.

The sample was then broken open and moisture content and particle specific gravity tests gave respectively 17 per cent and 2.73.

The specific gravity of the wax was 0.89. Determine the bulk density and unit weight, void ratio and degree of saturation.

Solution

Mass of soil = 683 g
Mass of wax = 690.6 − 683 = 7.6 g

$$\Rightarrow \quad \text{Volume of wax} = \frac{7.6}{0.89} = 8.55 \text{ ml}$$

$$\Rightarrow \quad \text{Volume of soil} = 350 - 8.6 = 341.4 \text{ ml}$$

$$\rho_b = \frac{683}{341.4} = 2 \text{ g/ml} = 2.0 \text{ Mg/m}^3$$

$$\gamma_b = 2 \times 9.81 = 19.6 \text{ kN/m}^3$$

$$\rho_d = \frac{2}{1.17} = 1.71 \text{ Mg/m}^3$$

Now

$$\frac{\rho_w G_s}{1 + e} = 1.71$$

$$\Rightarrow \quad e = \frac{2.73 - 1.71}{1.71} = 0.596$$

Now

$$\rho_b = 2.0 = \rho_w \frac{(G_s + eS_r)}{1 + e}$$

$$\Rightarrow \quad 1.596 \times 2.0 = 2.73 + 0.596 \times S_r$$
$$\Rightarrow \quad S_r = 77.0 \text{ per cent}$$

1.7.5 *Relative density* (D_r)

A granular soil generally has a large range into which the value of its void ratio may be fitted. If the soil is vibrated and compacted the particles are pressed close together and a minimum value of void ratio is obtained, but if the soil is loosely poured a maximum value of void ratio is obtained.

These maximum and minimum values can be obtained from laboratory tests and it is often convenient to relate them to the naturally occurring void ratio of the soil. This relationship is expressed as the relative density of the soil:

$$D_r = \frac{e_{max} - e}{e_{max} - e_{min}}$$

The theoretical maximum possible density of a granular soil must occur when $e = e_{min}$, i.e. when $D_r = 1.0$. Similarly the minimum possible density occurs when $e = e_{max}$ and $D_r = 0$. In practical terms this means that a loose granular soil will have a D_r value close to zero whilst a dense granular soil will have a D_r value close to 1.0.

1.8 Soil physical relations

A summary of the relations established in Section 1.7 is given below.

Water content $$w = \frac{W_w}{W_s} = \frac{M_w}{M_s}$$

Void ratio $$e = \frac{V_v}{V_s}$$

$$e = wG_s \quad \text{(saturated)}$$

$$e = \frac{wG_s}{S_r} \quad \text{(partially saturated)}$$

Porosity $$n = \frac{V_v}{V} = \frac{e}{1+e}$$

Degree of saturation $$S_r = \frac{V_w}{V_v}$$

Particle specific gravity $$G_s = \frac{W_s}{V_s\gamma_w} = \frac{M_s}{V_s\rho_w}$$

Bulk density $$\rho_b = \rho_w \frac{(G_s + eS_r)}{1+e}$$

Dry density $$\rho_d = \frac{\rho_w G_s}{1+e} = \frac{\rho_b}{1+w}$$

Saturated density $$\rho_{sat} = \rho_w \frac{(G_s + e)}{1+e}$$

Submerged density $$\rho' = \rho_w \frac{(G_s - 1)}{1+e}$$

Bulk unit weight $$\gamma_b = \gamma_w \frac{(G_s + eS_r)}{1+e}$$

Dry unit weight $$\gamma_d = \frac{\gamma_w G_s}{1+e} = \frac{\gamma_b}{1+w}$$

Saturated unit weight $$\gamma_{sat} = \gamma_w \frac{(G_s + e)}{1+e}$$

Submerged unit weight $$\gamma' = \gamma_w \frac{(G_s - 1)}{1+e}$$

Relative density $$D_r = \frac{e_{max} - e}{e_{max} - e_{min}}$$

Exercises

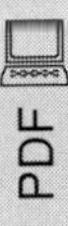

Exercise 1.1

The results of a sieve analysis on a soil were:

Sieve size (mm)	Mass retained (g)
50	0
37.5	15.5
20	17
14	10
10	11
6.3	33
3.35	114.5
1.18	63.3
0.6	18.2
0.15	17
0.063	10.5

The total mass of the sample was 311 g. Plot the particle size distribution curve and, from the inspection of this curve, determine the effective size and uniformity coefficient. Classify the soil.

Answer $D_{10} = 0.7$ mm; $D_{60} = 5.2$ mm. $C_u = 7.4$. 70 per cent gravel, 30 per cent sand. Well graded sandy GRAVEL – symbol GWS.

Exercise 1.2

Plot the particle size distribution curve for the following sieve analysis, given the sieve sizes and the mass retained on each. Classify the soil.

Sample 642 g. Retained on 425 μm sieve – 11 g, 300 μm sieve – 28 g, 212 μm sieve – 77 g, 150 μm sieve –173 g, 63 μm sieve – 321 g.

Answer By inspection of grading curve soil is a uniform SAND – symbol SPu. This is confirmed from the value of $C_u = 2.35$.

Exercise 1.3

A BS cone penetrometer test carried out on a sample of boulder clay gave the following results:

Cone penetration (mm)	15.9	17.1	19.4	20.9	22.8
Moisture content (%)	32.0	32.8	34.5	35.7	37.0

Determine the liquid limit of the soil.

Answer $w_L = 35$ per cent

Exercise 1.4

A liquid and plastic limit test gave the following results:

Test No.	1	2	3	4	PL	PL
Wet mass (g)	33.20	32.10	28.20	31.00	11.83	15.04
Dry mass (g)	28.20	26.50	22.40	23.90	11.25	14.07
Tin (g)	7.02	7.04	7.10	7.02	7.04	7.25
Penetration (mm)	14.5	17.0	20.9	22.7	–	–

Determine the plasticity index of the soil and classify the soil.

Answer 22, CI

If the natural moisture content was 28%, determine the liquidity index in the field.

Answer 0.64, CI

Exercise 1.5

A sand sample has a porosity of 35 per cent and the specific gravity of the particles is 2.73. What is its dry density and void ratio?

Answer $e = 0.54$, $\rho_d = 1.77$ Mg/m^3

Exercise 1.6

A sample of silty clay was found to have a volume of 14.88 ml, whilst its mass at natural moisture content was 28.81 g and the particle specific gravity was 2.7. Calculate the void ratio and degree of saturation if, after oven drying, the sample had a mass of 24.83 g.

Answer $e = 0.618$, $S_r = 70$ per cent

Exercise 1.7

A sample of moist sand was cut out of a natural deposit by means of a sampling cylinder. The volume of the cylinder was 478 ml; the weight of the sample alone was 884 g and 830 g after drying. The volume of the dried sample, when rammed tight into a graduated cylinder, was 418 ml and its volume, when poured loosely into the same cylinder, was 616 ml. If the particle specific gravity was 2.67, compute the relative density and the degree of saturation of the deposit.

Answer $D_r = 69$ per cent, $S_r = 32$ per cent

Exercise 1.8

In order to determine the density of a clay soil an undisturbed sample was taken in a sampling tube of volume 0.001 664 m^3.

The following data were obtained:

Mass of tube (empty) = 1.864 kg
Mass of tube and clay sample = 5.018 kg
Mass of tube and clay sample after drying = 4.323 kg

Calculate the moisture content, the bulk, and the dry densities. If the particle specific gravity was 2.69, determine the void ratio and the percentage saturation of the clay.

Answer $w = 28$ per cent, $\rho_d = 1.49\ Mg/m^3$, $\rho_b = 1.90\ Mg/m^3$, $S_r = 93$ per cent, $e = 0.82$

Chapter 2
Soil Water, Permeability and Flow

2.1 Subsurface water

This is the term used to define all water found beneath the Earth's surface. The main source of subsurface water is rainfall, which percolates downwards to fill up the voids and interstices. Water can penetrate to a considerable depth, estimated to be as much as 12 000 metres, but at depths greater than this, due to the large pressures involved, the interstices have been closed by plastic flow of the rocks. Below this level water cannot exist in a free state, although it is often found in chemical combination with the rock minerals, so that the upper limit of plastic flow within the rock determines the lower limit of subsurface water.

Subsurface water can be split into two distinct zones: saturation zone and aeration zone.

2.1.1 Saturation zone

This is the depth throughout which all the fissures, etc., are filled with water under hydrostatic pressure. The upper level of this water is known as the water table, phreatic surface or groundwater level, and water within this zone is called phreatic water or groundwater.

The water table tends to follow in a more gentle manner the topographical features of the surface above (Fig. 2.1). At groundwater level the hydrostatic pressure is zero, so another definition of water table is the level to which water will eventually rise in an unlined borehole.

The water table is not constant but rises and falls with variations of rainfall, atmospheric pressure, temperature, etc., whilst coastal regions are affected by tides.

When the water table reaches the surface, springs, lakes, swamps, and similar features can be formed.

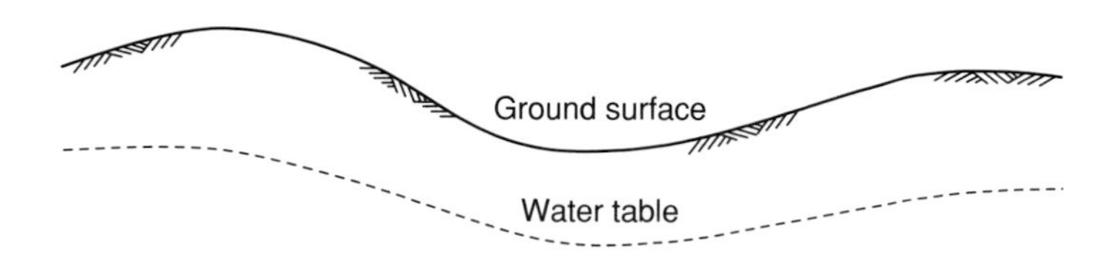

Fig. 2.1 Tendency of the water table to follow the earth's surface.

2.1.2 Aeration zone

Sometimes referred to as the vadose zone, this zone occurs between the water table and the surface, and can be split into three sections.

Capillary fringe

Owing to capillarity, water is drawn up above the water table into the interstices of the soil or rock. Water held in this manner is in a state of suction or negative pressure; its height depends upon the material, and in general the finer the voids the greater the capillary rise. In silts the rise can be as high as 2.5 m and in clays can reach twice that amount, as illustrated later in Section 2.19, which deals with capillarity.

Intermediate belt

As rainwater percolates downward to the water table a certain amount is held in the soil by the action of surface tension, capillarity, adsorption, chemical action, etc. The water retained in this manner is termed held water and is deep enough not to be affected by plants.

Soil belt

This zone is constantly affected by evaporation and plant transpiration. Moist soil in contact with the atmosphere either evaporates water or condenses water into itself until its vapour pressure is equal to atmospheric pressure. Soil water in atmospheric equilibrium is called hygroscopic water, whilst the moisture content (which depends upon relative humidity) is known as the hygroscopic water content.

The various zones are illustrated in Fig. 2.2.

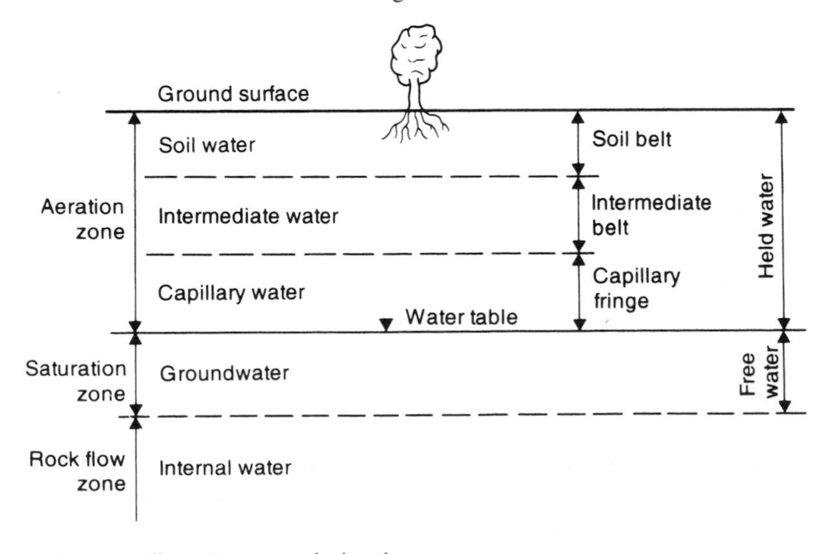

Fig. 2.2 Diagram illustrating types of subsurface water.

2.2 Flow of water through soils

The voids of a soil (and of most rocks) are connected together and form continuous passageways for the movement of water brought about by rainfall infiltration, transpiration of plants, unbalance of chemical energy, variation of intensity of dissolved salts, etc.

When rainfall falls on the soil surface, some of the water infiltrates the surface and percolates downward through the soil. This downward flow results from a gravitational force acting on the water. During flow, some of the water is held in the voids in the *aeration zone* and the remainder reaches the groundwater table and the *saturation zone*. In the aeration zone, flow is said to be *unsaturated*. Below the water table, flow is said to be *saturated*.

2.2.1 *Saturated flow*

The water within the voids of a soil is under pressure. This water, known as pore water, may be static or flowing. Water in saturated soil will flow in response to variations in hydrostatic head within the soil mass. These variations may be natural or induced by excavation or construction.

2.2.2 *Hydraulic or hydrostatic head*

The head of water acting at a point in a submerged soil mass is known as the hydrostatic head and is expressed by Bernoulli's equation:

Hydrostatic head = Velocity head + Pressure head + Elevation head

$$h = \frac{v^2}{2g} + \frac{p}{\gamma_w} + z$$

In seepage problems atmospheric pressure is taken as zero and the velocity is so small that the velocity head becomes negligible; the hydrostatic head is therefore taken as:

$$h = \frac{p}{\gamma_w} + z$$

Excess hydrostatic head

Water flows from points of high to points of low head. Hence flow will occur between two points if the hydrostatic head at one is less than the hydrostatic head at the other, and in flowing between the points the water experiences a head loss equal to the difference in head between them. This difference is known as the excess hydrostatic head.

2.2.3 Seepage velocity

The conduits of a soil are irregular and of small diameter – an average value of the diameter is $D_{10}/5$. Any flow quantities calculated by the theory of pipe flow must be in error and it is necessary to think in terms of an average velocity through a given area of soil rather than specific velocities through particular conduits.

If Q is the quantity of flow passing through an area A in time t, then the average velocity (v) is:

$$v = \frac{Q}{At}$$

This average velocity is sometimes referred to as the seepage velocity. In further work the term velocity will imply average velocity.

2.3 Darcy's law of saturated flow

In 1856 Darcy showed experimentally that a fluid's velocity of flow through a porous medium was directly related to the hydraulic gradient causing the flow, i.e.

$$v \propto i$$

where i = hydraulic gradient (the head loss per unit length), or

$$v = Ci$$

where C = a constant involving the properties of both the fluid and the porous material.

2.4 Coefficient of permeability (k)

In soils we are generally concerned with water flow: the constant C is determined from tests in which the permeant is water. The particular value of the constant C obtained from these tests is known as the coefficient of permeability and is given the symbol k.

It is important to realise that when a soil is said to have a certain coefficient of permeability this value only applies to water (at 20°C). If heavy oil is used as the permeant the value of C would be considerably less than k.

Temperature causes variation in k, but in most soils work this is insignificant.

Provided that the hydraulic gradient is less than 1.0, as is the case in most seepage problems, the flow of water through a soil is linear and Darcy's law applies, i.e.

$$v = ki$$

or

$$Q = Atki$$

or

$$q = Aki \left(\text{where } q = \text{quantity of unit flow} = \frac{Q}{t} \right)$$

From this latter expression a definition of k is apparent: the coefficient of permeability is the rate of flow of water per unit area of soil when under a unit hydraulic gradient.

BS 1377 specifies that the dimensions for k should be m/s and these dimensions are used in this chapter.

Whilst suitable for coarse grained soils, Swartzendruber (1961) showed that Darcy's law is not truly applicable to cohesive soils due to the departure from Newtonian flow (perfect fluid flow) and he therefore proposed a modified flow equation for such soils. Many workers maintain that these variations from Darcy's law are related to the adsorbed water in the soil system, with its much higher viscosity than free water, and also to the soil structure, which can cause small flows along the sides of the voids in the opposite direction to the main flow. Absorbed water is discussed in Section 2.19. Although these effects are not always negligible, the unmodified form of Darcy's law is invariably used in seepage problems as it has the great advantage of simplicity. It may be that, as work in this field proceeds, some form of modification may be adopted.

2.5 Determination of k in the laboratory

The constant head permeameter

The test is described in BS 1377: Part 5 and the apparatus is shown in Fig. 2.3. Water flows through the soil under a head which is kept constant by means of the overflow arrangement. The head loss, h, between two points along the length of the sample, distance l apart, is measured by means of a manometer (in practice there are more than just two manometer tappings).

From Darcy's law: $q = Aki$

The unit quantity of flow, $q = \frac{Q}{t}$

The hydraulic gradient, $i = \frac{h}{l}$

and A = area of sample

Hence k can be found from the expression

$$k = \frac{q}{Ai} \quad \text{or} \quad k = \frac{Ql}{tAh}$$

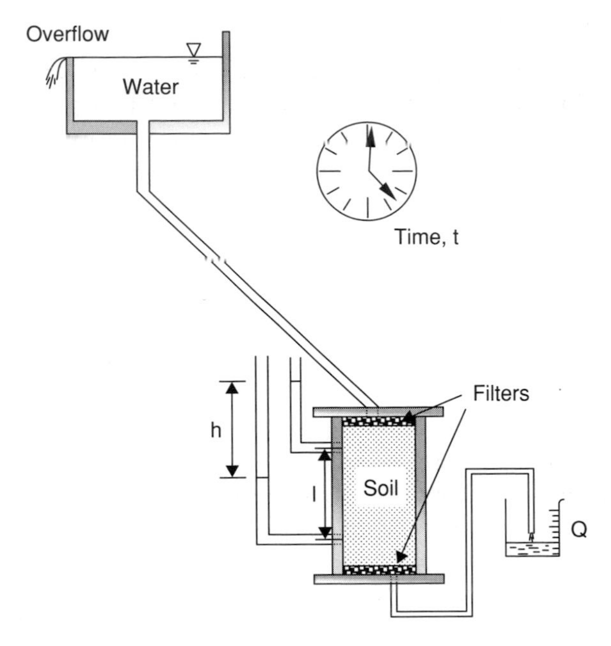

Fig. 2.3 The constant head permeameter.

A series of readings can be obtained from each test and an average value of k determined. The test is suitable for gravels and sand and could be used for many fill materials.

Excel

Example 2.1

In a constant head permeameter test the following results were obtained:

Duration of test	= 4.0 min
Quantity of water collected	= 300 ml
Head difference in manometer	= 50 mm
Distance between manometer tappings	= 100 mm
Diameter of test sample	= 100 mm

Determine the coefficient of permeability in m/s.

Solution

$$A = \frac{\pi \times 100^2}{4} = 7850 \text{ mm}^2 \qquad q = \frac{300}{4 \times 60} = 1.25 \text{ ml/s}$$

$$k = \frac{ql}{Ah} = \frac{1250 \times 100}{7850 \times 50} = 3.18 \times 10^{-1} \text{ mm/s} = 3.2 \times 10^{-4} \text{ m/s}$$

The falling head permeameter

A sketch of the falling head permeameter is shown Fig. 2.4. In this test, which is suitable for silts and some clays, the flow of water through the sample is measured at the inlet. The height, h_1, in the stand pipe is measured and the valve is then opened as a stop clock is started. After a measured time, t, the height to which the water level has fallen, h_2, is determined.

k is given by the formula:

$$k = 2.3 \frac{al}{At} \log_{10} \frac{h_1}{h_2}$$

where

A = cross-sectional area of sample
a = cross-sectional area of stand pipe
l = length of sample.

During the test, the water in the stand pipe falls from a height h_1 to a final height h_2. Let h be the height at some time, t.

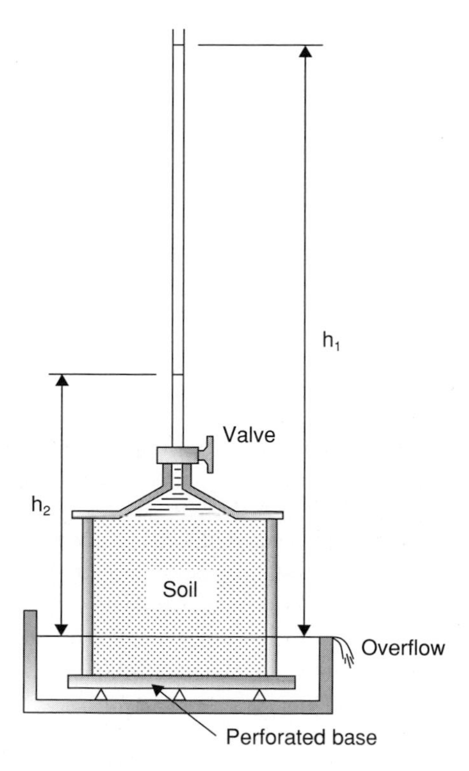

Fig. 2.4 The falling head permeameter.

Consider a small time interval, dt, and let the change in the level of h during this time be −dh (negative as it is a drop in elevation).

The quantity of flow through the sample in time dt = −adh and is given the symbol dQ. Now

$$dQ = Aki\,dt$$

$$= Ak\frac{h}{l}dt = -adh$$

or

$$dt = -\frac{al}{Ak}\frac{dh}{h}$$

Integrating between the test limits:

$$\int_0^t dt = -\frac{al}{Ak}\int_{h_1}^{h_2}\frac{1}{h}dh$$

i.e.

$$t = -\frac{al}{Ak}\ln\frac{h_2}{h_1} = \frac{al}{Ak}\ln\frac{h_1}{h_2}$$

or

$$k = \frac{al}{At}\ln\frac{h_1}{h_2}$$

$$= 2.3\frac{al}{At}\log_{10}\frac{h_1}{h_2}$$

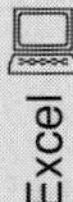

Example 2.2

An undisturbed soil sample was tested in a falling head permeameter. Results were:

Initial head of water in stand pipe	= 1500 mm
Final head of water in stand pipe	= 605 mm
Duration of test	= 281 s
Sample length	= 150 mm
Sample diameter	= 100 mm
Stand-pipe diameter	= 5 mm

Determine the permeability of the soil in m/s.

Solution

$$a = \frac{\pi \times 5^2}{4} = 19.67\ \text{mm}^2 \qquad A = \frac{\pi \times 100^2}{4} = 7854\ \text{mm}^2$$

$$\log_{10}\frac{h_1}{h_2} = \log_{10} 2.48 = 0.3945$$

$$k = \frac{2.3 \times 19.67 \times 150 \times 0.3945}{7854 \times 281} = 1.21 \times 10^{-3}\ \text{mm/s} = 1.2 \times 10^{-6}\ \text{m/s}$$

The hydraulic consolidation cell (Rowe cell)

The Rowe cell (described in Chapter 9) was developed for carrying out consolidation tests. The apparatus can also be used for determining the permeability of a soil. The test procedure is described in BS 1377: Part 6.

2.6 Determination of k in the field

The pumping out test

The pumping out test can be used to measure the average k value of a stratum of soil below the water table and is effective up to depths of about 45 m.

A casing of about 400 mm diameter is driven to bedrock or to impervious stratum. Observation wells of at least 35 mm diameter are put down on radial lines from the casing, and both the casing and the observation wells are perforated to allow easy entrance of water. The test consists of pumping water out from the central casing at a measured rate (q), and observing the resulting drawdown in groundwater level by means of the observation wells.

At least four observation wells, arranged in two rows at right angles to each other should be used although it may be necessary to install extra wells if the initial ones give irregular results. If there is a risk of fine soil particles clogging the observation wells then the wells should be surrounded by a suitably graded filter material. (The design of filters is discussed later in this chapter.)

It may be that the site boundary conditions, e.g. a river, canal or a steep sloping surface of impermeable subsurface rock, a fault or a dyke, do not allow the two rows of observation wells to be placed at right angles. In such circumstances the two rows of wells should be placed parallel to each other and at right angles to the offending boundary.

The minimum distance between the observation wells and the pumping well should be ten times the radius of the pumping well and at least one of the observation wells in each row should be at a radial distance greater than twice the thickness of the ground being tested.

In addition to the observation wells an additional standpipe inside the pumping well is desirable so that a reliable record of the drawdown of the well itself can be obtained.

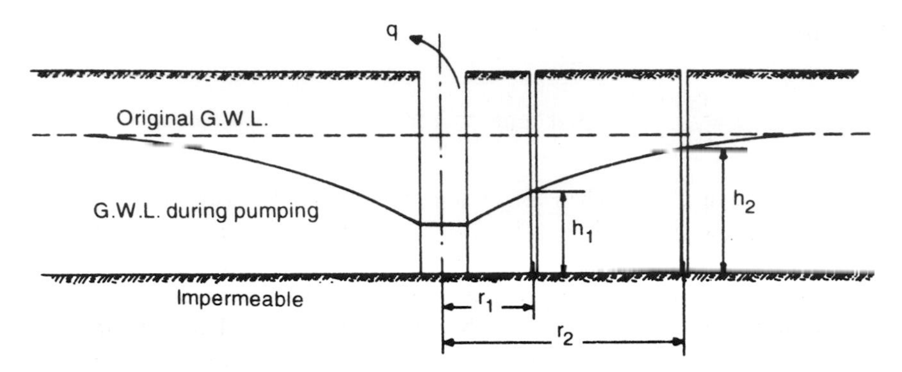

Fig. 2.5 The pumping out test.

Figure 2.5 illustrates conditions during pumping.

Consider an intermediate distance r from the centre of the pumping well and let the height of the GWL above the impermeable layer during pumping be h.

The hydraulic gradient, i, is equal to the slope of the

$$h - r \text{ curve} = \frac{\partial h}{\partial r}$$

where $2\pi rh$ = area of the walls of an imaginary cylinder of radius r and height h. Now

$$q = Aki = 2\pi rhk \frac{\partial h}{\partial r}$$

i.e.

$$q \frac{\partial r}{r} = k2\pi h \partial h$$

and, integrating between test limits:

$$q = \int_{r_1}^{r_2} \frac{1}{r} \partial r = k2\pi \int_{h_1}^{h_2} h \, \partial h$$

$$= k2\pi \left[\frac{h_2^2 - h_1^2}{2} \right]$$

i.e.

$$q \ln \frac{r_2}{r_1} = k\pi(h_2^2 - h_1^2)$$

or

$$k = \frac{q \ln r_2/r_1}{\pi(h_2^2 - h_1^2)} = \frac{2.3q \log_{10} r_2/r_1}{\pi(h_2^2 - h_1^2)}$$

Pumping tests can be expensive as they require the installation of both the pumping and the observation wells as well as suitable pumping and support equipment. Care must be taken in the design of a suitable test programme and, before attempting to carry out any pumping test, reliable data should be obtained about the subsoil profile, if necessary by means of boreholes specially sunk for the purpose.

Suction pumps can be used where the groundwater does not have to be lowered by more than about 5 m below the intake chamber of the pump but for greater depths submersible pumps are generally necessary.

Where a pumping test has been completed in which there are no observation well data, it is still possible to obtain a very rough estimate of k with the formula proposed by Logan (1964), $k \approx 1.22q/(s_w/h_o)$, where h_o is the thickness of confined ground and s_w is the corrected drawdown in the pumping well.

(The observed value of drawdown in the well has to be corrected for head loss through the well screens before being used in calculation of permeability. It is usual to reduce the observed value by 25 per cent to give s_w, unless the head losses from water entering the well are observed to be obviously much greater.)

Example 2.3

A 9.15 m thick layer of sandy soil overlies an impermeable rock. Groundwater level is at a depth of 1.22 m below the top of the soil. Water was pumped out of the soil from a central well at the rate of 5680 kg/min and the drawdown of the water table was noted in two observation wells. These two wells were on a radial line from the centre of the main well at distances of 3.05 and 30.5 m.

During pumping the water level in the well nearest to the pump was 4.57 m below ground level and in the furthest well was 2.13 m below ground level.

Determine an average value for the permeability of the soil in m/s.

Solution

$$q = 5680 \text{ kg/min} = 5.68 \text{ m}^3\text{/min} = 0.0947 \text{ m}^3\text{/s}$$

$$h_1 = 9.15 - 4.57 = 4.58 \text{ m} \qquad h_2 = 9.15 - 2.13 = 7.02 \text{ m}$$

$$k = \frac{q \ln r_2/r_1}{(h_2^2 - h_1^2)\pi} = \frac{0.0947 \times 2.3026}{28.3 \times \pi} = 2.45 \times 10^{-3} \text{ m/s}$$

The pumping in test

Where bedrock level is very deep or where the permeabilities of different strata are required the pumping in test can be used. A casing, perforated for a metre or so at its end, is driven into the ground. At intervals during the driving the rate of flow required to maintain a constant head in the casing is determined and a measure of the soil's permeability obtained.

2.7 Approximation of k

It is obvious that a soil's coefficient of permeability depends upon its porosity, which is itself related to the particle size distribution curve of the soil (a gravel is much more permeable than a clay). It would therefore seem possible to evaluate the permeability of a soil given its particle size distribution, and various formulae have indeed been produced.

A formula often used is the one produced by Hazen (1892), who stated that, for clean sands:

$$k = 10D_{10}^2 \text{ mm/s} = 0.01D_{10}^2 \text{ m/s}$$

where D_{10} = effective size in mm.

Other workers have proposed methods for evaluating the permeability of a soil from its pore size distribution, e.g. Marshall (1958) and Wise (1992).

It should be remembered that no formula is as good as an actual permeability test.

Typical values of permeability

Gravel	$> 10^{-1}$ m/s
Sands	10^{-1} to 10^{-5} m/s
Fine sands, coarse silts	10^{-5} to 10^{-7} m/s
Silts	10^{-7} to 10^{-9} m/s
Clays	$< 10^{-9}$ m/s

Example 2.4

Compute an approximate value for the coefficient of permeability for the soil in Example 1.1.

Solution

$$k = 0.01D_{10}^2 = 0.01 \times 0.17^2 = 2.9 \times 10^{-4} \text{ m/s}$$

2.8 General differential equation of flow

Figure 2.6 shows an elemental cube, of dimensions dx, dy and dz, in an orthotropic soil with an excess hydrostatic head h acting at its centre. (An *orthotropic* soil is a soil whose material properties are different in all directions.)

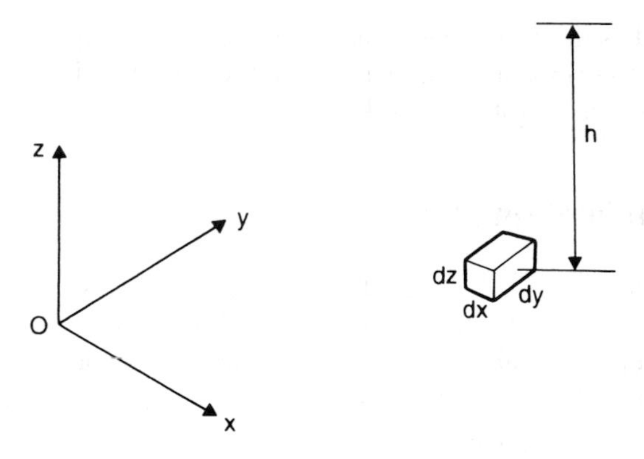

Fig. 2.6 Element in an orthotropic soil.

Let the coefficients of permeability in the coordinate directions x, y and z be k_x, k_y and k_z, respectively. Consider the component of flow in the x direction.

The component of the hydraulic gradient, i_x, at the centre of the element will be

$$i_x = -\frac{\partial h}{\partial x}$$

(Note that it is of negative sign as there is a head loss in the direction of flow.)

The rate of change of the hydraulic gradient i_x along the length of the element in the x direction will be:

$$\frac{\partial i_x}{\partial x} = -\frac{\partial^2 h}{\partial x^2}$$

Hence the gradient at the face of the element nearest the origin

$$= -\frac{\partial h}{\partial x} + \left(\frac{\partial i_x}{\partial x}\right)\left(\frac{-dx}{2}\right)$$

$$= -\frac{\partial h}{\partial x} + \frac{\partial^2 h}{\partial x^2}\frac{dx}{2}$$

From Darcy's law:

$$\text{Flow} = \text{Aki} = k_x\left(-\frac{\partial h}{\partial x} + \frac{\partial^2 h}{\partial x^2}\frac{dx}{2}\right)dy.dz \qquad (1)$$

The gradient at the face furthest from the origin is

$$-\frac{\partial h}{\partial x} + \left(\frac{\partial i_x}{\partial x}\right)\frac{dx}{2}$$

$$= -\frac{\partial h}{\partial x} - \frac{\partial^2 h}{\partial x^2}\frac{dx}{2}$$

Therefore

$$\text{Flow} = k_x\left(-\frac{\partial h}{\partial x} - \frac{\partial^2 h}{\partial x^2}\frac{dx}{2}\right)dy.dz \qquad (2)$$

Expressions (1) and (2) represent respectively the flow into and out of the element in the x direction, so that the net rate of increase of water within the element, i.e. the rate of change of the volume of the element, is (1)−(2).

Similar expressions may be obtained for flow in the y and z directions. The sum of the rates of change of volume in the three directions gives the rate of change of the total volume:

$$\left(\frac{k_x\partial^2 h}{\partial x^2} + \frac{k_y\partial^2 h}{\partial y^2} + \frac{k_z\partial^2 h}{\partial z^2}\right)dx.dy.dz$$

Under the laminar flow conditions that apply in seepage problems there is no change in volume and the above expression must equal zero:

$$\frac{k_x\partial^2 h}{\partial x^2} + \frac{k_y\partial^2 h}{\partial y^2} + \frac{k_z\partial^2 h}{\partial z^2} = 0$$

This is the general expression for three-dimensional flow.

In many seepage problems the analysis can be carried out in two dimensions, the y term usually being taken as zero so that the expression becomes:

$$\frac{k_x\partial^2 h}{\partial x^2} + \frac{k_z\partial^2 h}{\partial z^2} = 0$$

If the soil is isotropic, $k_x = k_z = k$ and the expression is:

$$\frac{\partial^2 h}{\partial x^2} + \frac{\partial^2 h}{\partial z^2} = 0$$

An *isotropic* soil is a soil whose material properties are the same in all directions.

It should be noted that these expressions only apply when the fluid flowing through the soil is incompressible. This is more or less the case in seepage problems when submerged soils are under consideration, but in partially saturated soils considerable volume changes may occur and the expressions are no longer valid.

2.9 Potential and stream functions

The Laplacian equation just derived can be expressed in terms of the two conjugate functions ϕ and ψ.

If we put

$$\frac{\partial\phi}{\partial x} = v_x = ki_x = -\frac{k\partial h}{\partial x} \quad \text{and} \quad \frac{\partial\phi}{\partial z} = v_z = -\frac{k\partial h}{\partial z}$$

then

$$\frac{\partial^2\phi}{\partial x^2} = -\frac{k\partial^2 h}{\partial x^2} \quad \text{and} \quad \frac{\partial^2\phi}{\partial z^2} = -\frac{k\partial^2 h}{\partial z^2}$$

hence

$$\frac{\partial^2\phi}{\partial x^2} + \frac{\partial^2\phi}{\partial z^2} = 0$$

Also, if we put

$$\frac{\partial^2\psi}{\partial x^2} = v_x = \frac{\partial\phi}{\partial z} \quad \text{and} \quad -\frac{\partial\psi}{\partial x} = v_z = \frac{\partial\phi}{\partial z}$$

then

$$\frac{\partial^2\psi}{\partial z^2} = \frac{\partial^2\phi}{\partial x\partial z} \quad \text{and} \quad \frac{\partial^2\psi}{\partial z^2} = -\frac{\partial^2\phi}{\partial x\partial z}$$

hence

$$\frac{\partial^2\psi}{\partial x^2} + \frac{\partial^2\psi}{\partial z^2} = 0$$

ϕ and ψ are known respectively as potential and stream functions. If ϕ is given a particular constant value then an equation of the form h = a constant can be derived (the equation of an equipotential line); if ψ is given a particular constant value then the equation derived is that of a stream or flow line.

Direct integration of these expressions in order to obtain a solution is possible for straightforward cases, and readers interested in this subject are referred to Harr (1962). Generally such integration cannot be easily carried out and a solution obtained by a graphical method in which a flow net is drawn has been used by engineers for many decades. Nowadays, however, much use is made of computer software to find the solution using numerical techniques such as the finite difference and finite element methods. Nevertheless, the method for drawing a flow net by hand is given in Section 2.13 for readers interested in learning the techniques involved.

2.10 Flow nets

The flow of water through a soil can be represented graphically by a flow net, a form of curvilinear net made up of a set of flow lines intersected by a set of equipotential lines.

Flow lines

The paths which water particles follow in the course of seepage are known as flow lines. Water flows from points of high to points of low head, and makes smooth curves when changing direction. Hence we can draw, by hand or by computer, a series of smooth curves representing the paths followed by moving water particles.

Equipotential lines

As the water moves along the flow line it experiences a continuous loss of head. If we can obtain the head causing flow at points along a flow line, then by joining up points of equal potential we obtain a second set of lines known as equipotential lines.

2.11 Hydraulic gradient

The potential drop between two adjacent equipotentials divided by the distance between them is known as the hydraulic gradient. It attains a maximum along a path normal to the equipotentials and in isotropic soil the flow follows the paths of the steepest gradients, so that flow lines cross equipotential lines at right angles.

Figure 2.7 shows a typical flow net representing seepage through a soil beneath a dam. The flow is assumed to be two dimensional, a condition that covers a large number of seepage problems encountered in practice.

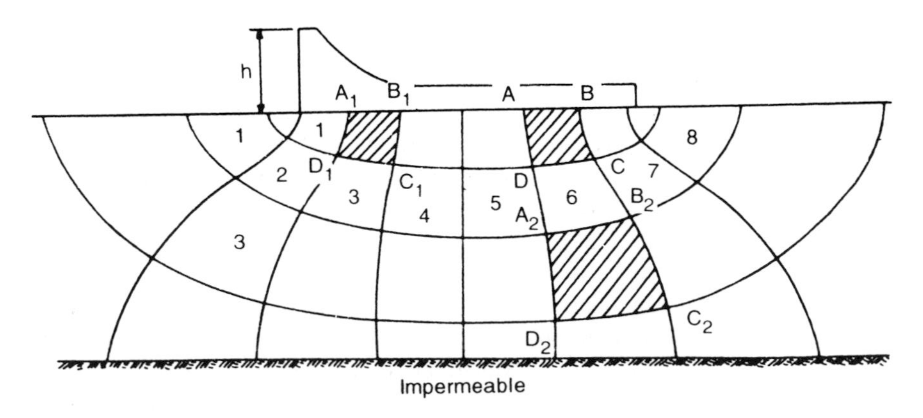

Fig. 2.7 Flow net for seepage beneath a dam.

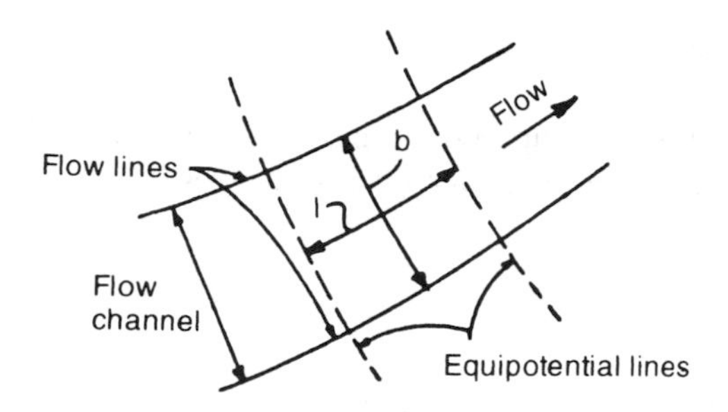

Fig. 2.8 Section of a flow net.

From Darcy's law q = Aki, so if we consider unit width of soil and if Δq = the unit flow through a flow channel (the space between adjacent flow lines), then:

$$\Delta q = b \times 1 \times k \times i = bki$$

where b = distance between the two flow lines.

In Fig. 2.7 the figure ABCD is bounded by the same flow lines as figure $A_1B_1C_1D_1$ and by the same equipotentials as figure $A_2B_2C_2D_2$. For any figure in the net $\Delta q = kib = k\Delta hb/l$, where

Δh = head loss between the two equipotentials
l = distance between the equipotentials (see Fig. 2.8).

Referring to Fig. 2.7:

$$\text{Flow through } A_1B_1C_1D_1 = \Delta q_1 = k\Delta h_1 \frac{b_1}{l_1}$$

$$\text{Flow through } A_2B_2C_2D_2 = \Delta q_2 = k\Delta h_2 \frac{b_2}{l_2}$$

$$\text{Flow through ABCD} = \Delta q = k\Delta h \frac{b}{l}$$

If we assume that the soil is homogeneous and isotropic then k is the same for all figures and it is possible to draw the flow net so that $b_1 = l_1$, $b_2 = l_2$, $b = l$. When we have this arrangement the figures are termed 'squares' and the flow net is a square flow net. With this condition:

$$\frac{b_1}{l_1} = \frac{b_2}{l_2} = \frac{b}{l} = 1.0$$

Since square ABCD has the same flow lines as $A_1B_1C_1D_1$,

$$\Delta q = \Delta q_1$$

Since square ABCD has the same equipotentials as $A_2B_2C_2D_2$,

$$\Delta h = \Delta h_2$$
$$\Rightarrow \quad \Delta q_2 = k\Delta h_2 = k\Delta h = \Delta q = \Delta q_1$$

i.e.

$$\Delta q = \Delta q_1 = \Delta q_2 \quad \text{and} \quad \Delta h = \Delta h_1 = \Delta h_2$$

Hence, in a flow net, where all the figures are square, there is the same quantity of unit flow through each figure and there is the same head drop across each figure.

No figure in a flow net can be truly square, but the vast majority of the figures do approximate to squares in that the four corners of the figure are at right angles and the distance between the flow lines, b, equals the distance between the equipotentials, l. As will be seen, a little imagination is sometimes needed when asserting that a certain figure is a square and some figures are definitely triangular in shape, but provided the flow net is drawn with a sensible number of flow channels (generally five or six) the results obtained will be within the range of accuracy possible. The more flow channels that are drawn the more the figures will approximate to true squares, but the apparent increase in accuracy is misleading and the extra work involved (if drawing by hand) perhaps twelve channels is not worthwhile.

2.12 Calculation of seepage quantities from a flow net

Let

N_d = number of potential drops
N_f = number of flow channels
h = total head loss
q = total quantity of unit flow.

Then

$$\Delta h = \frac{h}{N_d}; \quad \Delta q = \frac{q}{N_f}$$

$$\Delta q = k\Delta h \frac{b}{l} = k\Delta h \left(\text{as } \frac{b}{l} = 1\right)$$

$$\Rightarrow \quad k\frac{h}{N_d} = \frac{q}{N_f}$$

$$\Rightarrow \quad \text{Total unit flow per unit length } (q) = kh\frac{N_f}{N_d}$$

2.13 Drawing a flow net

A soft pencil, a rubber and a pair of dividers or compasses are necessary. The first step is to draw in one flow line, upon the accuracy of which the final correctness of the flow net depends. There are various boundary conditions that help to position this first flow line, including:

(i) Buried surfaces (e.g. the base of the dam, sheet piling), which are flow lines as water cannot penetrate into such surfaces.
(ii) The junction between a permeable and an impermeable material, which is also a flow line; for flow net purposes a soil that has a permeability of one-tenth or less the permeability of the other may be regarded as impermeable.
(iii) The horizontal ground surfaces on each side of the dam, which are equipotential lines.

The procedure is as follows

(a) Draw the first flow line and hence establish the first flow channel.
(b) Divide the first flow channel into squares. At first the use of compasses is necessary to check that in each figure b = 1, but after some practice this sketching procedure can be done by eye.
(c) Project the equipotentials beyond the first flow channel, which gives an indication of the size of the squares in the next flow channel.
(d) With compasses determine the position of the next flow line; draw this line as a smooth curve and complete the squares in the flow channel formed.
(e) Project the equipotentials and repeat the procedure until the flow net is completed.

As an example, suppose that it is necessary to draw the flow net for the conditions shown in Fig. 2.9a. The boundary conditions for this problem are shown in Fig. 2.9b, and the sketching procedure for the flow net is illustrated in Figs c, d, e and f of Fig. 2.9.

If the flow net is correct the following conditions will apply.

(i) Equipotentials will be at right angles to buried surfaces and the surface of the impermeable layer.
(ii) Beneath the dam the outermost flow line will be parallel to the surface of the impermeable layer.

After completing part of a flow net it is usually possible to tell whether or not the final diagram will be correct. The curvature of the flow lines and the direction of the equipotentials indicate if there is any distortion, which tends to be magnified as more of the flow net is drawn and gives a good indication of what was wrong with the first flow line. This line must now be redrawn in its corrected position and the procedure repeated again, amending the first flow line if necessary, until a satisfactory net is obtained.

Generally the number of flow channels, N_f, will not be a whole number, and in these cases an estimate is made as to where the next flow line would be if the impermeable layer was lower. The width of the lowest channel can then be found (in Fig. 2.9f, $N_f = 3.3$).

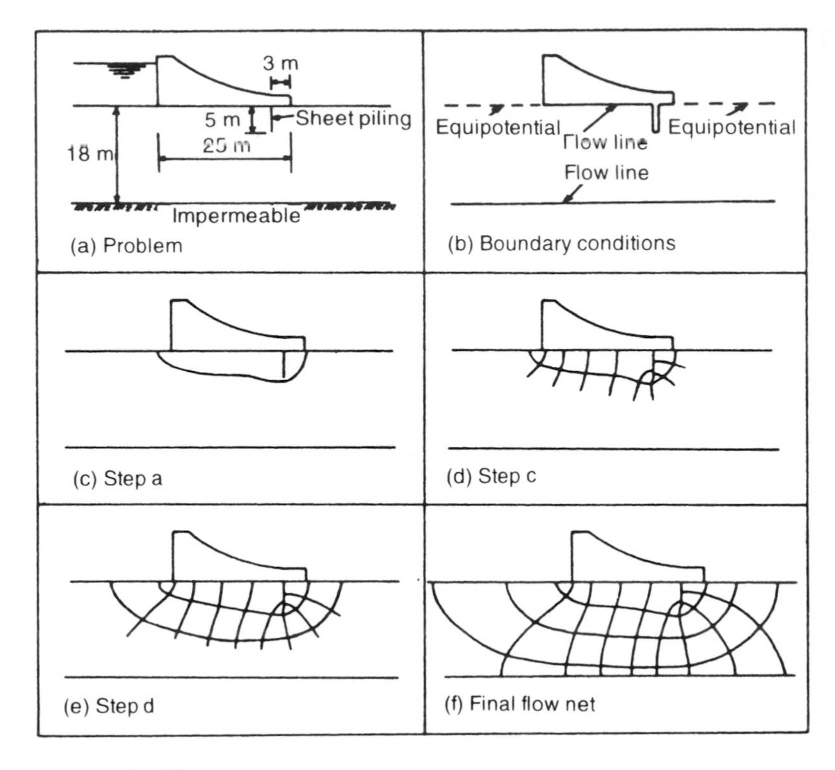

Fig. 2.9 Example of flow net construction.

Note In flow net problems we assume that the permeability of the soil is uniform throughout the soil's thickness. This is a considerable assumption and we see therefore that refinement in the construction of a flow net is unnecessary, since the difference between a roughly sketched net and an accurate one is small compared with the actual flow pattern in the soil and the theoretical pattern assumed.

Example 2.5

Using Fig. 2.9f, determine the loss through seepage under the dam in cubic metres per year if $k = 3 \times 10^{-6}$ m/s and the level of water above the base of the dam is 10 m upstream and 2 m downstream. The length of the dam perpendicular to the plane of seepage is 300 m.

Solution

From the flow net $N_f = 3.3$, $N_d = 9$

Total head loss (h) = 10 − 2 = 8 m

$$q/\text{metre length of dam} = kh\frac{N_f}{N_d} = 3 \times 10^{-6} \times 8 \times \frac{3.3}{9}$$

$$= 8.8 \times 10^{-6}\ \text{m}^3/\text{s}$$

$$\text{Total seepage loss per year} = 300 \times 8.8 \times 60 \times 60 \times 24 \times 365 \times 10^{-6}\ \text{m}^3$$

$$= 83\ 000\ \text{m}^3$$

2.14 Critical hydraulic gradient, i_c

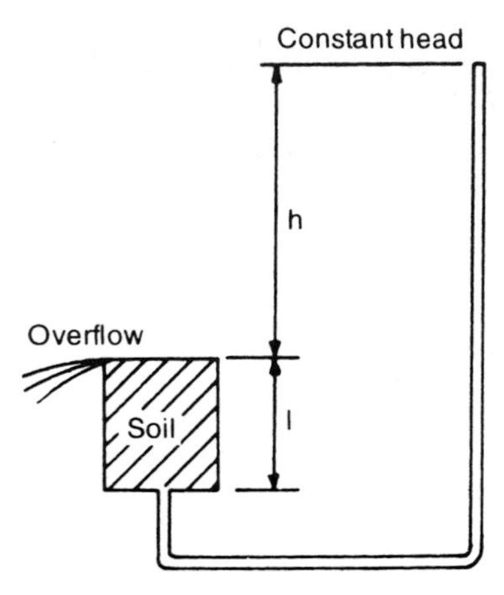

Fig. 2.10 Upward flow through a soil sample.

Figure 2.10 shows a sample of soil encased in a vessel of cross-sectional area A, with upward flow of water through the soil taking place under a constant head. The total head of water above the sample base = h + 1, and the head of water in the sample above the base = 1, therefore the excess hydrostatic pressure acting on the base of the sample = $\gamma_w h$.

If any friction between the soil and the side of the container is ignored, then the soil is on the point of being washed out when the downward forces equal the upward forces:

$$\text{Downward forces} = \text{Buoyant unit weight} \times \text{Volume}$$

$$= \gamma_w \frac{G_s - 1}{1 + e} Al$$

$$\text{Upward forces} = h\gamma_w A$$

i.e.

$$h\gamma_w A = \gamma_w \frac{G_s - 1}{1 + e} Al$$

or when

$$\frac{h}{l} = \frac{G_s - 1}{1 + e} = i_c$$

This particular value of hydraulic gradient is known as the critical hydraulic gradient and has an average value of about unity for most soils. It makes a material a quicksand, which is not a type of soil but a flow condition within the soil. Generally quicksand conditions occur in fine sands when the upward flow conditions achieve this state, but there is no theoretical reason why they should not occur in gravels (or any granular material) provided that the quantity of flow and the head are large enough. Other terms used to describe this condition are 'piping' or 'boiling', but piping will not occur in fine silts and clays due to cohesive forces holding the particles together; instead there can be a heave of a large mass of soil if the upward forces are large enough.

2.15 Seepage forces

Whenever water flows through a soil a seepage force is exerted (as in quicksands). In Fig. 2.10 the excess head h is used up in forcing water through the soil voids over a length l; this head dissipation is caused by friction and, because of the energy loss, a drag or force is exerted in the direction of flow.

The upward force $h\gamma_w A$ represents the seepage force, and in the case of uniform flow conditions it can be assumed to spread uniformly throughout the volume of the soil:

$$\frac{\text{Seepage force}}{\text{Unit volume of soil}} = \frac{h\gamma_w A}{Al} = i\gamma_w$$

This means that in an isotropic soil the seepage force acts in the direction of flow and has a magnitude $= i\gamma_w$ per unit volume.

2.16 Alleviation of piping

The risk of piping can occur in several circumstances, such as a cofferdam (Fig. 2.11a) or the downstream end of a dam (Fig. 2.11b).

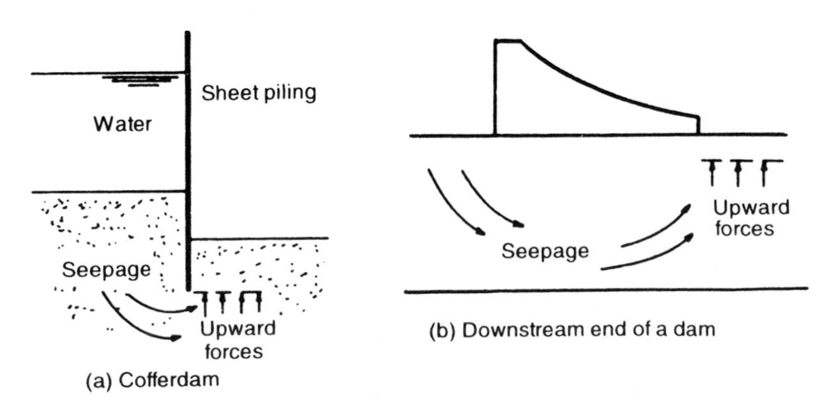

Fig. 2.11 Examples where piping can occur.

In order to increase the factor of safety against piping in these cases two methods can be adopted. The first procedure involves increasing the depth of pile penetration in Fig. 2.11a and inserting a sheet pile at the heel of the dam in Fig. 2.11b; in either case there is an increase in the length of the flow path for the water with a resulting drop in the excess pressure at the critical section. A similar effect is achieved by laying down a blanket of impermeable material for some length along the upstream ground surface.

The second procedure is to place a surcharge or filter apron on top of the downstream side, the weight of which increases the downward forces.

2.17 Design of filters

The design for a filter is largely empirical, but it must be fine enough to prevent soil particles being washed through it and yet coarse enough to allow the passage of water.

Terzaghi's rule

Terzaghi developed the following formulae:

D_{15} filter $> 4 \times D_{15}$ of base material
D_{15} filter $< 4 \times D_{85}$ of base material

The first equation ensures that the filter layer has a permeability several times higher than that of the soil it is designed to protect. The requirement of the second equation is to prevent piping within the filter. The ratio D_{15} (filter)/D_{85} (base) is known as the *piping ratio*.

Penman (1983) gave a simple explanation of the concept behind the Terzaghi rule, and the following section draws heavily from his material.

It is generally accepted that the size of an individual pore in a filter is determined by the size of the solid soil particles that both surround it and are in contact with each other. The Terzaghi rule assumes that, in general, a pore will not allow passage of a soil particle that is greater than one quarter of the average size of the surrounding filter grains. Penman points out that, if the filter grains were spheres, then they could be six times the diameter of the spherical particles that would be caught in the pores. He also refers to the work of Bertram (1940), who showed that the grain size of a uniform filter may be ten times that of a uniform sized soil that it protects.

Although Terzaghi's rule has proved extremely useful for several decades it is now seen to be conservative and has been modified by various workers. However, the basic idea of defining pore size by particle size was left unaltered until 1982 when a new approach to filter design was proposed by Vaughan and Soares who used the permeability of the filter to indicate pore size. A theoretical relationship between pore size and permeability can be written as:

$$k = Ad^2$$

where

d = pore dimension
A = a constant depending on other factors.

If the size of the particle that will just pass through the filter is represented by δ, then it can be expected that the permeability of a satisfactory filter could be found from a relationship of the form:

$$k = A\delta^{x}$$

Laboratory tests, in which clay flocs and fine quartz powders in suspension were passed through sand filters which had uniformity coefficient values between 2 and 6, indicated that a suitable form of the expression is:

$$k = 6.7 \times 10^{-6} \times \delta^{1.52}$$

This new approach is particularly relevant to the filter layers that protect the clay cores of earth dams as it is now possible to design filters to retain clay flocs. Until recently clay cores were surrounded by filters that were too coarse to retain flocs let alone clay particles so that, if there were a hydraulic fracture of the core, the flow of water towards the filter would tend to pull flocs of clay particles from the sides of the crack. If these were able to pass through the filter then the crack would continue to erode.

Pumping and observation wells used in pumping tests are perforated to allow the ingress of water. When it is likely that fine material will be washed into the well it should be surrounded by a layer of filter material. A commonly accepted rule for the choice of a suitable material to form such a filter is:

$$\frac{D_{85}\ \text{(filter)}}{\text{Hole diameter}} > 1.0$$

The required thickness of a filter layer depends upon the flow conditions and can be estimated with the use of Darcy's law of flow.

In addition to meeting these requirements filter material should be well graded, with a grading curve more or less parallel to the base material. All material should pass the 75 mm size sieve and not more than 5 per cent should pass the 0.063 mm size sieve. (See Example 2.7 and Fig. 2.13.)

Reversed filters

Protective filters are usually constructed in layers, each of which is coarser than the one below it, and for this reason they are often referred to as reversed filters. Even when there is no risk of piping, filters are often used to prevent erosion of foundation materials and they are extremely important in earth dams.

Example 2.6

An 8 m thick layer of silty clay is overlying a gravel stratum containing water under artesian pressure. A stand pipe was inserted into the gravel and water rose up the pipe to reach a level of 2 m above the top of the clay (Fig. 2.12).

The clay has a particle specific gravity of 2.7 and a natural moisture content of 30 per cent. The permeability of the silty clay is 3.0×10^{-8} m/s.

It is proposed to excavate 2 m into the soil in order to insert a wide foundation which, when constructed, will exert a uniform pressure of 100 kPa on to its supporting soil.

Determine: (a) the unit rate of flow of water through the silty clay in m^3 per year before the work commences; (b) how safe the foundation will be against heaving (i) at end of excavation (ii) after construction of the foundation.

Solution

(a) Assume that GWL occurs at top of clay.

Head of water in clay = 8 m
Head of water in gravel = 10 m
$\Rightarrow$ Head of water lost in clay = 2 m
q = Aki

Consider a unit area of 1 m^2 then:

$$q = 1 \times 3 \times 10^{-8} \times \frac{2}{8}$$

$$= 7.5 \times 10^{-9} \text{ m}^3/\text{s}$$
$$= 7.5 \times 10^{-9} \times 60 \times 60 \times 24 \times 365$$
$$= 0.237 \text{ m}^3/\text{year per m}^2 \text{ of surface area}$$

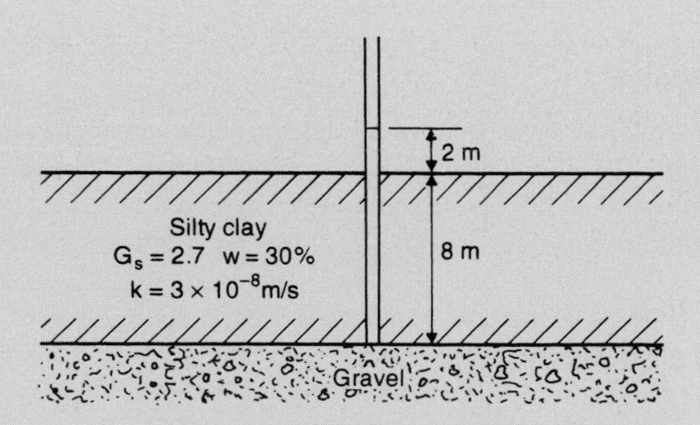

Fig. 2.12 Example 2.6.

(b) (i) $e = wG_s = 0.3 \times 2.7 = 0.81$

$$\gamma_{sat} = \gamma_w \frac{G_s + e}{1 + e} = 9.81 \frac{3.51}{1.81}$$

$$= 19.0 \text{ kN/m}^3$$

Height of clay left above gravel after excavation = 8 − 2 = 6 m
Upward pressure from water on base of clay = 10 × 9.81 = 98.1 kPa
Downward pressure of clay = 6 × 19 = 114 kPa

It is clear that the downward pressure exceeds the upward pressure and thus, on the face of it, the foundation will not be lifted by the buoyant effect of the upward-acting water pressure, i.e. it is *safe*. We can quantify how 'safe' the foundation is against buoyancy by introducing the term *factor of safety*, F:

$$\text{Factor of safety, F} = \frac{\text{Downward pressure}}{\text{Upward pressure}} = \frac{114}{98.1} = 1.16$$

(b) (ii) Downward pressure after construction = 114 + 100 = 214 kPa

$$\text{Factor of safety, F} = \frac{214}{98.1} = 2.18$$

i.e. the factor of safety against buoyant uplift is higher after construction.

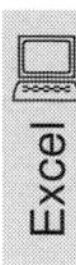

We can also assess the safety against buoyancy using the limit state design approach defined in Eurocode 7 (see Chapter 7). The solution to Example 2.6 when assessed in accordance with Eurocode 7 is available for download.

Units of pressure

The pascal is the stress value of one newton per square metre, 1.0 N/m^2, and is given the symbol Pa. In the example above, pressure has been expressed in kilopascals, kPa. Pressure could have equally been expressed in kN/m^2 as the two terms are synonymous.

1.0 kN/m^2 = 1.0 kPa
1.0 MN/m^2 = 1.0 MPa

Both terms are in common usage in the UK although the kPa is becoming the preferred term. Accordingly, this book has adopted the term kPa.

Example 2.7

Determine the approximate limits for a filter material suitable for the material shown in Fig. 2.13.

Solution

From the particle size distribution curve:

$$D_{15} = 0.01 \text{ mm}; \qquad D_{85} = 0.2 \text{ mm}$$

Using Terzaghi's method:

$$\text{Maximum size of } D_{15} \text{ for filter} = 4 \times D_{85} \text{ of base} = 4 \times 0.2 = 0.8 \text{ mm}$$
$$\text{Minimum size of } D_{15} \text{ for filter} = 4 \times D_{15} \text{ of base} = 4 \times 0.01 = 0.04 \text{ mm}$$

This method gives two points on the 15 per cent summation line. Two lines can be drawn through these points roughly parallel to the grading curve of the soil, and the space between them is the range of material suitable as a filter (Fig. 2.13).

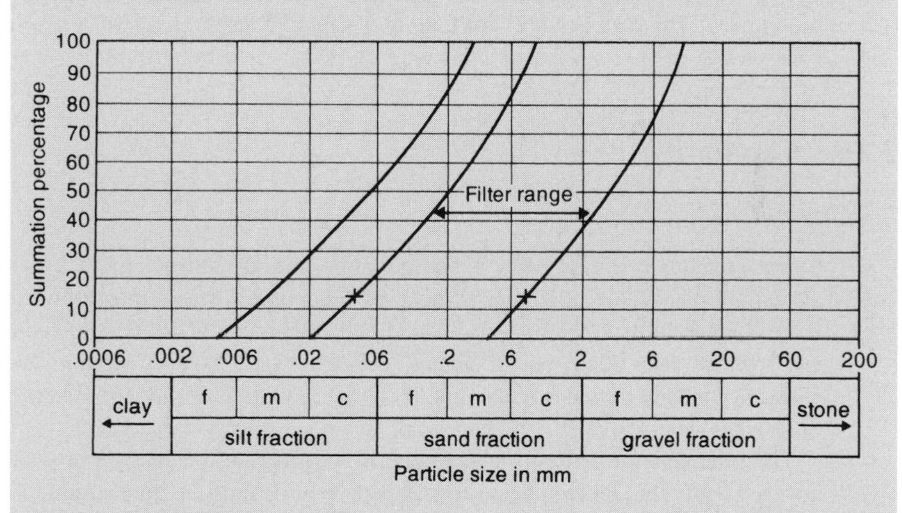

Fig. 2.13 Example 2.7.

2.18 Total and effective stress

The stress that controls changes in the volume and strength of a soil is known as the *effective stress*. In Chapter 1 it was seen that a soil mass consists of a collection of mineral particles with voids between them. These voids are filled with water, air and water, or air only (see Fig. 1.10).

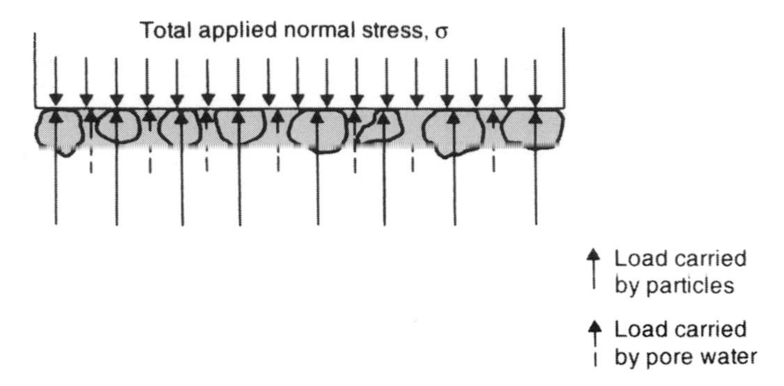

Fig. 2.14 Load carried by soil particles and pore water.

For the moment let us consider saturated soils only. When a load is applied to such a soil it will be carried by the water in the soil voids (causing an increase in the pore water pressure) or by the soil skeleton (in the form of grain to grain contact stresses), or else it will be shared between the water and the soil skeleton as illustrated in Fig. 2.14.

The portion of the total stress carried by the soil particles is known as the effective stress, σ'. The load carried by the water gives rise to an increase in the pore water pressure, u. The determination of total and effective stress in a soil is examined in Chapter 4.

2.19 Capillarity

Surface tension

Surface tension is the property of water that permits the surface molecules to carry a tensile force. Water molecules attract each other and, within a mass of water, these forces balance out. At the surface, however, the molecules are only attracted inwards and towards each other, which creates surface tension. Surface tension causes the surface of a body of water to attempt to contract into a minimum area: hence a drop of water is spherical.

The phenomenon is easily understood if we imagine the surface of water to be covered with a thin molecular skin capable of carrying tension. Such a skin, of course, cannot exist on the surface of a liquid, but the analogy can explain surface tension effects without going into the relevant molecular theories.

Surface tension is given the symbol T and can be defined as the force in newtons per millimetre length that the water surface can carry. T varies slightly with temperature, but this variation is small and an average value usually taken for the surface tension of water is 0.000 075 N/mm (0.075 N/m).

The fact that surface tension exists can be shown by the familiar laboratory experiment in which an open-ended glass capillary tube is placed in a basin of water subjected to atmospheric pressure; the rise of water within the tube is then observed. It is seen that the water wets the glass and the column of water within the tube reaches a definite height above the liquid in the basin.

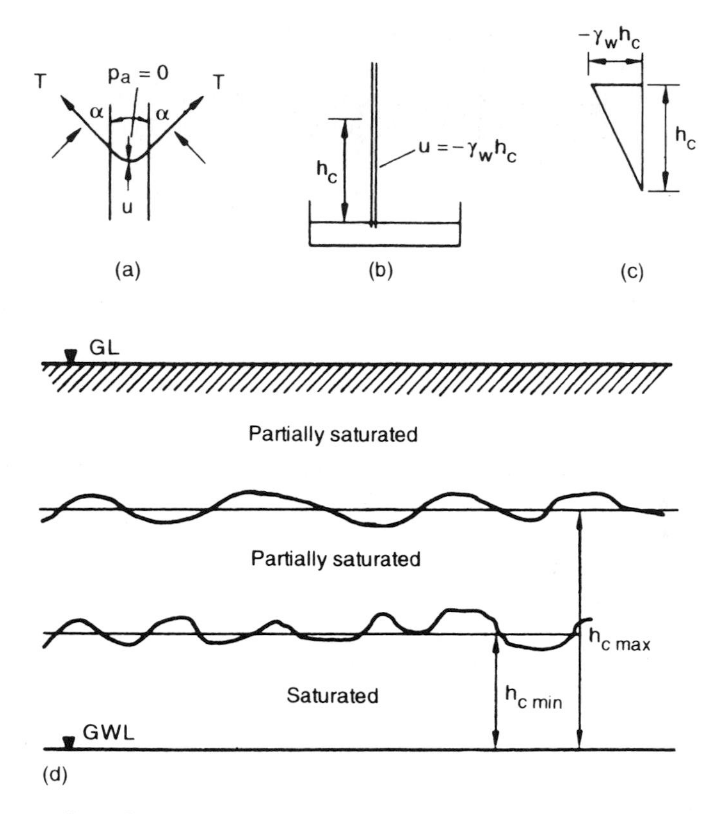

Fig. 2.15 Capillary effects.

The surface of the column forms a meniscus such that the curved surface of the liquid is at an angle α to the walls of the tube (Fig. 2.15a). The arrangement of the apparatus is shown in Fig. 2.15b.

The base of the column is at the same level as the water in the basin and, as the system is open, the pressure must be atmospheric. The pressure on the top surface of the column is also atmospheric. There are no externally applied forces that keep the column in position, which shows that there must be a tensile force acting within the surface film of the water.

Let

Height of water column = h_c
Radius of tube = r
Unit weight of water = γ_w

If we take atmospheric pressure as datum, i.e. the air pressure = 0, we can equate the vertical forces acting at the top of the column:

$$T 2\pi r \cos\alpha + u\pi r^2 = 0$$

$$\Rightarrow \quad u = \frac{-2T\cos\alpha}{r}$$

Hence, as expected, we see that u is negative; the water within the column is in a state of suction. The maximum value of this negative pressure is $\gamma_w h_c$ and occurs at the top of the column. The pressure distribution along the length of the tube is shown in Fig. 2.15c. It is seen that the water pressure gradually increases with loss of elevation to a value of 0 at the base of the column.

An expression for the height h_c can be obtained by substituting $u = -\gamma_w h_c$ in the above expression to yield:

$$h_c = \frac{2T \cos \alpha}{\gamma_w r}$$

From the two expressions we see that the magnitudes of both $-u$ and h_c increase as r decreases.

A further interesting point is that, if we assume that the weight of the capillary tube is negligible, then the only vertical forces acting are the downward weight of the water column supported by the surface tension at the top and the reaction at the base support of the tube. The tube must therefore be in compression. The compressive force acting on the walls of the tube will be constant along the length of the water column and of magnitude $2\pi T \cos \alpha$ (or $\pi r^2 h_c \gamma_w$).

It may be noted that for pure water in contact with clean glass which it wets, the value of angle α is zero. In this case the radius of the meniscus is equal to the radius of the tube and the derived formulae can be simplified by removing the term $\cos \alpha$.

With the use of the expression for h_c we can obtain an estimate of the theoretical capillary rise that will occur in a clay deposit. The average void size in a clay is as about 3 μm and, taking $\alpha = 0$, the formula gives $h_c = 5.0$ m. This possibly explains why the voids exposed when a sample of a clay deposit is split apart are often moist. However, capillary rises of this magnitude seldom occur in practice as the upward velocity of the water flow through a clay in the capillary fringe is extremely small and is often further restricted by adsorbed water films, which considerably reduce the free diameter of the voids.

Capillary effects in soil

The region within which water is drawn above the water table by capillarity is known as the capillary fringe. A soil mass, of course, is not a capillary tube system, but a study of theoretical capillarity enables one to determine a qualitative view of the behaviour of water in the capillary fringe of a soil deposit. Water in this fringe can be regarded as being in a state of negative pressure, i.e. at pressure values below atmospheric. A diagram of a capillary fringe appears in Fig. 2.15d.

The minimum height of the fringe, h_{cmin}, is governed by the maximum size of the voids within the soil. Up to this height above the water table the soil will be sufficiently close to full saturation to be considered as such.

The maximum height of the fringe h_{cmax} is governed by the minimum size of the voids. Within the range h_{cmin} to h_{cmax} the soil can be only partially saturated.

Terzaghi and Peck (1948) give an approximate relationship between h_{cmax} and grain size for a granular soil:

$$h_{cmax} = \frac{C}{eD_{10}} \text{ mm}$$

where C is a constant depending upon the shape of the grains and the surface impurities (varying from 10.0 to 50.0 mm^2) and D_{10} is the effective size expressed in millimetres.

Owing to the irregular nature of the conduits in a soil mass it is not possible, even approximately, to calculate moisture content distributions above the water table from the theory of capillarity. This is a problem of importance in highway engineering and is best approached by the concept of soil suction.

Contact moisture

Water in a moist sand occurs in the form of droplets between the points of contact of the individual grains and is therefore referred to as contact moisture. This water, retained by surface tension, holds the particles together and produces a resistance to applied stress resembling cohesion. The effect is temporary and will be destroyed if the sand is dried or flooded.

Contact moisture has two main effects, the first being to augment the strength of the sand and enable slopes to be at steeper angles than if dry or completely submerged. The second effect is the phenomenon known as bulking: a damp sand will not settle to the same volume as an equal weight of dry sand since the temporary cohesion prevents the grains from moving downwards, with the result that the volume of a damp sand may be 20 to 30 per cent more than for a dry sand similarly placed.

Adsorbed water

Clay minerals, due to their shape and crystalline structure, have surface forces that exceed gravity. These forces attract water molecules to the soil particles and hence, in cohesive soils, each particle is coated with a molecular film of water. This water, known as adsorbed water, has properties that differ considerably from ordinary water: its viscosity, density and boiling point are all higher than normal water and it does not freeze under frost action. It is generally believed that adsorbed water gives fine-grained soils their plastic properties.

2.20 Earth dams

Seepage patterns through an earth dam

As the upper flow line is subjected to atmospheric pressure, the boundary conditions are not completely defined and it is consequently difficult to sketch a flow net until this line has been located.

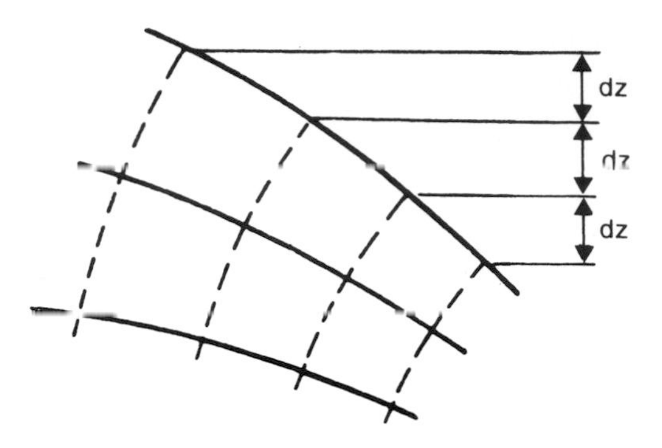

Fig. 2.16 Part of a flow net for an earth dam.

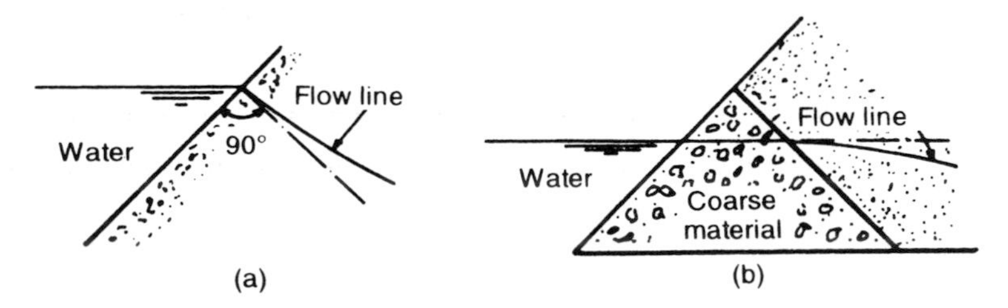

Fig. 2.17 Conditions at the start of an upper flow line.

Part of such a flow net is shown in Fig. 2.16. It has already been shown that the hydrostatic head at a point is the summation of velocity, pressure and elevation heads. As the top flow line is at atmospheric pressure the only type of head that can exist along it is elevational, so that between each successive point where an equipotential cuts an upper flow line there must be equal drops in elevation. This is the first of three conditions that must be satisfied by the upper flow line.

The second condition is that, as the upstream face of the dam is an equipotential, the flow line must start at right angles to it (see Fig. 2.17a), but an exception to this rule is illustrated in Fig. 2.17b where the coarse material is so permeable that the resistance to flow is negligible and the upstream equipotential is, in effect, the downstream face of the coarse material. The top flow line cannot be normal to this surface as water with elevation head only cannot flow upwards, so that in this case the flow line starts horizontally.

The third condition concerns the downstream end of the flow line where the water tends to follow the direction of gravity and the flow line either exits at a tangent to the downstream face of the dam (Fig. 2.18a) or, if a filter of coarse material is inserted, takes up a vertical direction in its exit into the filter (Fig. 2.18b).

Types of flow occurring in an earth dam

From Fig. 2.18 it is seen that an earth dam may be subjected to two types of seepage: when the dam rests on an impermeable base the discharge must occur on the surface of the downstream slope (the upper flow line for this case is shown in Fig. 2.19a),

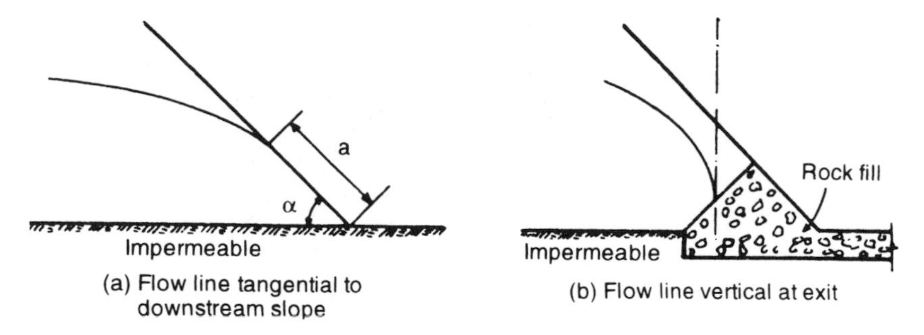

Fig. 2.18 Conditions at the downstream end of an upper flow line.

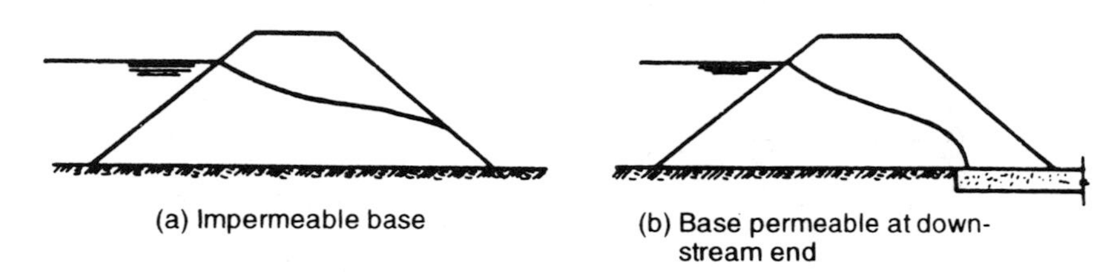

Fig. 2.19 Types of seepage through an earth dam.

whereas when the dam sits on a base that is permeable at its downstream end the discharge will occur within the dam (Fig. 2.19b). This is known as the underdrainage case. From a stability point of view underdrainage is more satisfactory since there is less chance of erosion at the downstream face and the slope can therefore be steeper but, on the other hand, seepage loss is smaller in dams resting on impermeable bases.

Parabolic solutions for seepage through an earth dam

In Fig. 2.20 is shown the cross-section of a theoretical earth dam, the flow net of which consists of two sets of parabolas. The flow lines all have the same focus, F, as do the equipotential lines. Apart from the upstream end, actual dams do not differ

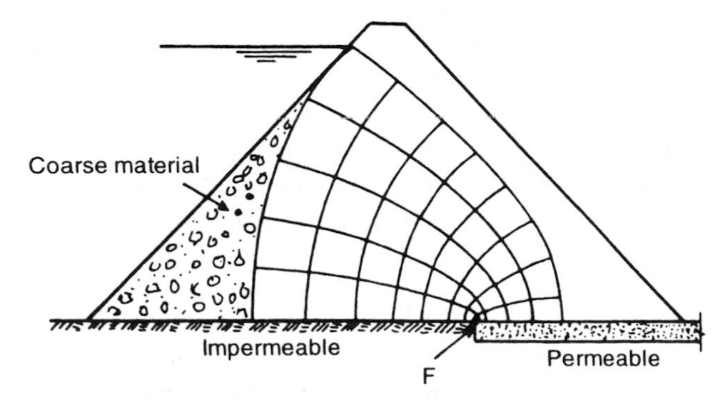

Fig. 2.20 Flow net for a theoretical earth dam.

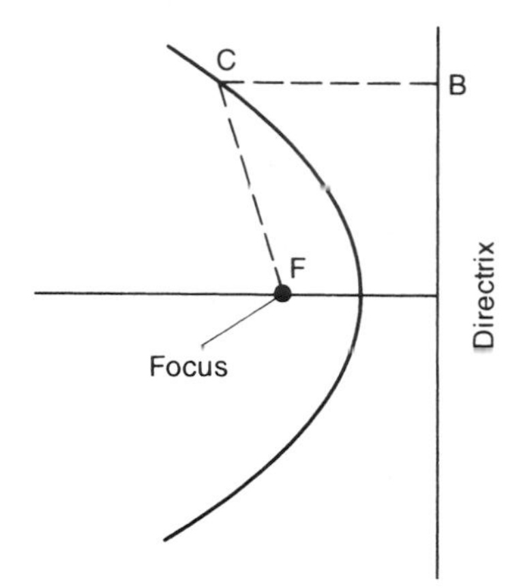

Fig. 2.21 The parabola.

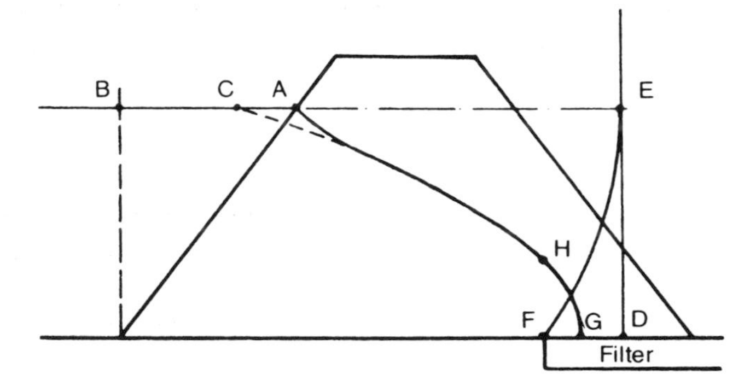

Fig. 2.22 Determination of upper flow line.

substantially from this imaginary example, so that the flow net for the middle and downstream portions of the dam are similar to the theoretical parabolas. (A parabola is a curve such that any point along it is equidistant from both a fixed point, called the focus, and a fixed straight line, called the directrix. In Fig. 2.21, FC = CB.)

The graphical method for determining the phreatic surface in an earth dam was evolved by Casagrande (1937) and involves the drawing of an actual parabola and then the correction of the upstream end. Casagrande showed that this parabola should start at the point C of Fig. 2.22 (which depicts a cross-section of a typical earth dam) where AC $\approx$ 0.3AB (the focus, F, is the upstream edge of the filter). To determine the directrix, draw, with compasses, the arc of the circle as shown, using centre C and radius CF; the vertical tangent to this arc is the directrix, DE. The parabola passing through C, with focus F and directrix DE, can now be constructed. Two points that are easy to establish are G and H, as FG = GD and FH = FD; other points can quickly be obtained using compasses. Having completed the parabola a correction is made as shown to its upstream end so that the flow line actually starts from A.

This graphical solution is only applicable to a dam resting on a permeable material. When the dam is sitting on impermeable soil the phreatic surface cuts the

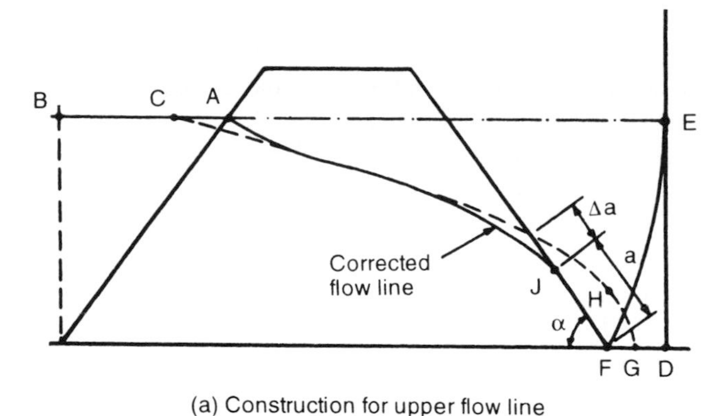

(a) Construction for upper flow line

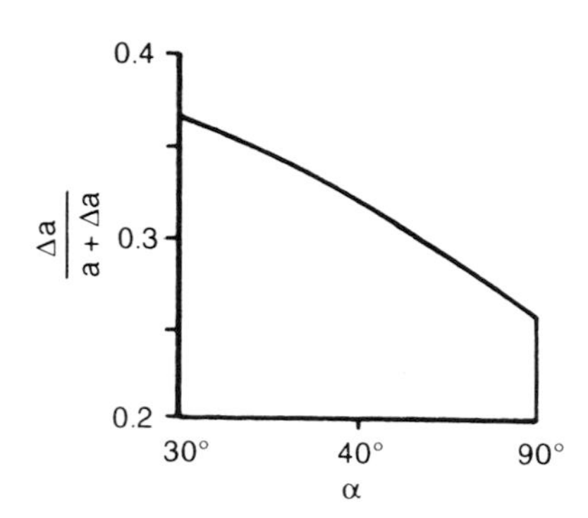

(b) Relationship between a and Δa (after Casagrande)

Fig. 2.23 Dam resting on an impermeable soil.

downstream slope at a distance (a) up the slope from the toe (Fig. 2.18a). The focus, F, is the toe of the dam, and the procedure is now to establish point C as before and draw the theoretical parabola (Fig. 2.23a). This theoretical parabola will actually cut the downstream face at a distance Δa above the actual phreatic surface; Casagrande established a relationship between a and Δa in terms of α, the angle of the downstream slope (Fig. 2.23b). In Fig. 2.23 the point J can thus be established and the corrected flow line sketched in as shown.

Tailings dams and lagoons

The population's increasing consumption of minerals and fossil fuels dictates that further exploitation of the world's resources must involve minimum wastage and must be carried out more efficiently and completely.

In line with this philosophy, the mining industry is now extracting mineral deposits that previously would not have been considered suitable. Old mines are now being remined in order to obtain minerals from deposits that were left undisturbed as their extraction was considered uneconomical when these mines were first in operation.

The extent of the problem can be appreciated by considering the copper industry. The quality of metal ore now being extracted is such that it contains some 0.3 to 0.4 per cent copper whereas, in the 1960s and 1970s, only deposits containing some 3 to 4 per cent of the metal were worked.

In order to extract the maximum amount of mineral from such low percentages it is necessary to crush the rock bearing ore to sand and silt sizes with the result that the mining industry now has the problem of disposing of large volumes of fine waste.

Penman (1985) points out that, for 1974 alone, world production of non-ferrous metals amounted to some 16×10^6 tons yet, by 1982, the annual production of tailings had increased to more than 5×10^9 tons. It is seen that mine waste material makes up the largest tonnage of material handled, considerably exceeding the amount of all other waste materials created by other industrial activities.

Generally mining waste material, left at the end of the extraction process, is wet so that the simplest and cheapest method for its disposal is to remove it hydraulically by pumping it through a pipeline to a suitable point of deposition.

The waste material is disgorged from the pipe into a lagoon which is retained by a tailings dam. Tailings dams can be of massive dimensions. The South African gold industry tends to use lagoons of rectangular plan with side lengths often more than a kilometre. With coalmining the discharge of tailings into a lagoon usually consists of slurry (untreated coal below the 2 mm size and soil together with crushed rock particles). With such a coal content the sediment that eventually fills a coal lagoon may have a commercial value as a low grade fuel.

Whether or not a lagoon will set up seepage forces in the tailings dam that supports it depends upon many factors. If the volume of water pumped into a lagoon is extremely large then it is possible for the whole base of the lagoon to become covered with water and, in theory, a flow net condition similar to that shown in Fig. 2.24a could be established. If no overflow arrangement is provided, and if the quantity of water is vast enough, then the level in the pond could rise until flow takes place through the embankment on the pond (Fig. 2.24b). A deposit of silt will very quickly form on the bottom of the pond and penetrate slightly into the spoil material (probably to about 0.3 m), resulting in the creation of a layer of low permeability material above the relatively high permeability of the tip.

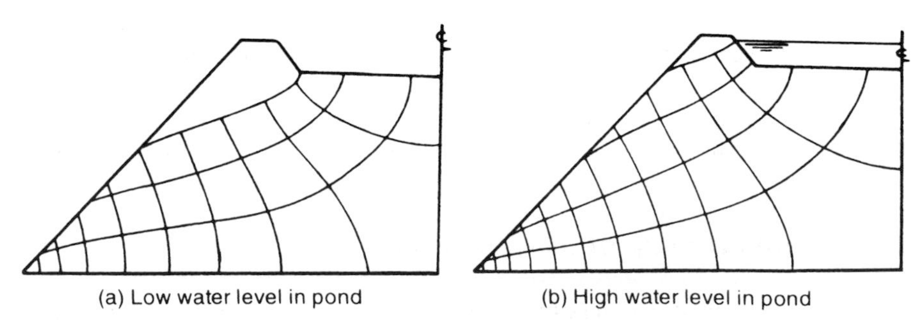

Fig. 2.24 Form of flow nets that may occur in a spoil heap if the quantity of effluent supplied is vast.

The seepage rate through this silt layer will be slow, and will decrease further as the thickness of the deposit is augmented. It may be that the quantity of water flowing through the silt will be insufficient to supply the spoil material with enough water to maintain a continuous flow condition and hydrostatic water pressures within the tip will consequently disappear. How long this effect will take to occur depends upon such factors as the particle sizes in the effluent, the permeability of the tip and the input quantity; any conclusions drawn from one lagoon will not necessarily apply to another.

If the input quantity is low then the bottom of the pond, in the initial stages, will not become covered over with water and silt will be deposited first at the inlet pipe and then gradually across the pond. Water will tend to run over this layer so that there may be a pool at the edge of the silt deposit, but there is little chance that this amount of water will set up seepage forces in the spoil heap. Eventually, when the bottom of the pond has silted over, water may cover a large area of the lagoon, but with the layer of low permeability material overlying the tip and the small quantity of water involved there is no chance of seepage forces developing, provided an overflow arrangement is installed.

Some lagoons are constructed on natural ground. If the input quantity is large enough and if the level of water is not controlled by an overflow, then seepage through the embankment can occur. The flow conditions in this case will be much the same as for an earth dam sitting on an impermeable base.

The situation of the phreatic surface cutting through the downstream slope must be avoided. It can lead to erosion, slips and back sapping that may be extremely dangerous, particularly in large tailings dams. If it is proposed that a lagoon is to be installed on the top of an existing spoil heap then, just as for an earth dam, the design should include drainage provision to ensure that the upper surface of any water flow is kept below the downstream slope.

It would appear that lagoons can only become hazardous from a seepage point of view when the quantity of effluent supplied in the initial stages is sufficient to set up continuous flow conditions through the tip before the silt deposit has been formed. The pumping of large quantities of mine water, with little solid material, into a newly excavated lagoon could give rise to this condition.

Another problem arising from lagoons is the risk of deterioration in the spoil heap material. Even though there are not continuous flow conditions a large quantity of water will percolate through a spoil heap supporting a lagoon and some gravel-like materials rapidly achieve the consistency of mud when immersed in water. In such cases a gradual change in the strength properties of the tip material may lead to a stability failure even after the lagoon has ceased to be used, so before a lagoon is installed it is advisable to check on the suitability of the material in the tip from a deterioration point of view.

2.21 The problem of stratification

Most loosely tipped deposits are probably isotropic, i.e. the value of permeability in the horizontal direction is the same as in the vertical direction. Present practice is to construct most spoil heaps, earth embankments and dams by spreading the soil in loose layers which are then compacted. This construction technique results in a greater value of permeability in the horizontal direction, k_x, than that in the vertical direction (the anisotropic condition). The value of k_z is usually 1/5 to 1/10 the value of k_x.

The general differential equation for flow was derived earlier in this chapter:

$$k_x \frac{\partial^2 h}{\partial x^2} + k_y \frac{\partial^2 h}{\partial y^2} + k_z \frac{\partial^2 h}{\partial z^2} = 0$$

For the two dimensional, i.e. anisotropic, case the equation becomes:

$$k_x \frac{\partial^2 h}{\partial x^2} + k_z \frac{\partial^2 h}{\partial z^2} = 0$$

Unless k_x is equal to k_z the equation is not a true Laplacian and cannot therefore be solved by a flow net.

To obtain a graphical solution the equation must be written in the form:

$$\frac{k_x}{k_z} \frac{\partial^2 h}{\partial z^2} + \frac{\partial^2 h}{\partial z^2} = 0$$

or

$$\frac{\partial^2 h}{\partial x_t^2} + \frac{\partial^2 h}{\partial z^2} = 0$$

where

$$\frac{1}{x_t^2} = \frac{k_x}{k_z} \cdot \frac{1}{x^2}$$

or

$$x_t^2 = x^2 \frac{k_z}{k_x}$$

i.e.

$$x_t = x \sqrt{\frac{k_z}{k_x}}$$

This equation is Laplacian and involves the two co-ordinate variables x_t and z. It can be solved by a flow net provided that the net is drawn to a vertical scale of z and a horizontal scale of

$$x_t = z \sqrt{\frac{k_z}{k_x}}$$

2.22 Calculation of seepage quantities in an anisotropic soil

This is exactly as before:

$$q = kh \frac{N_f}{N_d}$$

and the only problem is what value to use for k.

Using the transformed scale a square flow net is drawn and N_f and N_d obtained. If we consider a 'square' in the transformed flow net it will appear as shown in Fig. 2.25a. The same figure, drawn to natural scales (i.e. scale x = scale z), will appear as shown in Fig. 2.25b.

Let k′ be the effective permeability for the anisotropic condition. Then k′ is the operative permeability in Fig. 2.25a.

Hence, in Fig. 2.25a:

$$\text{Flow} = ak' \frac{\Delta h}{a} = k'\Delta h$$

and, in Fig. 2.25b:

$$\text{Flow} = ak_x \frac{\Delta h}{a \cdot \sqrt{\frac{k_x}{k_z}}} = \sqrt{k_x k_z}\ \Delta h$$

i.e. the effective permeability, $k' = \sqrt{k_x k_z}$.

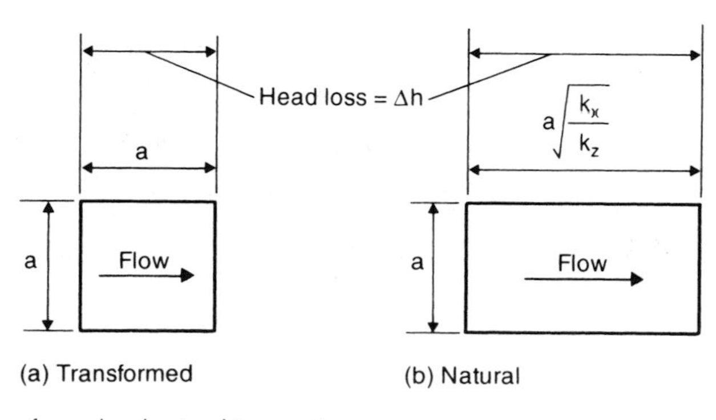

Fig. 2.25 Transformed and natural 'squares'.

Example 2.8

The cross-section of an earth dam is shown in Fig. 2.26a. Assuming that the water level remains constant at 35 m, determine the seepage loss through the dam. The width of the dam is 300 m, and the soil is isotropic with $k = 5.8 \times 10^{-7}$ m/s.

Solution

The flow net is shown in Fig. 2.26b, from it $N_f = 4.0$ and $N_d = 14$.

$$q/\text{metre width of dam} = 5.8 \times 35 \times \frac{4.0}{14} \times 60 \times 60 \times 24 \times 10^{-7}$$

$$= 5.0 \times 10^{-1} \text{ m}^3/\text{day}$$

$$\text{Total seepage loss per day} = 300 \times 5.0 \times 10^{-1}$$

$$= 150 \text{ m}^3/\text{day}$$

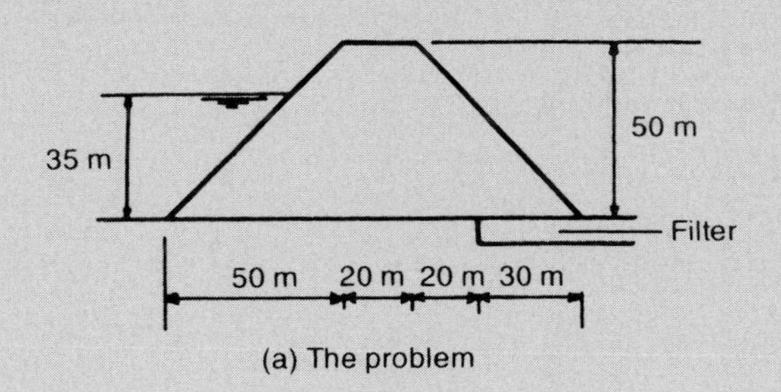

(a) The problem

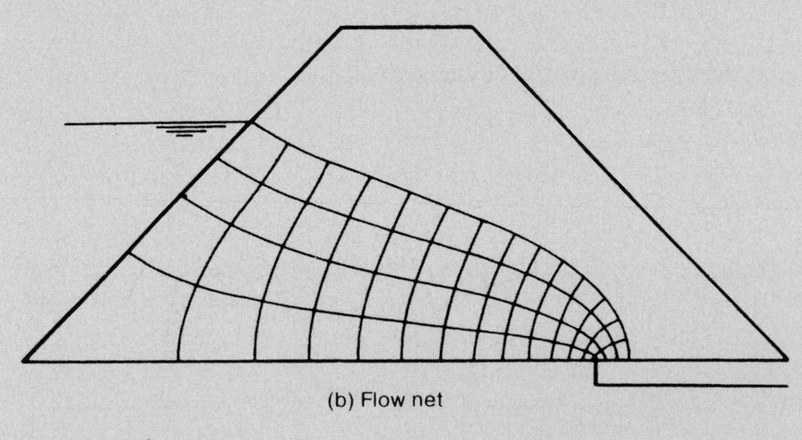

(b) Flow net

Fig. 2.26 Example 2.8.

Example 2.9

A dam has the same details as in Example 2.8 except that the soil is anisotropic with $k_x = 5.8 \times 10^{-7}$ m/s and $k_z = 2.3 \times 10^{-7}$ m/s.

Determine the seepage loss through the dam.

Solution

$$\text{Transformed scale for x direction } x_t = x\sqrt{\frac{k_z}{k_x}}$$

$$= x\sqrt{\frac{2.3}{5.8}}$$

$$= 0.63x$$

This means that, if the vertical scale is 1 : 500, then the horizontal scale is 0.63 : 500 or 1 : 794.

The flow net is shown in Fig. 2.27.

From the flow net, $N_f = 5.0$ and $N_d = 14$.

$$k' = \sqrt{k_x k_z} = \sqrt{5.8 \times 2.3} \times 10^{-7} = 3.65 \times 10^{-7} \text{ m/s}$$

$$\text{Total seepage loss} = 300 \times 3.65 \times 35 \times \frac{5.0}{14} \times 60 \times 60 \times 24 \times 10^{-7}$$

$$= 118 \text{ m}^3/\text{day}$$

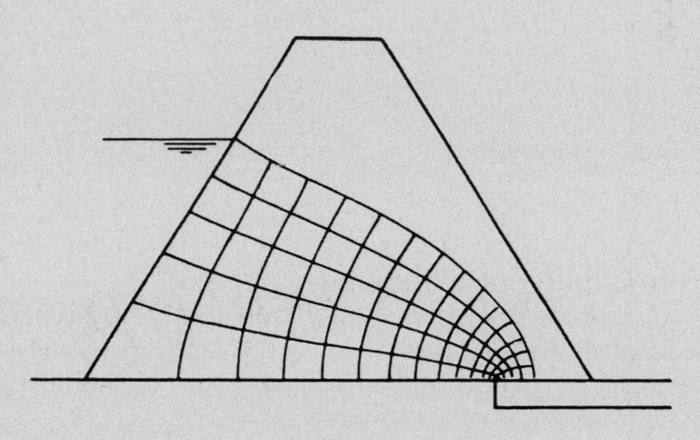

Fig. 2.27 Example 2.9.

Example 2.10

A dam has the same details as in Example 2.8, except that there is no filter drain at the toe.

Solution

The flow net is shown in Fig. 2.28, from it $N_f = 4.0$ and $N_d = 18$ (average). From the flow net it is also seen that $a + \Delta a = 22.4$ m. Now $\alpha = 45°$, and hence (according to Casagrande):

$$\frac{\Delta a}{a + \Delta a} = 0.34 \text{ (taken from Fig. 2.23b).}$$

Hence $\Delta a = 7.6$ m.

$$\text{Total seepage loss} = 300 \times 5.8 \times \frac{4}{18} \times 35 \times 60 \times 60 \times 24 \times 10^{-7}$$

$$= 117 \text{ m}^3\text{/day}$$

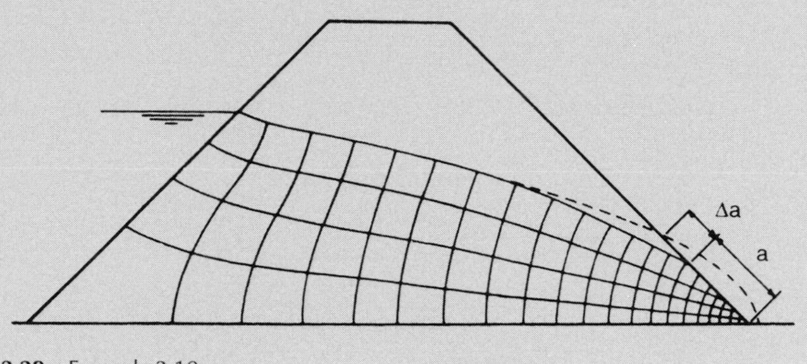

Fig. 2.28 Example 2.10.

2.23 Permeability of sedimentary deposits

A sedimentary deposit may consist of several different soils and it is often necessary to determine the average values of permeability in two directions, one parallel to the bedding planes and the other at right angles to them.

Let there be n layers of thicknesses $H_1, H_2, H_3, \ldots H_n$.
Let the total thickness of the layers be H.
Let $k_1, k_2, k_3, \ldots k_n$ be the respective coefficients of permeability for each individual layer.
Let the average permeability for the whole deposit be k_x for flow parallel to the bedding planes and k_z for flow perpendicular to this direction.

Consider flow parallel to the bedding planes:

$$\text{Total flow} = q = Ak_x i$$

where A = total area and i = hydraulic gradient.

This total flow must equal the sum of the flow through each layer, therefore:

$$Ak_x i = A_1k_1i + A_2k_2i + A_3k_3i + \cdots + A_nk_ni$$

Considering unit width of soil:

$$Hk_x i = i(H_1k_1 + H_2k_2 + H_3k_3 + \cdots + H_nk_n)$$

hence

$$k_x = \frac{H_1k_1 + H_2k_2 + H_3k_3 + \cdots + H_nk_n}{H}$$

Considering flow perpendicular to the bedding planes:

$$\text{Total flow} = q = Ak_z i = Ak_1i_1 = Ak_2i_2 = Ak_3i_3 = Ak_ni_n$$

Considering unit area:

$$q = k_z i = k_1i_1 = k_2i_2 = k_3i_3 = k_ni_n$$

Now

$$k_z i = k_z \frac{(h_1 + h_2 + h_3 + \cdots + h_n)}{H}$$

where h_1, h_2, h_3, etc., are the respective head losses across each layer.

Now

$$\frac{k_1h_1}{H_1} = q; \quad \frac{k_2h_2}{H_2} = q; \quad \frac{k_3h_3}{H_3} = q; \quad \frac{k_nh_n}{H_n} = q$$

$$\Rightarrow \quad h_1 = \frac{qH_1}{k_1}; \quad h_2 = \frac{qH_2}{k_2}; \quad h_3 = \frac{qH_3}{k_3}; \quad h_n = \frac{qH_n}{k_n}$$

$$\Rightarrow \quad \frac{k_z\left(\dfrac{qH_1}{k_1} + \dfrac{qH_2}{k_2} + \dfrac{qH_3}{k_3} + \cdots + \dfrac{qH_n}{k_n}\right)}{H} = q$$

hence

$$k_z = \frac{H}{\dfrac{H_1}{k_1} + \dfrac{H_2}{k_2} + \dfrac{H_3}{k_3} + \cdots + \dfrac{H_n}{k_n}}$$

Example 2.11

A three-layered soil system consisting of fine sand, coarse silt, and fine silt in horizontal layers is shown in Fig. 2.29.

Beneath the fine silt layer there is a stratum of water-bearing gravel with a water pressure of 155 kPa. The surface of the sand is flooded with water to a depth of 1 m.

Determine the quantity of flow per unit area in mm^3/s, and the excess hydrostatic heads at the sand/coarse silt and the coarse silt/fine silt interfaces.

Solution

$$k_z = \frac{12}{\frac{4}{2.0\times10^{-4}} + \frac{4}{4.0\times10^{-5}} + \frac{4}{2.0\times10^{-5}}} = 3.75\times10^{-5}\ \text{mm/s}$$

Taking the top of the gravel as datum:

Head of water due to artesian pressure = 15.5 m
Head of water due to groundwater = 3 × 4 + 1 = 13 m

Therefore excess head causing flow = 15.5 − 13 = 2.5 m.

$$\text{Flow} = q = Aki = 3.75 \times \frac{2.5}{12} \times 10^{-5} = 7.8 \times 10^{-6}\ \text{mm}^3/\text{s}$$

This quantity of flow is the same through each layer.

Excess head loss through fine silt:

$$\text{Flow} = 7.8\times10^{-6} = 2.0\times10^{-5} \times \frac{h}{4}$$

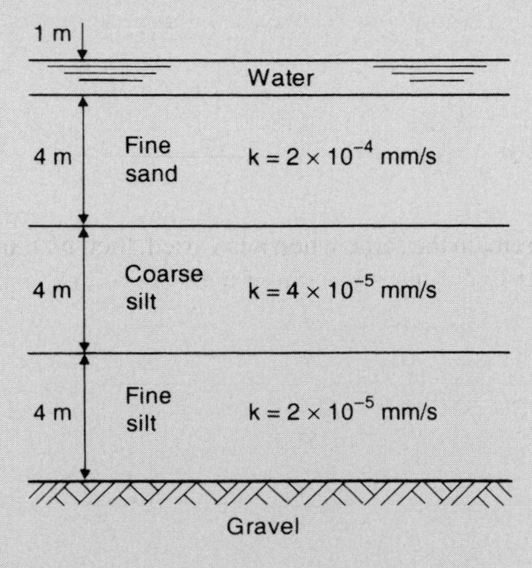

Fig. 2.29 Example 2.11.

Therefore

$$h = \frac{31.2 \times 10^{-6}}{2 \times 10^{-5}} = 1.56 \text{ m}$$

Excess head loss through coarse silt:

$$h = \frac{7.8 \times 10^{-6} \times 4}{4 \times 10^{-5}} = 0.78 \text{ m}$$

Excess head loss through fine sand:

$$h = \frac{7.8 \times 10^{-6} \times 4}{2 \times 10^{-4}} = 0.16 \text{ m}$$

Excess head at interface between fine and coarse silt

$$= 2.5 - 1.56 = 0.94 \text{ m}$$

Excess head at interface between fine sand and coarse silt

$$= 0.94 - 0.78 = 0.16 \text{ m}$$

2.24 Seepage through soils of different permeability

When water seeps from a soil of permeability k_1 into a soil of permeability k_2 the principle of the square flow net is no longer valid. If we consider a flow net in which the head drop across each figure, Δh, is a constant then, as has been shown, the flow through each figure is given by the expression:

$$\Delta q = k \, \Delta h \, \frac{b}{l}$$

If Δq is to remain the same when k is varied, then b/l must also vary. As an illustration of this effect consider the case of two soils with $k_1 = k_2/3$.

Then

$$\Delta q = k_1 \, \Delta h \, \frac{b_1}{l_1}$$

and

$$\Delta q = k_2 \, \Delta h \, \frac{b_2}{l_2} = 3k_1 \, \Delta h \, \frac{b_2}{l_2}$$

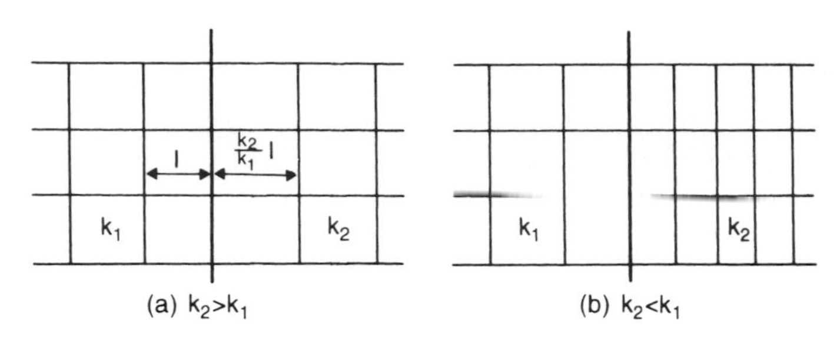

Fig. 2.30 Effect of variation of permeability on a flow net.

i.e.

$$\frac{b_1}{l_1} = 3\,\frac{b_2}{l_2}$$

If the portion of the flow net in the soil of permeability k_1 is square, then:

$$\frac{b_2}{l_2} = \frac{1}{3} \quad \text{or} \quad \frac{b_2}{l_2} = \frac{k_1}{k_2}$$

The effect on a flow net is illustrated in Fig. 2.30.

2.25 Refraction of flow lines at interfaces

An interface is the surface or boundary between two soils. If the flow lines across an interface are normal to it, then there will be no refraction and the flow net appears as

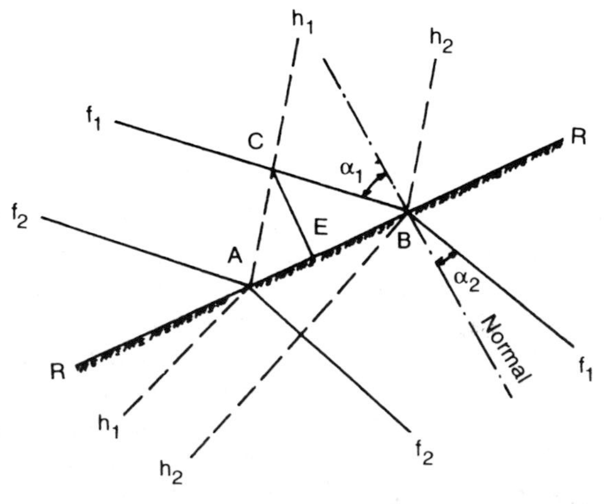

Fig. 2.31 Flow across an interface when the flow lines are at an angle to it.

shown in Fig. 2.30. When the flow lines meet the interface at some acute angle to the normal, then the lines are bent as they pass into the second soil.

In Fig. 2.31 let RR be the interface of two soils of permeabilities, k_1 and k_2. Consider two flow lines, f_1 and f_2, making angles to the normal of α_1 and α_2 in soils 1 and 2 respectively.

Let f_1 cut RR in B and f_2 cut RR in A.
Let h_1 and h_2 be the equipotentials passing through A and B respectively and let the head drop between them be Δh.

With uniform flow conditions the flow into the interface will equal the flow out. Consider flow normal to the interface.

In soil (1):

$$\text{Normal component of hydraulic gradient} = \frac{\text{head drop along CE}}{\text{CE}}$$

$$\text{Head drop from A to E} = \Delta h\,\frac{\text{AE}}{\text{CE}} = \text{Head drop from C to E}$$

$$\Rightarrow \quad q_1 = \text{AB}k_1\,\frac{\Delta h}{\text{CE}}\,\frac{\text{AE}}{\text{AB}} = k_1\,\Delta h\,\frac{\text{AE}}{\text{CE}} = \frac{k_1\,\Delta h}{\tan\alpha_1}$$

Similarly it can be shown that, in soil (2):

$$q_2 = \frac{k_2\,\Delta h}{\tan\alpha_2}$$

Now $q_1 = q_2$,

$$\Rightarrow \quad \frac{k_1}{k_2} = \frac{\tan\alpha_1}{\tan\alpha_2}$$

A flow net which illustrates the effect is illustrated in Fig. 2.32.

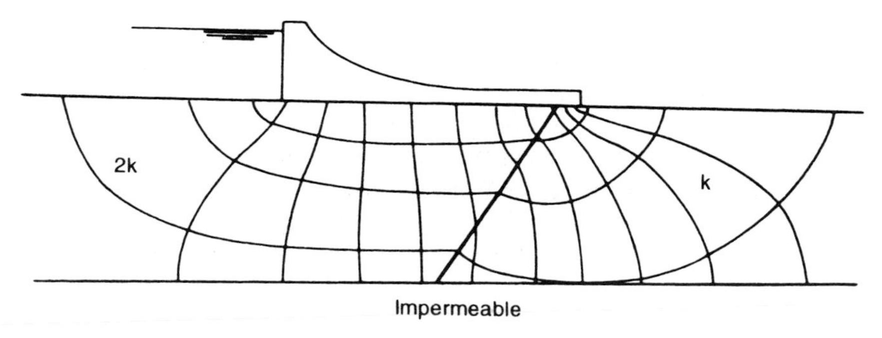

Fig. 2.32 Flow net for seepage through two soils of different permeabilities.

Exercises

Exercise 2.1

In a falling head permeameter test on a fine sand the sample had a diameter of 76 mm and a length of 152 mm with a stand pipe of 12.7 mm diameter. A stopwatch was started when h was 508 mm and read 19.6 s when h was 254 mm; the test was repeated for a drop from 254 mm to 127 mm and the time was 19.4 s.

Determine an average value for k in m/s.

Answer 1.5×10^{-4} m/s

Exercise 2.2

A sample of coarse sand 150 mm high and 55 mm in diameter was tested in a constant head permeameter. Water percolated through the soil under a head of 400 mm for 6.0 s and the discharge water had a mass of 400 g.

Determine k in m/s.

Answer 1.05×10^{-2} m/s

Exercise 2.3

In order to determine the average permeability of a bed of sand 12.5 m thick overlying an impermeable stratum, a well was sunk through the sand and a pumping test carried out. After some time the discharge was 850 kg/min and the drawdowns in observation wells 15.2 m and 30.4 m from the pump were 1.625 m and 1.360 m respectively. If the original water table was at a depth of 1.95 m below ground level, find the permeability of the sand (in m/s) and an approximate value for the effective grain size.

Answer $k = 6.7 \times 10^{-4}$ m/s, $D_{10} \approx 0.26$ mm

Exercise 2.4

A cylinder of cross-sectional area 2500 mm^2 is filled with sand of permeability 5.0 mm/s. Water is caused to flow through sand under a constant head using the arrangement shown in Fig. 2.33.

Determine the quantity of water discharged in 10 min.

Answer 9×10^6 mm^3

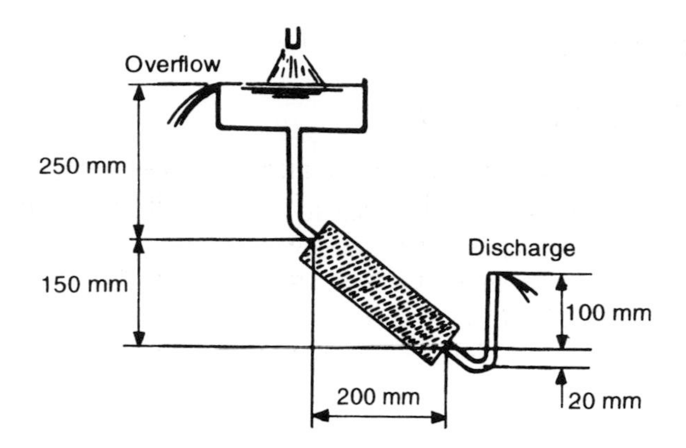

Fig. 2.33 Exercise 2.4.

Exercise 2.5

The specific gravity of particles of a sand is 2.54 and their porosity is 45 per cent in the loose state and 37 per cent in the dense state. What are the critical hydraulic gradients for these two states?

Answer 0.85, 0.98

Exercise 2.6

A large open excavation was made into a stratum of clay with a saturated unit weight of 17.6 kN/m^3. When the depth of the excavation reached 7.63 m the bottom rose, gradually cracked, and was flooded from below with a mixture of sand and water; subsequent borings showed that the clay was underlain by a bed of sand with its surface at a depth of 11.3 m.

Compute the elevation to which water would have risen from the sand into a drill hole before excavation was started.

Answer 6.45 m above top of sand

Exercise 2.7

A soil deposit consists of three horizontal layers of soil: an upper stratum A (1 m thick), a middle stratum B (2 m thick) and a lower stratum C (3 m thick). Permeability tests gave the following values:

Soil A 3×10^{-1} mm/s
Soil B 2×10^{-1} mm/s
Soil C 1×10^{-1} mm/s

Determine the ratio of the average permeabilities in the horizontal and vertical directions.

Answer 1.22

Beneath the deposit there is a gravel layer subjected to artesian pressure, the surface of the deposit coinciding with the groundwater level. Standpipes show that the fall in head across soil A is 150 mm. Determine the value of the water pressure in the gravel.

Answer 80 kPa

Chapter 3
Shear Strength of Soils

The property that enables a material to remain in equilibrium when its surface is not level is known as its shear strength. Soils in liquid form have virtually no shear strength and even when solid have shear strengths of relatively small magnitudes compared with those exhibited by steel or concrete.

To appreciate this section some knowledge of the relevant strength of materials is useful. A brief summary of this subject is set out below.

3.1 Friction

Consider a block of weight W resting on a horizontal plane (Fig. 3.1a). The vertical reaction, R, equals W, and there is consequently no tendency for the block to move. If a small horizontal force, H, is now applied to the block and the magnitude of H is such that the block still does not move, then the reaction R will no longer act vertically but becomes inclined at some angle, α, to the vertical.

By considering the equilibrium of forces, first in the horizontal direction and then in the vertical direction, it is seen that:

Horizontal component of R = H = R sin α
Vertical component of R = W = R cos α (Fig. 3.1b)

The angle α is called the angle of obliquity and is the angle that the reaction on the plane of sliding makes with the normal to that plane. If H is slowly increased in magnitude a stage will be reached at which sliding is imminent; as H is increased the

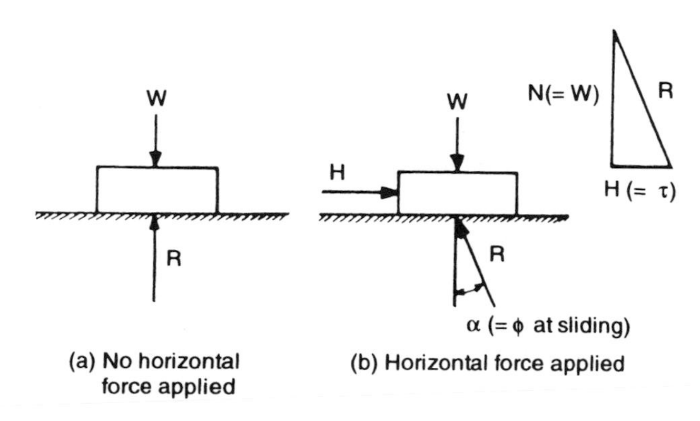

Fig. 3.1 Friction.

value of α will also increase until, when sliding is imminent, α has reached a limiting value, ϕ. If H is now increased still further the angle of obliquity, ϕ, will not become greater and the block, having achieved its maximum resistance to horizontal movement, will move (ϕ is known as the angle of friction). The frictional resistance to sliding is the horizontal component of R and, as can be seen from the triangle of forces in Fig. 3.1b, equals N tan ϕ where N equals the normal force on the surface of sliding (in this case N = W).

As α only achieves the value ϕ when sliding occurs, it is seen that the frictional resistance is not constant and varies with the applied load until movement occurs. The term tan ϕ is known as the coefficient of friction.

3.2 Complex stress

When a body is acted upon by external forces then any plane within the body will be subjected to a stress that is generally inclined to the normal to the plane. Such a stress has both a normal and a tangential component and is known as a *compound*, or *complex*, *stress* (Fig. 3.2).

Principal plane

A plane that is acted upon by a normal stress only is known as a principal plane, there is no tangential, or shear, stress present. As is seen in the next section dealing with principal stress, only three principal planes can exist in a stressed mass.

Principal stress

The normal stress acting on a principal plane is referred to as a principal stress. At every point in a soil mass, the applied stress system that exists can be resolved into three principal stresses that are mutually orthogonal. The principal planes corresponding to these principal stresses are called the major, intermediate and minor principal planes and are so named from a consideration of the principal stresses that act upon them. The largest principal stress, σ_1, is known as the major principal stress and acts on the major principal plane. Similarly the intermediate principal stress, σ_2, acts on the intermediate principal plane whilst the smallest principal stress, σ_3, called the minor principal stress, acts on the minor principal plane. Critical stress values and obliquities generally occur on the two planes normal to the intermediate plane so that the effects of σ_2 can be ignored and a two-dimensional solution is possible.

σ = σ_n + τ

Fig. 3.2 Complex stress.

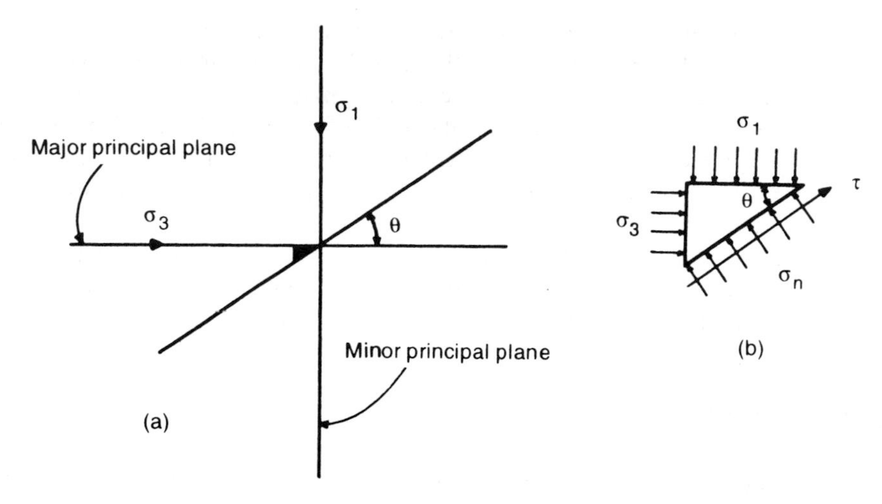

Fig. 3.3 Stress induced by two principal stresses, σ_1 and σ_3, on a plane inclined at θ to σ_3.

3.3 The Mohr circle diagram

Figure 3.3a shows a major principal plane, acted upon by a major principal stress, σ_1, and a minor principal plane, acted upon by a minor principal stress, σ_3.

By considering the equilibrium of an element within the stressed mass (Fig. 3.3b) it can be shown that on any plane, inclined at angle θ to the direction of the major principal plane, there is a shear stress, τ, and a normal stress, σ_n. The magnitudes of these stresses are:

$$\tau = \frac{\sigma_1 - \sigma_3}{2} \sin 2\theta$$

$$\sigma_n = \sigma_3 + (\sigma_1 - \sigma_3) \cos^2 \theta$$

These formulae lend themselves to graphical representation, and it can be shown that the locus of stress conditions for all planes through a point is a circle (generally called a Mohr circle). In order to draw a Mohr circle diagram a specific convention must be followed, all normal stresses (including principal stresses) being plotted along the axis OX while shear stresses are plotted along the axis OY. For most cases the axis OX is horizontal and OY is vertical, but the diagram is sometimes rotated to give correct orientation. The convention also assumes that the direction of the major principal stress is parallel to axis OY, i.e. the direction of the major principal plane is parallel to axis OX.

To draw the diagram, first lay down the axes OX and OY, then set off OA and OB along the OX axis to represent the magnitudes of the minor and major principal stresses respectively, and finally construct the circle with diameter AB. This circle is the locus of stress conditions for all planes passing through the point A, i.e. a plane passing through A and inclined to the major principal plane at angle θ cuts the

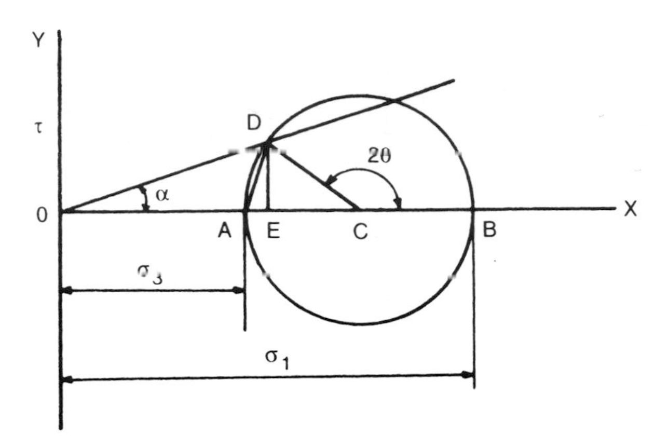

Fig. 3.4 Mohr circle diagram.

circle at D. The coordinates of the point D are the normal and shear stresses on the plane (Fig. 3.4).

$$\begin{aligned}
\text{Normal stress} = \sigma_n = \text{OE} = \text{OA} + \text{AE} &= \sigma_3 + \text{AD} \cos\theta \\
&= \sigma_3 + \text{AB} \cos^2\theta \\
&= \sigma_3 + (\sigma_1 - \sigma_3)\cos^2\theta \\
\text{Shear stress} = \tau = \text{DE} &= \text{DC} \sin(180^\circ - 2\theta) \\
&= \text{DC} \sin 2\theta \\
&= \frac{\sigma_1 - \sigma_3}{2} \sin 2\theta
\end{aligned}$$

In Fig. 3.4, OE and DE represent the normal and shear stress components of the complex stress acting on plane AD. From the triangle of forces ODE it can be seen that this complex stress is represented in the diagram by the line OD, whilst the angle DOB represents the angle of obiquity, α, of the resultant stress on plane AD.

Limit conditions

It has been stated that the maximum shearing resistance is developed when the angle of obliquity equals its limiting value, ϕ. For this condition the line OD becomes a tangent to the stress circle, inclined at angle ϕ to axis OX (Fig. 3.5).

An interesting point that arises from Fig. 3.5 is that the failure plane is not the plane subjected to the maximum value of shear stress. The criterion of failure is maximum obliquity, not maximum shear stress. Hence, although the plane AE in Fig. 3.5 is subjected to a greater shear stress than the plane AD, it is also subjected to a larger normal stress and therefore the angle of obliquity is less than on AD, which is the plane of failure.

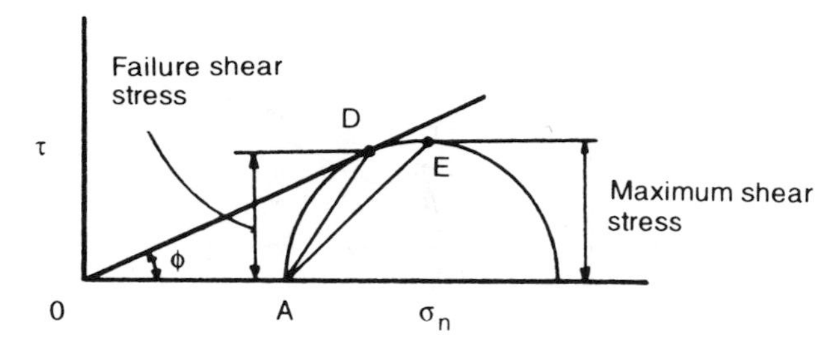

Fig. 3.5 Mohr circle diagram for limit shear resistance.

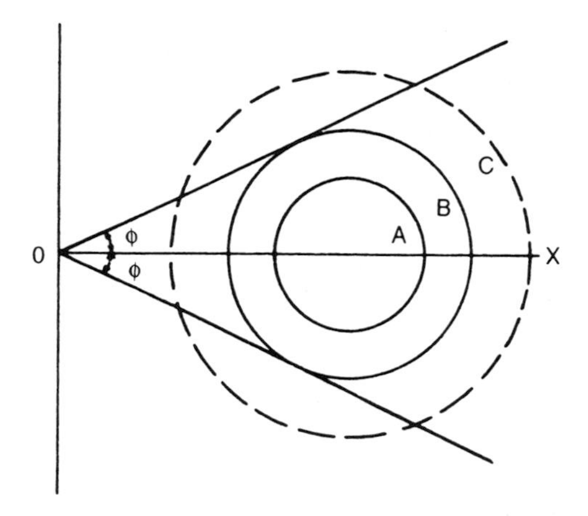

Fig. 3.6 Mohr strength envelope.

Strength envelopes

If ϕ is assumed constant for a certain material, then the shear strength of the material can be represented by a pair of lines passing through the origin, O, at angles $+\phi$ and $-\phi$ to the axis OX (Fig. 3.6). These lines comprise the Mohr strength envelope for the material.

In Fig. 3.6 a state of stress represented by circle A is quite stable as the circle lies completely within the strength envelope. Circle B is tangential to the strength envelope and represents the condition of incipient failure, since a slight increase in stress values will push the circle over the strength envelope and failure will occur. Circle C cannot exist as it is beyond the strength envelope.

Relationship between ϕ and θ

In Fig. 3.7, $\angle DCO = 180° - 2\theta$.

In triangle ODC: $\angle DOC = \phi$, $\angle ODC = 90°$, $\angle OCD = 180° - 2\theta$. These angles summate to 180°, i.e.

$$\phi + 90° + 180° - 2\theta = 180°$$

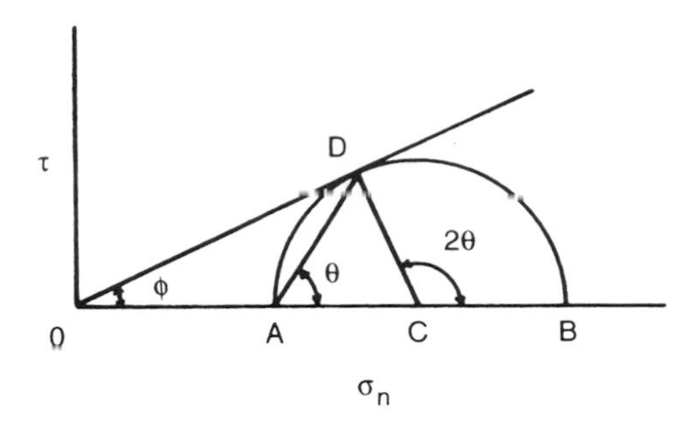

Fig. 3.7 Relationship between ϕ and θ.

hence

$$\theta = \frac{\phi}{2} + 45°$$

Example 3.1

On a failure plane in a purely frictional mass of dry sand the total stresses at failure were: shear = 3.5 kPa; normal = 10.0 kPa.

Determine (a) by calculation and (b) graphically the resultant stress on the plane of failure, the angle of shearing resistance of the soil, and the angle of inclination of the failure plane to the major principal plane.

Solution

(a) By calculation

The soil is frictional, therefore the strength envelope must go through the origin. The failure point is represented by point D in Fig. 3.8a with coordinates (10, 3.5).

$$\text{Resultant stress} = \text{OD} = \sqrt{3.5^2 + 10^2} = 10.6 \text{ kPa}$$

$$\tan\phi = \frac{3.5}{10} = 0.35$$

$$\Rightarrow \phi = 19.3°$$

$$\theta = \frac{\phi}{2} + 45° = 54.6°$$

(b) Graphically

The procedure (Fig. 3.8b) is first to draw the axes OX and OY and then, to a suitable scale, set off point D with coordinates (10, 3.5); join OD (this is the strength envelope). The stress circle is tangential to OD at the point D; draw line DC perpendicular to OD to cut OX in C, C being the centre of the circle.

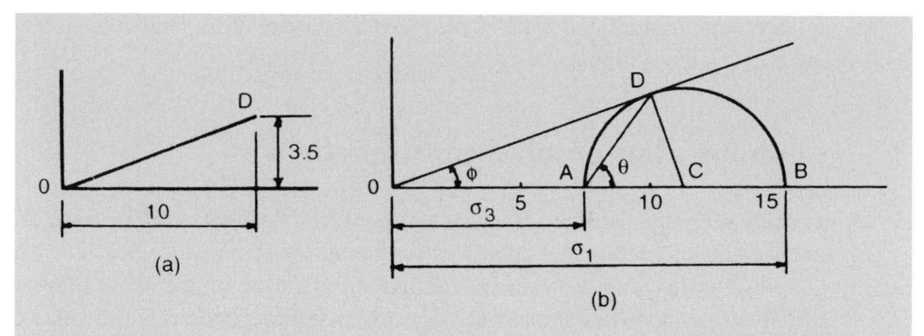

Fig. 3.8 Example 3.1.

With centre C and radius CD draw the circle establishing the points A and B on the x-axis.

By scaling, OD = resultant stress = 10.6 kPa. With protractor, $\phi = 19°$; $\theta = 55°$.

Note From the diagram we see that

$$OA = \sigma_3 = 7.6 \text{ kPa}$$
$$OB = \sigma_1 = 15 \text{ kPa}$$

3.4 Cohesion

It is possible to make a vertical cut in silts and clays and for this cut to remain standing, unsupported, for some time. This cannot be done with a dry sand which, on removal of the cutting implement, will slump until its slope is equal to an angle known as the *angle of repose*. In silts and clays, therefore, some other factor must contribute to shear strength. This factor is called *cohesion* and results from the mutual attraction existing between fine particles that tends to hold them together in a solid mass without the application of external forces. In terms of the Mohr diagram this means that the strength envelope for the soil, for undrained conditions, no longer goes through the origin but intercepts the shear stress axis (see Fig. 3.9). The value of

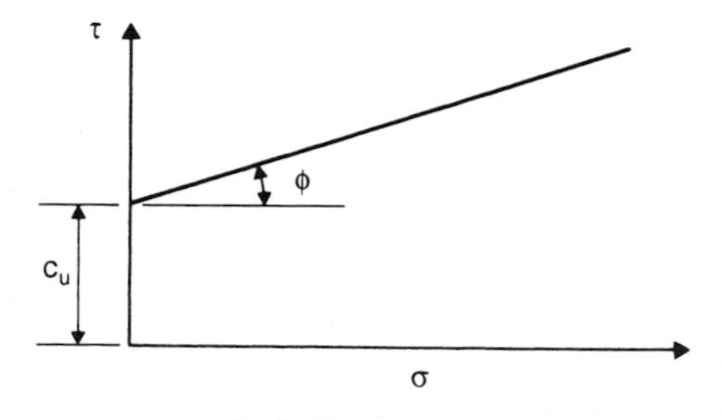

Fig. 3.9 A cohesive coil subjected to undrained conditions and zero total normal stress will still exhibit a shear stress, c_u.

the intercept, to the same scale as σ_n, gives a measure of the unit cohesion available and is given the symbols c or c_u.

3.5 Coulomb's law of soil shear strength

It can be seen that the shear resistance offered by a particular soil is made of the two components of friction and cohesion. Frictional resistance does not have a constant value but varies with the value of normal stress acting on the shear plane whereas cohesive resistance has a constant value which is independent of the value of σ_n. In 1776 Coulomb suggested that the equation of the strength envelope of a soil could be expressed by the straight line equation

$$\tau_f = c + \sigma \tan \phi$$

where

τ_f = shear stress at failure, i.e. the shear strength
c = unit cohesion
σ = total normal stress on failure plane
ϕ = angle of shearing resistance.

The equation gave satisfactory predictions for sands and gravels, for which it was originally intended, but it was not so successful when applied to silts and clays. The reasons for this are now well known and are that the drainage conditions under which the soil is operating together with the rate of the applied loading have a considerable effect on the amount of shearing resistance the soil will exhibit. None of this was appreciated in the 18th century, and this lack of understanding continued more or less until 1925 when Terzaghi published his theory of effective stress.

Note It should be noted that there are other factors that affect the value of the angle of shearing resistance of a particular soil. They include the effects of such items as the amount of friction between the soil particles, the shape of the particles and the degree of interlock between them, the density of the soil, its previous stress history, etc.

Effective stress, σ'

The principle of effective stress was introduced on Chapter 2. Terzaghi first presented the concept of effective stress in 1925 and again, in 1936, at the First International Conference on Soil Mechanics and Foundation Engineering, at Harvard University. He showed, from the results of many soil tests, that when an undrained saturated soil is subjected to an increase in applied normal stress, $\Delta\sigma$, the pore water pressure within the soil increases by Δu, and the value of Δu is equal to the value of $\Delta\sigma$. This increase in u caused no measurable changes in either the volumes or the strengths of the soils tested, and Terzaghi therefore used the term *neutral stress* to describe u, instead of the now more popular term *pore water pressure*.

Terzaghi concluded that only part of an applied stress system controls measurable changes in soil behaviour and this is the balance between the applied stresses and the neutral stress. He called these balancing stresses the *effective stresses*.

If a soil mass is subjected to the action of compressive forces applied at its boundaries then the stresses induced within the soil at any point can be estimated by the theory of elasticity, described in Chapter 4. For most soil problems, estimations of the values of the principal stresses σ_1, σ_2 and σ_3 acting at a particular point are required. Once these values have been obtained, the values of the normal and shear stresses acting on any plane through the point can be computed.

At any point in a saturated soil each of the three principal stresses consists of two parts:

(1) u, the neutral pressure acting in both the water *and* in the solid skeleton in every direction with equal intensity;
(2) the balancing pressures $(\sigma_1 - u)$, $(\sigma_2 - u)$ and $(\sigma_3 - u)$.

As explained above, Terzaghi's theory is that only the balancing pressures, i.e. the effective principal stresses, influence volume and strength changes in saturated soils:

Principal effective stress = Principal normal stress − Pore water pressure

i.e.

$$\sigma_1' = \sigma_1 - u, \text{ etc.}$$

where the prime represents 'effective stress'.

Terzaghi explained that if a saturated soil fails by shear, the normal stress on the plane of failure, σ, also consists of the neutral stress, u, and an effective stress which led to the equation known to all soils engineers:

$$\sigma' = \sigma - u$$

This equation has stood the test of time and is accepted as applicable to all saturated soils. The problem of a stress state equation for *unsaturated* soils is discussed in Chapter 12.

3.6 Modified Coulomb's law

Shear strength depends upon effective stress and not total stress. Coulomb's equation must therefore be modified in terms of effective stress and becomes:

$$\tau_f = c' + \sigma' \tan \phi'$$

where

c' = unit cohesion, with respect to effective stresses
σ' = effective normal stress acting on failure plane
ϕ' = angle of shearing resistance, with respect to effective stresses.

It is seen that, dependent upon the loading and drainage conditions, it is possible for a clay soil to exhibit purely frictional shear strength (i.e. to act as a '$c' - 0$' or 'ϕ'' soil), when it is loaded under drained conditions or to exhibit only cohesive strength (i.e. to act as a '$\phi = 0$' or 'c_u' soil) when it is loaded under undrained conditions. Obviously, at an interim stage the clay can exhibit both cohesion and frictional resistance (i.e. to act as a '$c' - \phi'$' soil). The same situation also applies to granular soils.

3.7 The Mohr–Coulomb yield theory

Over the years various yield theories have been proposed for soils. The best known ones are: the Tresca theory, the von Mises theory, the Mohr–Coulomb theory and the critical state theory. The first three theories have been described by Bishop (1966) and the critical state theory by Schofield and Wroth (1968) and by Muir Wood (1991).

Only the Mohr–Coulomb theory is discussed in this chapter. The theory does not consider the effect of strains or volume changes that a soil experiences on its way to failure; nor does it consider the effect of the intermediate principal stress, σ_2. Nevertheless satisfactory predictions of soil strength are obtained and, as it is simple to apply, the Mohr–Coulomb theory is widely used in the analysis of most practical problems which involve soil strength.

The Mohr strength theory is really an extension of the Tresca theory, which in turn was probably based on Coulomb's work – hence the title. The theory assumes that the difference between the major and minor principal stresses is a function of their sum, i.e. $(\sigma_1 - \sigma_3) = f(\sigma_1 + \sigma_3)$. Any effect due to σ_2 is ignored.

The Mohr circle has been discussed earlier in this chapter and a typical example of a Mohr circle diagram is shown in Fig. 3.10. The intercept on the shear stress axis of the strength envelope is the intrinsic pressure, i.e. the strength of the material when under zero normal stress. As we know, this intercept is called cohesion in soil mechanics and given the symbol c.

In Fig. 3.10:

$$\sin\phi = \frac{DC}{O'C} = \frac{\frac{1}{2}(\sigma_1 - \sigma_3)}{k + \frac{1}{2}(\sigma_1 + \sigma_3)} = \frac{\sigma_1 - \sigma_3}{2k + \sigma_1 + \sigma_3}$$

Hence

$$\sigma_1 - \sigma_3 = 2k\sin\phi + (\sigma_1 + \sigma_3)\sin\phi$$

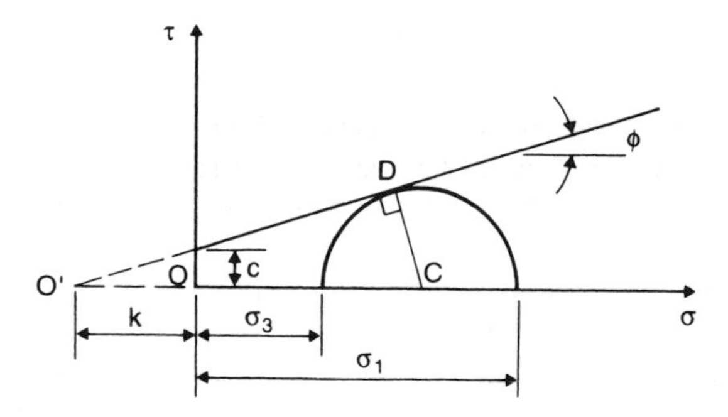

Fig. 3.10 Mohr circle diagram.

Now

$$k = c \cot \phi$$
$$\Rightarrow \quad (\sigma_1 - \sigma_3) = 2c \cos \phi + (\sigma_1 + \sigma_3) \sin \phi$$

which is the general form of the Mohr–Coulomb theory.

The equation can be expressed in terms of either total stress (as shown) or effective stress:

$$(\sigma_1' - \sigma_3') = 2c' \cos \phi' + (\sigma_1' + \sigma_3') \sin \phi'$$

3.8 Determination of the shear strength parameters

The shear strength of a soil is controlled by the effective stress that acts upon it and it is therefore obvious that a geotechnical analysis involving the operative strength of a soil should be carried out in terms of the effective stress parameters ϕ' and c′. This is the general rule and, as you would expect, there is at least one exception. The case of a fully saturated clay subjected to undrained loading is much more simple to analyse using total stress values and ϕ_u and c than with an effective stress approach. As will be illustrated in later chapters, such a situation can arise in both slope stability and bearing capacity problems.

It is seen therefore that both the values of the undrained parameters, ϕ_u and c_u (or ϕ and c), and of the drained parameters, ϕ' and c′ (or ϕ_d and c_d) are generally required. They are obtained from the results of laboratory tests carried out on representative samples of the soil with loading and drainage conditions approximating to those in the field where possible. The tests in general use are the direct shear box test, the triaxial test and the unconfined compression test, an adaptation of the triaxial test.

3.8.1 The direct shear box test

The apparatus consists of a brass box, split horizontally at the centre of the soil specimen. The soil is gripped by perforated metal grilles, behind which porous discs can be placed if required to allow the sample to drain (see Fig. 3.11).

The usual plan size of the sample is $60 \times 60\ \text{mm}^2$, but for testing granular materials such as gravel or stony clay it is necessary to use a larger box, generally $300 \times 300\ \text{mm}^2$, although even greater dimensions are sometimes used.

A vertical load is applied to the top of the sample by means of weights. As the shear plane is predetermined in the horizontal direction the vertical load is also the normal load on the plane of failure. Having applied the required vertical load a shearing force is gradually exerted on the box from an electrically driven screw-jack. The shear force is measured by means of a load transducer connected to a computer.

By means of another transducer (fixed to the shear box) it is possible to determine the strain of the test sample at any point during shear:

$$\text{Strain} = \frac{\text{Movement of box}}{\text{Length of sample}}$$

The load reading is taken at fixed displacements, and failure of the soil specimen is indicated by a sudden drop in the magnitude of the reading or a levelling off in successive readings. In most cases the computer plots a graph of the shearing force against strain as the test continues. Failure of the soil is visually apparent from a turning point in the graph.

The apparatus can be used for both drained and undrained tests, although undrained tests on silts and sands are not possible.

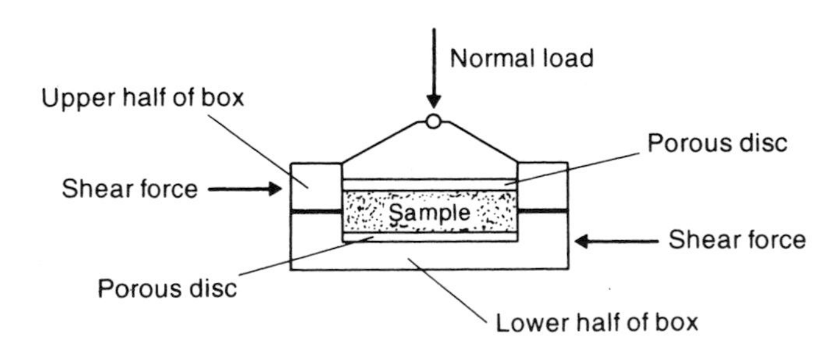

Fig. 3.11 Diagrammatic sketch of the shear box apparatus.

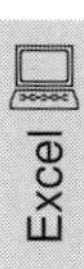

Example 3.2

Undrained shear box tests were carried out on a series of soil samples with the following results:

Test no.	Total normal stress (kPa)	Total shear stress at failure (kPa)
1	100	98
2	200	139
3	300	180
4	400	222

Determine the cohesion and the angle of friction of the soil, with respect to total stress.

Solution

In this case both the normal and the shear stresses at failure are known, so there is no need to draw stress circles and the four failure points may simply be plotted. These points must lie on the strength envelope and the best straight line through the points will establish it (Fig. 3.12).

From the plot, $c_u = 55$ kPa; $\phi_u = 23°$.

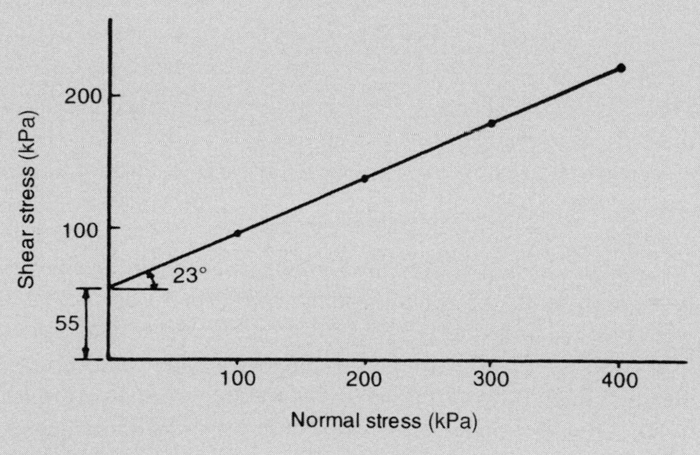

Fig. 3.12 Example 3.2.

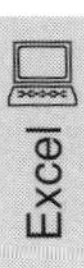

Example 3.3

The following results were obtained from an undrained shear box test carried out on a set of undisturbed soil samples:

Normal load (kN)	0.2	0.4	0.8
Strain (%)	Shearing force (N)		
0	0	0	0
1	21	33	45
2	46	72	101
3	70	110	158
4	89	139	203
5	107	164	248
6	121	180	276
7	131	192	304
8	136	201	330
9	138	210	351
10	138	217	370
11	137	224	391
12	136	230	402
13		234	410
14		237	414
15		236	416
16			417
17			417
18			415

The cross-sectional area of the box was 3600 mm^2 and the test was carried out in a fully instrumented shear box apparatus.

Determine the strength parameters of the soil in terms of total stress.

Solution

The plot of load transducer readings against strain is shown in Fig. 3.13a. From this plot the maximum readings for normal loads of 0.2, 0.4 and 0.8 kN were 138, 237 and 417 kN.

For this particular case the maximum readings could obviously have been obtained directly from the tabulated results, but viewing the plots is sometimes useful to demonstrate whether one of the sets of readings differs from the other two.

The shear stress at each maximum load reading is calculated.

Normal load (kN)	Normal stress (kPa)	Shear force (N)	Shear stress (kPa)
0.2	$\frac{0.2 \times 10^6}{3600} = 56$	138	$\frac{0.138 \times 10^6}{3600} = 38$
0.4	111	237	66
0.8	222	417	116

The plot of shear stress to normal stress is given in Fig. 3.13b.

The total stress envelope is obtained by drawing a straight line through the three points. The strength parameters are: $\phi_u = 25°$; $c_u = 13$ kPa.

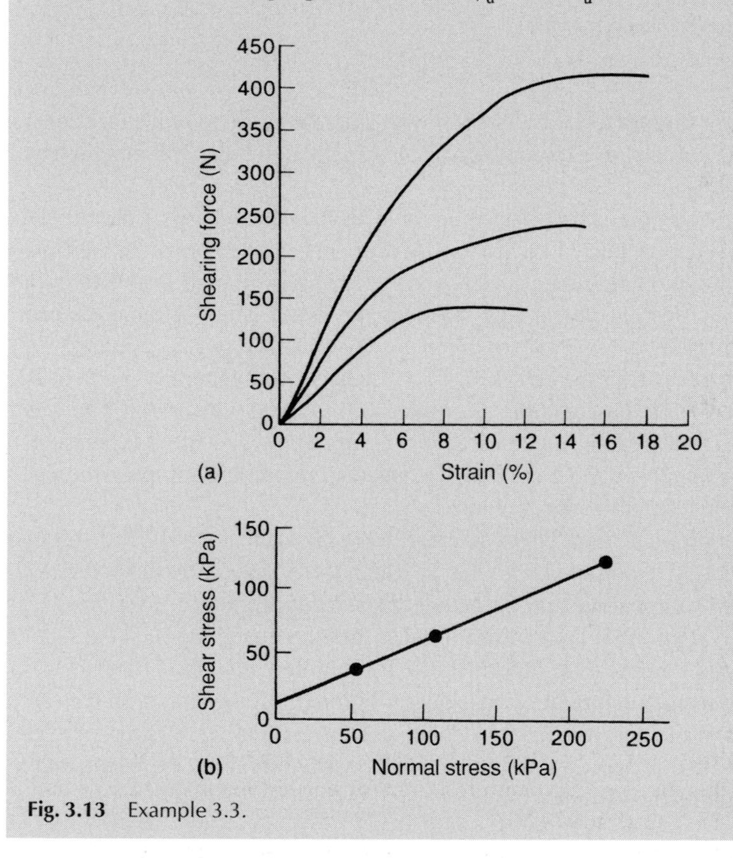

Fig. 3.13 Example 3.3.

3.8.2 *The triaxial test*

As its name implies this test (Fig. 3.14) subjects the soil specimen to three compressive stresses at right angles to each other, one of the three stresses being increased until the sample fails in shear. Its great advantage is that the plane of shear failure is not predetermined as in the shear box test.

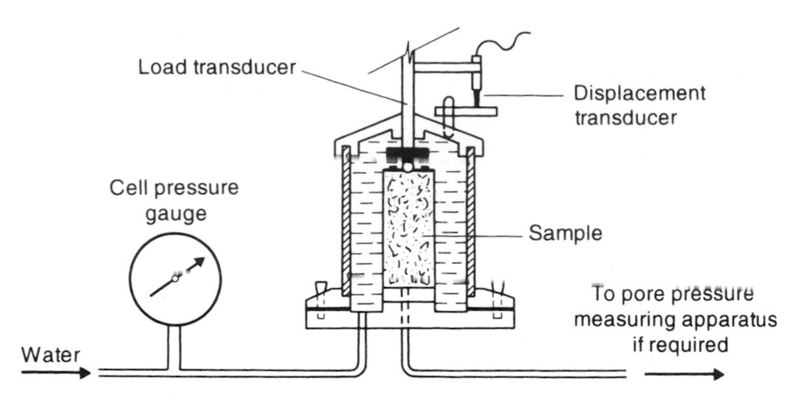

Fig. 3.14 The triaxial apparatus.

The soil sample test is cylindrical with a height equal to twice its diameter. In the UK the usual sizes are 76 mm high by 38 mm diameter and 200 mm high by 100 mm diameter.

The test sample is first placed on the pedestal of the base of the triaxial cell and a loading cap is placed on its top. A thin rubber membrane is then placed over the sample, including the pedestal and the loading cap, and made watertight by the application of tight rubber ring seals, known as 'O' rings, around the pedestal and the loading cap.

The upper part of the cell, which is cylindrical and generally made of Perspex, is next fixed to the base and the assembled cell is filled with water. The water is then subjected to a predetermined value of pressure, known as the cell pressure, which is kept constant throughout the length of the test. It is this water pressure that subjects the sample to an all-round pressure.

The additional axial stress is created by an axial load applied through a load transducer, in a similar way to that in which the horizontal shear force is applied in the shear box apparatus. By the action of an electric motor the axial load is gradually increased at a constant rate of strain and as the axial load is applied the sample suffers continuous compressive deformation. The amount of this vertical deformation is obtained from a deformation transducer. Throughout the test, until the sample fails, readings of the deformation transducer and corresponding readings of axial load are taken. With this data the computer plots the variation of the axial load on the sample against its vertical strain.

Determination of the additional axial stress

From the load transducer it is possible at any time during the test to determine the additional axial load that is being applied to the sample.

During the application of this load the sample experiences shortening in the vertical direction with a corresponding expansion in the horizontal direction. This means that the cross-sectional area of the sample varies, and it has been found that very little error is introduced if the cross-sectional area is evaluated on the assumption that the volume of the sample remains unchanged during the test. In other words the cross-sectional area is found from:

$$\text{Cross-sectional area} = \frac{\text{Volume of sample}}{\text{Original length} - \text{Vertical deformation}}$$

Principal stresses

The intermediate principal stress, σ_2, and the minor principal stress, σ_3, are equal and are the radial stresses caused by the cell pressure, p_c. The major principal stress, σ_1, consists of two parts: the cell water pressure acting on the ends of the sample and the additional axial stress from the load transducer, q. To ensure that the cell pressure acts over the whole area of the end cap, the bottom of the plunger is drilled so that the pressure can act on the ball seating.

From this we see that the triaxial test can be considered as happening in two stages (Fig. 3.15), the first being the application of the cell water pressure (p_c, i.e. σ_3), while the second is the application of a deviator stress (q, i.e. $[\sigma_1 - \sigma_3]$).

A set of at least three samples is tested. The deviator stress is plotted against vertical strain and the point of failure of each sample is obtained. The Mohr circles for each sample are then drawn and the best common tangent to the circles is taken as the strength envelope (Fig. 3.16). A small curvature occurs in the strength envelope of most soils, but this effect is slight and for all practical work the envelope can be taken as a straight line.

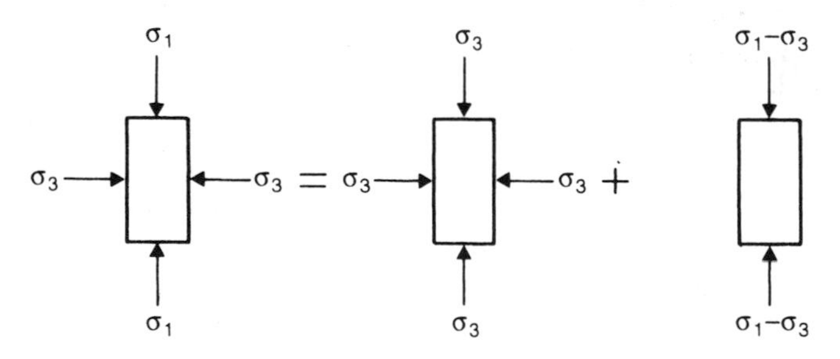

Fig. 3.15 Stresses in the triaxial test.

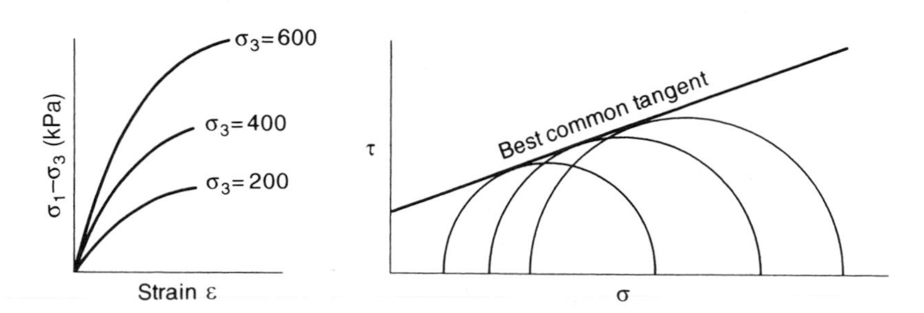

Fig. 3.16 Typical triaxial test results.

Types of failure

Not all soil samples will fail in pure shear; there are generally some barrelling effects as well. In a sample that fails completely by barrelling there is no definite failure point, the deviator stress simply increasing slightly with strain. In this case an arbitrary value of the failure stress is taken as the stress value at 20 per cent strain (see Fig. 3.17).

Note In the past, soil laboratories made use of dial gauges to measure displacement, and proving rings to measure applied loads. Some laboratories still use such equipment, and any reader interested in an explanation and examples of their use is guided to the 6th, or earlier, editions of this book.

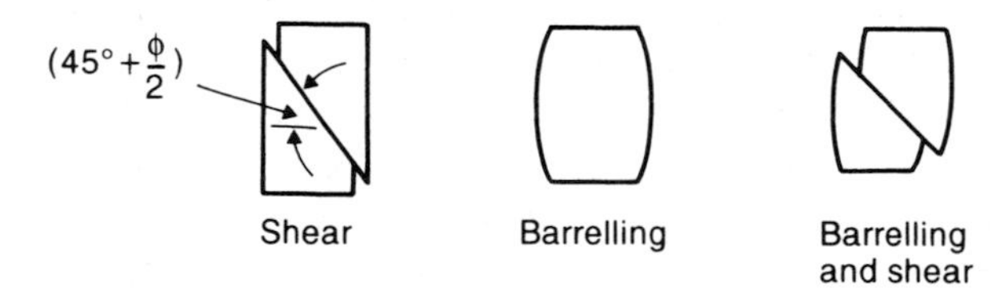

Fig. 3.17 Types of failure in the triaxial test.

3.8.3 The unconfined compression test

In this test (Fig. 3.18) no all-round pressure is applied to the soil specimen and the results obtained give a measure of the *unconfined* compressive strength of the soil. The test is only applicable to cohesive soils and, although not as popular as the triaxial test, it is used where a rapid result is required. An electric motor within

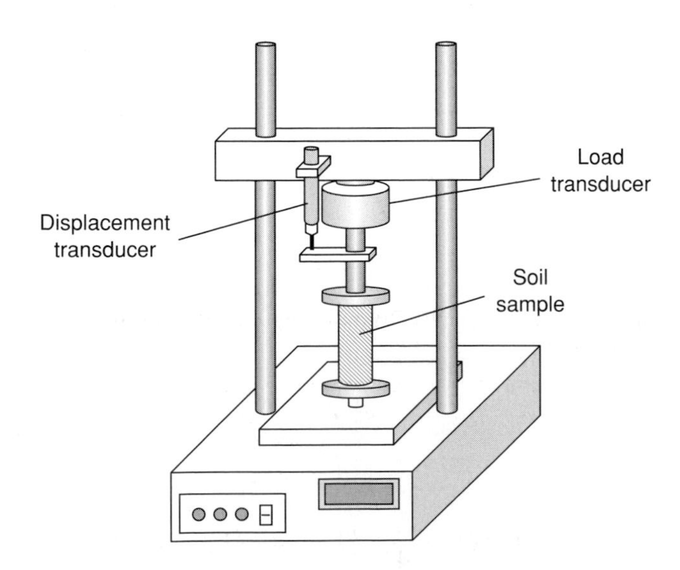

Fig. 3.18 The unconfined compression test.

the base unit drives the platen supporting the specimen upwards and the load carried by the soil is recorded by the load transducer. The vertical strain is recorded by a displacement transducer and the load–displacement curve is plotted on a PC connected to the system. The load and strain readings at failure are used to give a direct measure of the unconfined compressive strength of the soil.

3.9 Determination of the total stress parameters ϕ_u and c_u

The undrained shear test

The simplest method to determine values for the total shear strength parameters of a soil is to subject suitable samples of the soil to this test. In the test the soil sample is prevented from draining during shear and is therefore sheared immediately after the application of the normal load (in the shear box) or immediately after the application of the cell pressure (in the triaxial apparatus). A sample can be tested in 15 minutes or less, so that there is no time for any pore pressures developed to dissipate or to distribute themselves evenly throughout the sample. Measurements of pore water pressure are therefore not possible and the results of the test can only be expressed in terms of total stress.

The unconfined compression apparatus is only capable of carrying out an undrained test on a clay sample with no radial pressure applied. The test takes about a minute.

Undrained tests on silts and sands are not possible in the shear box.

Example 3.4

The following results were obtained from a series of undrained triaxial tests carried out on undisturbed samples of a compacted soil:

Cell pressure (kPa)	Additional axial load at failure (N)
200	342
400	388
600	465

Each sample, originally 76 mm long and 38 mm in diameter, experienced a vertical deformation of 5.1 mm.

Draw the strength envelope and determine the Coulomb equation for the shear strength of the soil in terms of total stresses.

Solution

$$\text{Volume of sample} = \frac{\pi}{4} \times 38^2 \times 76 = 86\ 193\ \text{mm}^3$$

Therefore cross-sectional area at failure $= \dfrac{86\,193}{76 - 5.1} = 1216\ \text{mm}^2$.

Cell pressure σ_3 (kPa)	Deviator stress $(\sigma_1 - \sigma_3)$ (kPa)	Major principal stress σ_1 (kPa)
200	$\dfrac{0.342 \times 10^6}{1216} = 281$	481
400	$\dfrac{0.388 \times 10^6}{1216} = 319$	719
600	$\dfrac{0.465 \times 10^6}{1216} = 382$	982

The Mohr circles for total stress and the strength envelope are shown in Fig. 3.19. From the diagram $\phi_u = 7°$; $c_u = 100$ kPa.

Coulomb's equation is:

$$c_u + \sigma \tan \phi_u = 100 + \sigma \tan 7° = 100 + 0.123\sigma\ \text{kPa}$$

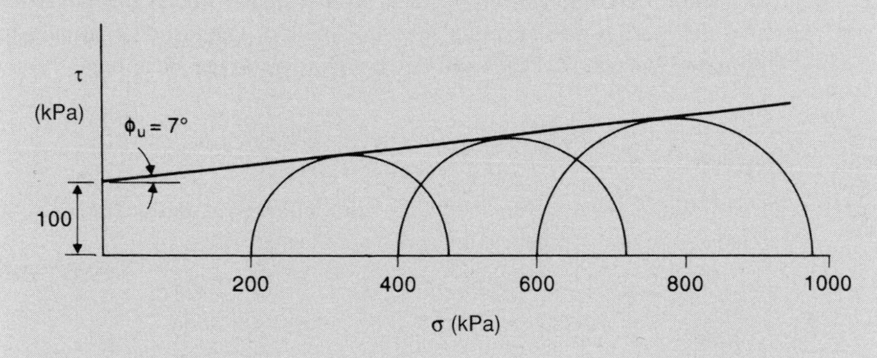

Fig. 3.19 Example 3.4.

Example 3.5

A sample of clay was subjected to an undrained triaxial test with a cell pressure of 100 kPa and the additional axial stress necessary to cause failure was found to be 188 kPa. Assuming that $\phi_u = 0°$, determine the value of additional axial stress that would be required to cause failure of a further sample of the soil if it was tested undrained with a cell pressure of 200 kPa.

Solution

The first step is to draw the stress circle that represents the conditions for the first test, i.e. $\sigma_3 = 100$ kPa and $\sigma_1 = 188 + 100 = 288$ kPa. The circle is shown in

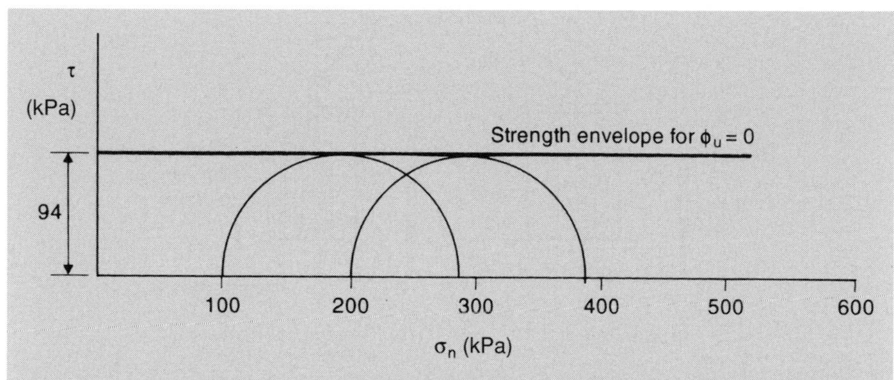

Fig. 3.20 Example 3.5.

Fig. 3.20 and the strength envelope representing the condition that $\phi_u = 0°$ is now drawn as a horizontal line tangential to the stress circle. The next step is to draw the stress circle with $\sigma_3 = 200$ kPa and tangential to the strength envelope. Where this circle cuts the normal stress axis it gives the value of σ_1, which is seen to be 388 kPa.

The additional axial stress required for failure = $(\sigma_1 - \sigma_3) = (388 - 200) =$ 188 kPa.

It can be seen from the figure that $c_u = 94$ kPa. This value can be obtained numerically, from the result of either test, if it is remembered that:

$$c_u = \frac{\sigma_1 - \sigma_3}{2} \quad \text{when} \quad \phi_u = 0°$$

3.10 Determination of the effective stress parameters ϕ' and c'

There are two relevant triaxial tests.

3.10.1 *The drained test*

A porous disc is placed on the pedestal before the test sample is placed in position so that water can drain out from the soil. The triaxial cell is then assembled, filled with water and pressurised. The cell pressure creates a pore water pressure within the soil sample and the apparatus is left until the sample has consolidated, i.e. until the pore water pressure has been dissipated by water seeping out through the porous disc into the burette (see Fig. 3.21). This process usually takes about a day but is quicker if a porous disc is installed beneath the loading cap and joined to the pedestal disc by connecting strips of vertical filter paper placed on the outside of the sample but within the rubber membrane. During this consolidation stage the water level in a burette half full of water and connected to the base of the sample is monitored. When the water level stops rising then the point of full consolidation has been reached.

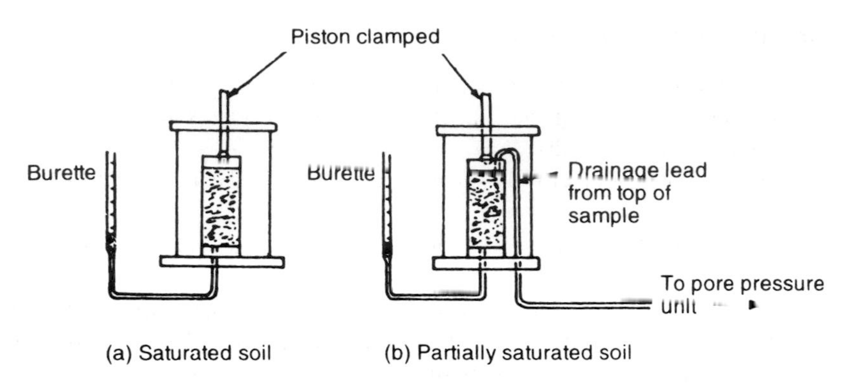

Fig. 3.21 Alternative arrangements for consolidation of test samples.

An alternative method (sometimes preferable with a partially saturated soil) is to allow drainage into a burette from one end of the sample and to connect a pore pressure measuring device to the other. When the pore water pressure reaches zero the sample is consolidated.

When consolidation has been completed the sample is sheared by applying a deviator stress at such a low rate of strain that any pore water pressures induced in the sample have time to dissipate through the porous discs. In this test the pore water pressure is therefore always zero and the effective stresses are consequently equal to the applied stresses.

The main drawback of the drained test is the length of time it takes, with the attendant risk of testing errors: an average test time for a clay sample is about three days but with some soils a test may last as long as two weeks.

3.10.2 The consolidated undrained test

This is the most common form of triaxial test used in soils laboratories to determine c' and ϕ'. It has the advantage that the shear part of the test can be carried out in only two to three hours.

The sample is consolidated exactly as for the drained test, but at this stage the drainage connection is shut off and the sample is sheared under undrained conditions. The application of the deviator stress induces pore water pressures (which are measured), and the effective deviator stress is then simply the total deviator stress less the pore water pressure.

Although the sample is sheared undrained, the rate of shear must be slow enough to allow the induced pore water pressures to distribute themselves evenly throughout the sample. For most soils a strain rate of 0.05 mm/min is satisfactory, which means that the majority of samples can be sheared in under three hours.

Note: With respect to total stress (i.e. the undrained parameters) c_u and ϕ_u are occasionally written as c and ϕ, while with respect to effective stresses (i.e. the drained parameters) c' and ϕ' are occasionally written as c_d and ϕ_d.

Testing with back pressures

It should be noted that, with some soils, the reduction of the pore water pressure to atmospheric during the consolidation stage of a triaxial test on a saturated soil sample can cause air dissolved in the water to come out of solution. If this happens, the sample is no longer fully saturated and this can affect the results obtained during the shearing part of the test.

To maintain a state of occlusion in the pore water, i.e. the state where air can no longer exist in a free state but only in the form of bubbles, its pressure can be increased by applying a pressure (known as a back pressure) to the water in the burette. (The soil water can still drain into the burette.) The back pressure ensures that air does not come out of solution and, by applying the same increase in pressure to the value of the cell pressure, the effective stress situation is unaltered.

The technique can also be used to create full saturation during the consolidation and shearing of partially saturated natural or remoulded soils for both the drained and consolidated undrained triaxial tests. In these cases, back pressure values often as high as 650 kPa are necessary in order to achieve full saturation.

Example 3.6

A series of drained triaxial tests were performed on a soil. Each test was continued until failure and the effective principal stresses for the tests were:

Test no.	σ'_3 (kPa)	σ'_1 (kPa)
1	200	570
2	300	875
3	400	1162

Plot the relevant Mohr stress circles and hence determine the strength envelope of the soil with respect to effective stress.

Solution

The Mohr circle diagram is shown in Fig. 3.22. The circles are drawn first and then, by constructing the best common tangent to these circles, the strength envelope is obtained.

In this case it is seen that the soil is cohesionless as there is no cohesive intercept.

By measurement, $\phi' = 29°$.

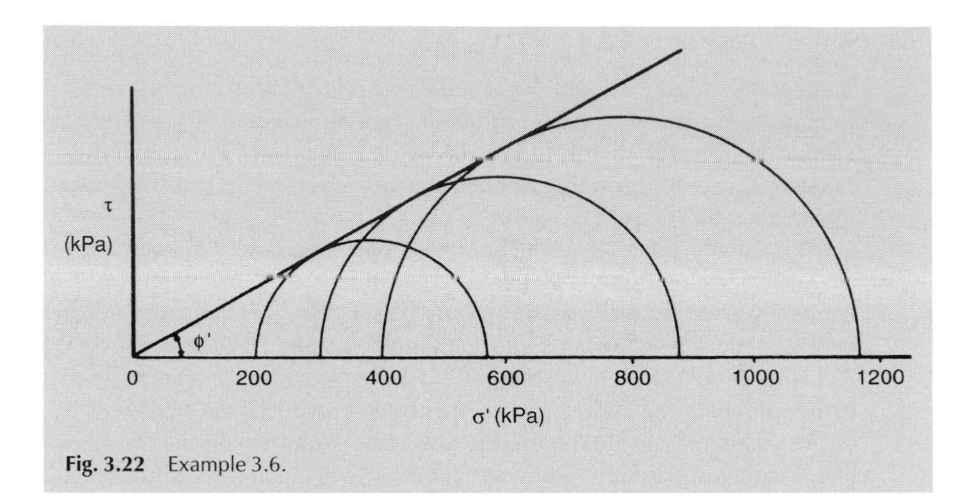

Fig. 3.22 Example 3.6.

Example 3.7

A series of undisturbed samples from a normally consolidated clay was subjected to consolidated undrained tests.

The results were:

Cell pressure (kPa)	Deviator stress at failure (kPa)	Pore water pressure at failure (kPa)
200	118	110
400	240	220
600	352	320

Plot the strength envelope of the soil (a) with respect to total stresses and (b) with respect to effective stresses.

Solution

The two Mohr circle diagrams are shown in Fig. 3.23. The total stress circles are obtained as previously described and are shown with full lines. To determine an effective stress circle it is necessary to subtract the pore water pressure for that circle from each of the principal stresses, e.g. for a cell pressure of 200 kPa the major principal total stress was 200 + 118 = 318 kPa. The pore water pressure was 110 kPa.

$$\sigma_3' = 200 - 110 = 90 \text{ kPa}; \ \sigma_1' = 318 - 110 = 208 \text{ kPa}$$

The values of ϕ_u and ϕ' can be obtained from Fig. 3.23 by direct measurement. Alternatively, knowing that both strength envelopes go through the origin, the

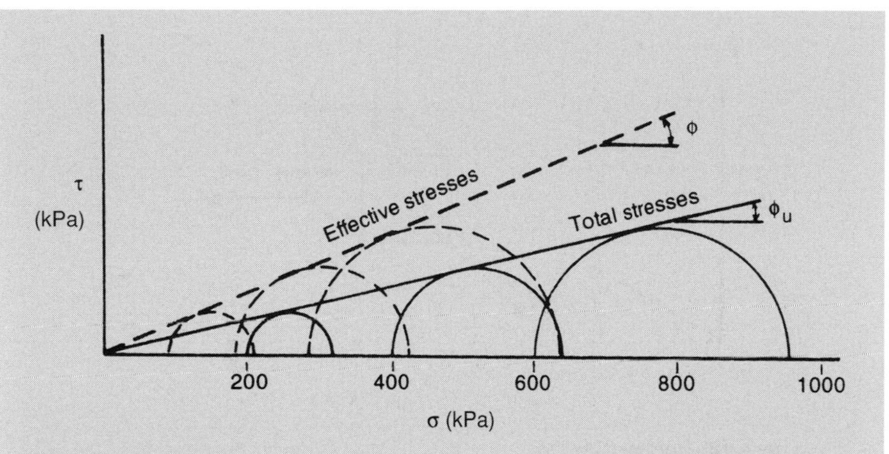

Fig. 3.23 Example 3.7.

values can be obtained from the Mohr–Coulomb equation. Consider the Mohr stress circles created when the cell pressure, $\sigma_3 = 200$ kPa:

$$\sin \phi' = \frac{\sigma_1' - \sigma_3'}{\sigma_1' + \sigma_3'} = \frac{118}{90 + 208} = 0.396, \quad \text{i.e.} \quad \phi' = 23.3°$$

$$\sin \phi_u = \frac{\sigma_1 - \sigma_3}{\sigma_1 + \sigma_3} = \frac{118}{(118 + 200) + 200} = 0.228, \quad \text{i.e.} \quad \phi_u = 13.2°$$

3.11 The pore pressure coefficients A and B

These coefficients were proposed by Skempton in 1954 and are now almost universally accepted. The relevant theory is set out below.

$$\text{Volumetric strain} = \frac{\text{Change in volume}}{\text{Original volume}} = \frac{-\Delta V}{V}$$

(ΔV is negative when dealing with compressive stresses as is the general case in soil mechanics.)

Consider an elemental cube of unit dimensions and acted upon by compressive principal stresses σ_1, σ_2 and σ_3 (Fig. 3.24).

On horizontal plane (2,3):

$$\text{Compressive strain} = \frac{\sigma_1}{E}$$

$$\text{Lateral strain from stresses } \sigma_2 \text{ and } \sigma_3 = -\left(\frac{\mu\sigma_2}{E} + \frac{\mu\sigma_3}{E}\right)$$

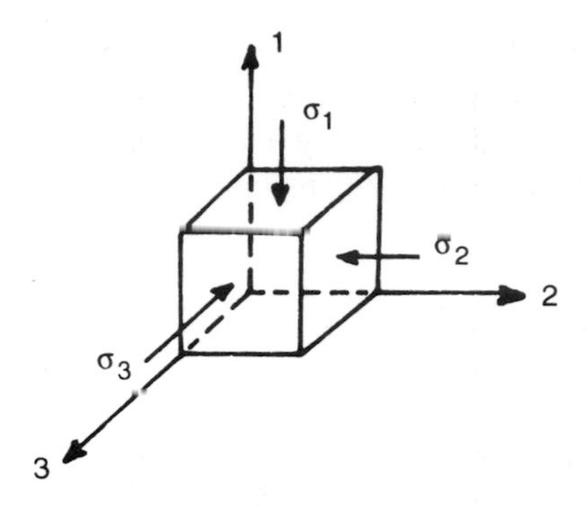

Fig. 3.24 Compressive principal stresses.

where μ = Poisson's ratio.

i.e. total strain on this plane $= \frac{\sigma_1}{E} - \frac{\mu}{E}(\sigma_2 + \sigma_3)$.

Similarly, strains on the other two planes are:

$$\frac{\sigma_2}{E} - \frac{\mu}{E}(\sigma_3 + \sigma_1)$$

$$\frac{\sigma_3}{E} - \frac{\mu}{E}(\sigma_1 + \sigma_2)$$

Now it can be shown that, no matter what the stresses on the faces of the cube, the volumetric strain is equal to the sum of the strains on each face.

$$-\frac{\Delta V}{V} = \frac{(\sigma_1 + \sigma_2 + \sigma_3)}{E} - \frac{2\mu}{E}(\sigma_1 + \sigma_2 + \sigma_3)$$

i.e.

$$\frac{-\Delta V}{V} = \frac{1 - 2\mu}{E}(\sigma_1 + \sigma_2 + \sigma_3)$$

Compressibility of a material is the volumetric strain per unit pressure, i.e. for a soil skeleton,

$$C_c = \frac{-\Delta V}{V} \text{ per unit pressure increase}$$

Average pressure increase $= \frac{1}{3}(\sigma_1 + \sigma_2 + \sigma_3)$. Therefore, for a perfectly elastic soil:

$$C_c = \frac{3(1 - 2\mu)}{E} \frac{(\sigma_1 + \sigma_2 + \sigma_3)}{(\sigma_1 + \sigma_2 + \sigma_3)} = \frac{3(1 - 2\mu)}{E}$$

Consider a sample of saturated soil subjected to an undrained triaxial test. The applied stress system for this test has already been discussed (Fig. 3.15). The pore water pressure, u, produced during the test will be made up of two parts corresponding to the application of the cell pressure and the deviator stress.

Let

u_a = pore pressure due to σ_3
u_d = pore pressure due to $(\sigma_1 - \sigma_3)$.

If we consider the effects of small total pressure increments $\Delta\sigma_3$ and $\Delta\sigma_1$ then $\Delta\sigma_3$ will cause a pore pressure change Δu_a and $\Delta\sigma_1 - \Delta\sigma_3$ will cause a pore pressure change Δu_d.

Effect of $\Delta\sigma_3$

When an all-around pressure is applied to a saturated soil and drainage is prevented the proportions of the applied stress carried by the pore water and by the soil skeleton depend upon their relative compressibilities:

$$\text{Compressibility of the soil } C_C = \frac{-\Delta V}{V\Delta\sigma_3}$$

$$\text{Compressibility of the pore water} = C_v = \frac{-\Delta V_v}{V_v \Delta u_a}$$

Consider a saturated soil of initial volume V.
Then volume of pore water = nV where n = porosity.
Assume a change in total ambient stress = $\Delta\sigma_3$.

Assume that the change in effective stress caused by this total stress increment is $\Delta\sigma'_3$ and that the corresponding change in pore water pressure is Δu_a. Then,

$$\text{Decrease in volume of soil skeleton} = C_c V\Delta\sigma'_3$$

and

$$\text{Decrease in volume of pore water} = C_v nV\Delta u_a$$

With no drainage these changes must be equal:

i.e.

$$C_c V\Delta\sigma'_3 = C_v nV\Delta u_a$$

$$\Rightarrow \quad \Delta\sigma'_3 = \frac{nC_v}{C_c}\Delta u_a$$

Now

$$\Delta\sigma_3' = \Delta\sigma_3 - \Delta u_a$$

$$\Rightarrow \frac{nC_v}{C_c}\Delta u_a = \Delta\sigma_3 - \Delta u_a$$

or

$$\Delta u_a\left(1 + \frac{nC_v}{C_c}\right) = \Delta\sigma_3$$

$$\Rightarrow \Delta u_a = \frac{\Delta\sigma_3}{1 + \dfrac{nC_v}{C_c}}$$

i.e.

$$\Delta u_a = B\Delta\sigma_3 \quad \text{where} \quad B = \frac{1}{1 + \dfrac{nC_v}{C_c}}$$

The compressibility of water is of the order of 1.63×10^{-7} kPa.

Typical results from soil tests are given in Table 3.1 and show that, for all saturated soils, B can be taken as equal to 1.0 for practical purposes.

Table 3.1 Compression of saturated soils.

Soil type	Soft clay	Stiff clay	Compact silt	Loose sand	Dense sand
n (%)	60	37	35	46	43
C_c (m^2/kN)	4.79×10^{-4}	3.35×10^{-5}	9.58×10^{-5}	2.87×10^{-5}	1.44×10^{-5}
B	0.9998	0.9982	0.9994	0.9973	0.9951

Effect of $\Delta\sigma_1 - \Delta\sigma_3$

Increase in effective stresses:

$$\Delta\sigma_1' = (\Delta\sigma_1 - \Delta\sigma_3) - \Delta u_d$$

$$\Delta\sigma_2' = \Delta\sigma_3' = -\Delta u_d$$

Change in volume of soil skeleton, $\Delta V_c = -C_c V(\Delta\sigma_1' + 2\Delta\sigma_3')$

i.e.

$$\Delta V_c = -V\frac{C_c}{3}[(\Delta\sigma_1 - \Delta\sigma_3) - 3\Delta u_d]$$

Now

$$\Delta V_v = -C_v n \Delta u_d V \text{ and } \Delta V_c \text{ must equal } \Delta V_v$$

$$\Rightarrow \quad \frac{1}{3} C_c(\Delta\sigma_1 - \Delta\sigma_3) - C_c \Delta u_d = C_v n \Delta u_d$$

or

$$\Delta u_d (C_c + nC_v) = \frac{1}{3} C_c(\Delta\sigma_1 - \Delta\sigma_3)$$

$$\Rightarrow \quad \Delta u_d = \frac{1}{1 + \dfrac{nC_v}{C_c}} \frac{1}{3} (\Delta\sigma_1 - \Delta\sigma_3)$$

$$= B \times \frac{1}{3} (\Delta\sigma_1 - \Delta\sigma_3)$$

Now

$$\Delta u = \Delta u_a + \Delta u_d$$

$$\Rightarrow \quad \Delta u = B\left[\Delta\sigma_3 + \frac{1}{3}(\Delta\sigma_1 - \Delta\sigma_3)\right]$$

Generally a soil is not perfectly elastic and the above expression must be written in the form:

$$\Delta_u = B[\Delta\sigma_3 + A(\Delta\sigma_1 - \Delta\sigma_3)]$$

where A is a coefficient determined experimentally.

The expression is often written in the form:

$$\Delta u = B\Delta\sigma_3 + \bar{A}(\Delta\sigma_1 - \Delta\sigma_3) \quad \text{where} \quad \bar{A} = AB$$

$\bar{A}$ and B can be obtained directly from the undrained triaxial test. As has been shown, for a saturated soil $B = 1.0$ and the above expression must be.

$$\Delta u = \Delta\sigma_3 + A(\Delta\sigma_1 - \Delta\sigma_3)$$

Values of A

For a given soil, A varies with both the stress value and the rate of strain, due mainly to the variation of Δu_d with the deviator stress. The value of Δu_d under a particular stress system depends upon such factors as the degree of saturation and whether the soil is normally consolidated or overconsolidated. The value of A must be quoted for some specific point, e.g. at maximum deviator stress or at maximum effective stress ratio (σ_1'/σ_3'); at maximum deviator stress it can vary from 1.5 (for a highly sensitive clay) to −0.5 (for a heavily overconsolidated clay).

3.12 The triaxial extension test

In the normal triaxial test the soil sample is subjected to an all-around water pressure and fails under an increasing axial load. This is known as a compression test in which $\sigma_1 > \sigma_2 = \sigma_3$.

When the cohesive intercept, c′, is equal to zero, as is the case for drained granular soils, silts and normally consolidated clays, then the relevant form of the Mohr–Coulomb equation is:

$$\sigma_1 - \sigma_3 = \sigma_1 \sin\phi + \sigma_3 \sin\phi$$

i.e.

$$\sigma_{1f}(\text{max}) = \sigma_{3f}\,\frac{1+\sin\phi}{1-\sin\phi}$$

where σ_{1f} and σ_{3f} are the respective stresses at failure.

It is possible to fail the sample in axial tension by first subjecting it to equal pressures σ_1 and σ_3 and then gradually reducing σ_1 below the value of σ_3 until failure occurs. This test is known as an extension test and the Mohr–Coulomb expression becomes:

$$\sigma_{1f}(\text{min}) = \sigma_{3f}\,\frac{1-\sin\phi}{1+\sin\phi}, \quad \text{where} \quad \sigma_1 < \sigma_2 = \sigma_3$$

The Mohr circle diagram showing the maximum and minimum values of σ_1 for a fixed value of σ_3 is shown in Fig. 3.25. In the triaxial compression test the stress state is $\sigma_1 > \sigma_2 = \sigma_3$, and in the triaxial extension test the stress state is $\sigma_1 < \sigma_2 = \sigma_3$.

The symbols used in Fig. 3.25 might be confusing to a casual observer. Strictly speaking, for the extension test, σ_{1f}(min) should really be given the symbol σ_{3f} and its accompanying σ_{3f} given the symbol σ_{1f}. In order to avoid this sort of confusion between major and minor principal stresses it has become standard practice to designate the axial effective stress as σ_a' and the radial effective stress as σ_r'.

A comprehensive survey of techniques used in the triaxial test was prepared by Bishop and Henkel (1962). Although the book was published more than 40 years ago it is still regarded as a standard reference for specialist triaxial tests.

For the standard triaxial tests discussed in this chapter, fuller descriptions can be found in BS 1377, and are given by Head (1992).

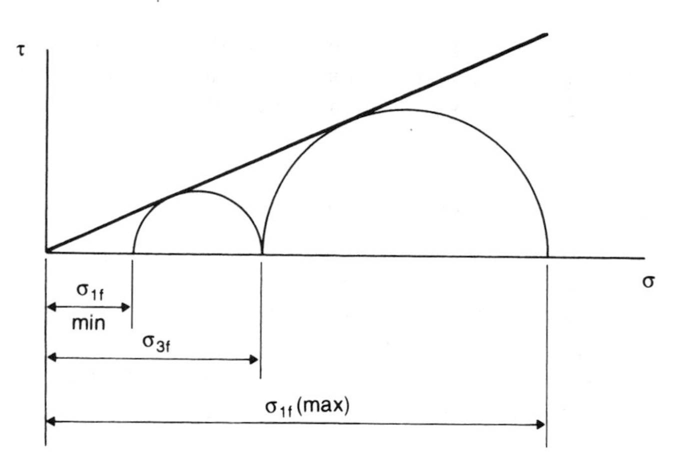

Fig. 3.25 Mohr circle diagram for triaxial compression and tension tests.

Example 3.8

A series of consolidated undrained triaxial tests were carried out on undisturbed samples of an overconsolidated clay.

Results were:

Cell pressure (kPa)	Deviator stress at failure (kPa)	Pore water pressure at failure (kPa)
100	410	−65
200	520	−10
400	720	80
600	980	180

(i) Plot the strength envelope for the soil (a) with respect to total stresses, and (b) with respect to effective stresses.

(ii) If the preconsolidation to which the clay had been subjected was 800 kPa, plot the variation of the pore pressure parameter A_f with the overconsolidation ratio.

Solution

The Mohr circle diagrams are shown in Fig. 3.26a. When a pore pressure is negative the principle of effective stress still applies, i.e. $\sigma' = \sigma - u$; for a cell pressure of 100 kPa, $\sigma_1 = 510$ and $u = -65$, so that

$$\sigma'_3 = 100 - (-65) = 165 \text{ kPa} \quad \text{and} \quad \sigma'_1 = 510 - (-65) = 575 \text{ kPa}$$

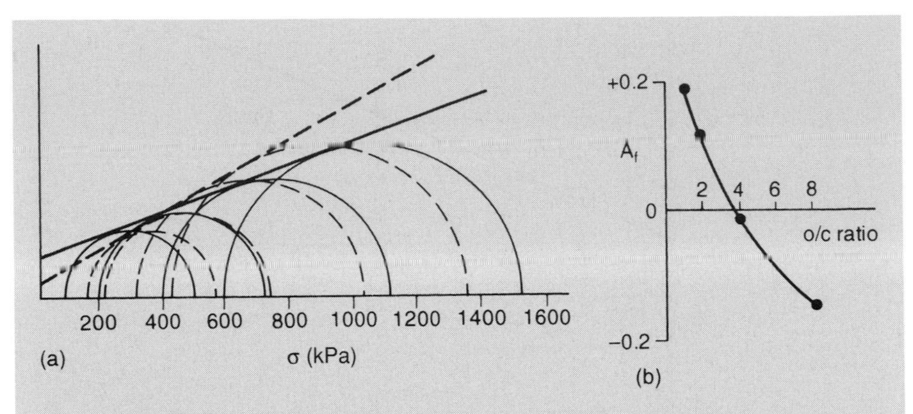

Fig. 3.26 Example 3.8.

After consolidation in a consolidated undrained test (i.e. when shear commences) the soil is saturated, B = 1, and hence the pore pressure coefficient $\bar{A} = A$.

σ_3	o/c ratio	$A = \dfrac{\Delta u_d}{\Delta\sigma_1 - \Delta\sigma_3}$
100	8	−65/410 = −0.146
200	4	−0.02
400	2	0.111
600	1.33	0.185

The results are shown plotted in Fig 3.26b.

Example 3.9

The following results were obtained from an undrained triaxial test on a compacted soil sample using a cell pressure of 300 kPa. Before the application of the cell pressure the pore water pressure within the sample was zero.

Strain (%)	σ_1 (kPa)	u (kPa)
0.0	300	120
2.5	500	150
5.0	720	150
7.5	920	120
10.0	1050	80
15.0	1200	10
20.0	1250	−60

(i) Determine the value of the pore pressure coefficient B and state whether or not the soil was saturated.

(ii) Plot the variation of deviator stress with strain.

(iii) Plot the variation of the pore pressure coefficient A with strain.

Solution

(i)

$$B = \frac{\Delta u_a}{\Delta \sigma_3} = \frac{120}{300} = 0.4$$

The soil was partially saturated as B was less than 1.0.

Strain (%)	Δu_d	$(\Delta\sigma_1 - \Delta\sigma_3)$	$\bar{A}$	$A\left(=\frac{\bar{A}}{B}\right)$
2.5	30	200	$\frac{300}{600} = 0.15$	0.375
5.0	30	420	0.071	0.178
7.5	0	620	0	0
10.0	−40	750	−0.053	−0.132
15.0	−110	900	−0.122	−0.304
20.0	−180	950	−0.188	−0.470

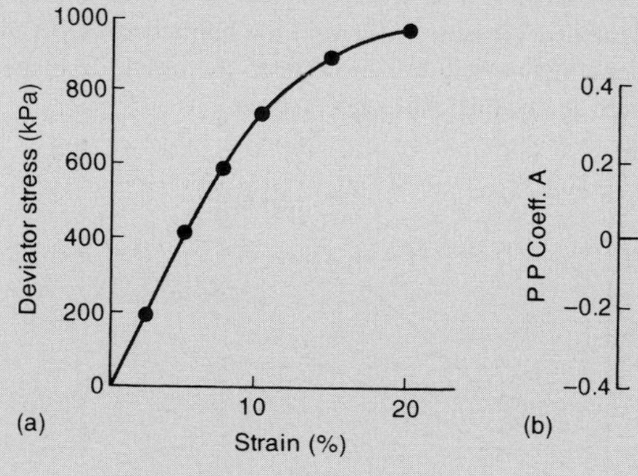

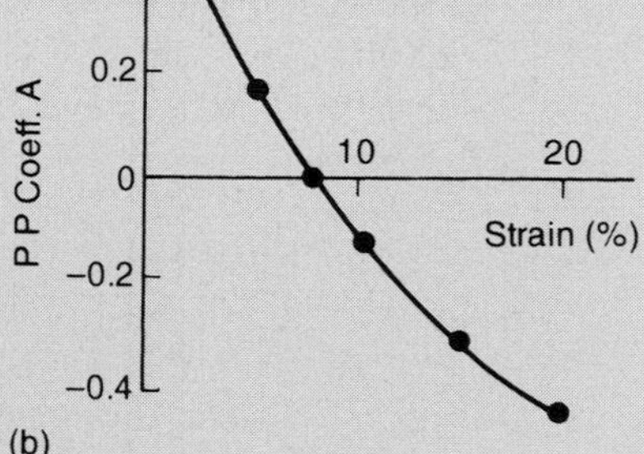

Fig. 3.27 Example 3.9.

3.13 Behaviour of soils under shear

Before discussing this important subject the following definitions must be established.

- *Overburden*: The overburden pressure at a point in a soil mass is simply the weight of the material above it. The effective overburden is the pressure from this material less the pore water pressure due to the height of water extending from the point up to the water table.
- *Normally consolidated clay*: Clay which, at no time in its history, has been subjected to pressures greater than its existing overburden pressure.
- *Overconsolidated clay*: Clay which, during its history, has been subjected to pressures greater than its existing overburden pressure. One cause of overconsolidation is the erosion of material that once existed above the clay layer. Boulder clays are overconsolidated, as the many tons of pressure exerted by the mass of ice above them has been removed.
- *Preconsolidation pressure*: The maximum value of pressure exerted on an overconsolidated clay before the pressure was relieved.
- *Overconsolidation ratio*: The ratio of the value of the effective preconsolidation pressure to the value of the presently existing effective overburden pressure. A normally consolidated clay has an OCR = 1.0 whilst an overconsolidated clay has an OCR > 1.0.

3.13.1 *Type of soil*

Sands and other granular materials

Unless drainage is deliberately prevented, a shear test on a sand will be a drained one as the high value of permeability makes consolidation and drainage virtually instantaneous.

A sand can be tested either dry or saturated. If dry there will be no pore water pressures and the intergranular pressure will equal the applied stress; if the sand is saturated, the pore water pressure will be zero due to the quick drainage, and the intergranular pressure will again equal the applied stress.

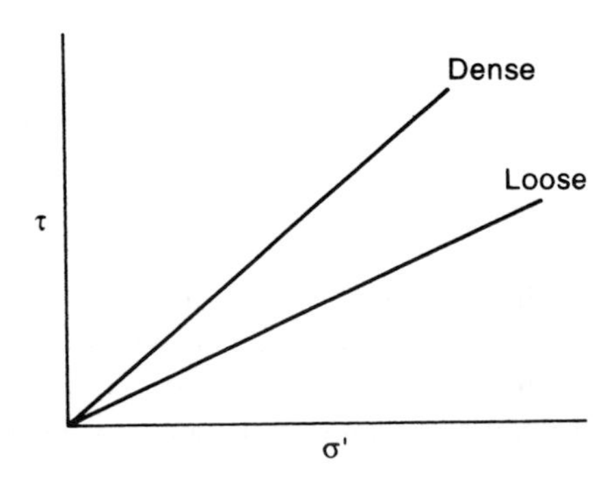

Fig. 3.28 Strength envelope of a granular material, showing the greater shear resistance of a dense sand.

A dense sand tends to dilate (increase in volume) during shear whereas a loose sand tends to decrease in volume, and if the movement of pore water is restricted the shear strength of the sand will be affected: a dense sand will have negative pore pressures induced in it, causing an increase in shear strength, while a loose sand will have positive pore pressures induced with a corresponding reduction in strength. A practical application of this effect occurs when a pile is driven into sand, the load on the sand being applied so suddenly that, for a moment, the water it contains has no time to drain away. The density at which there is no increase or decrease in shear strength when the sand is maintained at constant volume is called the critical density of the sand.

Saturated cohesive soils

These soils are defined as saturated clays and silts in either their natural or a remoulded state.

Unsaturated cohesive soils

Until the late 1980s, it was felt that both the value of the shear strength and the volume change characteristics of an unsaturated soil could be considered as functions of a single effective stress, in a similar manner to that for a saturated soil.

This theory has now been discarded as it has been found that the strength and volume changes in an unsaturated soil are governed by the two forms of the different stress paths that the soil experienced in reaching its relevant final pattern of applied stresses (see Chapter 12).

3.13.2 Undrained shear

The shear strength of a soil, if expressed in terms of total stress, corresponds to Coulomb's Law, i.e.

$$\tau_f = c_u + \sigma \tan \phi_u$$

where

c_u = unit cohesion of the soil, with respect to total stress
ϕ_u = angle of shearing resistance of soil, with respect to total stress
σ = total normal stress on plane of failure.

For saturated cohesive soils tested in undrained shear it is generally found that τ_f has a constant value being independent of the value of the cell pressure σ_3 (see Fig. 3.29). The main exception to this finding is a fissured clay.

Hence, we can say that $\phi_u = 0$ when a saturated cohesive soil is subjected to undrained shear. Hence:

$$\tau = c_u = \tfrac{1}{2}(\sigma_1 - \sigma_3)$$

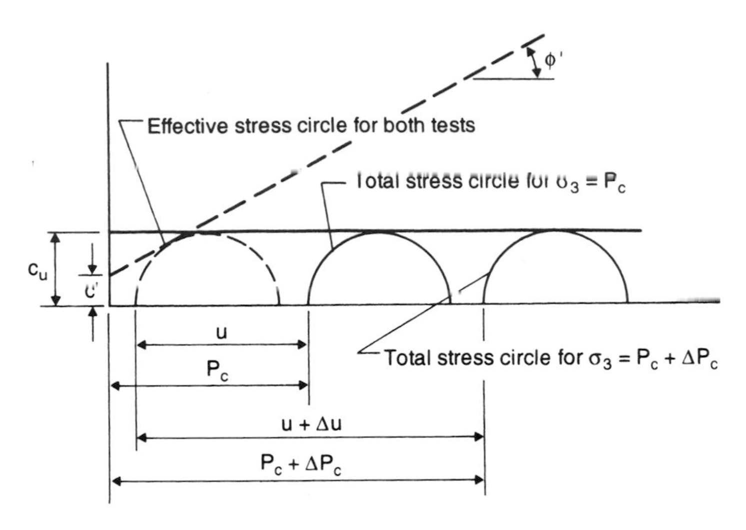

Fig. 3.29 Strength envelope for a saturated cohesive soil subjected to an undrained shear test.

Because of this, the term c_u is referred to as the undrained shear strength of the soil. As will be seen later, the value of c_u is used in slope stability analyses when it can be assumed that $\phi_u = 0$ and the value of c_u can be obtained on site by the simple and economical unconfined compression set.

If we wish to think of the results of an undrained test in terms of effective stress we should consider the nature of the test. In the standard compression undrained triaxial test, the soil sample is placed in the triaxial cell, the drainage connection is removed, the cell pressure is applied and the sample is immediately sheared by increasing the axial stress. Any pore water pressures generated throughout the test are not allowed to dissipate.

If, for a particular undrained shear test carried out at a cell pressure p_c, the pore water pressure generated at failure is u then the effective stresses at failure are:

$$\sigma_1' = \sigma_1 - u; \qquad \sigma_3' = \sigma_3 - u = p_c - u$$

Remembering that, in a saturated soil, the pore pressure parameter $B = 1.0$ it is seen that if the test is repeated using a cell pressure of $p_c + \Delta p_c$ the value of the undrained strength of the soil will be exactly as that obtained from the first test because the increase in the cell pressure, Δp_c, will induce an increase in pore water pressure, Δu, of the same magnitude ($\Delta u = \Delta p_c$). The effective stress circle at failure will therefore be the same as for the first test (Fig. 3.29), the soil acting as if it were purely cohesive. It is therefore seen that there can only be one effective stress circle at failure, independent of the cell pressure value, in an undrained shear test on a saturated soil.

3.13.3 Drained and consolidated undrained shear

The triaxial forms of these shear tests have already been described. It is generally accepted that, for all practical purposes, the values obtained for the drained parameters, c' and ϕ', from either test are virtually the same.

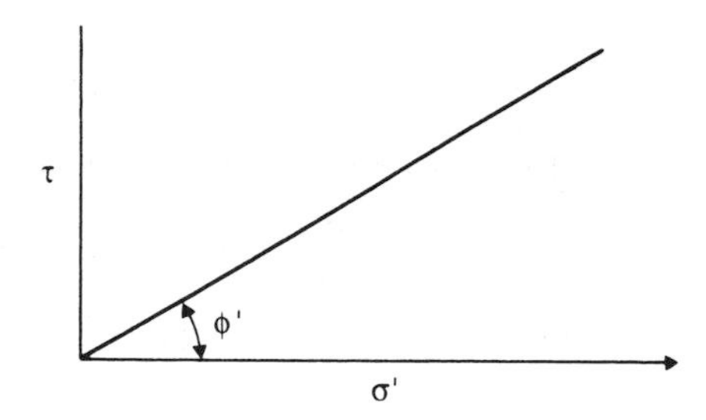

Fig. 3.30 Strength envelope for a normally consolidated clay subjected to a drained shear test.

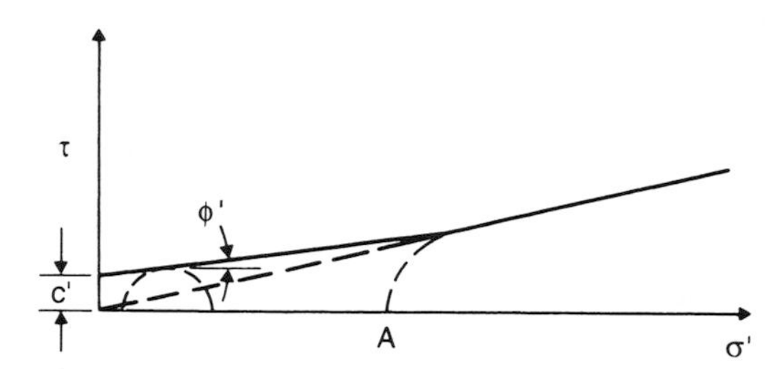

Fig. 3.31 Strength envelope for an overconsolidated soil subjected to a drained shear test.

The c′ value for normally consolidated clays is negligible and can be taken as zero in virtually every situation. A normally consolidated clay therefore has an effective stress strength envelope similar to that shown in Fig. 3.30 and, under drained conditions, will behave as if it were a frictional material.

The effective stress envelope for an overconsolidated clay is shown in Fig. 3.31. Unless unusually high cell pressures are used in the triaxial test the soil will be sheared with a cell pressure less than its preconsolidation pressure value. The resulting strength envelope is slightly curved with a cohesive intercept c′. As the curvature is very slight it is approximated to a straight line inclined at ϕ' to the normal stress axis.

In Fig. 3.31 the point A represents the value of cell pressure that is equal to the preconsolidation pressure. At cell pressures higher than this the strength envelope is the same as for a normally consolidated clay, the value of ϕ' being increased slightly. If this line is projected backwards it will pass through the origin.

Owing to the removal of stresses during sampling, even normally consolidated clays will have a slight degree of overconsolidation and may give a small c′ value, usually so small that it is difficult to measure and has little importance.

The shearing characteristics of silts are similar to those of normally consolidated clays.

The behaviour of saturated normally consolidated and overconsolidated clays in undrained shear is illustrated in Fig. 3.32 which illustrates the variations of both deviator stress and pore water pressure during shear.

An overconsolidated clay is considerably stronger at a given pressure than it would be if normally consolidated, and also tends to dilate during shear whereas a

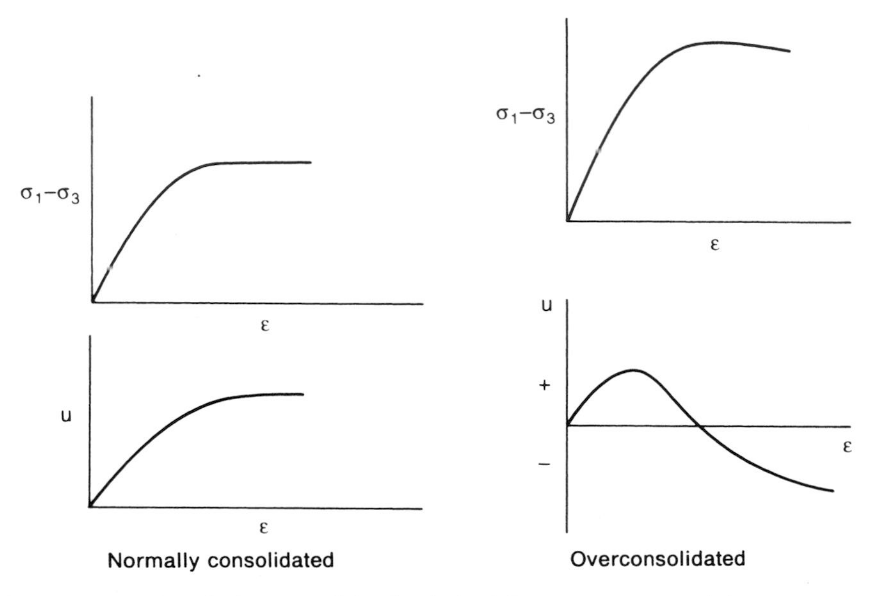

Fig. 3.32 Typical results from consolidated undrained shear tests on saturated clays.

normally consolidated clay will consolidate. Hence when an overconsolidated clay is sheared under undrained conditions negative pore water pressures are induced, the effective stress is increased, and the undrained strength is much higher than the drained strength – the exact opposite to a normally consolidated clay.

If an excavation is made through overconsolidated clay the negative pressures set up give an extremely high undrained strength, but these pore pressures gradually dissipate and the strength falls by as much as 60 or 80 per cent to the drained strength. A well-known example of overconsolidated clay is London clay, which when first cut will stand virtually unsupported to a height of 7.5 m. It does not remain stable for long, and so great is the loss in strength that there have been cases of retaining walls built to support it being pushed over.

Several case histories of retaining wall failures of this type are given in Clayton (1993).

3.14 Variation of the pore pressure coefficient A

An important effect of overconsolidation is its effect on the pore pressure parameter A. With a normally consolidated clay the value of A at maximum deviator stress, A_f, is virtually the same in a consolidated undrained test no matter what cell pressure is used, but with an overconsolidated clay the value of A_f falls off rapidly with increasing overconsolidation ratio (Fig. 3.33).

Overconsolidation ratio is the ratio of preconsolidation pressure divided by the cell pressure used in the test. When the overconsolidation ratio is 1.0 the soil is normally consolidated.

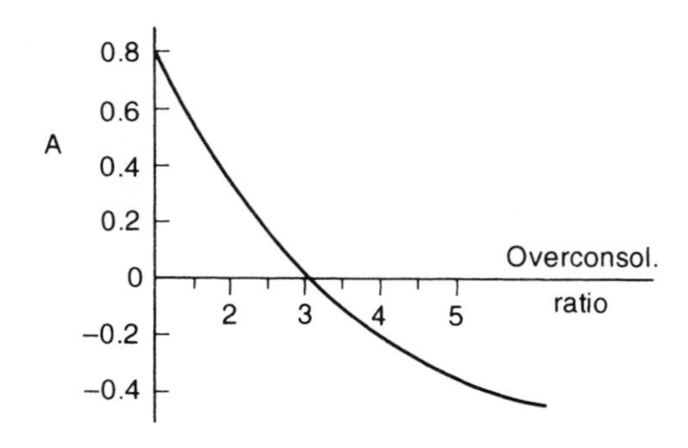

Fig. 3.33 Effects of overconsolidation on the pore pressure coefficient A.

3.15 Operative strengths of soils

For the solution of most soil mechanics problems the peak strength parameters can be used, i.e. the values corresponding to maximum deviator stress. The actual soil strength that applies *in situ* is dependent upon the type of soil, its previous stress history, the drainage conditions, the form of construction and the form of loading. Obviously the shear tests chosen to determine the soil strength parameters to be used in a design should reflect the conditions that will actually prevail during and after the construction period.

Set out below is a brief guide as to the variation of strength properties of different soils.

Sand and gravels

These soils have high values of permeability, and any excess pore water pressures generated within them are immediately dissipated. For all practical purposes these soils operate in the drained state. The appropriate strength parameter is therefore ϕ', with $c' = 0$.

In granular soils the value of ϕ' is highly dependent upon the density of the soil and, as it is difficult to obtain inexpensive undisturbed soil samples, its value is generally estimated from the results of *in situ* tests.

In the UK the standard penetration test (see Chapter 8) is the one most used and a very approximate relationship between the blow count N and the angle of internal friction ϕ' is shown in Fig. 3.34. It should be noted that the corrected value for N, i.e. N′ described in Chapter 8, can be used in conjunction with Fig. 3.34, and that the value obtained approximates to ϕ_t, the peak triaxial angle obtained from drained tests.

Other factors, besides the value of N, such as the type of minerals, the effective size, the grading and the shape of the particles are acknowledged to have an effect on the value of ϕ', but in view of the rough-and-ready method used to determine the value of N, any attempt at refinement seems unrealistic.

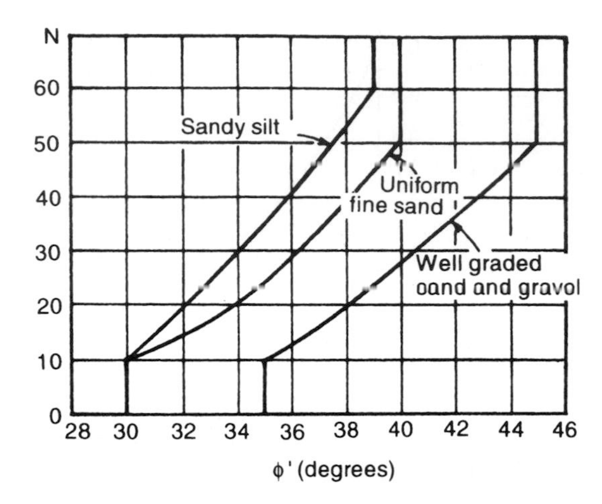

Fig. 3.34 Relationship between N and ϕ'.

Silts

These soils rarely occur in a pure form in the UK and are generally mixed with either sand or clay. It is therefore usually possible to classify silty soils as being either granular or clayey. When there is a reasonable amount of clay material within the soil there should be little difficulty in obtaining undisturbed samples for strength evaluation. With sandy silts, estimated values for ϕ' can be obtained from the results of the standard penetration test.

Clays

Owing to the low permeability of these soils, any excess pore water pressures generated within them will not dissipate immediately. The first step in any design work is to determine whether the clay is normally consolidated or overconsolidated.

Soft or normally consolidated clay

A clay with an undrained shear strength, c_u, of not more than 40 kPa is classified as a soft clay and will be normally consolidated (or lightly overconsolidated). Such clays, when subjected to undrained shear, tend to develop positive pore water pressures (Fig. 3.32), so that during and immediately after construction the strength of the soil is at its minimum value.

After completion of the construction, over a period of time, the soil will achieve its drained condition and will then be at its greatest strength.

Overconsolidated clay

With these soils any pore water pressures generated during shear will be negative. This means simply that the clay is at its strongest during and immediately after construction. The weakest strength value will occur once the soil achieves its fully drained state, the operative strength parameters then being c′ and ϕ'.

3.16 Space diagonal and octahedral plane

As will be seen in Chapter 13, there are occasions when we must think in terms of three-dimensional stress systems. In order to do this, it is general practice to use the stress space formed between the three principal stress axes, $0\sigma_1$, $0\sigma_1$, $0\sigma_3$. For example in Fig. 3.35a, point P represents the three-dimensional stress state $(\sigma_1, \sigma_2, \sigma_3)$.

If we consider all the points where $\sigma_1 = \sigma_2 = \sigma_3$, it is found that they lie on a straight line which passes through the origin and makes the same angle ($\cos^{-1} 1/\sqrt{3}$) with each of the three axes. This line is known as the hydrostatic stress axis or the space diagonal. A plane that is normal to the space diagonal is known as an octahedral plane. Obviously there are an infinite number of octahedral planes on a space diagonal but we are usually only interested in the one corresponding to the stress system being considered.

The normal stress acting at point P on the octahedral plane is given the symbol σ_{oct} and is called the octahedral normal stress. The shear stress acting on the octahedral plane at point P is given the symbol τ_{oct} and is called the octahedral shear stress. The expressions for σ_{oct} and τ_{oct} are derived in most stress analysis textbooks and are:

$$\sigma_{oct} = \frac{(\sigma_1 + \sigma_2 + \sigma_3)}{3}$$

$$\tau_{oct} = \frac{1}{3}\sqrt{[(\sigma_1 - \sigma_2)^2 - (\sigma_2 - \sigma_3)^2 - (\sigma_3 - \sigma_1)^2]}$$

By inserting $\sigma_1' = \sigma_1 - u$, etc., it can easily be shown that $\sigma_{oct}' = \sigma_{oct}$ and $\tau_{oct}' = \tau_{oct} - u$.

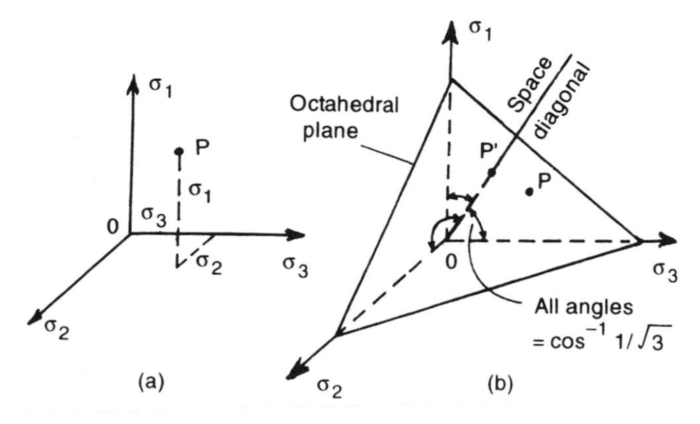

Fig. 3.35 Space diagonal and octahedral plane.

To express triaxial test results the formulae must be changed to:

$$\sigma_{oct} = \frac{(\sigma_a + 2\sigma_r)}{3}$$

$$\tau_{oct} = \frac{\sqrt{2}}{3}(\sigma_1 - \sigma_3)$$

where

σ_r = radial, or cell, pressure
σ_a = axial stress on sample due to applied loading
$\sigma_1 = \sigma_a$ = for compression tests and $\sigma_1 = \sigma_r$ for extension tests.

Consider again the point P and consider the octahedral plane passing through it. This plane will cut the space diagonal at the point P′ such that P′ represents the stress system $(\sigma_P, \sigma_P, \sigma_P)$ where

$$\sigma_P = \frac{(\sigma_1 + \sigma_2 + \sigma_3)}{3}$$

It is seen therefore that it is often convenient to divide a general stress system into two components:

(i) the hydrostatic component, $\sigma_P = (\sigma_1 + \sigma_2 + \sigma_3)/3$ (point P′ in Fig. 3.35b);
(ii) a deviator stress component accounting for the remainder (the distance PP′ in Fig. 3.35b).

The magnitude of the distance PP′ can be found as follows:

OP, the stress tensor of P (i.e. the length of OP) is:

$$OP = \sqrt{\sigma_1^2 + \sigma_2^2 + \sigma_3^2}$$

OP′, the stress tensor of P′ = $\sqrt{3\sigma_P^2}$.

As the space diagonal is at right angles to the octahedral plane, OPP′ is a right-angled triangle and PP′ = $\sqrt{OP^2 - OP'^2}$. Hence

$$PP'^2 = \frac{1}{\sqrt{3}}\sqrt{[(\sigma_1 - \sigma_2)^2 + (\sigma_2 - \sigma_3)^2 + (\sigma_3 - \sigma_1)^2]}$$

$$= \sqrt{3}\,\tau_{oct}$$

Note As σ_{oct} is also the mean value of the three principal stresses of the applied stress system, then $\sigma_{oct} = \sigma_P$.

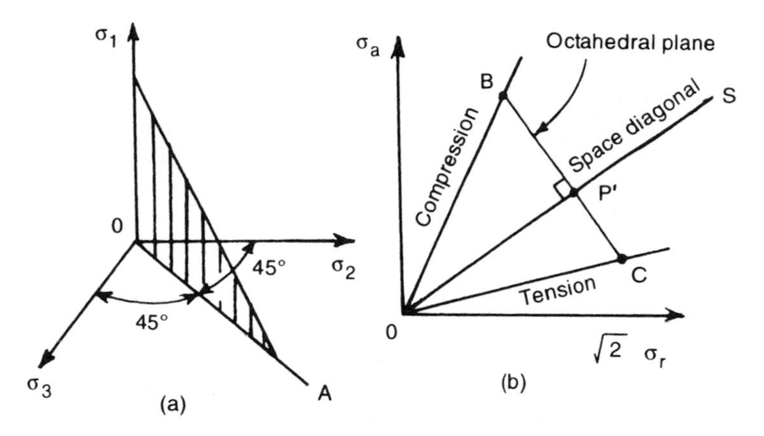

Fig. 3.36 Mohr–Coulomb theory applied to triaxial test.

Application to the triaxial test

The stress systems applied in the triaxial test have already been discussed:

The compression test:	$\sigma_1 > \sigma_2 = \sigma_3$	or	$\sigma_1' > \sigma_2' = \sigma_3'$
The extension test:	$\sigma_1 < \sigma_2 = \sigma_3$	or	$\sigma_1' < \sigma_2' = \sigma_3'$

The Mohr circle diagrams for the tests are illustrated in Fig. 3.25.

Representing σ_1 by the symbol σ_a and representing σ_2 and σ_3 by the symbol σ_r the plane $0\sigma_1 A$ is designated as $0\sigma_a A$ and its normal view is shown in Fig. 3.36b. On this diagram values of σ_a and σ_a' appear to a normal scale but the projected values of σ_r and σ_r' scale $\sqrt{2}\,\sigma_r$ and $\sqrt{2}\sigma_r'$ respectively. The two boundaries for the Mohr–Coulomb theory can be drawn on the diagram, their equations, in terms of effective stress, being:

For axial compression:

$$\sigma_a' = \frac{1 + \sin\phi'}{1 - \sin\phi'}$$

For axial tension:

$$\sigma_a' = \frac{1 - \sin\phi'}{1 + \sin\phi'}$$

No point that represents a stress state outside these two boundaries can exist, as the soil would have suffered plastic failure.

As plane $0\sigma_a A$ represents the condition $\sigma_2 = \sigma_3$ then the space diagonal can be drawn upon it. This line, OS, can be drawn quickly by selecting some suitable value for σ_a and then plotting the point represented by $\sqrt{2}\,(\sigma_a)$, σ_a. This point must lie on the space diagonal, which can now be positioned by drawing a line from the origin and through the point. Any line drawn at right angles to OS represents an octahedral plane. In Fig. 3.36 the octahedral plane drawn is for the stress system acting at compression failure (point B) and also for the stress system acting at tension failure (point C). The intersection point of the space diagonal and octahedral plane is the point P′, which represents the hydrostatic stress system.

Example 3.10

A saturated normally consolidated clay has the following properties: $c' = 0$; $\phi' = 40°$. A sample of the soil is to be subjected to a drained compression triaxial test with a cell pressure of 250 kPa.

Plot the yield surface of the Mohr–Coulomb theory, as viewed on the corresponding octahedral plane.

Solution

$$\frac{1 + \sin\phi'}{1 - \sin\phi'} = 4.60; \qquad \frac{1 - \sin\phi'}{1 + \sin\phi'} = 0.217$$

The test is a compression test, therefore effective axial stress at failure is

$$\sigma'_a = 4.60 \times 250 = 1150 \text{ kPa}$$

Procedure is to first plot the two failure boundaries on the $0\sigma_a A$ plane. As $c' = 0$, these boundaries will go through the origin and there is therefore only the need to establish one point on each boundary to find their positions.

On the compression boundary we have the point $\sigma'_r = 250$; $\sigma'_a = 1150$ kPa, which must be plotted as the point $(\sqrt{2} \times 250, 1150) = (354, 1150)$. This point is shown in Fig. 3.37a as point A.

On the extension boundary, for $\sigma'_r = 250$ kPa, $\sigma'_a = 0.217 \times 250 = 54.3$ kPa and this is plotted as point B, with coordinates (354, 54) in Fig. 3.37a.

For the space diagonal we can select any value for σ_p but, so that we can establish P′, we will select $(\sigma_p, \sigma_p, \sigma_p)$ where

$$\sigma_p = \frac{\sigma'_a + 2\sigma'_r}{3}$$

$$= \frac{1150 + (2 \times 250)}{3} = 550 \text{ kPa}$$

therefore to plot P′ we use the coordinates (778, 550).

Drawing a line from the origin through P′ establishes the position of the space diagonal, OS. If from A we draw a line through P′ to cut the extension failure

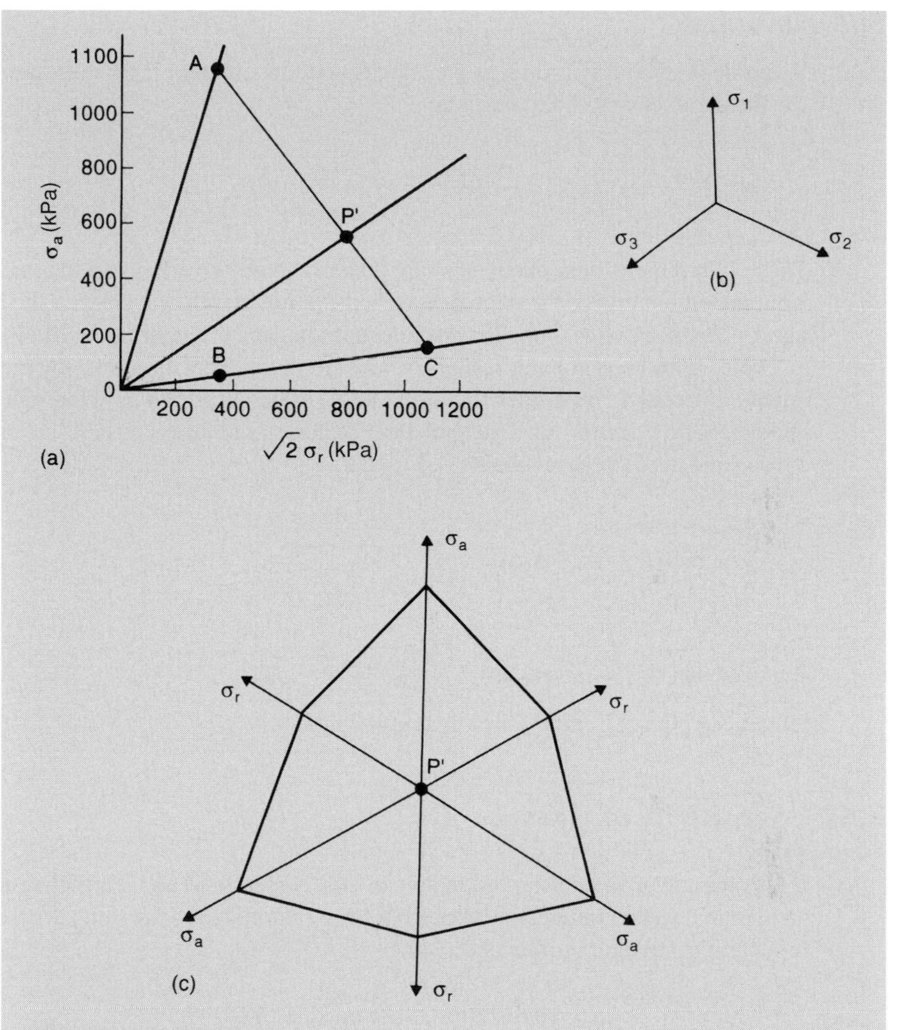

Fig. 3.37 Example 3.10.

boundary at C then the line AC represents the octahedral plane corresponding to the compressive failure conditions. (It can be checked by measurement that AC is at right angles to OS, as it should be.)

Fig. 3.37b shows the projection onto the octahedral plane of the three principal stress axes $0\sigma_1$, $0\sigma_2$ and $0\sigma_3$. (This is the view seen when looking down the space diagonal.)

If we adopt the convention that σ' is no longer greater or less than σ'_2 or σ'_3 and that each principal stress can be the greatest or the smallest value then we can assume that σ'_r can be σ'_1 or σ'_2 or σ'_3 to give the Mohr–Coulomb yield surface shown in Fig. 3.37c.

Stress invariants

When dealing with two-dimensional stress systems, modern textbooks now tend to use the parameters s, s′ and t, where:

$$s = \frac{\sigma_1 + \sigma_3}{2}; \qquad s' = \frac{\sigma_1' + \sigma_3'}{2}; \qquad t = \frac{\sigma_1 - \sigma_3}{2}$$

These two-dimensional parameters are actually stress invariants in that, no matter what the orientation of the reference axes, the parameters always define the centres, s and s′, and the radius t (or t′), of the relevant total and effective stress Mohr circles.

There is no need for the reader to become involved with stress invariants but anyone interested could refer to Smith (1971), who described how the roots of the characteristic equation of a general three-dimensional stress system are the first, second and third stress invariants J_1, J_2 and J_3:

$$J_1 = \sigma_1 + \sigma_2 + \sigma_3$$
$$J_2 = \sigma_1\sigma_2 + \sigma_2\sigma_3 + \sigma_3\sigma_1$$
$$J_3 = \sigma_1\sigma_2\sigma_3$$

It is seen that $\sigma_{oct} = \frac{1}{3}J_1$ and that $\tau_{oct} = \frac{1}{3}\sqrt{2J_1^2 - 3J_2}$.

Obviously both σ_{oct} and τ_{oct} are stress invariants.

3.17 Sensitivity of clays

If the strength of an undisturbed sample of clay is measured and it is then re-tested at an identical water content, but after it has been remoulded to the same dry density, a reduction in strength is often observed.

$$\text{Sensitivity} = S_t = \frac{\text{Undisturbed, undrained strength}}{\text{Remoulded, undrained strength}}$$

Normally consolidated clays tend to have sensitivity values varying from 5 to 10 but certain clays in Canada and Scandinavia have sensitivities as high as 100 and are referred to as quick clays. Sensitivity can vary, slightly, depending upon the moisture content of the clay. Generally, overconsolidated clays have negligible sensitivity, but some quick clays have been found to be overconsolidated. A classification of sensitivity appears in Table 3.2.

Thixotropy

Some clays, if kept at a constant moisture content, regain a portion of their original strength after remoulding, with time (Skempton and Northley, 1952). This property is known as thixotropy.

Table 3.2 Sensitivity classification.

S_t	Classification
1	insensitive
1–2	low
2–4	medium
4–8	sensitive
8–16	extra sensitive
>16	quick (can be up to 150)

Liquidity index I_L

The definition of this index has already been given in Chapter 1:

$$I_L = \frac{w - w_P}{I_P}$$

where w is the *in situ* water content.

This index probably more usefully reflects the properties of plastic soil than the generally used consistency limits w_P and w_L. Liquid and plastic limit tests are carried out on remoulded soil in the laboratory, but the same soil, in its *in situ* state (i.e. undisturbed), may exhibit a different consistency at the same moisture content as the laboratory specimen, due to sensitivity effects. It does not necessarily mean, therefore, that a soil found to have a liquid limit of 50 per cent will be in the liquid state if its *in situ* water content w is also 50 per cent.

If w is greater than the test value of w_L then I_L is >1.0 and it is obvious that if the soil were remoulded it would be transformed into a slurry. In such a case the soil is probably an unconsolidated sediment with an undrained shear strength, c_u, in the order of 15–50 kPa.

Most cohesive soil deposits have I_L values within the range 1.0–0.0. The lower the value of w, the greater the amount of compression that must have taken place and the nearer I_L will be to zero.

If w is less than the test value of the plastic limit then $I_L < 0.0$ and the soil cannot be remoulded (as it is outside the plastic range). In this case the soil is most likely a compressed sediment. Soil in this state will have a c_u value varying from 50 to 250 kPa.

3.18 Activity of a clay

Apart from their value in soil classification, the w_L and w_P values of a plastic soil give an indication of the types and amount of the clay minerals present in the soil. The three major clay mineral groups have been briefly described in Chapter 1, although there are, of course, many variations.

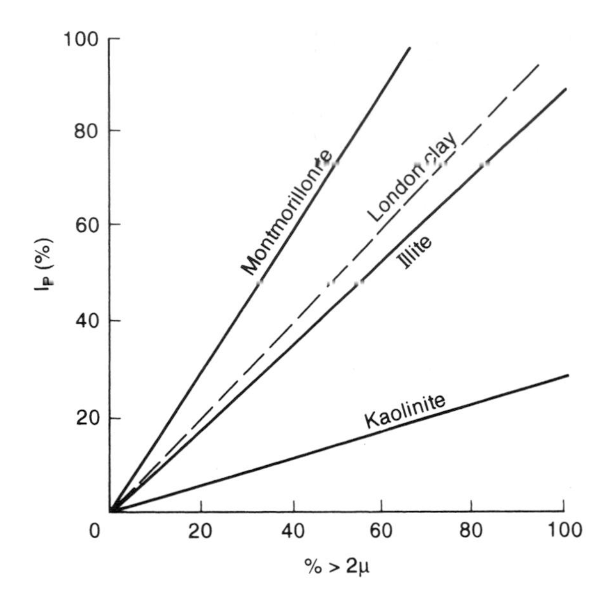

Fig. 3.38 Relationship between I_P and clay percentage (After Skempton, 1953).

It has been found that, for a given soil, the plasticity index increases in proportion to the percentage of clay particles in the soil. Indeed, if a group of soils is examined and their I_P values are plotted against their clay percentages, a straight line, passing through the origin, is obtained.

If a soil sample is taken and its clay percentage artificially varied, a relationship between I_P and clay percentage can be obtained. Each soil will have its own straight line because, although in two differing soils the percentages of clay may be the same, they will contain different minerals.

The relationship between montmorillonite, illite, kaolinite and the plasticity index is shown in Fig. 3.38. Note that $1\mu = 1$ micron $= 0.001$ mm.

The plot of London clay is also shown on the figure and, from its position, it is seen that the mineral content of this soil is predominantly illite. London Clay has a clay fraction of about 46 per cent and consists of illite (70%), kaolinite (20%) and montmorillonite (10%). The remaining fraction of 54 per cent consists of silt (quartz, feldspar and mica: 44%) and sand (quartz and feldspar: 10%).

In Fig. 3.38 the slope of the line is the ratio

$$\frac{I_P}{\%\ \text{clay}}$$

Skempton (1953) defined this ratio as the *activity* of the clay. Clays with large activities are called active clays and exhibit plastic properties over a wide range of water content values.

3.19 Residual strength of soil

In an investigation concerning the stability of a clay slope, the normal procedure is to take representative samples, conduct shear tests, establish the strength parameters c′ and ϕ' from the peak values of the tests, and conduct an effective stress analysis. For this analysis the shear strength of the soil can be expressed by the equation:

$$s_f = c' + \sigma' \tan \phi'$$

There have been many cases of slips in clay slopes which have afforded a means of checking this procedure. Obviously when a slope slipped its factor of safety was 1.0 and, knowing the mass of material involved and the location of the slip plane, it is possible to deduce the value of the average shear stress on the slip plane, $\bar{s}$, at the time failure occurred. It has often been found that s is considerably less than s_f, especially with slopes that have been in existence for some years, and Professor Skempton, in his Rankine lecture of 1964, presented a comprehensive study of the subject suppored by case records.

Figure 3.39a shows a typical stress-to-strain relationship obtained in a drained shear test on a clay. Normal practice is to stop the test as soon as the peak strength has been reached, but if the test is continued it is found that as the strain increases the shear strength decreases and finally levels out. This constant stress value is termed the residual strength, s_r, of the clay. The strength envelopes from the two sets of strength values are shown in Fig. 3.39b, it will be seen that the cohesive intercept, based on residual strength, is so negligible that it may be taken as zero.

For clays, residual strength tests can be carried out in a shear box with a large travel of about 150 mm so that the sample can be continually strained in the one direction. The normal shear box can be used provided that it is capable of reversing its direction at the end of the travel. The reduction down to zero and back to its original value of applied stress at the point of reversal can be assumed as occurring over zero strain. The total displacement of the sample is taken to be the length of travel of the box times the number of reversals.

The reversable shear box has become a standard piece of laboratory equipment but it is now believed that, due to the absence of a large one-directional displacement, the values obtained for s_r tend to be on the high side.

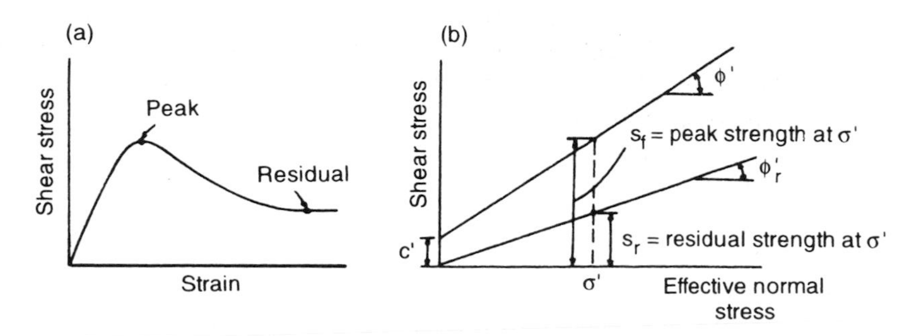

Fig. 3.39 The residual strength of clays (after Skempton, 1964).

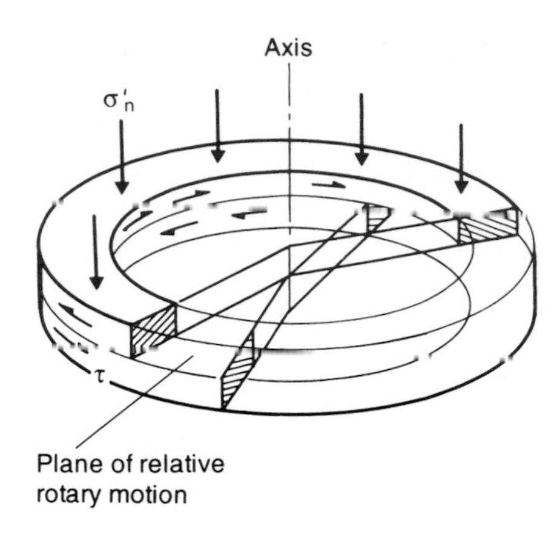

Fig. 3.40 Ring shear test sample (after Bishop *et al.*, 1971).

A more acceptable test can be carried out in the ring shear apparatus, which was developed originally by Hvorslev (and others independently) in about 1934. A thin annular soil specimen is sheared by clamping it between two metal discs, which are then rotated in opposite directions. The apparatus did not become popular, mainly because of the concentration at the time on the study of peak values, so readily obtained from the triaxial test, but probably also because the ring shear apparatus was complicated and it took a long time to carry out a test.

As a result of Skempton's work interest in the determination of soil strength after large displacement was re-established and, in 1971, Bishop *et al.* redeveloped the ring shear apparatus (Fig. 3.40), which is now considered as the most reliable means for determining residual strengths of cohesive soils.

Residual strength of clays

The reduction from peak to residual strength in clays is considered to result primarily from the formation of extremely thin layers of fine particles orientated in the direction of shear; these particles would originally have been in a random state of orientation and must therefore have had a greater resistance to shear than when they became parallel to each other in the shear direction.

The development of residual strength in a soil is a continuous process. If at a particular point the soil is stressed beyond its peak strength, its strength will decrease and additional stress will be transmitted to other points in the soil; these likewise becoming overstressed and decreasing in strength, the failure process continues once it has started (unless the slope slips) until the strength at every point along the potential slip surface has been reduced to residual strength.

Clays, especially overconsolidated deposits, contain fissures, such as those in London clay which occur some 150–200 mm apart; these fissures are already established points of weakness, the strength between their contact surfaces probably being about residual. An important feature of fissures is that they can tend to act as stress concentrators at their edges, leading to overstressing beyond the peak strength and hence to a progressive strength decrease.

Tests carried out by Skempton indicate that the residual strength of clay under a particular effective stress is the same, whether the clay was normally or overconsolidated. Hence in any clay layer, provided the particles are the same, the value of ϕ'_r will be constant.

Residual strength of silts and silty clays

From a study of case records Skempton showed that the value of ϕ'_r decreases with increasing clay percentage. Sand-sized particles, being roughly spherical in shape, cannot orientate themselves in the same way as flakey clay particles and when they are present in silts or clays the residual strength becomes greater as the percentage of sand increases.

Residual strength of sands

Shear tests on sand indicate that the stress–displacement curve for the loose and dense states are as shown in Fig. 3.41. The residual strength is seen to correspond to the peak strength of the loose density and is usually reached fairly quickly in one travel of the shear box, succeeding reversals having little effect.

Residual factor, R

In the slips investigated by Skempton, some were found to have an $\bar{s}$ value corresponding to s_r and some an $\bar{s}$ value lying between and s_f and s_r. The use of the term residual factor, R, was therefore suggested, where R is the proportion of the total slip surface in the clay along which strength has fallen to the residual value:

$$R = \frac{s_f - \bar{s}}{s_f - s_r}$$

If there is no reduction in strength, $\bar{s} = s_f$ and $R = 0.0$, but if there is a complete reduction in strength then $\bar{s} = s_r$ and $R = 1.0$. The work so far carried out on residual

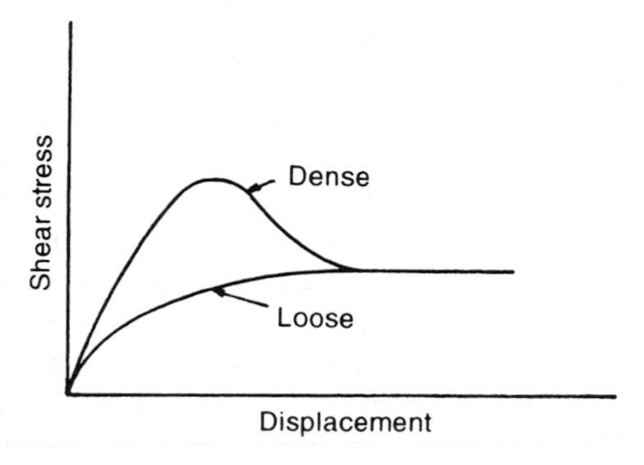

Fig. 3.41 Stress–displacement characteristics of sands.

strength has involved existing slopes and cuttings, for which Skempton's findings may be summarised as follows:

Unfissured clays: $R \approx 0.0$
Pre-existing slip: $R \sim 1.0$
Fissured clays: R varies from 0.0 to 1.0

Indications are that R increases with time, but there is at present no way of predicting its value from soil tests and it is not known if residual strengths can become evident in compact material; standard practice is to base stability analyses on peak soil strengths. If an earth embankment settles unevenly fissures can develop within it, but bearing pressures are generally kept within reasonable limits. With coal spoil heaps higher bearing capacities are often used which could lead to larger settlements. Such tips may also be subjected to mining subsidence (sometimes of several metres) and it does not seem impossible for fissuring to occur under these conditions. If there is fissuring then a potential slip surface will tend to travel through this weakened zone (which may have a strength closer to the loose than the compacted density), leading to a reduction in stability.

Exercises

Exercise 3.1

A soil sample is tested to failure in a drained triaxial test using a cell pressure of 200 kPa. The effective stress parameters of the soil are known to be $\phi' = 29°$ and $c' = 0$.

Determine the inclination of the plane of failure, with respect to the direction of the major principal stress, and the magnitudes of the stresses that will act on this plane. What is the maximum value of shear stress that will be induced in the soil?

Answer Failure plane inclined at 30.5° to major principal stress.
Effective normal stress on failure plane = 310 kPa.
Shear stress on failure plane = 170 kPa.
Maximum shear stress = 190 kPa.

Exercise 3.2

A soil has an effective angle of shearing resistance, ϕ', of 20° and an effective cohesion, c', of 20 kPa. What would you expect the value of the vertical stress would be at failure if the soil is subjected to:

(a) a drained triaxial extension test with a cell pressure of 250 kPa.
(b) a drained triaxial compression test with the same cell pressure?

Answer (a) 95 kPa, (b) 567 kPa

Exercise 3.3

An undisturbed soil sample, 110 mm in diameter and 220 mm in height, was tested in a triaxial machine. The sample sheared under an additional axial load of 3.35 kN with a vertical deformation of 21 mm. The failure plane was inclined at 50° to the horizontal and the cell pressure was 300 kPa.

(i) Draw the Mohr circle diagram representing the above stress conditions and from it determine:
  (a) Coulomb's equation for the shear strength of the soil, in terms of total stress;
  (b) the magnitude and obliquity of the resultant stress on the failure plane.

(ii) A further undisturbed sample of the soil was tested in a shear box under the same drainage conditions as used for the previous test. If the area of the box was 3600 mm^2 and the normal load was 500 N what would you expect the failure shear stress to have been?

Answer (i) (a) $\tau = 80 + 0.1763\ \sigma$kPa
(b) 465 kPa; 20.5°
(ii) 105 kPa

Exercise 3.4

The following results were obtained from a drained triaxial test on a soil:

Cell pressure (kPa)	Additional effective axial stress at failure (kPa)
200	700
400	855
600	1040

Determine the cohesion and angle of friction of the soil with respect to effective stresses.

Answer $\phi' = 17.5°$, $c' = 192$ kPa

Exercise 3.5

Undisturbed samples were taken from a compacted fill material and subjected to consolidated undrained triaxial tests. Results were:

Cell pressure (kPa)	Additional axial stress at failure (kPa)	Pore water pressure at failure (kPa)
200	650	50
400	770	200
600	880	350

Determine the values of the cohesion and the angle of internal friction with regard to (a) total stresses and (b) effective stresses.

Answer $c_u = 220$ kPa, $\phi_u = 13°$; $c' = 110$ kPa, $\phi' = 26°$

Exercise 3.6

An undrained triaxial test carried out on a compacted soil gave the following results:

Strain (%)	Deviator stress (kPa)	Pore water pressure (kPa)
0	0	240
1	240	285
2	460	300
3	640	270
4	840	200
5	950	160
7.5	1100	110
10.0	1150	75
12.5	1170	55
15.0	1150	50

The cell pressure was 400 kPa, and before its application the pore water pressure in the sample was zero.

(i) Determine the value of the pore pressure coefficient B.
(ii) Plot deviator stress (total) against strain.
(iii) Plot pore water pressure against strain.
(iv) Plot the variation of the pore pressure coefficient A with strain.

Answer (i) 0.6
(iv) Value of A at maximum deviator stress = −0.275.

Exercise 3.7

The following results were obtained from an undrained shear box test carried out on a set of undisturbed soil samples. The apparatus made use of a proving ring to measure the shear forces.

Normal load (kN)	0.2	0.4	0.6
Strain (%)	Proving ring dial gauge readings (no. of divisions)		
1	8.5	16.5	28.0
2	16.0	27.0	39.0
3	22.5	34.9	46.8
4	27.5	39.9	52.3
5	31.3	45.0	56.6
6	33.4	46.0	59.7
7	33.4	47.6	61.7
8	33.4	47.6	62.7
9		47.6	62.7
10			62.7

The cross-sectional area of the box was 3600 mm^2 and one division of the proving ring dial gauge equalled 0.01 mm. The calibration of the proving ring was 0.01 mm deflection equalled 8.4 N.

Determine the strength parameters of the soil in terms of total stress.

Answer $\phi_u = 32°$; $c_u = 43$ kPa

Chapter 4
Elements of Stress Analysis

4.1 Stress–strain relationships

Before commencing a study of the material in this chapter it is best to become familiar with the main terms used to describe the stress–strain relationships of a material. It is useful to begin by examining a typical stress–strain plot obtained for a metal (Fig. 4.1).

Results such as those indicated in the figure would normally be obtained by subjecting a specimen of the metal to a tensile test and plotting the values of tensile strain against the nominal values of tensile stress, as the stress–strain relationship obtained is equally applicable in tension or compression in the case of a metal.

Note Nominal stress = actual load/original cross-sectional area of specimen, i.e. no allowance is made for reduction in area, due to necking, as the load is increased.

From the plot it is seen that in the early stages of loading, up to point B, the stress is proportional to the strain. Unloading tests can also demonstrate that, up to the point A, the metal is elastic in that it will return to its original dimensions if the load is removed. The limiting stress at which elasticity effects are not quite complete is known as the elastic limit, represented by point A. The limiting stress at which linearity between stress and strain ceases is known as the limit of proportionality, point B.

In most metals points A and B occur so close together that they are generally assumed to coincide, i.e. elastic limit is assumed equal to the limit of proportionality.

Point C in Fig. 4.1a represents the yield point, i.e. the stress value at which there is a sudden drop of load, as illustrated, or the stress value at which there is a continuing extension with no further significant increase in load.

Figure 4.1a can be approximated to Fig. 4.1b which represents the ideal elastic–plastic material. In this diagram, point 1 represents the limit of elasticity

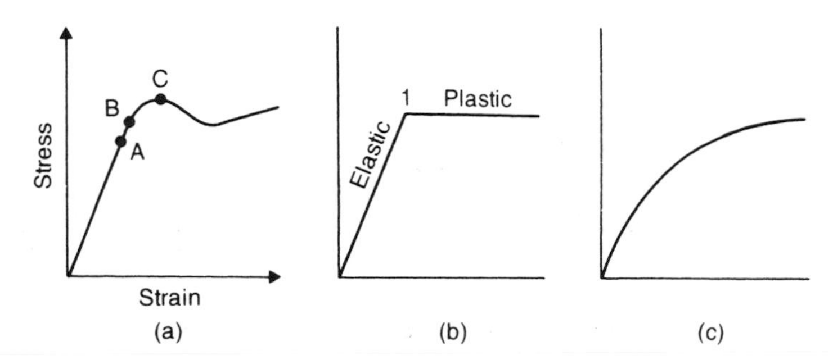

Fig. 4.1 Stress–strain relationships.

and proportionality and the point at which plastic behaviour occurs. The form of the compressive stress–strain relationships typical for all types of soil up to their peak values is as shown in Fig. 4.1c.

It is seen that the stress–strain relationship of a soil is never linear and, in order to obtain solutions, the designer is forced either to assume the idealised conditions of Fig. 4.1b or to solve a particular problem directly from the results of tests that subject samples of the soil to conditions that closely resemble those that are expected to apply *in situ*.

In most soil problems the induced stresses are either low enough to be well below the yield stress of the soil and it can be assumed that the soil will behave elastically (e.g. immediate settlement problems), or they are high enough for the soil to fail by plastic yield (bearing capacity and earth pressure problems), where it can be assumed that the soil will behave as a plastic material.

With soils, even further assumptions must be made if one is to obtain a solution. Generally it is assumed that the soil is both homogeneous and isotropic. As with the assumption of perfect elasticity these theoretical relationships do not apply in practice but can lead to realistic results when sensibly applied.

4.2 The state of stress at a point within a soil mass

A major problem in geotechnical analysis is the estimation of the state of stress at a point at a particular depth in a soil mass.

A load acting on a soil mass, whether internal, due to its self-weight, or external, due to a load applied at the boundary, creates stresses within the soil. If we consider an elemental cube of soil at the point considered then a solution by elastic theory is possible. Each plane of the cube is subjected to a stress, σ, acting normal to the plane, together with a shear stress, τ, acting parallel to the plane. There are therefore a total of six stress components acting on the cube (see Fig. 4.2a). Once the values

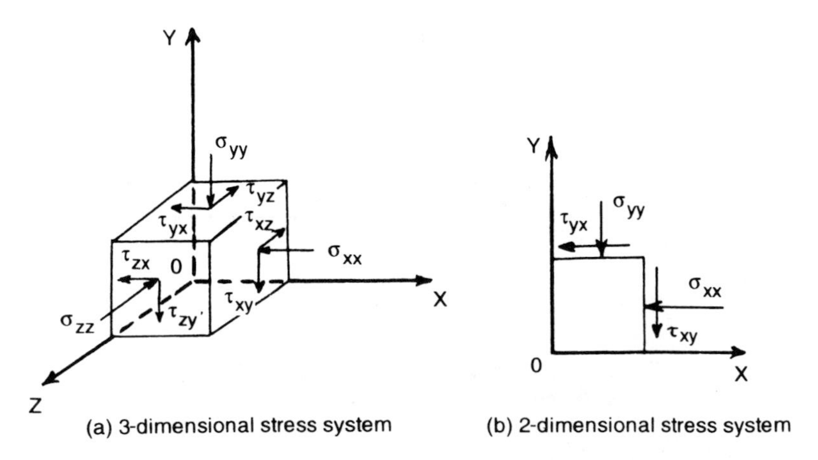

Fig. 4.2 Three- and two-dimensional stress states.

of these components are determined then they can be compounded to give the magnitudes and directions of the principal stresses acting at the point considered (see Timoshenko and Goodier, 1951).

Many geotechnical structures operate in a state of plane strain, i.e. one dimension of the structure is large enough for end effects to be ignored and the problem can be regarded as one of two dimensions. The two-dimensional stress state is illustrated in Fig. 4.2b.

4.3 Stresses induced by the self-weight of the soil

These stresses subject the elemental cube to vertical stress only. They cannot create shear stresses under a level surface.

Example 4.1

A 3 m layer of sand, of saturated unit weight 18 kN/m^3, overlies a 4 m layer of clay, of saturated unit weight 20 kN/m^3. If the groundwater level occurs within the sand at 2 m below the ground surface, determine the total and effective vertical stresses acting at the centre of the clay layer. The sand above groundwater level may be assumed to be saturated. Take $\gamma_w = 10$ kN/m^3.

Solution

For this sort of problem it is usually best to draw a diagram to represent the soil conditions (see Fig. 4.3).

Total vertical stress at centre of clay = Total weight of soil above

$$\sigma_v = 2 \text{ m saturated clay} + 3 \text{ m saturated sand}$$

$$= 2 \times 20 + 3 \times 18 = 94 \text{ kPa}$$

Effective stress = Total stress − Water pressure

$$\sigma'_v = 94 - 10(2 + 1) = 64 \text{ kPa}$$

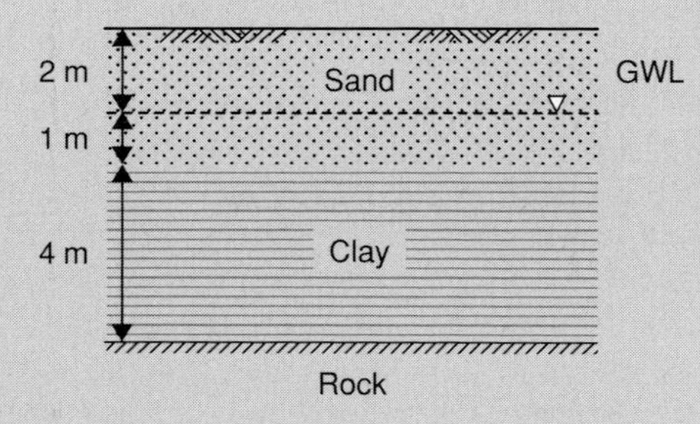

Fig. 4.3 Example 4.1.

4.4 Stresses induced by applied loads

Uniform loading over wide area

In the case of a uniform loading spread over a large area it can be assumed that the same value of vertical stress is induced at the same depth throughout the soil.

Example 4.2

Details of the subsoil conditions at a site are shown in Fig. 4.4 together with details of the soil properties. The ground surface is subjected to a uniform loading of 60 kPa and the groundwater level is 1.2 m below the upper surface of the silt. It can be assumed that the gravel has a degree of saturation of 50 per cent and that the silt layer is fully saturated. Take $\gamma_w = 10$ kN/m^3.

Determine the vertical effective stress acting at a point 1 m above the silt/rock interface.

Solution

$$\text{Bulk unit weight of gravel} = \gamma_w \frac{G_s + eS_r}{1 + e} = 10\,\frac{2.65 + 0.65 \times 0.5}{1 + 0.65}$$

$$= 18.03 = 18 \text{ kN/m}^3$$

$$\text{Saturated weight of silt} = \gamma_w \frac{G_s + e}{1 + e} = 10\,\frac{2.58 + 0.76}{1 + 0.76}$$

$$= 18.98 = 19 \text{ kN/m}^3$$

Effective vertical stress at 1 m above silt/rock interface

$$= \begin{bmatrix}\text{Uniform pressure}\\ \text{applied at ground}\\ \text{surface}\end{bmatrix} + \begin{bmatrix}\text{Total pressure}\\ \text{due to weight}\\ \text{of soils}\end{bmatrix} - [\text{Water pressure}]$$

$$= 60 + (1.8 \times 18 + 4.2 \times 19) - 3 \times 10$$

$$= 142.2 \text{ kPa}$$

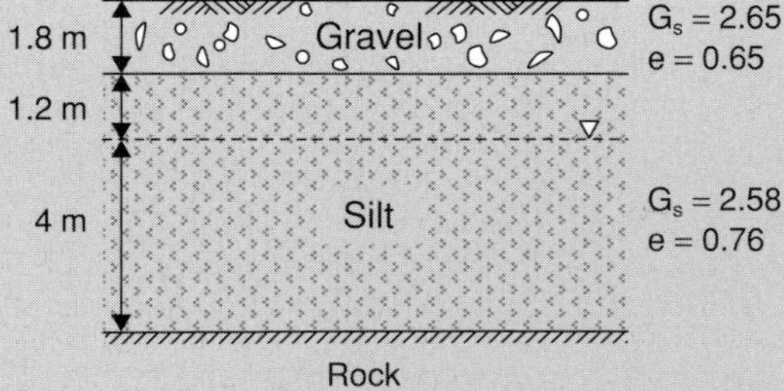

Fig. 4.4 Example 4.2.

Point load (Boussinesq Theory, 1885)

The simplest case of applied loading has been illustrated in Example 4.2. However, most loads are applied to soil through foundations of finite area so that the stresses induced within the soil directly below a particular foundation are different from those induced within the soil at the same depth but at some radial distance away from the centre of the foundation.

The determination of the stress distributions created by various applied loads has occupied researchers for many years. The basic assumption used in all their analyses is that the soil mass acts as a continuous, homogeneous and elastic medium. The assumption of elasticity obviously introduces errors but it leads to stress values that are of the right order and are suitable for most routine design work.

Some more modern methods of settlement analysis, such as those proposed by Lambe (1964, 1967), necessitate determining the increments of both major and minor principal stresses, but Jürgenson (1934) has prepared stress tables based on the elastic theory that can be very helpful in this and other aspects of stress analysis. A description of many solutions has also been prepared by Poulos and Davis (1974).

In most foundation problems, however, it is only necessary to be acquainted with the increase in vertical stresses (for settlement analysis) and the increase in shear stresses (for shear strength analysis).

Boussinesq (1885) evolved equations that can be used to determine the six stress components that act at a point in a semi-infinite elastic medium due to the action of a vertical point load applied on the horizontal surface of the medium.

His expression for the increase in vertical stress is:

$$\Delta\sigma_z = \frac{3Pz^3}{2\pi(r^2 + z^2)^{\frac{5}{2}}}$$

where

P = concentrated load

$r = \sqrt{x^2 + y^2}$ (see Fig. 4.5).

The expression has been simplified to:

$$\Delta\sigma_z = K\frac{P}{z^2}$$

where K is an influence factor.

Values of K against values of r/z are shown in Fig. 4.5.

Example 4.3

A concentrated load of 400 kN acts on the surface of a soil.

Determine the vertical stress increments at points directly beneath the load to a depth of 10 m.

Solution

For points below the load r = 0 and at all depths r/z = 0, whilst from Fig. 4.5 it is seen that K = 0.48.

z (m)	z^2	$\frac{P}{z^2}$	$\Delta\sigma_z = K\frac{P}{z^2}$ (kPa)
0.5	0.25	1600.0	768.0
1.0	1.00	400.0	192.0
2.5	6.25	64.0	30.7
5.0	25.00	16.0	7.7
7.5	56.25	7.1	3.4
10.0	100.00	4.0	1.9

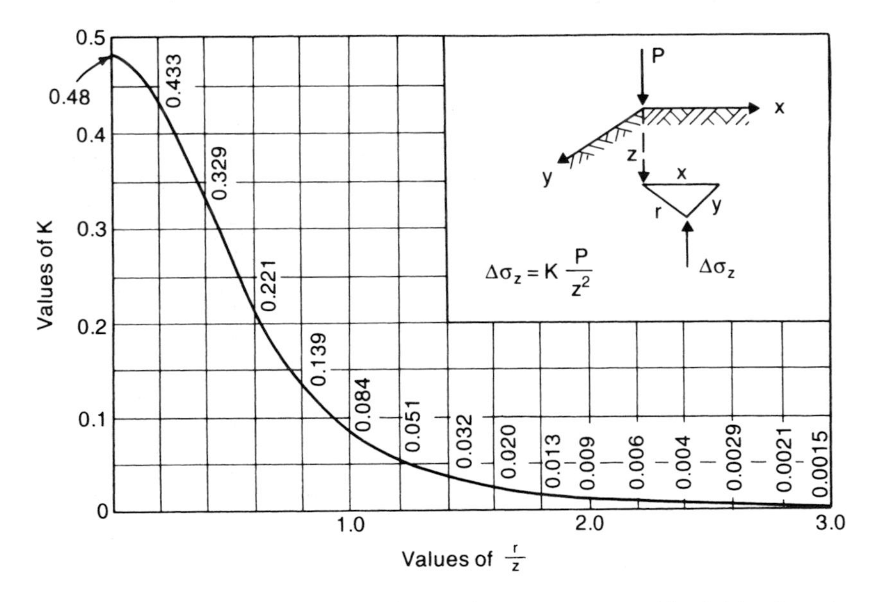

Fig. 4.5 Influence coefficients for vertical stress from a concentrated load (after Boussinesq, 1885).

This method is only applicable to a point load, which is a rare occurrence in soil mechanics, but the method can be extended by the principle of superposition to cover the case of a foundation exerting a uniform pressure on the soil. A plan of the foundation is prepared and this is then split into a convenient number of geometrical sections. The force due to the uniform pressure acting on a particular section is assumed to be concentrated at the centroid of the section, and the vertical stress increments at the point to be analysed due to all the sections are now obtained. The total vertical stress increment at the point is the summation of these increments.

Uniform rectangular load (Steinbrenner's method, 1934)

If a foundation of length L and width B exerts a uniform pressure, p, on the soil then the vertical stress increment due to the foundation at a depth z below one of the corners is given by the expression:

$$\sigma_z = pI_\sigma$$

where I_σ is an influence factor depending upon the relative dimensions of L, B and z.

I_σ can be evaluated by the Boussinesq theory and values of this factor (which depend upon the two coefficients m = B/z and n = L/z) were prepared by Fadum in 1948 (Fig. 4.6).

With the use of this influence factor the determination of the vertical stress increment at a point under a foundation is very much simplified, provided that the foundation can be split into a set of rectangles or squares with corners that meet over the point considered.

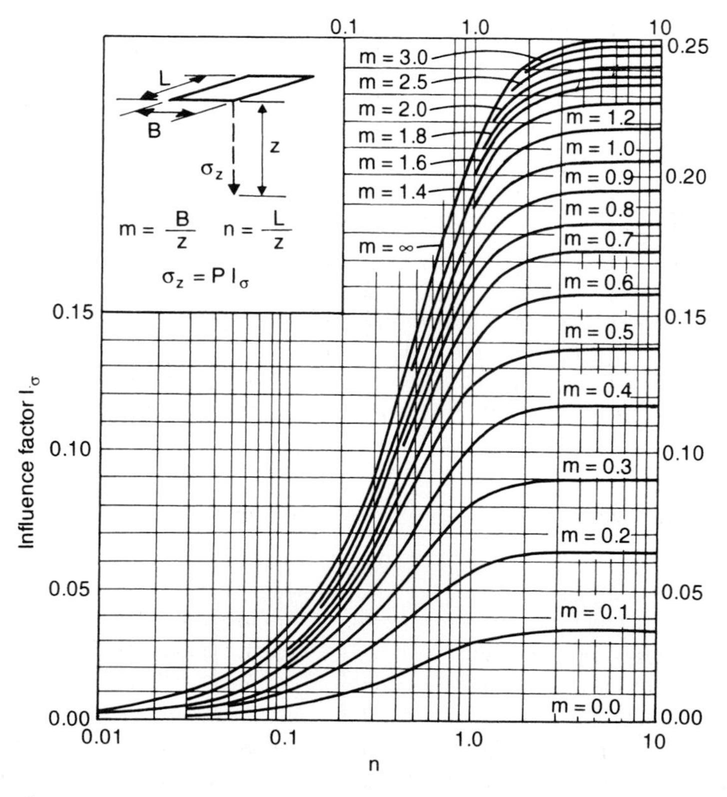

Fig. 4.6 Influence factors for the vertical stress beneath the corner of a rectangular foundation (after Fadum, 1948).

Example 4.4

A 4.5 m square foundation exerts a uniform pressure of 200 kPa on a soil. Determine (i) the vertical stress increments due to the foundation load to a depth of 10 m below its centre and (ii) the vertical stress increment at a point 3 m below the foundation and 4 m from its centre along one of the axes of symmetry.

Solution

(i) The square foundation can be divided into four squares whose corners meet at the centre O (Fig. 4.7a).

z (m)	$m = \frac{B}{z}$	$n = \frac{L}{z}$	I_σ	$4I_\sigma$	σ_z (kPa)
2.5	0.9	0.9	0.163	0.652	130
5.0	0.45	0.45	0.074	0.296	59
7.5	0.3	0.3	0.04	0.16	32
10.0	0.23	0.23	0.025	0.1	20

(ii) This example illustrates how the method can be used for points outside the foundation area (Fig. 4.7b). The foundation is assumed to extend to the point K (Fig. 4.7c) and is now split into two rectangles, AEKH and HKFD.

For both rectangles:

$$m = \frac{B}{z} = \frac{2.25}{3} = 0.75; \qquad n = \frac{L}{z} = \frac{6.25}{3} = 2.08$$

From Fig. 4.6, $I_\sigma = 0.176$, therefore $\sigma_z = 0.176 \times 2 \times 200 = 70.4$ kPa.

The effect of rectangles BEGK and KGCF must now be subtracted.

For both rectangles:

$$m = \frac{2.25}{3} = 0.75; \qquad n = \frac{1.75}{3} = 0.58$$

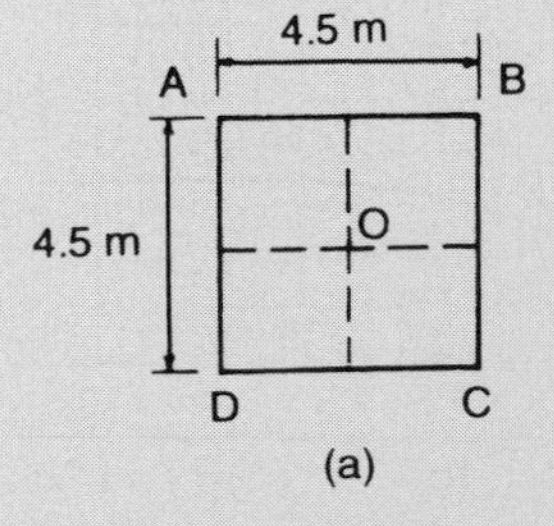

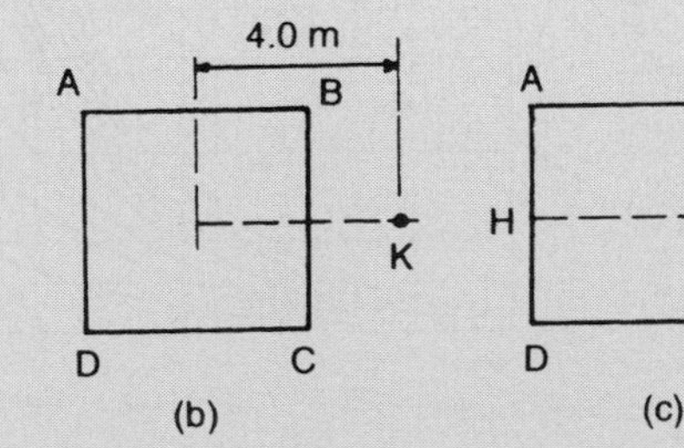

Fig. 4.7 Example 4.4.

From Fig. 4.6, $I_\sigma = 0.122$ (strictly speaking m is 0.58 and n is 0.75, but m and n are interchangeable in Fig. 4.6). Hence:

$$\sigma_z = 0.122 \times 2 \times 200 = 48.8 \text{ kPa}$$

Therefore the vertical stress increment due to the foundation

$$= 70.4 - 48.8 = 21.6 \text{ kPa}$$

Circular foundations can also be solved by Steinbrenner's method, and according to Jürgenson (1934) the stress effects from such a foundation may be found approximately by assuming that it is the same as for a square foundation of the same area.

Example 4.5

A circular foundation of diameter 100 m exerts a uniform pressure on the soil of 450 kPa. Determine the vertical stress increments for depths up to 200 m below its centre.

Solution

$$\text{Area of foundation} = \frac{\pi \times 100^2}{4} = 7850 \text{ m}^2$$

Length of side of square foundation of same area $= \sqrt{7850} = 88.6$ m. This imaginary square can be divided into four squares as in Example 4.4(i). Length of sides of squares = 44.3 m.

z (m)	$n = m = \frac{B}{z}$	I_σ	$4I_\sigma$	σ_z (kPa)
10	4.43	0.248	0.992	446
25	1.77	0.221	0.884	398
50	0.89	0.16	0.64	288
100	0.44	0.071	0.284	128
150	0.3	0.04	0.16	72
200	0.22	0.024	0.096	43

4.5 Influence charts for vertical stress increments

It may not be possible to employ Fadum's method for irregularly shaped foundations, and a numerical solution is then only possible by the use of Boussinesq's coefficients, K, and the principle of superposition.

Computer software for calculating the vertical stress increments beneath irregularly shaped foundations is widely available and nowadays such software is routinely used to determine the stress values. Historically, however, the Newmark chart (Newmark, 1942) was used to determine the values and any reader interested in the use of the Newmark chart is guided to the seventh, or earlier, editions of this book for a full and detailed explanation.

4.6 Bulbs of pressure

If points of equal vertical pressure are plotted on a cross-section through the foundation, diagrams of the form shown in Figs 4.8a and 4.8b are obtained.

These diagrams are known as bulbs of pressure and constitute another method of determining vertical stresses at points below a foundation that is of regular shape, the bulb of pressure for a square footing being obtainable approximately by assuming that it has the same effect on the soil as a circular footing of the same area.

In the case of a rectangular footing the bulb pressure will vary at cross-sections taken along the length of the foundation, but the vertical stress at points below the centre of such a foundation can still be obtained from the charts in Fig. 4.8 by either (i) assuming that the foundation is a strip footing or (ii) determining σ_z values for both the strip footing case and the square footing case and combining them by proportioning the length of the two foundations.

From a bulb of pressure one has some idea of the depth of soil affected by a foundation. Significant stress values go down roughly to 2.0 times the width of the foundation, and Fig. 4.9 illustrates how the results from a plate loading test may give quite misleading results if the proposed foundation is much larger: the soft layer of

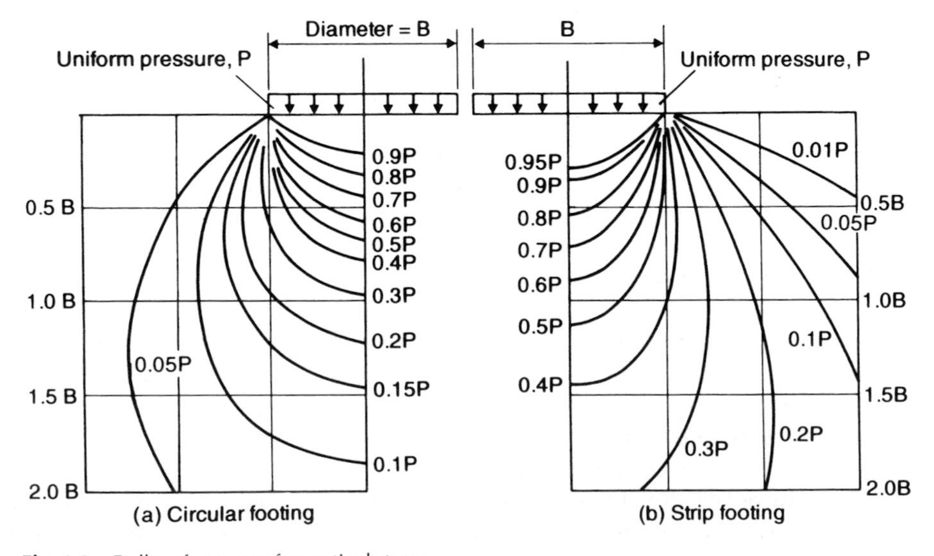

Fig. 4.8 Bulbs of pressure for vertical stress.

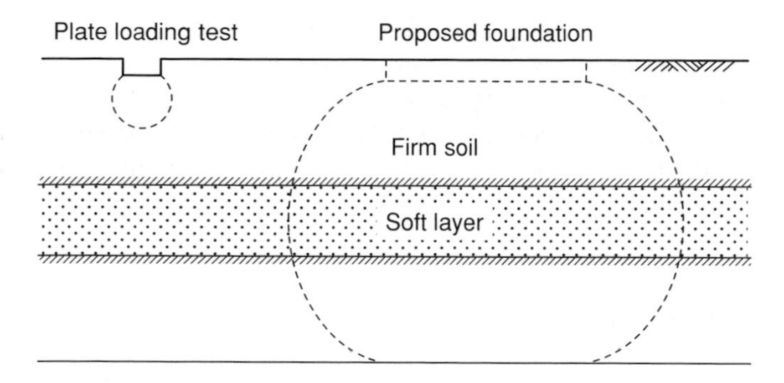

Fig. 4.9 Illustration of how a plate loading test may give misleading results.

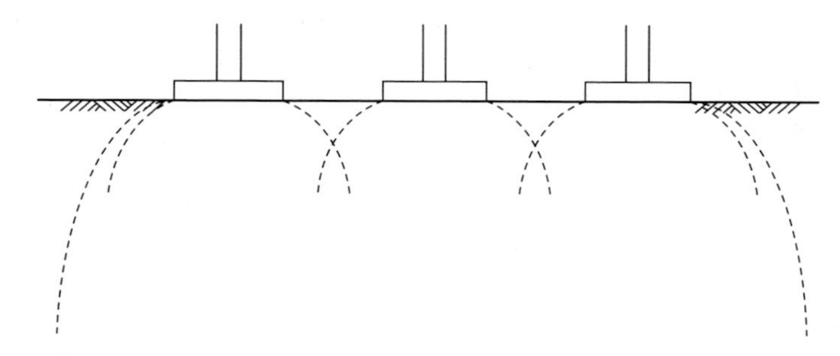

Fig. 4.10 Overlapping of pressure bulbs.

soil in the diagram is unaffected by the plate loading test but would be considerably stressed by the foundation.

Boreholes in a site investigation should therefore be taken down to a depth at least 1.5 times the width of the proposed foundation or until rock is encountered, whichever is the lesser.

Small foundations will act together as one large foundation (Fig. 4.10) unless the foundation are at a greater distance apart (c/c) than five times their width, which is not usual. Boreholes for a building site investigation should therefore be taken down to a depth of approximately 1.5 times the width of the proposed building.

4.7 Shear stresses

In normal foundation design procedure it is essential to check that the shear strength of the soil will not be exceeded. The shear stress developed by loads from foundations of various shapes can be calculated. Jürgenson obtained solutions for the case of a circular footing and for the case of a strip footing (Fig. 4.11). It may be noted that, in the case of a strip footing, the maximum stress induced in the soil is p/π, this value occurring at points lying on a semicircle of diameter equal to the foundation width B. Hence the maximum shear stress under the centre of a continuous foundation occurs at a depth of B/2 beneath the centre.

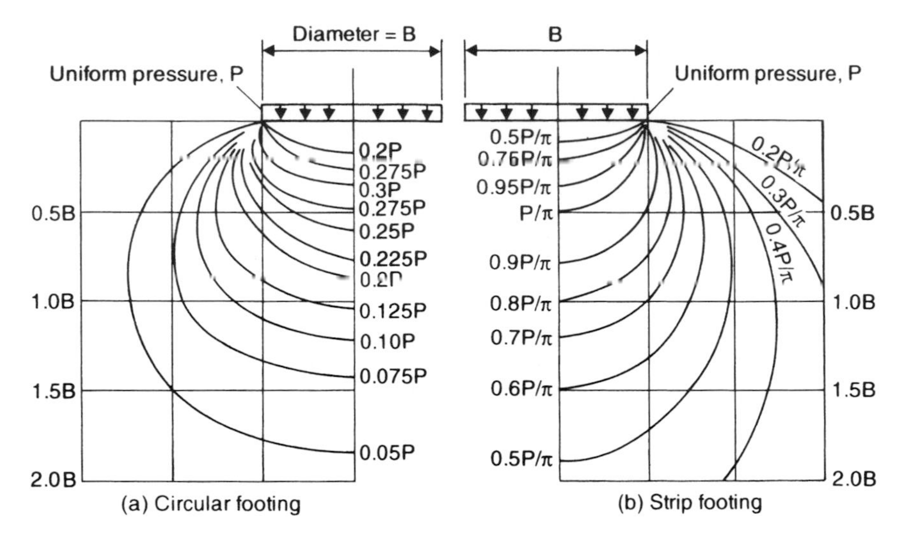

Fig. 4.11 Pressure bulbs of shear stress (after Jürgenson, 1934).

Shear stresses under a rectangular foundation

It is sometimes necessary to evaluate the shear stresses beneath a foundation in order to determine a picture of the likely overstressing in the soil.

Unfortunately a large number of foundations are neither circular nor square but rectangular, but Figs 4.11a and 4.11b can be used to give a rough estimate of shear stress under the centre of a rectangular footing.

Example 4.6

A rectangular foundation has the dimensions 15 m × 5 m and exerts a uniform pressure on the soil of 600 kPa. Determine the shear stress induced by the foundation beneath the centre at a depth of 5 m.

Solution

Strip footing

$$\frac{z}{B} = \frac{5}{5} = 1.0$$

From Fig. 4.11b:

$$\tau = \frac{0.81 \times 600}{\pi} = 155 \text{ kPa}$$

For a square footing:

$$\text{Area} = 5 \times 5 = 25\ \text{m}^2$$

Diameter of circle of same area:

$$\sqrt{\frac{25 \times 4}{\pi}} = 5.64\ \text{m}$$

Hence the shear stress under a 5 m square foundation can be obtained from the bulb of pressure of shear stress for a circular foundation of diameter 5.64 m.

$$\frac{z}{B} = \frac{5}{5.64} = 0.89$$

From Fig. 4.11a:

$$\tau = 0.2 \times 600 = 120\ \text{kPa}$$

These values can be combined if we proportion them to the respective areas (or lengths):

$$\tau = 120 + (155 - 120)\frac{15}{15 + 5} = 146\ \text{kPa}$$

The method is approximate but it does give an indication of the shear stress values.

4.8 Contact pressure

Contact pressure is the actual pressure transmitted from the foundation to the soil. In all the foregoing discussions it has been assumed that this contact pressure value, p, is uniform over the whole base of the foundation, but a uniformly loaded foundation will not necessarily transmit a uniform contact pressure to the soil. This is only possible if the foundation is perfectly flexible. The contact pressure distribution of a rigid foundation depends upon the type of soil beneath it. Figures 4.12a and 4.12b show the form of contact pressure distribution induced in a cohesive soil (a) and in a cohesionless soil (b) by a rigid, uniformly loaded, foundation.

On the assumption that the vertical settlement of the foundation is uniform, it is found from the elastic theory that the stress intensity at the edges of a foundation on cohesive soils is infinite. Obviously local yielding of the soil will occur until the resultant distribution approximates to Fig. 4.12a.

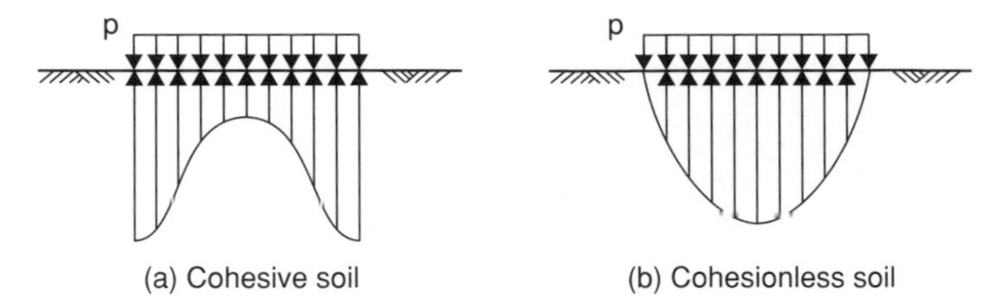

Fig. 4.12 Contact pressure distribution under a rigid foundation loaded with a uniform pressure, p.

For a rigid surface footing sitting on sand the stress at the edges is zero as there is no overburden to give the sand shear strength, whilst the pressure distribution is roughly parabolic (Fig. 4.12b). The more the foundation is below the surface of the sand the more shear strength there is developed at the edges of the foundation, with the result that the pressure distribution tends to be more uniform.

In the case of cohesive soil, which is at failure when the whole of the soil is at its yield stress, the distribution of the contact pressure again tends to uniformity.

A reinforced concrete foundation is neither perfectly flexible nor perfectly rigid, the contact pressure distribution depending upon the degree of rigidity. This pressure distribution should be considered when designing for the moments and shears in the foundation, but in order to evaluate shear and vertical stresses below the foundation the assumption of a uniform load inducing a uniform pressure is sufficiently accurate.

Contact pressures of a spoil heap

A spoil heap, even if compacted, is flexible as far as the supporting soil is concerned. Most existing heaps are of non-uniform section and the stresses induced in the soil below the tip can be approximately determined by superposition in which the tip is divided into a set of equivalent layers. Each layer is assumed to act in turn on the surface of the soil, and the total induced stresses are obtained by addition (Fig. 4.13). The method can be extended to include earth embankments.

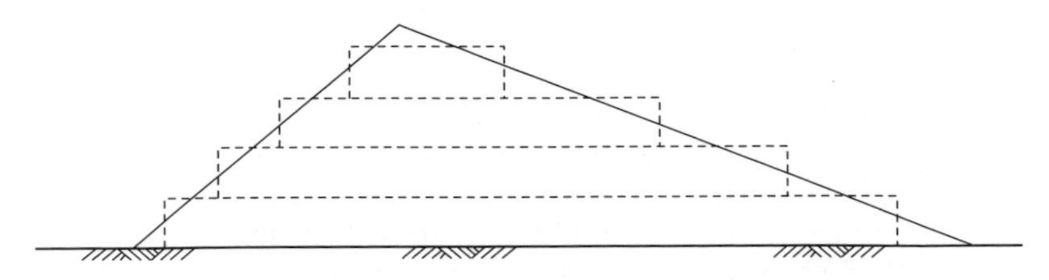

Fig. 4.13 Method for determining subsoil stresses beneath a spoil heap (the effect of slope is of course three-dimensional).

Exercises

Exercise 4.1

A raft foundation subjects its supporting soil to a uniform pressure of 300 kPa. The dimensions of the raft are 6.1 m by 15.25 m. Determine the vertical stress increments due to the raft at a depth of 4.58 m below it (i) at the centre of the raft and (ii) at the central points of the long edges.

Answer (i) 192 kPa, (ii) 132 kPa

Exercise 4.2

A concentrated load of 85 kN acts on the horizontal surface of a soil. Plot the variation of vertical stress increments due to the load on horizontal planes at depths of 1 m, 2 m and 3 m directly beneath it.

Answer Fig. 4.14.

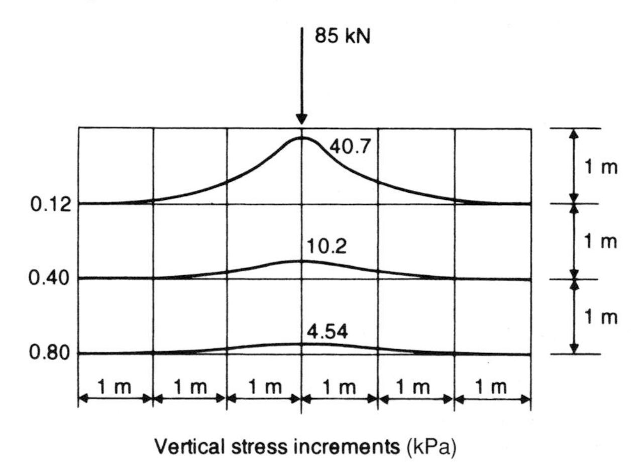

Fig. 4.14 Exercise 4.2.

Exercise 4.3

The plan of a foundation is given in Fig. 4.15a. The uniform pressure on the soil is 40 kPa. Determine the vertical stress increment due to the foundation at a depth of 5 m below the point X, using Fig. 4.6.

Note In order to obtain a set of rectangles whose corners meet at a point, a section of the foundation area is sometimes included twice and a correction made. For this particular problem the foundation area must be divided into six rectangles (Fig. 4.15b); the effect of the shaded portion will be included twice and must therefore be subtracted once.

Answer 11.1 kPa

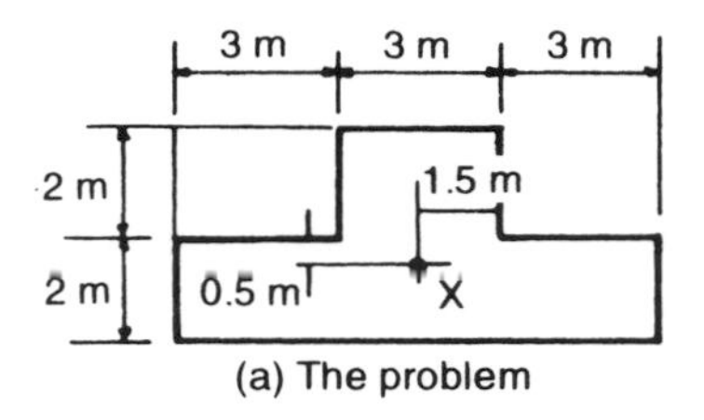

(a) The problem

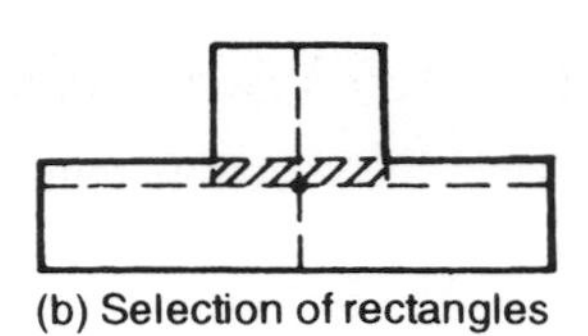
(b) Selection of rectangles

Fig. 4.15 Exercise 4.3.

Exercise 4.4

A load of 500 kN is uniformly distributed through a pad foundation of dimensions 1.0 m × 1.5 m. Determine the increase in vertical stress at a depth of 2.0 m below one corner of the foundation.

Answer 36 kPa

Chapter 5
Stability of Slopes

5.1 Granular materials

Soils such as gravel and sand are collectively referred to as granular soils and normally exhibit only a frictional component of strength. A potential slip surface in a slope of granular material will be planar and the analysis of the slope is relatively simple. However, most soils exhibit both cohesive and frictional strength and pure granular soils are fairly infrequent. Nevertheless a study of granular soils affords a useful introduction to the later treatment of soil slopes that exhibit both cohesive and frictional strength.

Figure 5.1 illustrates an embankment of granular material with an angle of shearing resistance, ϕ, and with its surface sloping at angle β to the horizontal.

Consider an element of the embankment of weight W:

$$\text{Force parallel to slope} = \text{W} \sin \beta$$
$$\text{Force perpendicular to slope} = \text{W} \cos \beta$$

For stability,

$$\text{Sliding forces} = \frac{\text{Restraining forces}}{\text{Factor of safety (F)}}$$

i.e.

$$\text{W} \sin \beta = \frac{\text{W} \cos \beta \tan \phi}{\text{F}}$$

$$\Rightarrow \quad \text{F} = \frac{\tan \phi}{\tan \beta}$$

For limiting equilibrium (F = 1), $\tan \beta = \tan \phi$, i.e. $\beta = \phi$.

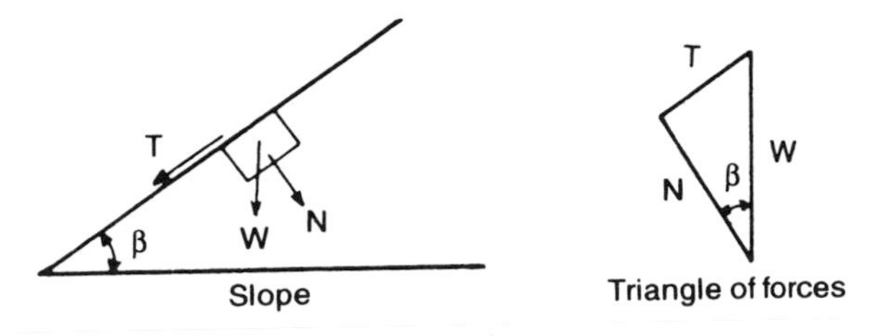

Fig. 5.1 Forces involved in a slope of granular material.

From this it is seen that (a) the weight of a material does not affect the stability of the slope, (b) the safe angle for the slope is the same whether the soil is dry or submerged, and (c) the embankment can be of any height.

Failure of a submerged sand slope can occur, however, if the water level of the retained water falls rapidly while the water level in the slope lags behind, since seepage forces are set up in this situation.

5.1.1 Seepage forces in a granular slope subjected to rapid drawdown

In Fig. 5.2a the level of the river has dropped suddenly due to tidal effects. The permeability of the soil in the slope is such that the water in it cannot follow the water level changes as rapidly as the river, with the result that seepage occurs from the high water level in the slope to the lower water level of the river. A flow net can be drawn for this condition and the excess hydrostatic head for any point within the slope can be determined.

Assume that a potential failure plane, parallel to the surface of the slope, occurs at a depth z and consider an element within the slope of weight W. Let the excess pore water pressure induced by seepage be u at the mid point of the base of the element.

$$\text{Normal reaction } N = W\cos\beta$$

$$\text{Normal stress } \sigma = \frac{W\cos\beta}{1} = \frac{W\cos^2\beta}{b} \qquad \left(\text{since } 1 = \frac{b}{\cos\beta}\right)$$

$$\text{Normal effective stress } \sigma' = \frac{W\cos^2\beta}{b} - u$$

$$= \frac{\gamma z b\cos^2\beta}{b} - u = \gamma z\cos^2\beta - u$$

(Where γ = the average unit weight of the whole slice, it is usually taken that the whole slice is saturated.)

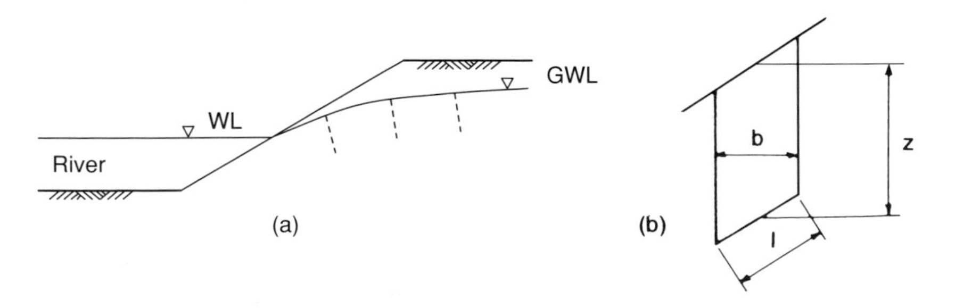

Fig. 5.2 Seepage due to rapid drawdown.

Tangential force = W sin β

$$\Rightarrow \quad \text{Tangential shear stress, } \tau = \frac{W \sin \beta}{1} = \gamma z \sin \beta \cos \beta$$

Ultimate shear strength of soil = $\sigma' \tan \phi = \tau F$

$$\Rightarrow \quad \gamma z \sin \beta \cos \beta = (\gamma z \cos^2 \beta - u) \frac{\tan \phi}{F}$$

$$\Rightarrow \quad F = \left(\frac{\cos \beta}{\sin \beta} - \frac{u}{\gamma z \sin \beta \cos \beta} \right) \tan \phi$$

$$= \left(1 - \frac{u}{\gamma z \cos^2 \beta} \right) \frac{\tan \phi}{\tan \beta}$$

This expression may be written as:

$$F = \left(1 - \frac{r_u}{\cos^2 \beta} \right) \frac{\tan \phi}{\tan \beta}$$

where

$$r_u = \frac{u}{\gamma_z}$$

5.1.2 *Pore pressure ratio*

The ratio, at any given point, of the pore water pressure to the weight of the material acting on unit area above it is known as the pore pressure ratio and is given the symbol r_u. (See also Section 5.5.1.)

Flow parallel to the surface and at the surface

The flow net for these special conditions is illustrated in Fig. 5.3.

If we consider the same element as before, the excess pore water head, at the centre of the base of the element, is represented by the height h_w in Fig. 5.3. In the figure, AB = z cos β and h_w = AB cos β. Hence, $h_w = z \cos^2 \beta$, so that excess pore water pressure at the base of the element = $\gamma_w z \cos^2 \beta$.

$$\Rightarrow \quad r_u = \frac{u}{\gamma z} = \frac{\gamma_w z \cos^2 \beta}{\gamma z} = \frac{\gamma_w}{\gamma} \cos^2 \beta$$

The equation for F becomes:

$$F = \left(1 - \frac{\gamma_w}{\gamma} \right) \frac{\tan \phi}{\tan \beta} = \left(\frac{\gamma - \gamma_w}{\gamma} \right) \frac{\tan \phi}{\tan \beta} = \frac{\gamma' \tan \phi}{\gamma_{sat} \tan \beta}$$

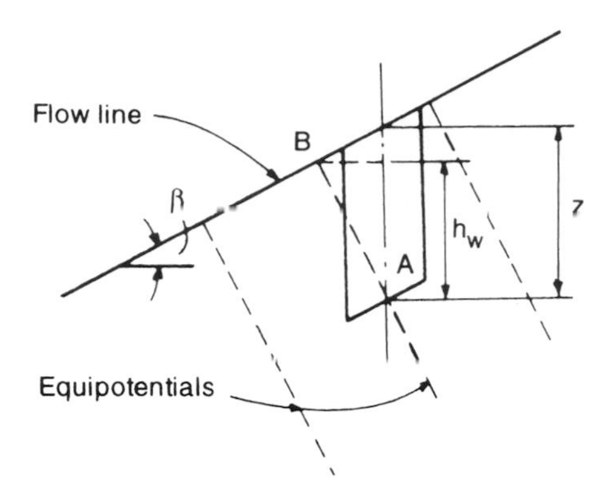

Fig. 5.3 Flow net when flow is parallel and at the surface.

Example 5.1

A granular soil has a saturated unit weight of 18.0 kN/m^3 and an angle of shearing resistance of 30°. A slope is to be made of this material. If the factor of safety is to be 1.25, determine the safe angle of the slope (i) when the slope is dry or submerged and (ii) if seepage occurs at and parallel to the surface of the slope.

Solution

(i) When dry or submerged:

$$F = \frac{\tan\phi}{\tan\beta} \quad \Rightarrow \quad \tan\beta = \frac{0.5774}{1.25} = 0.462$$

$$\Rightarrow \quad \beta = 25°$$

(ii) When flow occurs at and parallel to the surface:

$$F = \frac{\gamma' \tan\phi}{\gamma_{sat} \tan\beta} \quad \Rightarrow \quad \tan\beta = \frac{(18 - 9.81) \times 0.5774}{1.25 \times 18} = 0.210$$

$$\Rightarrow \quad \beta = 12°$$

Seepage more than halves the safe angle of slope in this particular example.

5.2 Soils with two strength components

Failures in embankments made from soils that possess both cohesive and frictional strength components tend to be rotational, the actual slip surface approximating to the arc of a circle (Fig. 5.4).

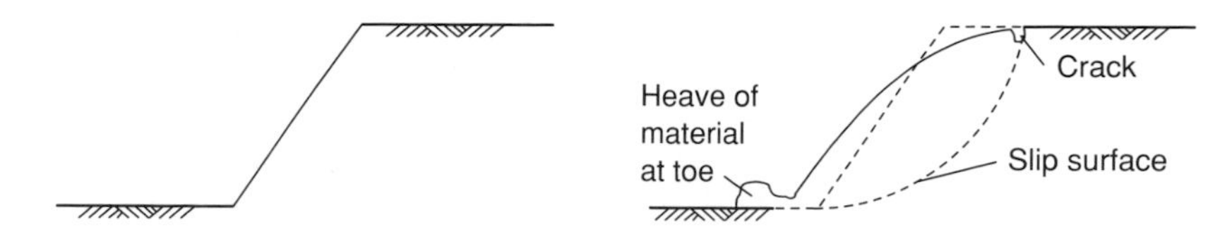

Fig. 5.4 Typical rotational slip in a cohesive soil.

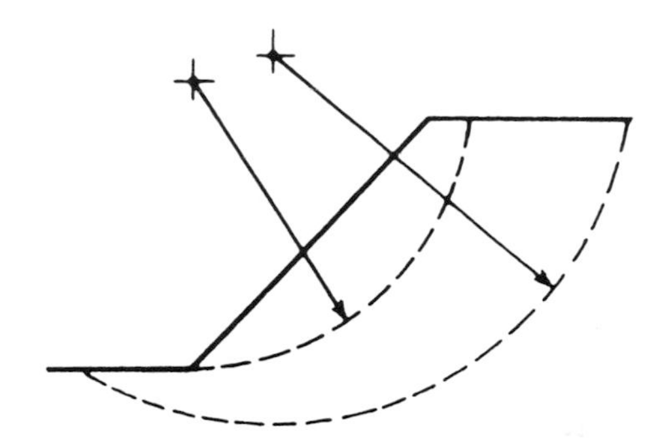

Fig. 5.5 Example of two possible slip surfaces.

5.3 Methods of investigating slope stability

Contemporary methods of investigation are based on (a) assuming a slip surface and a centre about which it rotates, (b) studying the equilibrium of the forces acting on this surface, and (c) repeating the process until the worst slip surface is found as illustrated in Fig. 5.5. The worst slip surface is that surface which yields the lowest factor of safety, F, where F is the ratio of the restoring moment to the disturbing moment, each moment considered about the centre of rotation. The methods of assessing stability using this moment equilibrium approach are described in the next few sections. Alternatively, if stability assessment is to be performed in accordance with Eurocode 7, the strength parameters of the soil are first divided by partial factors, and stability is then confirmed by checking the GEO limit state (see Section 5.7).

5.4 Total stress analysis

This analysis, often called the $\phi_u = 0$ analysis, is intended to give the stability of an embankment immediately after its construction. At this stage it is assumed that the soil in the embankment has had no time to drain and the strength parameters used in the analysis are the ones representing the undrained strength of the soil (with respect to total stresses), which are found from either the unconfined compression test or an undrained triaxial test without pore pressure measurements.

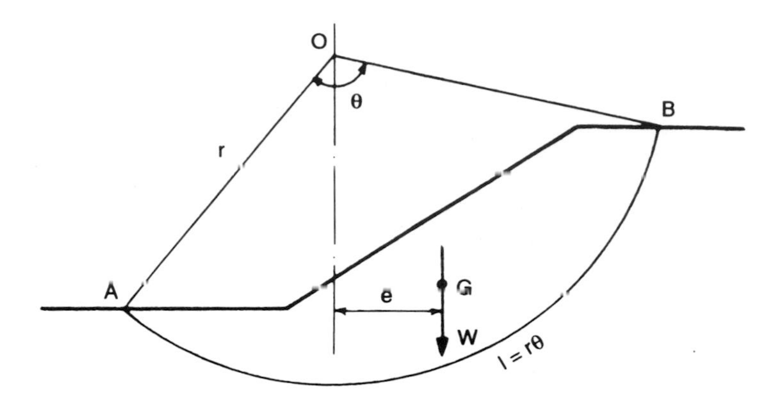

Fig. 5.6 Total stress analysis.

Consider in Fig. 5.6 the sector of soil cut off by arc AB of radius r. Let W equal the weight of the sector and G the position of its centre of gravity. As $\phi_u = 0°$, shear strength component = c_u.

Taking moments about O, the centre of rotation:

$$We = c_u lr = c_u r\theta r = c_u r^2\theta \text{ for equilibrium}$$

$$F = \frac{\text{Restoring moment}}{\text{Disturbing moment}} = \frac{c_u r^2\theta}{We}$$

The position of G is not needed, and it is only necessary to ascertain where the line of action of W is. This can be obtained by dividing the sector into a set of vertical slices and taking moments of area of these slices about a convenient vertical axis.

5.4.1 *Effect of tension cracks*

With a slip in a cohesive soil there will be a tension crack at the top of the slope (Fig. 5.7) along which no shear resistance can develop. In a purely cohesive soil the depth of the crack, h_c, is given by the following formula (see Chapter 6):

$$h_c = \frac{2c_u}{\gamma}$$

The effect of the tension crack is to shorten the arc AB to AB′. If the crack is to be allowed for, the angle θ' must be used instead of θ in the formula for F, and the full weight W of the sector is still used in order to compensate for any water pressures that may be exerted if the crack fills with rain water.

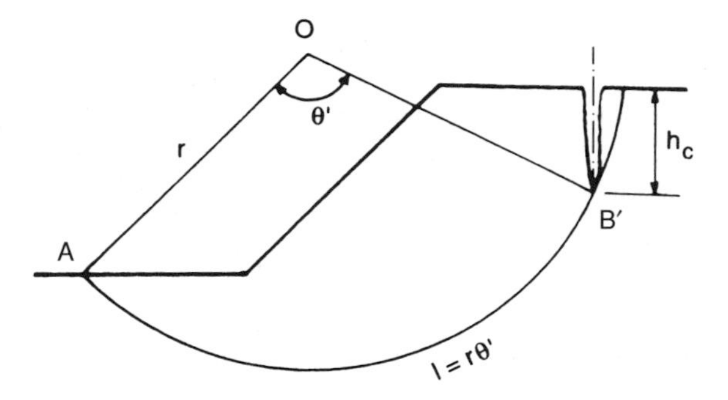

Fig. 5.7 Tension crack in a cohesive soil.

Example 5.2

Figure 5.8 gives details of an embankment made of cohesive soil with $\phi_u = 0$ and $c_u = 20$ kPa. The unit weight of the soil is 19 kN/m^3.

For the trial circle shown, determine the factor of safety against sliding. The weight of the sliding sector is 346 kN acting at an eccentricity of 5 m from the centre of rotation. What would the factor of safety be if the shaded portion of the embankment were removed? In both cases assume that no tension crack develops.

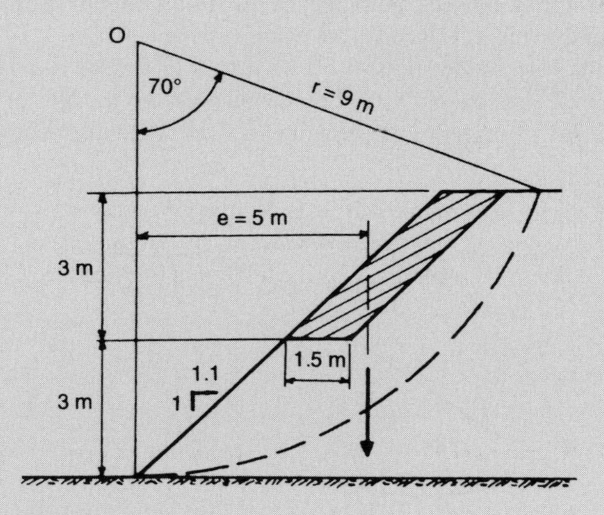

Fig. 5.8 Example 5.2.

Solution

Disturbing moment = 346 × 5 = 1730 kN m

Restoring moment = $c_u r^2 \theta = 20 \times 9^2 \times \frac{70}{180} \times \pi = 1980$ kN m

$\Rightarrow \quad F = \frac{1980}{1730} = 1.14$

Area of portion removed = 1.5 × 3 = 4.5 m^2

Weight of portion removed = 4.5 × 19 = 85.5 kN

Eccentricity from O = $3.3 + \frac{3.3 + 1.5}{2} = 5.7$ m

Relief of disturbing moment = 5.7 × 85.5 = 488 kN m

$\Rightarrow \quad F = \frac{1980}{1730 - 488} = 1.6$

5.4.2 The Swedish method of slices analysis

With partially saturated soils the undrained strength envelope is no longer parallel to the normal stress axis and the soil has a value of both ϕ and c.

The total stress analysis can be adapted to cover this case by assuming a slip circle procedure and dividing the sector into a suitable number of vertical slices, the stability of one such slice being considered in Fig. 5.9 (the lateral reactions on the two vertical sides of the wedge, L_1 and L_2, are assumed to be equal).

At the base of the slice set off its weight to some scale. Draw the direction of its normal component, N, and by completing the triangle of forces determine its magnitude, together with the magnitude of the tangential component T.

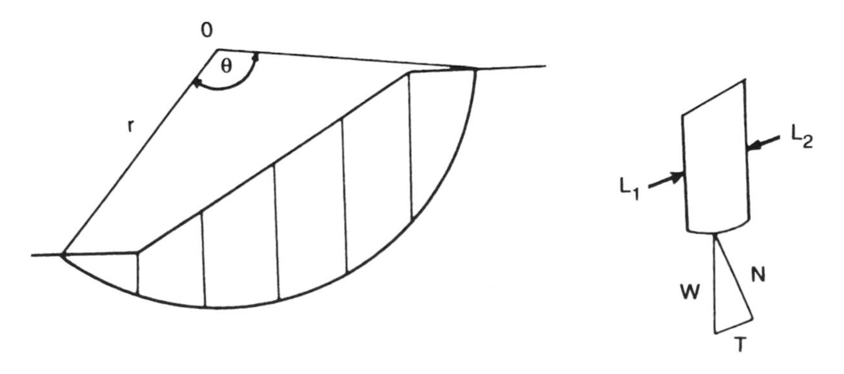

Fig. 5.9 The Swedish method of slices.

Taking moments about the centre of rotation, O:

Disturbing moment = $r\Sigma T$
Restoring moment = $r(cr\theta + \Sigma N \tan\phi)$

Hence

$$F = \frac{cr\theta + \Sigma N \tan\phi}{\Sigma T}$$

The effect of a tension crack can again be allowed for, and in this case (see Chapter 6):

$$h_c = \frac{2c}{\gamma}\tan\left(45^\circ + \frac{\phi}{2}\right)$$

5.4.3 Location of the most critical circle

The centre of the most critical circle was traditionally found by trial and error, various slip circles being analysed and the minimum factor of safety eventually obtained. Nowadays slope stability computer programs are routinely used to rapidly find the centre of the critical slip circle using iterative procedures. Nonetheless, an explanation of one traditional approach is offered in this section.

The procedure is suggested in Fig. 5.10. The centre of each trial circle is plotted and the F value for the circle is written alongside it. After several points have thus been established it is possible to draw 'contours' of F values, which are roughly elliptical so that their centre indicates where the centre of the critical circle will be. Note that the value of F is more sensitive to horizontal movements of the circle's centre than to vertical movements.

To determine a reasonable position for the centre of a first trial slip circle is not easy, but a study of the various types of slips that can occur is helpful (it should

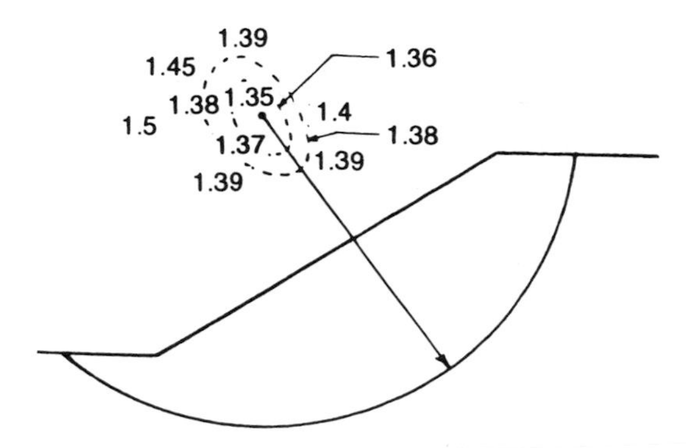

Fig. 5.10 Method for determining the centre of the critical circle.

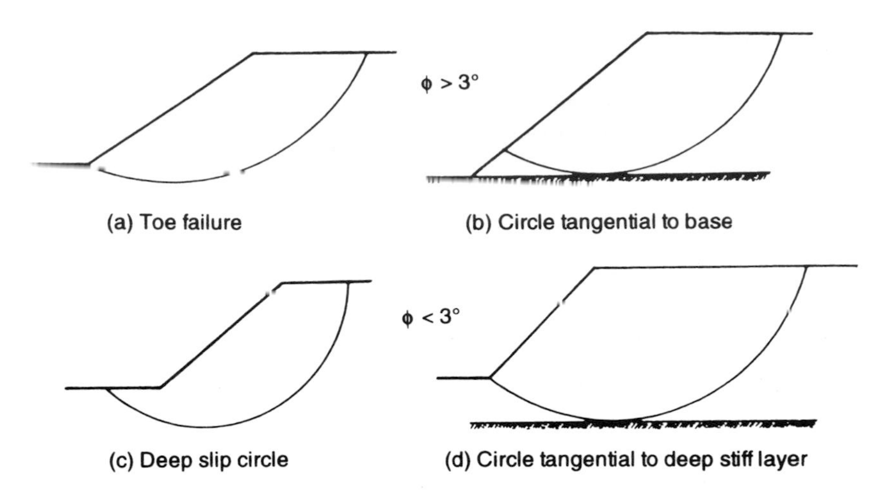

Fig. 5.11 Types of slip failures.

be remembered, however, that the following considerations apply to homogeneous soils). In the case of soils with angles of shearing resistance that are not less than 3°, the critical slip circle is invariably through the toe – as it is for any soil (no matter what its ϕ value) if the angle of slope exceeds 53° (Fig. 5.11a). An exception to this rule occurs when there is a layer of relatively stiff material at the base of the slope, which will cause the circle to be tangential to this layer (Fig. 5.11b).

For cohesive soils with little angle of friction the slip circle tends to be deeper and usually extends in front of the toe (Fig. 5.11c); this type of circle can of course be tangential to a layer of stiff material below the embankment which limits the depth to which it would have extended (Fig. 5.11d).

In the case of a slope made out of homogeneous cohesive soil it is possible to determine directly the centre of the critical circle by a method that Fellenius proposed in 1936 (Fig. 5.12); the centre of the circle is the intersection of two lines set off from the bottom and top of the slope at angles α and β respectively (Fellenius's values for α and β are given in the table below).

Slope	Angle of slope	Angle α	Angle β
1:0.58	60°	29°	40°
1:1	45°	28°	37°
1:1.5	33.79°	26°	35°
1:2	26.57°	25°	35°
1:3	18.43°	25°	35°
1:5	11.32°	25°	37°

This technique is not applicable in its original form to frictional cohesive soils but has been adapted by Jumikis (1962) to suit them, provided that they are homogeneous

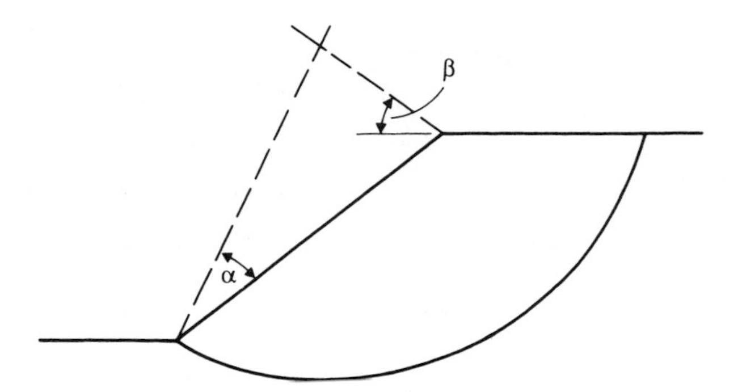

Fig. 5.12 Fellenius' construction for the centre of the critical circle.

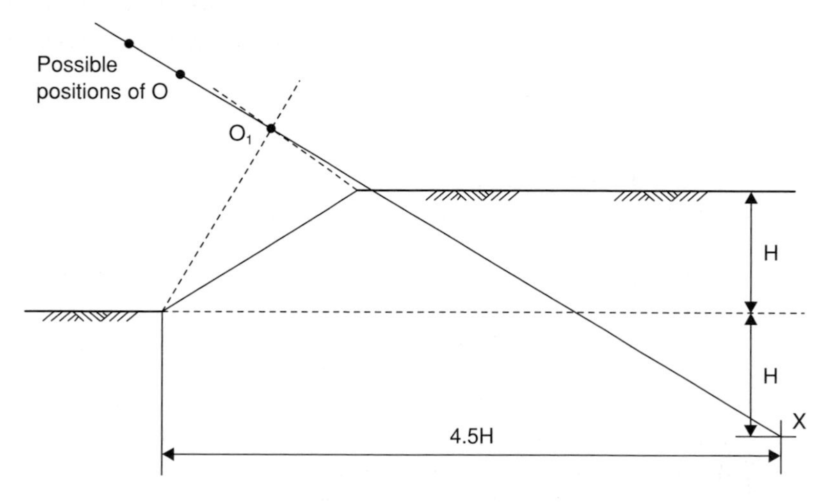

Fig. 5.13 Construction for the centre of the critical circle for a c–ϕ soil.

(Fig. 5.13). It is necessary first to obtain the centre of the Fellenius circle, O_1, as before, after which a point X is established such that X is 2H below the top of the slope and a distance of 4.5H horizontally away from the toe of the slope (H = the vertical height of the slope). The centre of the critical circle, O, lies on the line XO_1 extended beyond O_1, the distance of O beyond O_1 becoming greater as the angle of friction increases.

Such a method can only be used as a means of obtaining a set of sensibly positioned trial slip circles. When the slope is irregular or when there are pore pressures in the soil, conditions are no longer homogeneous and the method becomes less reliable.

Example 5.3

The embankment in Fig. 5.14 is made up from a soil with $\phi = 20°$ and c = 20 kPa. The soil on which the embankment sits has a ϕ of 7° and c = 75 kPa. For both soils $\gamma = 19.3$ kN/m^3.

Determine the factors of safety for the two slip circles shown.

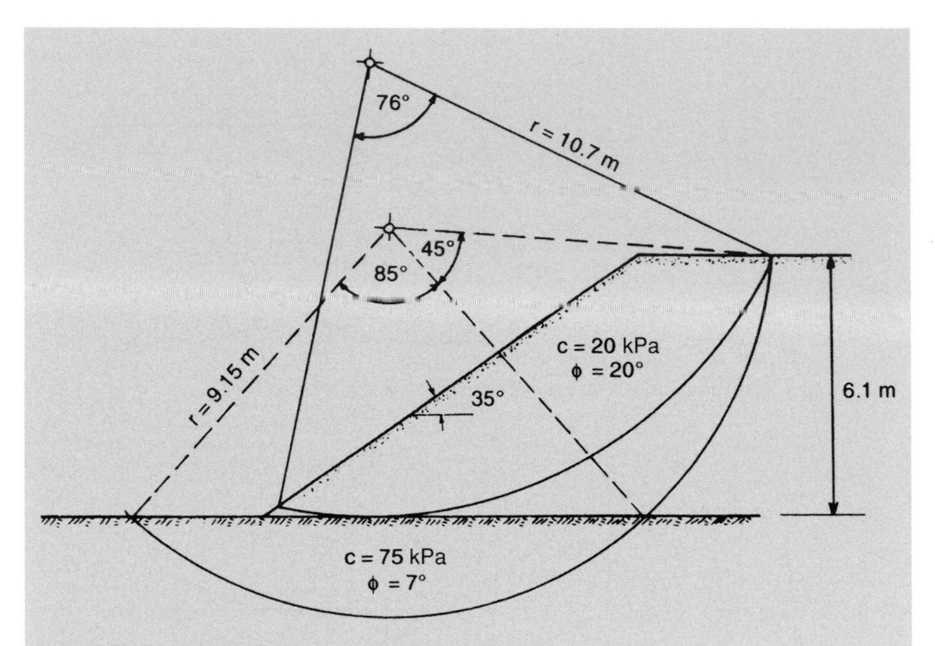

Fig. 5.14 Example 5.3.

Solution

This example is the classic case of an embankment resting on a stiff layer. The slip circle tangential to the lower layer (Fig. 5.15) will give a lower factor of safety, the example being intended to illustrate this effect.

The sliding sector of soil is conveniently divided into four equal vertical slices. To determine the area of a particular slice its mid-height is multiplied by its breadth, and then the weight of the slice is obtained (unit weight × area) and set off as a vector below it. The triangle of forces for the normal and tangential components is then drawn.

The procedure is repeated for each slice, after which the algebraic sum of the tangential forces and the numerical sum of the normal forces is obtained and F evaluated.

The calculations are best set out in tabular form.

Slice no.	Area (m^2)	Weight W (kN)	Normal component N (kN)	Tangential component T (kN)
1	3.7	71	71	−7
2	8.7	168	163	42
3	11.6	224	191	116
4	7.7	148	104	106
			$\Sigma N = 529$	$\Sigma T = 257$

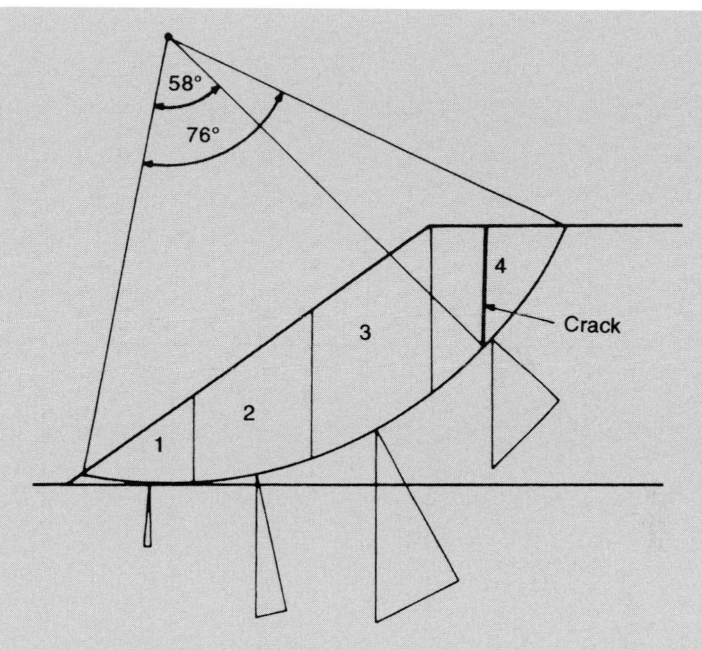

Fig. 5.15 Example 5.3: slip circle tangential to lower layer.

$$\Sigma N \tan \phi = 529 \times 0.364 = 192 \text{ kN}$$

$$cr\theta = 20 \times 10.7 \times \frac{76}{180} \times \pi = 284 \text{ kN}$$

$$F = \frac{cr\theta + \Sigma N \tan \phi}{\Sigma T} = \frac{284 + 192}{257} = 1.85$$

If a tension crack is allowed for:

$$h_c = \frac{2c}{\gamma} \tan\left(45^\circ + \frac{\phi}{2}\right) = \frac{40}{19.3} \times 1.43 = 2.96 \text{ m}$$

θ becomes 58°

$$\Rightarrow \quad cr\theta = 20 \times 10.7 \times \frac{58}{180} \times \pi = 217 \text{ kN}$$

$$F = \frac{192 + 217}{257} = 1.59$$

The deep slip circle is shown in Fig. 5.16.

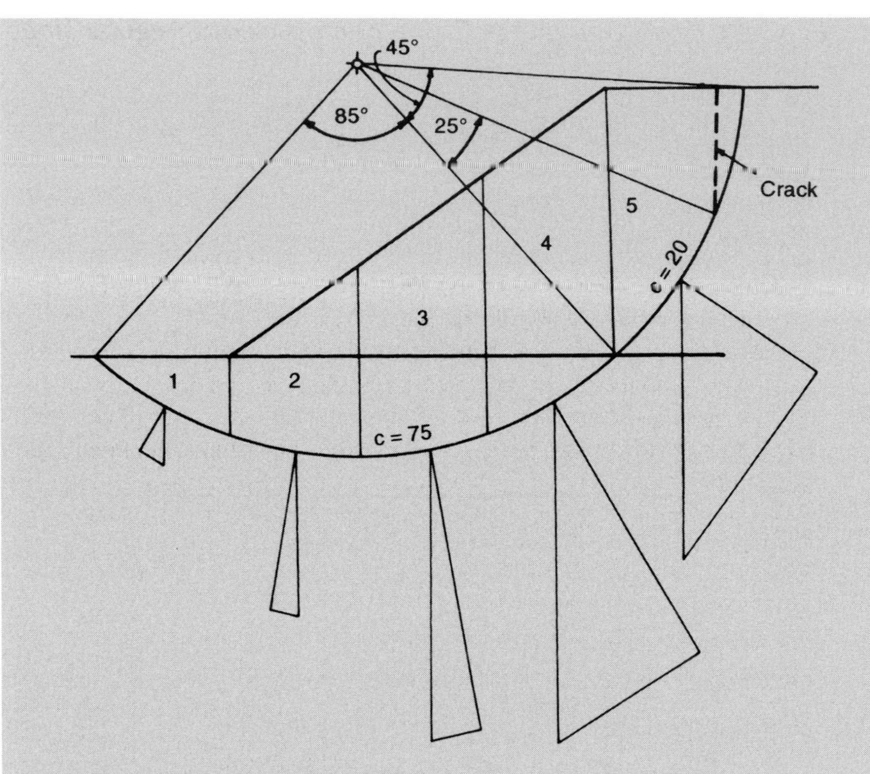

Fig. 5.16 Example 5.3: deep slip circle.

Slice no.	Area (m^2)	Weight W (kN)	Normal component N (kN)	Tangential component T (kN)
1	3.7	71	61	−36
2	9.7	187	184	−33
3	16.6	320	316	52
4	19.2	370	322	186
5	14.3	276	162	224
				$\Sigma T = 393$

ΣN Upper layer = 162 kN
Lower layer = 883 kN

$$\Sigma N \tan \phi = 162 \times 0.364 + 883 \times 0.123 = 169 \text{ kN}$$

$$cr\theta = 75 \times 9.15 \times \frac{85}{180} \times \pi + 20 \times 9.15 \times \frac{45}{180} \times \pi = 1163 \text{ kN}$$

$$F = \frac{1163 + 168}{393} = 3.39$$

If a tension crack is allowed for, F becomes 2.95.

5.4.4 Rapid determination of F for a homogeneous, regular slope

It can be shown that for two similar slopes made from two different soils the ratio $c_m/\gamma H$ is the same for each slope provided that the two soils have the same angle of friction. The ratio $c_m/\gamma H$ is known as the stability number and is given the symbol N, where c_m = cohesion mobilised with regard to total stress, γ = unit weight of soil, and H = vertical height of embankment.

For any type of soil, the critical circle always passes through the toe when $\beta > 53°$. In theory, when $\phi = 0°$ (in practice when $\phi < 3°$) and $\beta > 53°$, the critical slip circle can extend to a considerable depth (Fig. 5.11c).

Taylor (1948) prepared two curves that relate the stability number to the angle of the slope: the first (Fig. 5.17) is for the general case of a c–ϕ soil whilst the second (Fig. 5.18) is for a soil with $\phi = 0°$ with a layer of stiff material or rock at a depth DH below the top of the embankment. D is known as the depth factor, and depending upon its value the slip circle will either emerge at a distance nH in front of the toe or pass through the toe (using the dashed lines the value of n can be obtained from the curves).

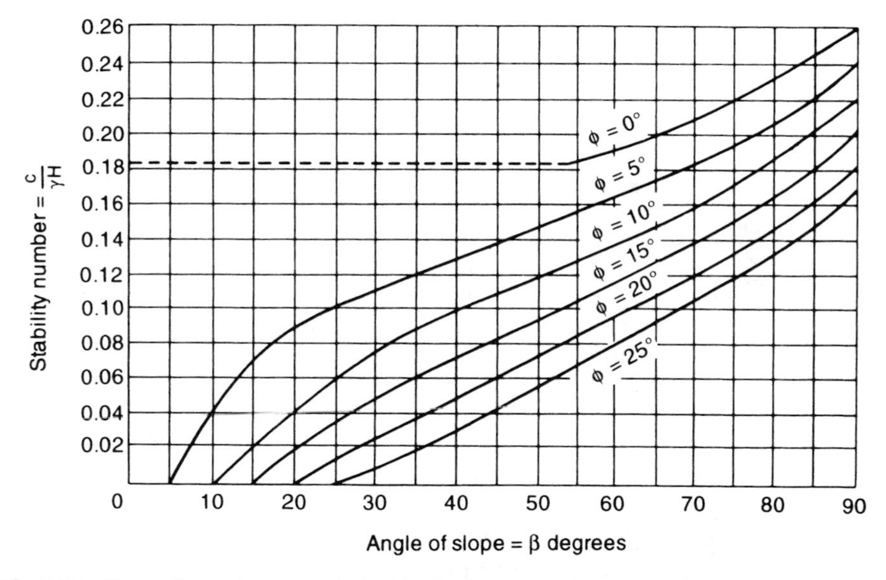

Fig. 5.17 Curves for total stress analysis (After Taylor (1948); for $\phi = 0°$ and $\beta < 53°$, use Fig. 5.18).

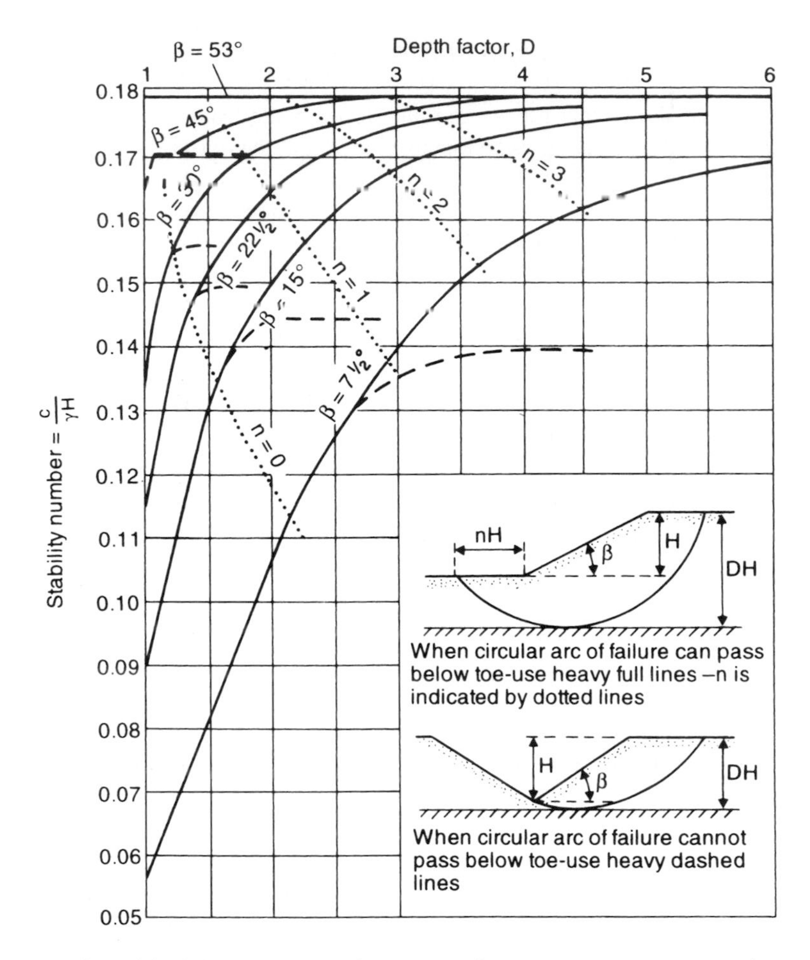

Fig. 5.18 Effect of depth limitation on Taylor's curves ($\beta < 53°$, use Fig. 5.17). Note: $c/\gamma H = 0.181$ at $D = \infty$ for all β values.

Example 5.4

An embankment has a slope of 1 vertical to 2 horizontal. The properties of the soil are: $c = 25$ kPa, $\phi = 20°$, $\gamma = 16$ kN/m^3, and $H = 31$ m.

Using Taylor's charts, determine the F value for the slope.

Solution

From the charts it is seen that a slope with $\phi = 20°$ and an inclination of 26.6° has a stability number of 0.017. This means that, if the factor of safety for friction was unity, c_m, the cohesion which must be mobilised would be found from the expression:

$$\frac{c_m}{\gamma H} = 0.017 \quad \text{i.e.} \quad c_m = 16 \times 31 \times 0.017 = 8.43 \text{ kPa}$$

$$\Rightarrow \quad \text{Factor of safety, with respect to cohesion} = \frac{25}{8.43} = 2.96$$

This is not the factor of safety used in slope stability, which is:

$$F = \frac{\text{Shear strength}}{\text{Disturbing shear}} \quad \text{i.e.} \quad F = \frac{c + \sigma \tan \phi}{\tau}$$

This safety factor applies equally to cohesion and to friction. F can be found by successive approximations:

$$\tau = \frac{c}{F} + \frac{\sigma \tan \phi}{F}$$

so try $F = 1.5$:

$$\frac{\tan \phi}{F} = \frac{0.364}{1.5} = 0.242 = \text{tangent of angle of } 13.5°$$

Use this value of ϕ to establish a new N value from the charts:

$$N = 0.047$$

$$\Rightarrow \quad c_m = 0.047 \times 16 \times 31 = 23.3 \text{ kPa}$$

$$\Rightarrow \quad F \text{ (for c)} = \frac{25}{23.3} = 1.07$$

Try $F = 1.3$:

$$\frac{\tan \phi}{F} = \frac{0.364}{1.3} = 0.28 \qquad (\phi = 15.75°)$$

From the charts $N = 0.036 \quad \Rightarrow \quad F_c = \frac{25}{17.8} = 1.4$

Try $F = 1.35$:

$$\frac{\tan \phi}{F} = \frac{0.364}{1.35} = 0.27 \qquad (\phi = 15°)$$

From the charts $N = 0.037 \quad \Rightarrow \quad F_c = \frac{25}{18.3} = 1.37 \qquad (= F_\phi)$

Factor of safety for slope = 1.35

Example 5.5

Slope = 1 vertical to 4 horizontal, $c = 12.5$ kPa, $H = 31$ m, $\phi = 20°$, $\gamma = 16$ kN/m^3. Find the F value of the slope.

Solution

Angle of slope = 14°, so obviously the slope is safe as it is less than the angle of friction. With this case N from the charts = 0.

The procedure is identical with Example 5.4.

Try F = 1.5:

$$\frac{\tan\phi}{F} = \frac{0.364}{1.5} = 0.24 \qquad (\phi = 13.5°)$$

$$\text{From the charts } N = 0.005 \quad \Rightarrow \quad F_c = \frac{12.5}{0.005 \times 31 \times 16} = 5.04$$

Try F = 2.0:

$$\frac{\tan\phi}{F} = \frac{0.364}{2.0} = 0.182 \qquad (\phi = 10.25°)$$

$$\text{From the charts } N = 0.016 \quad \Rightarrow \quad F_c = \frac{12.5}{7.95} = 1.57$$

Try F = 1.9:

$$\frac{\tan\phi}{F} = \frac{0.364}{1.9} = 0.192 \qquad (\phi = 11°)$$

$$\text{From the charts } N = 0.013 \quad \Rightarrow \quad F_c = \frac{12.5}{6.45} = 1.94 \quad \text{(acceptable)}$$

Factor of safety for slope = 1.9

5.5 Effective stress analysis

The methods for analysing a slip circle that have been discussed so far can be used to give an indication of the factor of safety immediately after construction has been completed, but they are not applicable in the case of an existing embankment if water pressures are present. However, if an analysis is carried out in terms of effective stress then it can be used to determine F after drainage has occurred or for any intermediate value of r_u between undrained and drained, such an analysis affording a better estimate for stability immediately after construction than the total stress methods.

Before this system can be examined, the determination of the pore pressure ratio, r_u, must be considered.

5.5.1 *Pore pressure ratio, r_u*

The prediction of pore pressures in an earth dam or an embankment has been discussed by Bishop (1954). There are two main types of problem: those in which the value of the pore water pressure depends upon the magnitude of the applied stresses (e.g. during the rapid construction of an embankment), and those where the value of the pore water pressure depends upon either the groundwater level within the embankment or the seepage pattern of water impounded by it.

Rapid construction of an embankment

The pore pressure at any point in a soil mass is given by the expression:

$$u = u_0 + \Delta u$$

where

u_0 = initial value of pore pressure before any stress change
Δu = change in pore pressure due to change in stress.

From Chapter 3:

$$\Delta u = B[\Delta\sigma_3 + A(\Delta\sigma_1 - \Delta\sigma_3)]$$

Skempton (1954) showed that the ratio of the pore pressure change to the change in the total major principal stress gives another pore pressure coefficient $\bar{B}$:

$$\frac{\Delta u}{\Delta\sigma_1} = \bar{B} = B\left[\frac{\Delta\sigma_3}{\Delta\sigma_1} = A\left(1 - \frac{\Delta\sigma_3}{\Delta\sigma_1}\right)\right]$$

The coefficient $\bar{B}$ can be used to determine the magnitude of pore pressures set up at any point in an embankment if it is assumed that no drainage occurs during construction (a fairly reasonable thesis if the construction rate is rapid). Now

$$\gamma_u = \frac{u}{\gamma z}$$

i.e.

$$r_u = \frac{u_0}{\gamma z} + \frac{\bar{B}\Delta\sigma_1}{\gamma z}$$

A reasonable assumption to make for the value of the major principal stress is that it equals the weight of the material above the point considered. Hence

$$\Delta\sigma_1 = \gamma z \qquad \text{and} \qquad r_u = \frac{u_0}{\gamma z} + \bar{B}$$

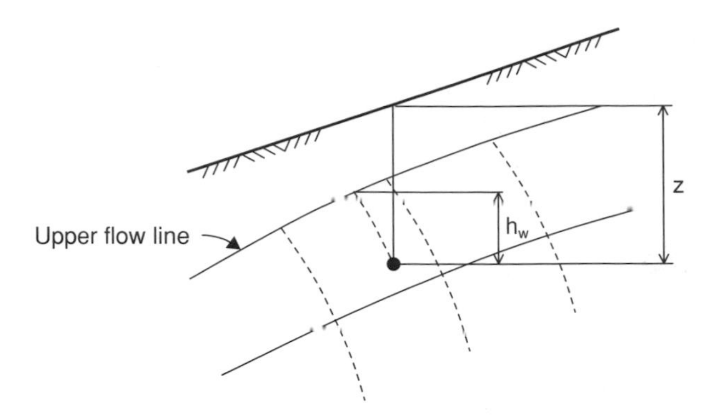

Fig. 5.19 Determination of excess head at a point on a flow net.

For soils placed at or below optimum moisture content, u_0 is small and can even be negative, its effect is of little consequence and may be ignored so that the analysis for stability at the end of construction is often determined from the relationship $r_u = \bar{B}$.

The pore pressure coefficient $\bar{B}$ is determined from a special stress path test known as a dissipation test, which has been described by Bishop and Henkel (1962). Briefly, a sample of the soil is inserted in a triaxial cell and subjected to increases in the principal stresses $\Delta\sigma_1$ and $\Delta\sigma_3$ of magnitudes approximating to those expected in the field. The resulting pore pressure is measured and $\bar{B}$ obtained.

Steady seepage

It is easy to determine r_u from a study of the flow net (Fig. 5.19). The procedure is to trace the equipotential through the point considered up to the top of the flow net, so that the height to which water would rise in a standpipe inserted at the point is h_w. Since $u = \gamma_w h_w$:

$$r_u = \frac{h_w \gamma_w}{\gamma z}$$

Rapid drawdown

In the case of lagoons a sudden drawdown in the level of the slurry is unlikely, but the problem is important in the case of a normal earth dam. Bishop (1954) considered the case of the upstream face of a dam subjected to this effect, the slope having a rock fill protection as shown in Fig. 5.20. A simplified expression for u under these conditions is obtained by the following calculation:

$$u = u_0 + \Delta u$$

and

$$u_0 = \gamma_w(h_w + h_r + h_c - h')$$

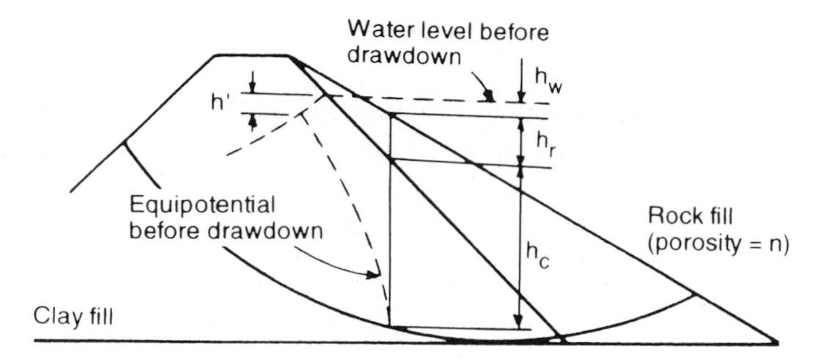

Fig. 5.20 Upstream dam face subjected to sudden drawdown (after Bishop, 1954).

If it is assumed that the major principal stress equals the weight of material, then the initial total major principal stress is given by the expression:

$$\sigma_{1_0} = \gamma_c h_c + \gamma_r h_r + \gamma_w h_w$$

where γ_c and γ_r are the saturated unit weights of the clay and the rock. The final total major principal stress, after drawdown, will be:

$$\sigma_{1_F} = \gamma_c h_c + \gamma_{dr} h_r$$

where γ_{dr} equals the drained unit weight of the rock fill.

$$\begin{aligned}\text{Change in major principal stress} &= \sigma_{1_F} - \sigma_{1_0} \\ &= h_r(\gamma_{dr} - \gamma_r) - \gamma_w h_w\end{aligned}$$

i.e.

$$\Delta\sigma_1 = -\gamma_w n h_r - \gamma_w h_w$$

Note Porosity of rock fill, $n = V_v/V$ or, when we consider unit volume, $n = V_v$. Hence $(\gamma_{dr} - \gamma_r) = -\gamma_w n$.

$$\Rightarrow \quad \Delta u = -\bar{B}(\gamma_w n h_r + \gamma_w h_w)$$

The pore pressure coefficient $\bar{B}$ can be obtained from a laboratory test but standard practice is to assume, conservatively, that $\bar{B} = 1.0$. In this case

$$\Delta u = -\gamma_w (n h_r + h_w)$$

and the expression for u becomes

$$u = \gamma_w [h_c + h_r(1 - n) - h']$$

5.5.2 *Measurements of* in situ *pore water pressures*

For any important structures the theoretical evaluations of pore pressures must be checked against actual values measured in the field. These measurements can be obtained by an instrument known as a piezometer, of which there are four basic types: open standpipe, hydraulic, pneumatic and electric.

Open standpipe

For fully saturated soils of high permeability, such as sands and gravels, the water pressure can be obtained from the water level in an open standpipe placed in a borehole. The borehole must be sealed at its top to prevent the ingress of surface water (see Fig. 14.4).

5.5.3 *Piezometers*

For soils of medium to low permeability, open standpipes cannot be used because of the large time lag involved. In this case *pneumatic* or *vibrating wire* piezometers may be employed. A review of these piezometers has been given by Penman (1961) and their use within boreholes has been described by Mikkelsen and Green (2003).

The pneumatic piezometer is usually operated within a borehole and contains a flexible diaphragm housed within a protective casing and connected to a sensor at the ground surface via twin pneumatic tubes. The outer aspect of the diaphragm is in contact with the saturated soil and is pushed inwards within the housing as a result of the pore water pressure acting on it. A flow of nitrogen gas is passed from the sensor to the inner aspect of the diaphragm until the point that the diaphragm is forced back outwards. At this point the gas pressure is equal to the pore water pressure and this value is simply read from the sensor unit.

The vibrating wire piezometer also incorporates a diaphragm within a protective housing operated within a borehole. The diaphragm is connected to a tensioned steel wire and the water pressure acting on the diaphragm causes a change in tension of the wire. An electromagnetic coil is used to excite the wire, which then vibrates at its natural frequency. The vibration of the wire, which is dependent on its tension, generates a frequency signal that is transmitted to a readout device at the surface and then converted to display the pore water pressure.

5.5.4 *Effective stress analysis by Bishop's method*

Bishop's conventional method

The effective stress methods of analysis now in general use were evolved by Bishop (1955). Figure 5.21 illustrates a circular failure arc, ABCD, and shows the forces on a vertical slice through the sliding segment.

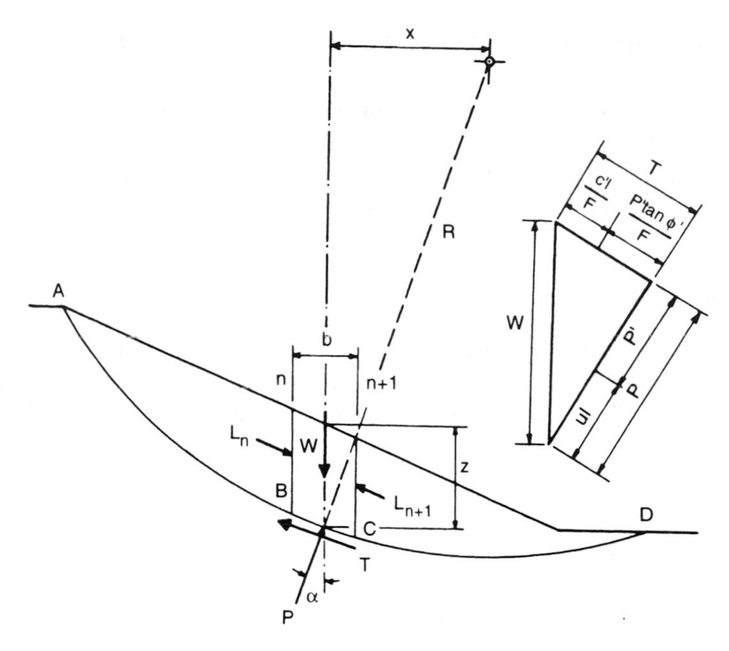

Fig. 5.21 Effective stress analysis: forces acting on a vertical side.

Let L_n and L_{n+1} equal the lateral reactions acting on sections n and n + 1 respectively. The difference between L_n and L_{n+1} is small and the effect of these forces can be ignored with little loss in accuracy.

Let the other forces acting on the slice be:

W = weight of slice
P = total normal force acting on base of slice
T = shear force acting on base of slice

and the other notation is:

z = height of slice
b = breadth of slice
l = length of BC (taken as straight line)
α = angle of between P and the vertical
x = horizontal distance from centre of slice to centre of rotation, O.

In terms of effective stress, the shear strength mobilised is

$$\tau = \frac{c' + (\sigma_n - u) \tan \phi'}{F}$$

Total normal stress on base of slice:

$$\sigma_n = \frac{P}{l}$$

i.e.

$$\tau = \frac{1}{F}\left[c' + \left(\frac{P}{l} - u\right)\tan\phi'\right]$$

Shear force acting on base of slice, $T = \tau l$

For equilibrium, Disturbing moment = Restoring moment

i.e.

$$\Sigma Wx = \Sigma TR = \Sigma \tau l R$$

$$= \frac{R}{F}\Sigma[c'l + (P - ul)\tan\phi']$$

$$\Rightarrow \quad F = \frac{R}{\Sigma Wx}\Sigma[c'l + (P - ul)\tan\phi']$$

If we ignore the effects of L_n and L_{n+1} the only vertical force acting on the slice is W. Hence

$$P = W\cos\alpha$$

$$\Rightarrow \quad F = \frac{R}{\Sigma Wx}\Sigma[c'l + (W\cos\alpha - ul)\tan\phi']$$

Putting $x = R\sin\alpha$:

$$F = \frac{1}{\Sigma W\sin\alpha}\Sigma[c'l + (W\cos\alpha - ul)\tan\phi']$$

If we express u in terms of the pore pressure ratio, r_u:

$$u = r_u\,\gamma z = r_u\frac{W}{b}$$

Now

$$b = l\cos\alpha \qquad \Rightarrow \quad u = \frac{r_u W}{l\cos\alpha} = \frac{r_u W}{l}\sec\alpha$$

$$\Rightarrow \quad F = \frac{1}{\Sigma W\sin\alpha}\Sigma[c'l + W(\cos\alpha - r_u\sec\alpha)\tan\phi']$$

This formula gives a solution generally known as the conventional method which allows rapid determination of F when sufficient slip circles are available to permit the determination of the most critical. For analysing the stability of an existing tip it should prove perfectly adequate.

In the case of an earth embankment F is usually considered to be satisfactory if it is not less than 1.25, and for economic reasons it should not be designed for a greater F value than 1.5.

Bishop's rigorous method

The formula for the conventional method of analysis can give errors of up to 15 per cent in the value of F obtained, although the error is on the safe side since it gives a lower value than is the case. In the construction of new embankments and earth dams, however, this error can lead to unnecessarily high costs, and it becomes particularly pronounced with a deep slip circle where the variations of α over the slip length are large.

Returning to the equation:

$$F = \frac{R}{\Sigma Wx} \Sigma[c'l + (P - ul) \tan \phi']$$

Let the normal effective force, $(P - ul) = P'$.

Resolving forces vertically:

$$W = P \cos \alpha + T \sin \alpha$$

Now

$$P = P' + ul$$

and

$$T = \frac{1}{F}(c'l + P' \tan \phi']$$

$$W = ul \cos \alpha + P' \cos \alpha + \frac{P' \tan \phi'}{F} \sin \alpha + \frac{c'l}{F} \sin \alpha$$

$$= ul \cos \alpha + \frac{c'l \sin \alpha}{F} + P'\left(\cos \alpha + \frac{\tan \phi'}{F} \sin \alpha\right)$$

$$= l\left(u \cos \alpha + \frac{c'}{F} \sin \alpha\right) + P'\left(\cos \alpha + \frac{\tan \phi'}{F} \sin \alpha\right)$$

$$\Rightarrow \quad P' = \frac{W - l\left(u \cos \alpha + \frac{c'}{F} \sin \alpha\right)}{\cos \alpha + \frac{\tan \phi' \sin \alpha}{F}}$$

slice no.	z (m)	b (m)	$N = \gamma zb$	α°	$\sin\alpha$	$W\sin\alpha$ (1)	$c'b$ (2)	$W(1-r_u)\tan\phi'$ (3)	2 + 3 (4)	$\sec\alpha$	$\tan\alpha$	(5) $\dfrac{\sec\alpha}{1+\dfrac{\tan\phi'\tan\alpha}{F}}$		4 × 5 (6)	
												F=	F=	F=	F=

Fig. 5.22 Example spreadsheet template for slope stability calculation.

Substituting P′ for (P – ul) in the original equation:

$$F = \frac{R}{\Sigma Wx}\sum[c'l + (P - ul)\tan\phi']$$

$$F = \frac{R}{\Sigma Wx}\sum\left\{c'l + \left[\frac{\left(W - ul\cos\alpha - \dfrac{c'l}{F}\sin\alpha\right)\tan\phi'}{\cos\alpha + \dfrac{\tan\phi'\sin\alpha}{F}}\right]\right\}$$

and substituting

$$x = R\sin\alpha,\ b = l\cos\alpha \text{ and } \frac{ub}{W} = \frac{u}{\gamma z} = r_u$$

$$F = \frac{1}{\Sigma W\sin\alpha}\sum\left[(c'b + W(1 - r_u)\tan\phi')\frac{\sec\alpha}{1 + \dfrac{\tan\phi'\tan\alpha}{F}}\right]$$

When working by hand the final analysis of forces acting on a vertical slice is best carried out by tabulating the calculations. However, in most design offices, slope stability problems are now computerised (Fig. 5.22).

Excel

Example 5.6

The cross-section of an earth dam sitting on an impermeable base is shown in Fig. 5.23. The stability of the downstream slope is to be investigated using the slip circle shown and given the following information:

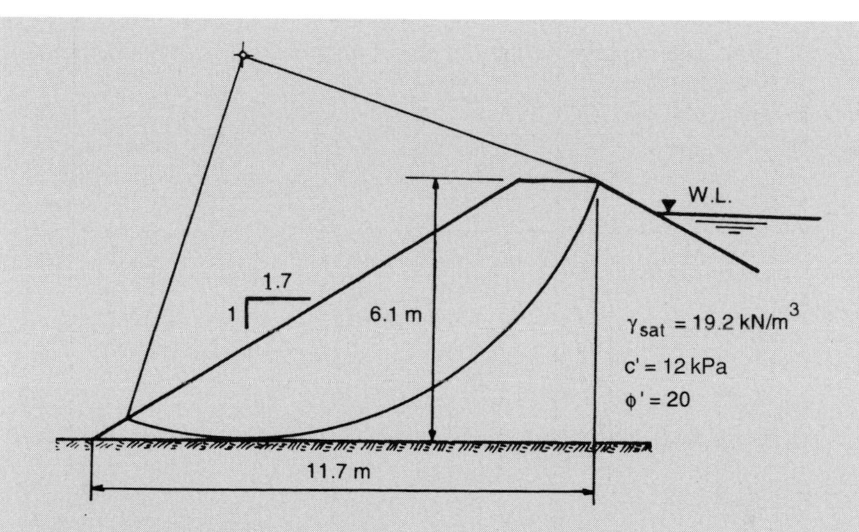

Fig. 5.23 Example 5.6.

$\gamma_{sat} = 19.2 \text{ kN/m}^3$
$c' = 12 \text{ kPa}$
$\phi' = 20°$
$R = 9.15 \text{ m}$
Angle subtended by arc of slip circle, $\theta = 89°$

For this circle determine the factor of safety (a) with the conventional method and (b) with the rigorous method.

Solution

The earth dam is drawn to scale on graph paper. The first step in the analysis is to divide the sliding sector into a suitable number of slices and determine the pore pressure ratio at the mid-point of the base of each slice.

The phreatic surface must be drawn, using the method of Casagrande. A rough form of the flow net must then be established so that the equipotentials through the centre points of each slice can be inserted. Five slices is a normal number (Fig. 5.24).

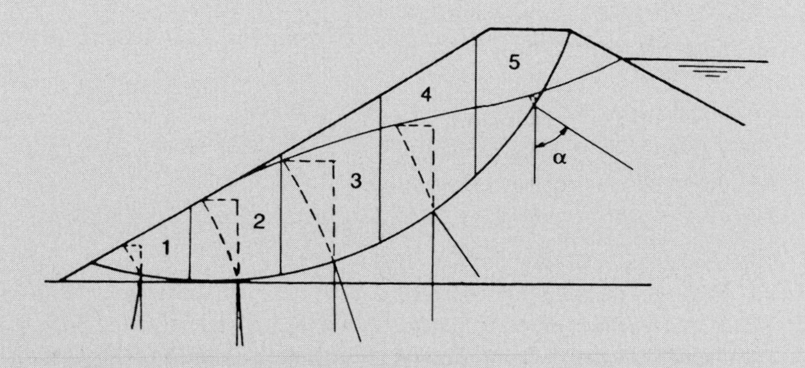

Fig. 5.24 Working diagram for Example 5.6.

The determination of the r_u values is required for both methods and will be considered first.

Slice no.	h_w (m)	u (kPa)	z (m)	r_u
1	0.654	6.42	0.95	0.352
2	1.958	19.21	2.44	0.41
3	2.440	23.90	3.32	0.370
4	2.020	19.82	3.50	0.295
5	0.246	2.41	1.74	0.072

The calculations for the conventional method are set out in Fig. 5.25a:

$$\theta = 89°$$

$$\Rightarrow \quad c'l = c'R\theta = 12 \times 9.15 \times \frac{\pi}{180} \times 89 = 170.6 \text{ kN}$$

$$F = \frac{106.5 + 170.6}{207.9} = 1.33$$

The rigorous method calculations are set out in Fig. 5.25b. With the first approximation:

$$F = \frac{297.5}{207.9} = 1.43$$

This value was obtained by assuming a value for F of 1.5 in the expression:

$$\frac{\sec \alpha}{1 + \left(\dfrac{\tan \phi' \tan \alpha}{F}\right)}$$

of column (5).

Columns 5 and 6 are now recalculated using F = 1.43 and a revised value of F is obtained:

$$F = \frac{295.8}{207.9} = 1.42$$

This is approximately equal to the assumed value of 1.43 and is taken as correct. Thus the factor of safety of the slope is 1.42.

Had the assumed and derived values of F not been approximately equal, the iterative procedure could have been repeated once again to find an improved value of F, as can be demonstrated through the *Example 5.6.xls* spreadsheet.

Slice	z (m)	b (m)	W (kN)	α (°)	cos α	sec α	h_w (m)	r_u	cos $\alpha - r_u$ sec α	W(cos $\alpha - r_u$ sec α) × tan ϕ'	sin α	W(sin α)
1	0.95	2.35	42.9	−10	0.985	1.015	0.654	0.352	0.985	15.4	−0.174	−7.4
2	2.44	2.35	110.1	4	0.998	1.002	1.958	0.410	0.587	23.5	0.070	7.7
3	3.32	2.35	149.8	20	0.940	1.064	2.440	0.376	0.540	29.4	0.342	51.2
4	3.50	2.35	157.9	35	0.819	1.221	2.020	0.295	0.459	26.4	0.574	90.6
5	1.74	2.35	78.5	57	0.545	1.836	0.246	0.072	0.412	11.8	0.839	65.8
										Σ106.5		Σ207.9

(a) Conventional method

						(1)	(2)	(3)	(4)			(5)		(6)	
Slice	z (m)	b (m)	W (kN)	α (°)	sin α	W sin α	c′b	W(1 − r_u) × tan ϕ'	(2) + (3)	sec α	tan α	$\dfrac{\sec\alpha}{1+\dfrac{\tan\phi'\tan\alpha}{F}}$		(4) × (5)	
												F = 1.5	F = 1.43	F = 1.5	F = 1.43
1	0.95	2.35	42.9	−10	−0.17	−7.4	22.6	10.1	38.3	1.015	−0.18	1.061	1.063	40.6	40.7
2	2.44	2.35	110.1	4	0.07	7.7	22.6	23.6	51.8	1.002	0.07	0.986	0.985	51.1	51.1
3	3.32	2.35	149.8	20	0.34	51.2	22.6	34.1	62.2	1.064	0.36	0.978	0.974	60.9	60.6
4	3.50	2.35	157.9	35	0.57	90.6	22.6	40.5	68.7	1.221	0.70	1.043	1.036	71.7	71.2
5	1.74	2.35	78.5	57	0.84	65.8	22.6	26.5	54.7	1.836	1.54	1.337	1.319	73.1	72.2
						Σ207.9								Σ297.5	295.8

(b) Rigorous method

Fig. 5.25 Example 5.6

Example 5.7

Figure 5.26 gives details of the cross-section of an embankment. The soil has the following properties: $\phi' = 35°$, $c' = 10$ kPa, $\gamma = 16$ kN/m^3.

For the slip circle shown, determine the factor of safety for the following values of r_u: 0.2, 0.4 and 0.6.

Plot the variation of F with r_u.

Solution

The calculations were based on the rigorous method and are shown in Figs. 5.27a–c.

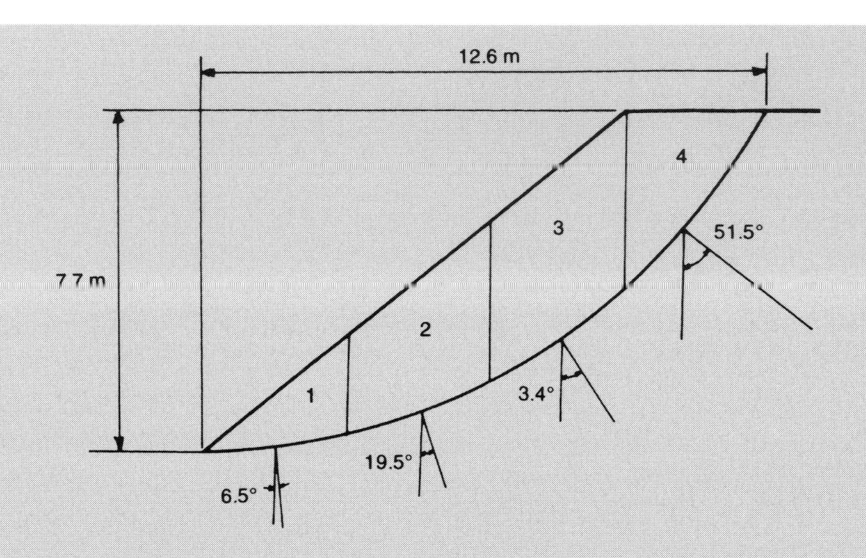

Fig. 5.26 Example 5.7(a).

$r_u = 0.2$

						(1)	(2)	(3)	(4)			(5)		(6)	
Slice	z (m)	b (m)	W (kN)	α (°)	sin α	W sin α	c'b	W(1 − r_u) × tan φ'	(2) + (3)	sec α	tan α	$\frac{\sec\alpha}{1+\frac{\tan\phi'\tan\alpha}{F}}$		(4) × (5)	
												F = 1.5	F = 1.47	F = 1.5	F = 1.47
1	1.00	3.15	50.4	7	0.113	5.7	31.5	28.2	59.7	1.006	0.114	0.956	0.955	57.1	57.0
2	3.08	3.15	155.2	20	0.334	51.8	31.5	87.0	118.5	1.061	0.354	0.910	0.908	107.8	107.5
3	4.00	3.15	201.6	34	0.559	112.7	31.5	112.9	144.4	1.206	0.675	0.917	0.913	132.5	131.8
4	2.70	3.15	136.1	52	0.783	106.5	31.5	76.2	107.7	1.606	1.257	1.012	1.004	109.1	108.2
						Σ276.8								Σ406.5	404.6

$$F = \frac{406.5}{276.8} = 1.47 \qquad F = \frac{404.6}{276.8} = 1.46$$

$r_u = 0.4$

(3)	(4)	(5)		(6)	
W(1 − r_u) × tan φ'	(2) + (3)	$\frac{\sec\alpha}{1+\frac{\tan\phi'\tan\alpha}{F}}$		(4) × (5)	
		F = 1.3	F = 1.17	F = 1.3	F = 1.17
21.2	52.7	0.948	0.942	49.9	49.6
65.2	96.7	0.891	0.875	86.2	84.7
84.7	116.2	0.885	0.859	102.8	99.9
57.2	88.7	0.958	0.917	84.9	81.3
				Σ323.9	315.4

$$F = \frac{323.9}{276.8} = 1.17 \qquad F = \frac{315.4}{276.8} = 1.14$$

$r_u = 0.6$

(3)	(4)	(5)		(6)	
W(1 − r_u) × tan φ'	(2) + (3)	$\frac{\sec\alpha}{1+\frac{\tan\phi'\tan\alpha}{F}}$		(4) × (5)	
		F = 1.0	F = 0.86	F = 1.0	F = 0.86
14.1	45.6	0.932	0.921	42.5	42.0
43.5	75.0	0.850	0.823	63.7	61.7
56.5	88.0	0.819	0.778	72.1	68.5
38.1	69.6	0.854	0.793	59.5	55.2
				Σ237.8	227.4

$$F = \frac{237.8}{276.8} = 0.86 \qquad F = \frac{227.4}{276.8} = 0.82$$

Fig. 5.27 Example 5.7(b).

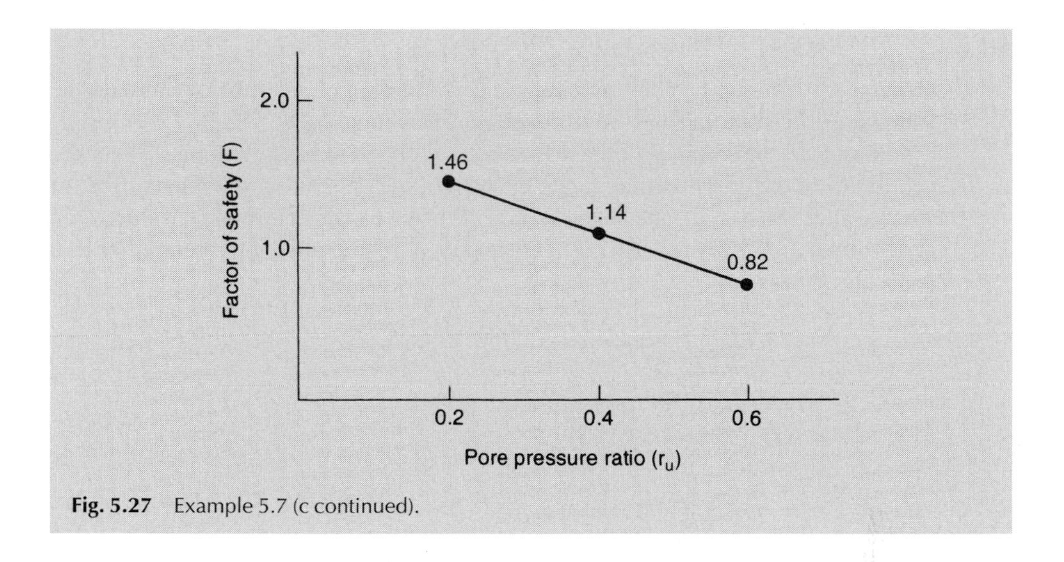

Fig. 5.27 Example 5.7 (c continued).

5.5.5 Rapid determination of F for a homogeneous, regular slope with a constant pore pressure ratio

If on a trial slip circle the value of F is determined for various values of r_u and the results plotted, a linear relationship is found between F and r_u (see Example 5.7). The usual values of r_u encountered in practice range from 0.0 to 0.7 and it has been established that this linear relationship between F and r_u applies over this range. The factor of safety, F, may therefore be determined from the expression:

$$F = m - nr_u$$

in which m is the factor of safety with respect to total stresses (i.e. when no pore pressures are assumed) and n is the coefficient which represents the effect of the pore pressures on the factor of safety. These terms m and n are known as stability coefficients and were evolved by Bishop and Morgenstern (1960); they depend upon $c'/\gamma H$ (the stability number with c' equalling cohesion with respect to effective stress), cot β (the cotangent of the slope angle, e.g. a 5:1 slope means 5 horizontal to 1 vertical), and ϕ' (the angle of friction with respect to effective stresses).

Bishop and Morgenstern prepared charts of m and n for three sets of $c'/\gamma H$ values (0.0, 0.025, 0.05), which are reproduced at the end of this chapter and cover slopes from 2:1 (26.5°) to 5:1 (11.5°). Extrapolation, within reason, is possible for a case outside this range.

Graphs to cover depth factor values, D, up to 1.5 were produced for $c'/\gamma H = 0.05$, but for the other two cases D values greater than 1.25 were not calculated as such values are not critical in these instances. As in Taylor's analysis, the effect of tension cracks has not been included. O'Connor and Mitchell (1977) extended the work of Bishop and Morgenstern to include $c'/\gamma H = 0.075$ to 0.150.

Determination of an average value for r_u

Generally r_u will not be constant over the cross-section of an embankment and the following procedure can be used to determine an average value.

In Fig. 5.28 the stability of the downstream slope is to be determined. From the centre line of the cross-section divide the base of the dam into a suitable number of vertical slices (a, b, c, d), and on the centre line of each slice determine r_u values for a series of points as shown. Then the average pore pressure ratio on the centre line of a particular slice is

$$r_u = \frac{h_1 r_{u1} + h_2 r_{u2} + h_3 r_{u3} + \cdots}{h_1 + h_2 + h_3 + \cdots}$$

The average r_u for whole cross-section

$$= \frac{A_a r_{ua} + A_b r_{ub} + A_c r_{uc} + \cdots}{A_a + A_b + A_c + \cdots}$$

where A_a = area of the slice a, and r_{ua} = average r_u value in slice a.

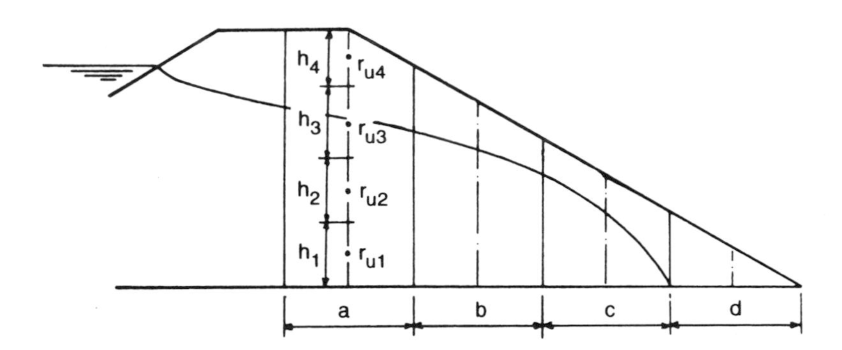

Fig. 5.28 Determination of average r_u value.

Example 5.8

An embankment has a slope of 1 vertical to 2 horizontal. The properties of the soil are: $c' = 25$ kPa, $\phi' = 20°$, $\gamma = 16$ kN/m^3. The height of the embankment is 31 m and the average pore pressure ratio is 0.4.

Using Bishop and Morgenstern's charts, determine the factor of safety for the slope.

Solution

On the charts of n values are plotted a series of dotted lines labelled r_{ue}. The authors have shown that if the relevant r_{ue} value is less than the actual design r_u then the set of charts with the next highest depth factor should be used. The procedure therefore becomes as follows.

Calculate $c'/\gamma H$ and, using the chart with D = 1.0, check that r_u is less than r_{ue}. If it is, then this is the correct chart to use: if r_u is not less than r_{ue} select the next chart (D = 1.25). In the case of $c'/\gamma H = 0.05$, r_{ue} should again be checked and if r_u is greater than this value then the chart for D = 1.5 should be used.

In the example (which is Example 5.4 with an r_u value):

$$\frac{c'}{\gamma H} = \frac{25}{16 \times 31} = 0.05$$

Select the chart with D = 1.0

$r_{ue} = 0.64 \Rightarrow$ this chart is acceptable ($r_{ue} > r_u$)
m = 1.39 (compare with Taylor's method = 1.35)
n = 1.07
$\Rightarrow$ F = 1.39 − (0.4 × 1.07) = 0.96

Example 5.9

Slope = 1 vertical to 4 horizontal, $c' = 12.5$ kPa, $\phi' = 20°$, $\gamma = 16$ kN/m^3, H = 31 m, $r_u = 0.35$. Find F.

Solution

This is Example 5.5 with an r_u value.

$$\frac{c'}{\gamma H} = \frac{12.5}{16 \times 31} = 0.025$$

Using D = 1.0, $r_{ue} = 0.0$ ($r_{ue} \not> r_u$).

Use chart for D = 1.25:

m = 1.97 (compare with Taylor's method = 1.9)
n = 1.78
F = 1.97 − (0.35 × 1.78) = 1.35

Example 5.10

Slope angle = 24°, $c' = 20$ kPa, $\gamma = 21$ kN/m^3, $\phi' = 30°$, $r_u = 0.3$, H = 50 m. Find F.

Solution

Slope (24°) = 1 vertical to 2.25 horizontal

$$\frac{c'}{\gamma H} = \frac{20}{21 \times 50} = 0.019$$

Chart $\frac{c'}{\gamma H} = 0.025$; D = 1.0

$m = 1.76$
$n = 1.68$
$F = 1.76 - (0.3 \times 1.68) = 1.26$

Chart $\frac{c'}{\gamma H} = 0.00$

$m = 1.3$
$n = 1.54$
$F = 1.3 - (0.3 \times 1.54) = 0.84$

The actual F value can be found by interpolation:

$$F = 0.84 + (1.26 - 0.84)\,\frac{0.019}{0.025}$$

$$= 0.84 + 0.32 = 1.16$$

Note It is preferable to calculate the two F values and then interpolate rather than to determine the m and n values by interpolation.

5.6 Planar failure surfaces

There are many occasions when the potential failure surface of a slope is more realistically represented by a straight line, or a series of straight lines, rather than the arc of a circle. It has already been established in this chapter that slip surfaces in granular soils are planar and examples of how planar failure surfaces can be created in other types of soil will now be given.

5.6.1 Planar translational slip

Quite often the surface of an existing slope is underlain by a plane of weakness lying parallel to it. This potential failure surface (often caused by downstream creep under alternating winter–summer conditions) generally lies at a depth below the surface that is small when compared with the length of the slope.

Owing to the comparative length of the slope and the depth to the failure surface we can generally assume that end effects are negligible and that the factor of safety of the slope against slip can be determined from the analysis of a wedge or slice of the material, as for the granular slope.

Consider Fig. 5.29. Angle of slope = β, depth to failure surface = z, width of slice = b, and weight of slice, W = γzb/unit width.

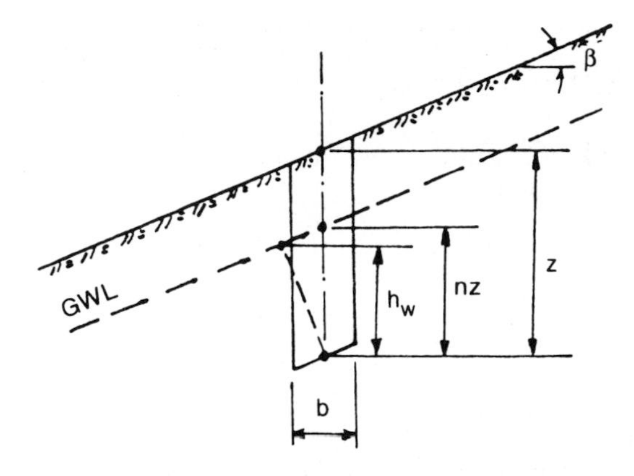

Fig. 5.29 Planar translational slip.

Let the groundwater lever be parallel to the surface at a constant height above the failure plane = nz.

Then excess hydrostatic head at midpoint of base of slice,

$$h_w = nz \cos^2 \beta$$

i.e.

$$u = \gamma_w nz \cos^2 \beta$$

Tangential force = W sin β

i.e.

$$\tau = \frac{\gamma zb \sin \beta}{b} \cos \beta = \gamma z \sin \beta \cos \beta$$

and

$$\tau_f = c' + (\sigma - u) \tan \phi'$$

Now

$$(\sigma - u) = \frac{\gamma zb \cos^2 \beta}{b} - u = \gamma z \cos^2 \beta - \gamma_w nz \cos^2 \beta$$

and

$$F = \frac{\tau_f}{\tau} = \frac{c' + (\gamma z - \gamma_w nz) \cos^2 \beta \tan \phi'}{\gamma z \sin \beta \cos \beta}$$

$$= \frac{c'}{\gamma z \sin \beta \cos \beta} + \frac{\gamma - n\gamma_w}{\gamma} \times \frac{\tan \phi'}{\tan \beta}$$

Note When $c' = 0$ and $n = 1.0$, we obtain the same expression as derived for a granular slope:

$$F = \frac{\gamma'}{\gamma_{sat}} \times \frac{\tan \phi'}{\tan \beta}$$

5.6.2 *Wedge failure*

When the potential sliding mass of the soil is bounded by two or three straight lines we have a wedge failure. Wedge failures can be brought about by a variety of geological conditions, and one example is shown in Fig. 5.30a with the design approximation illustrated in Fig. 5.30b. No doubt the reader can think of other situations.

The form of construction within an earth structure can also dictate that any stability failure will be of a wedge type. One example is that of an earth dam with a sloping impermeable core (Fig. 5.30c). Various wedge failure patterns could be assumed for the purpose of analysis. One such form is illustrated in Fig. 5.30d.

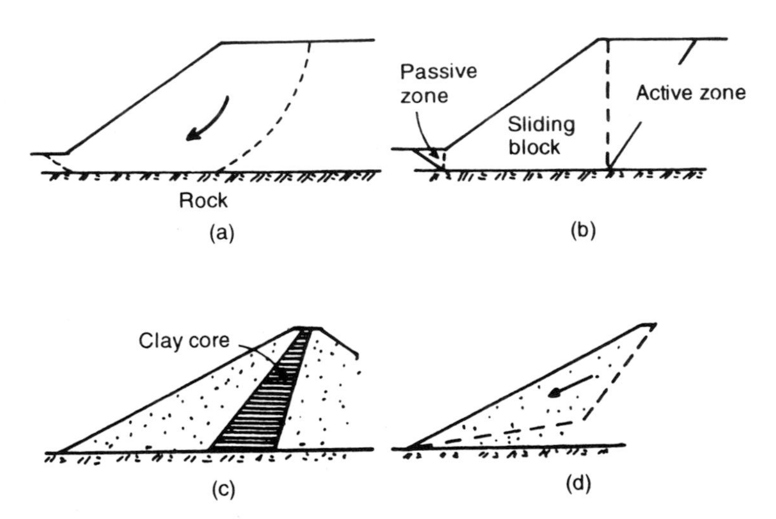

Fig. 5.30 Sliding block or wedge failure.

Example 5.11

The cross-section of a sloping core dam is shown in Fig. 5.31a and a probable form of wedge failure (based on Sultan and Seed, 1967).

Using the suggested failure shape determine the factor of safety of the dam.

Relevant properties: rock fill: $\phi = 40°$, $\gamma = 18$ kN/m^3; core: $c = 80$ kPa.

Solution

The forces acting on wedges (1) and (2) are shown in Fig. 5.31b. In the diagram ϕ_m = angle of friction mobilised in the rock fill and c_m = cohesion mobilised in the core.

$T_1 = c_m BE$ kN/m run of dam; $\quad T_2 = N_2 \tan \phi_m$ kN/m run of dam

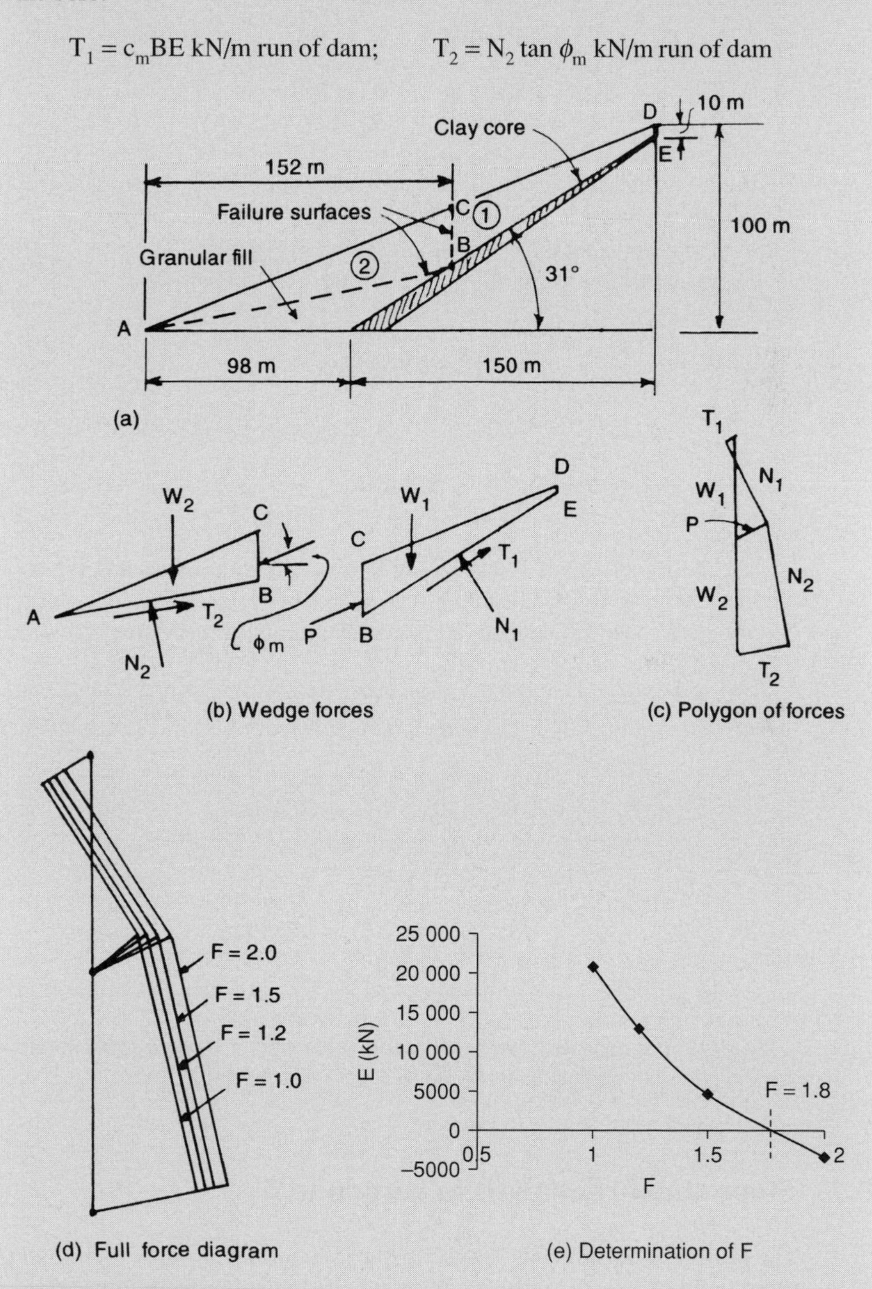

Fig. 5.31 Example 5.11.

The procedure starts by selecting suitable F values and evaluating the corresponding values for ϕ_m and c_m:

$F = 1.0 \quad c_m = 80\ \text{kPa}; \quad \phi_m = 40°$

$F = 1.2 \quad c_m = 66.7\ \text{kPa}; \quad \phi_m = \tan^{-1}\dfrac{\tan\phi}{1.2} = 35°$

$F = 1.5 \quad c_m = 53.3\ \text{kPa}; \quad \phi_m = 29°$

$F = 2.0 \quad c_m = 40\ \text{kPa}; \quad \phi_m = 23°$

For each value of F a polygon of forces for slice (1) is drawn and the force P obtained. Using this value for P the polygon of forces for wedge (2) is now drawn to give the total force diagram shown in Fig. 5.31c.

A typical set of calculations (for F = 2.0) is set out below.

$$W_1 = \left(\frac{10 + 29}{2}\right) 96 \times 18 = 33\,696\ \text{kN}$$

$$T_1 = 40 \times 112 = 4480\ \text{kN} \qquad \text{(112 m is the base length of wedge 1)}$$

$$W_2 = \frac{29}{2} \times 152 \times 18 = 39\,672\ \text{kN}; \qquad \phi_m = 23°$$

As the directions of P and T_2 are known, the value obtained for N_2 will be correct. The error of closure, E, can therefore be assessed by comparing the value of T_2 determined from the force diagram with the value calculated from $N_2 \tan\phi_m$.

The force diagrams for the four chosen values for F are shown superimposed on each other in Fig. 5.31d. The corresponding values of N_2 and T_2 are set out below.

F	N_2 (kN)	T_2 (kN)	$N_2 \tan\phi_m$ (kN)	E (kN)
1.0	42 750	15 080	35 871	20 791
1.2	42 500	16 700	29 759	13 059
1.5	42 250	18 800	23 420	4 620
2.0	41 500	21 000	17 616	–3 384

By plotting F against E we see that the value of E = 0 when F = 1.8. Hence, for the wedge failure surfaces chosen, F = 1.8 (Fig. 5.31e).

5.7 Slope stability analysis to Eurocode 7

The principles of Eurocode 7, described in Section 7.4.2, apply to the assessment of slope stability and the reader is advised to refer ahead to that section whilst studying the following few pages.

The overall stability of slopes is covered in Section 11 of Eurocode 7, and the GEO limit state is the principal state that is considered. The procedure to check overall stability uses the methods described earlier in this chapter and applies to both the undrained and drained states. Partial factors are applied to the characteristic values of the soils shear strength parameters, and to the representative values of the actions (e.g. weights of slices) to obtain the design values.

For circular failure surfaces, during a method of slices analysis, the weight of a single slice can contribute to the disturbing moment or it may contribute to the restoring moment, as illustrated by Example 5.6 (Fig. 5.24). In that example, slices 2–5 contribute to the disturbing moment and slice 1 contributes to the restoring moment. This follows from the particular choice of position of the centre of the slip circle: had the centre of the slip circle been in a different location the directions of the moment of each slice might have been different.

As explained in Section 7.4.2, an action is considered as either favourable or unfavourable. However, as we have just seen, the choice of the position of the centre of the circle influences whether the weight of a slice would be favourable or unfavourable and because of this it is impossible to know from the outset whether an action will be favourable or unfavourable. When using Design Approach 1 Combination 1, the partial factors (see Table 7.1) are different for favourable and unfavourable actions, whereas for Combination 2 the partial factors are the same (= 1.0). To this end, when using Design Approach 1 for circular failure surfaces, Combination 2 will almost always be used to check the overall stability.

The GEO limit state requirement is satisfied if the design effect of the actions (E_d) is less than or equal to the design resistance (R_d), i.e. $E_d \leq R_d$. Here E_d is the design value of the disturbing moment (or force, for planar failure surfaces) and R_d is the design value of the restoring moment (or force). By representing the ratio of the restoring moment (or force) to the disturbing moment (or force) as the over-design factor, Γ, it is seen that the limit state requirement is satisfied if $\Gamma \geq 1$.

Excel

Example 5.12

Check the GEO limit state for the earth dam described in Example 5.6 using Design Approach 1, Combination 2.

Solution

From Table 7.1, the relevant partial factors are: $\gamma_{\phi'} = 1.25$; and $\gamma_{c'} = 1.25$.

The design shear strength parameters c'_d and ϕ'_d are determined:

$$c'_d = \frac{c'}{\gamma_{c'}} = \frac{12}{1.25} = 9.6\ \text{kPa}$$

$$\phi'_d = \tan^{-1}\left(\frac{\tan\phi'}{\gamma_{\phi'}}\right) = \tan^{-1}\left(\frac{\tan 20^\circ}{1.25}\right) = 16.2^\circ$$

(a) Conventional method: Using c'_d and ϕ'_d the conventional analysis is performed and the calculations are set out in Fig. 5.32.

Slice	z (m)	b (m)	W (kN)	α (°)	cos α	sec α	h_w (m)	r_u	cos $\alpha - r_u$ sec α	W(cos $\alpha - r_u$ sec α) × tan ϕ'	sin α	W(sin α)
1	0.95	2.35	42.9	−10	0.985	1.015	0.654	0.352	0.985	12.3	−0.174	−7.4
2	2.44	2.35	110.1	4	0.998	1.002	1.958	0.410	0.587	18.8	0.070	7.7
3	3.32	2.35	149.8	20	0.940	1.064	2.440	0.376	0.540	23.6	0.342	51.2
4	3.50	2.35	157.9	35	0.819	1.221	2.020	0.295	0.459	21.1	0.574	90.6
5	1.74	2.35	78.5	57	0.545	1.836	0.246	0.072	0.412	9.4	0.839	65.8
										Σ85.2		Σ207.9

Fig. 5.32 Example 5.12: Conventional method.

						(1)	(2)	(3)	(4)			(5)		(6)	
Slice	z (m)	b (m)	W (kN)	α (°)	sin α	W sin α	c′b	W(1 − r_u) × tan ϕ'	(2) + (3)	sec α	tan α	$\dfrac{\sec\alpha}{1+\dfrac{\tan\phi'\tan\alpha}{F}}$		(4) × (5)	
												F = 1.5	F = 1.17	F = 1.5	F = 1.17
1	0.95	2.35	42.9	−10	−0.17	−7.4	22.6	8.1	30.7	1.015	−0.18	1.051	1.062	32.2	32.5
2	2.44	2.35	110.1	4	0.07	7.7	22.6	18.9	41.5	1.002	0.07	0.989	0.985	41.0	40.9
3	3.32	2.35	149.8	20	0.34	51.2	22.6	27.2	49.8	1.064	0.36	0.994	0.976	49.5	48.6
4	3.50	2.35	157.9	35	0.57	90.6	22.6	32.4	55.0	1.221	0.70	1.075	1.040	59.1	57.2
5	1.74	2.35	78.5	57	0.84	65.8	22.6	21.2	43.8	1.836	1.54	1.414	1.328	61.9	58.1
						Σ207.9								Σ243.7	237.3

Fig. 5.33 Example 5.12: Rigorous method.

$$c'_d \times l = 9.6 \times 9.15 \times \frac{\pi}{180} \times 89 = 136.4\ \text{kN}$$

$$\text{Over-design factor, } \Gamma = \frac{85.2 + 136.4}{207.9} = 1.07$$

Since $\Gamma > 1$, the GEO limit state requirement is satisfied.

(*b*) *Rigorous method*: The calculations are set out in Fig. 5.33.

$$\text{Over-design factor, } \Gamma = \frac{\Sigma(6)}{\Sigma(1)} = \frac{237.3}{207.9} = 1.14$$

that is, the GEO limit state requirement is satisfied.

Example 5.13

Consider an infinite slope, constructed at an angle of $\beta = 20°$ to the horizontal using the granular soil described in Example 5.1. Check the GEO limit state against failure along a plane parallel and 2 m beneath the surface of the slope, using Design Approach 1.

Solution

(i) Combination 1 (partial factor sets A1 + M1 + R1)

From Table 7.1, the relevant partial factors are: $\gamma_{G;dst} = 1.35$; $\gamma_{\phi'} = 1.0$.

From Section 5.1 it was seen that the sliding force = W sin β

$$\begin{aligned}\text{Design sliding force, } G_{dst;d} &= \gamma \times z \times \sin\beta \times \gamma_{G;dst}\\ &= 18 \times 2 \times \sin 20° \times 1.35\\ &= 16.6\ \text{kPa}\end{aligned}$$

No other actions contribute to the sliding force so $E_d = G_{dst;d} = 16.6$ kPa.

$$\phi'_d = \tan^{-1}\left(\frac{\tan\phi'}{\gamma_{\phi'}}\right) = \tan^{-1}\left(\frac{\tan 30°}{1.0}\right) = 30°$$

$$\begin{aligned}\text{Design resisting force, } R_d &= \gamma \times z \times \cos\beta \times \tan\phi'_d\\ &= 18 \times 2 \times \cos 20° \times \tan 30°\\ &= 19.5\ \text{kPa}\end{aligned}$$

Therefore the limit state requirement is satisfied since $E_d \leq R_d$.

(ii) Combination 2 (partial factor sets A2 + M2 + R1)

From Table 7.1: $\gamma_{G;dst} = 1.0$; $\gamma_{\phi'} = 1.25$.

$$\begin{aligned}\text{Design sliding force, } G_{d;dst} &= \gamma \times z \times \sin\beta \times \gamma_{G;dst}\\ &= 18 \times 2 \times \sin 20° \times 1.0\\ &= 12.3\ \text{kPa}\end{aligned}$$

that is, $E_d = 12.3$ kPa.

$$\phi'_d = \tan^{-1}\left(\frac{\tan\phi}{\gamma_{\phi'}}\right) = \tan^{-1}\left(\frac{\tan 30°}{1.25}\right) = 24.8°$$

$$\begin{aligned}\text{Design resisting force, } R_d &= \gamma \times z \times \cos\beta \times \tan\phi'_d\\ &= 18 \times 2 \times \cos 20° \times \tan 24.8°\\ &= 15.6\ \text{kPa}\end{aligned}$$

Once again, the limit state requirement is satisfied since $E_d \leq R_d$.

Exercises

Exercise 5.1

A proposed cutting is to have the dimensions shown in Fig. 5.34. The soil has the following properties: $\phi = 15°$, c = 13.5 kPa, $\gamma = 19.3$ kN/m^3.

Determine the factor of safety against slipping for the slip circle shown (i) ignoring tension cracks and (ii) allowing for a tension crack.

Answer (i) 1.7, (ii) 1.6

Exercise 5.2

Investigate the stability of the embankment shown in Fig. 5.35. The embankment consists of two soils, both with bulk densities of 19.3 kN/m^3; the upper

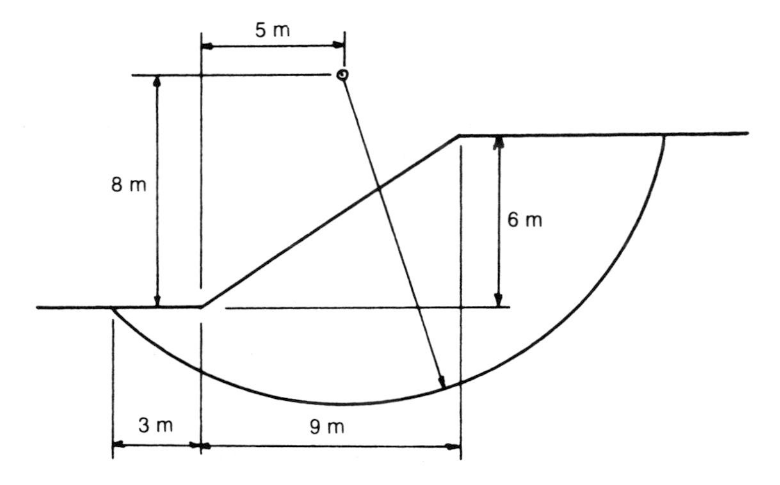

Fig. 5.34 Exercise 5.1.

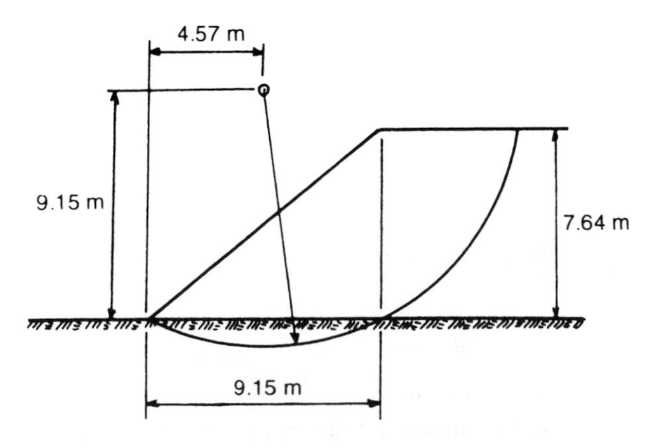

Fig. 5.35 Exercise 5.2.

soil has c = 7.2 kPa and $\phi = 30°$, whilst the lower soil has c = 32.5 kPa and $\phi = 0°$.

Analyse the slip circle shown (ignore tension cracks).

Answer F = 1.2

Excel

Exercise 5.3

The surface of a granular soil mass is inclined at 25° to the horizontal. The soil is saturated throughout with a water content of 15.8 per cent, particle specific gravity of 2.65 and an angle of internal friction of 38°.

At a depth of 1.83 m water is seeping through the soil parallel to the surface.

Determine the factor of safety against slipping on a plane parallel to the surface of the soil at a vertical depth of 3.05 m below the surface.

Answer 1.37

What would be the factor of safety for the same plane if the level of the seeping water lifted up to the surface of the soil?

Answer 0.91

Exercise 5.4

Using Taylor's curves, determine the factor of safety for the following slopes (assume D = 1.0):

$$H = 30.5 \text{ m}, \beta = 40°, c = 10.8 \text{ kPa}, \phi = 35°, \gamma = 14.4 \text{ kN/m}^3$$

Answer Approximately 1.25

$$H = 15.25 \text{ m}, \beta = 20°, c = 24.0 \text{ kPa}, \phi = 0°, \gamma = 19.3 \text{ kN/m}^3$$

Answer 0.8

$$H = 22.8 \text{ m}, \beta = 30°, c = 9.6 \text{ kPa}, \phi = 25°, \gamma = 16.1 \text{ kN/m}^3$$

Answer 1.2

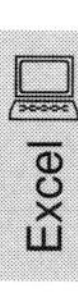

Exercise 5.5

During the analysis of a trial slip circle on a soil slope the rotating mass of the soil was divided into five vertical slices as shown in Fig. 5.36. The position of the GWT is also shown.

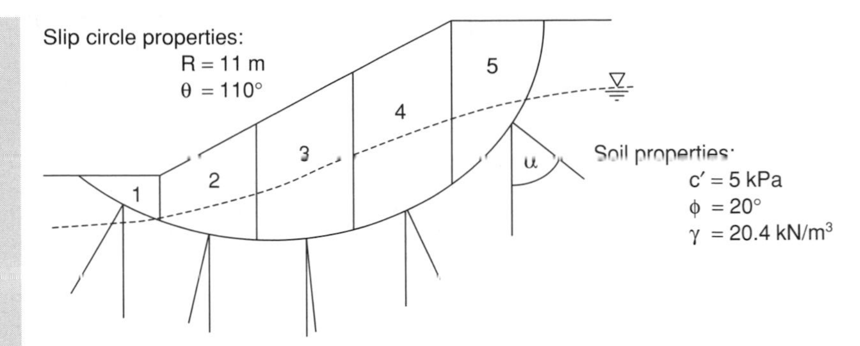

Fig. 5.36 Exercise 5.5.

The following measurements were made:

Slice	b (m)	z (m)	h_w (m)	α (degrees)
1	2.8	1.0	0	–20
2	3.4	3.0	0.8	–12
3	3.4	5.0	2.0	4
4	3.4	6.0	2.4	18
5	3.4	3.4	0.6	48

Determine the factor of safety of the slope using Bishop's conventional method of analysis.

Answer 1.70

Exercise 5.6

In the stability analysis of an earth embankment the slip circle shown in Fig. 5.37 was used and the following figures obtained:

Slice no.	Breadth b (m)	Weight W (kN)	α (degrees)
1	5.65	372	–26
2	5.65	656	–7
3	5.65	1070	12
4	5.65	1220	30
5	5.65	686	54

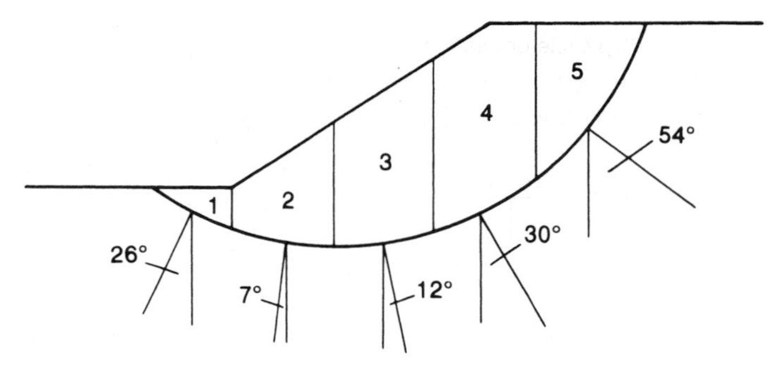

Fig. 5.37 Exercise 5.6.

With the values, and using the conventional method, determine the safety factor of the slope at the end of construction assuming the pore pressure ratio to be 0.45 and the cohesion of soil and the angle of friction (with regard to effective stresses) to be 19.1 kPa and 25°, respectively. The radius of the slip circle was 18.8 m and the angle subtended at the centre of the circle was 100°.

Answer 1.15

Exercise 5.7

Using the slip circle shown in Fig. 5.38, determine the F values for $r_u = 0.4$, 0.6 and 0.8. Plot r_u against F. $\gamma = 23.3$ kN/m^3, $c' = 17.1$ kPa, $\phi' = 37.5°$.

Answer By rigorous method

F	1.5	1.0	0.5
r_u	0.4	0.6	0.8

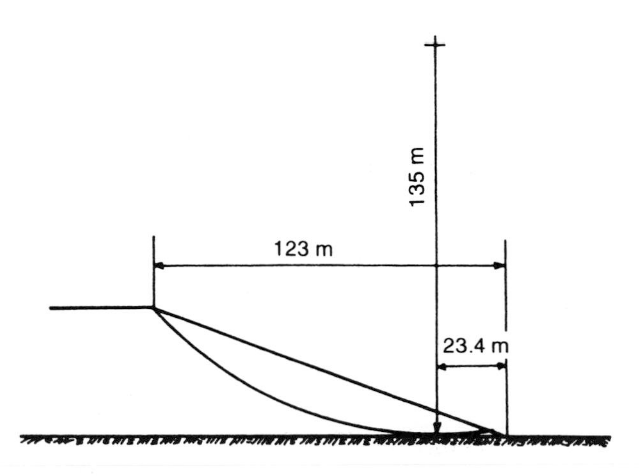

Fig. 5.38 Exercise 5.7.

Exercise 5.8

Using Bishop and Morgenstern's charts, determine the factors of safety for the following slopes:

(i) $r_u = 0.5$, $c' = 5.37$ kPa, $\phi' = 40°$, $\gamma = 14.4$ kN/m^3, H = 15.2 m, slope = 3:1.

Answer 1.6

(ii) $r_u = 0.3$, $c' = 7.2$ kPa, $\phi' = 39°$, $\gamma = 12.8$ kN/m^3, H = 76.4 m, slope = 2:1.

Answer 1.19

(iii) $r_u = 0.5$, $c' = 20$ kPa, $\gamma = 17.7$ kN/m^3, $\phi' = 25°$, H = 25 m, angle of slope = 20°.

Answer 1.2

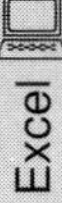

Exercise 5.9

Check the GEO limit state requirement for stability for the embankment described in Exercise 5.6 using Eurocode 7 Design Approach 1, Combination 2.

Answer $\Gamma = 0.92$

Stability coefficients for earth slopes

Adapted from originals by A. W. Bishop and N. Morgenstern.

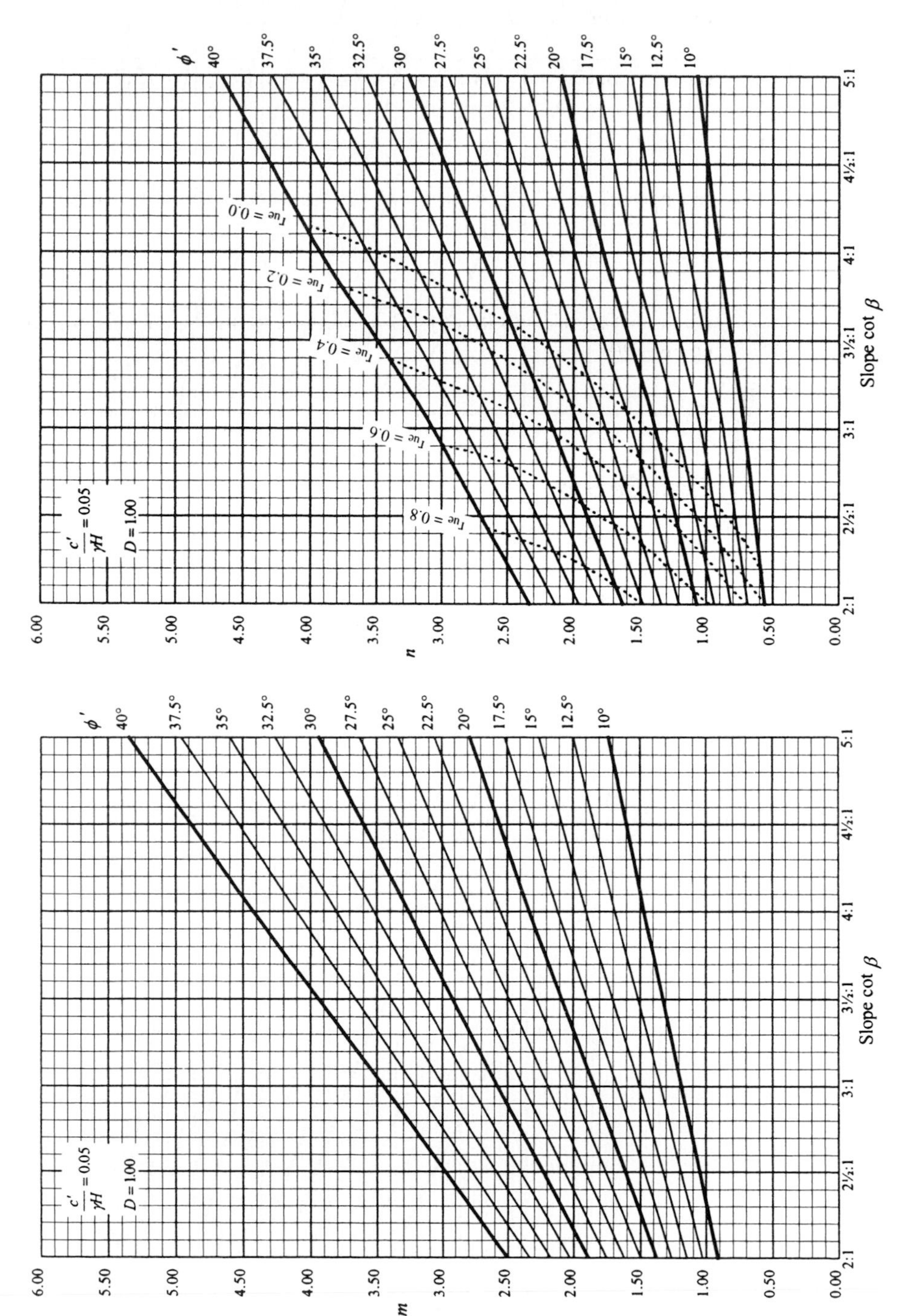

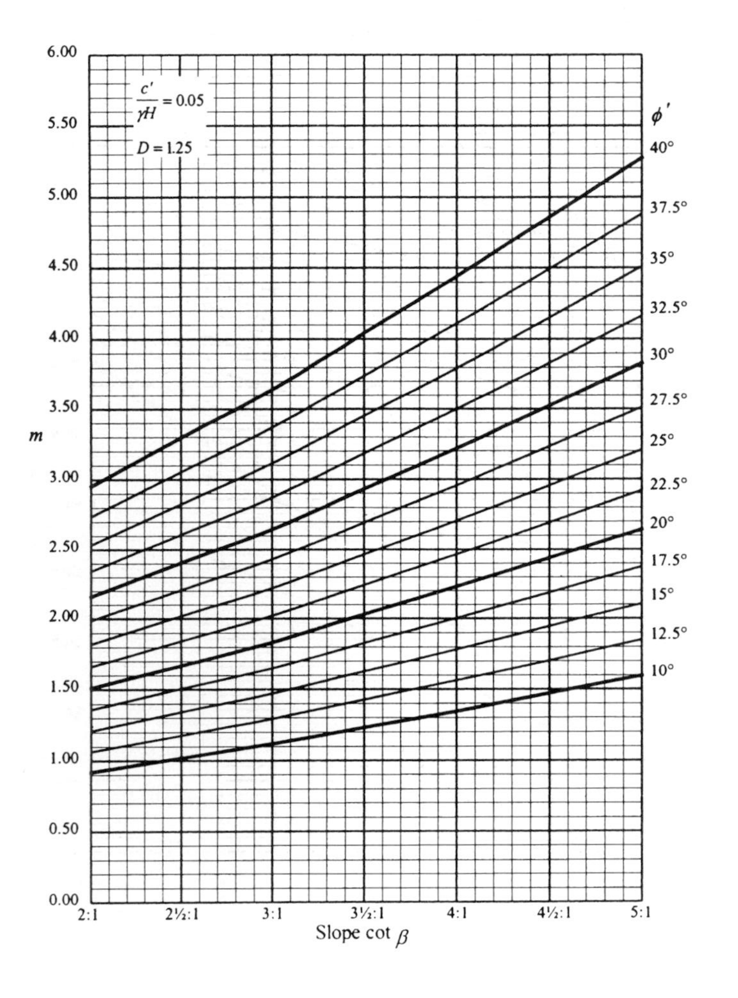
c'/γH = 0.05
D = 1.25
φ'
40°
37.5°
35°
32.5°
30°
27.5°
25°
22.5°
20°
17.5°
15°
12.5°
10°
m
6.00
5.50
5.00
4.50
4.00
3.50
3.00
2.50
2.00
1.50
1.00
0.50
0.00
2:1
2½:1
3:1
3½:1
4:1
4½:1
5:1
Slope cot β

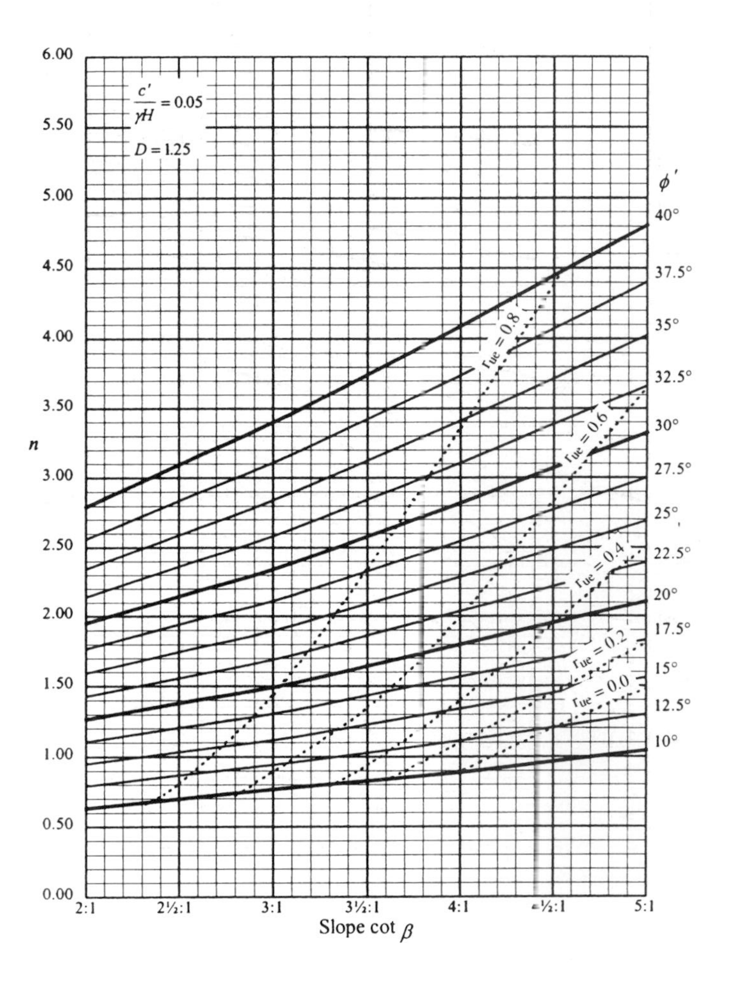
c'/γH = 0.05
D = 1.25
φ'
40°
37.5°
35°
32.5°
30°
27.5°
25°
22.5°
20°
17.5°
15°
12.5°
10°
r_ue = 0.8
r_ue = 0.6
r_ue = 0.4
r_ue = 0.2
r_ue = 0.0
n
6.00
5.50
5.00
4.50
4.00
3.50
3.00
2.50
2.00
1.50
1.00
0.50
0.00
2:1
2½:1
3:1
3½:1
4:1
4½:1
5:1
Slope cot β

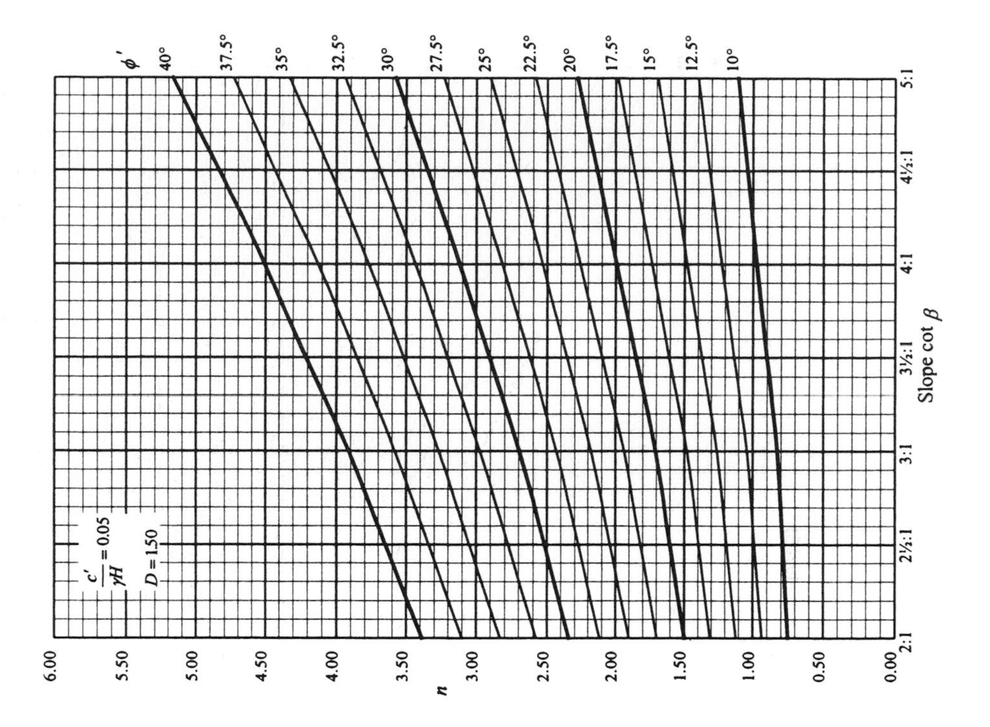
$\frac{c'}{\gamma H} = 0.05$
$D = 1.50$
ϕ'
40°
37.5°
35°
32.5°
30°
27.5°
25°
22.5°
20°
17.5°
15°
12.5°
10°
n
6.00
5.50
5.00
4.50
4.00
3.50
3.00
2.50
2.00
1.50
1.00
0.50
0.00
2:1
2½:1
3:1
3½:1
4:1
4½:1
5:1
Slope cot β

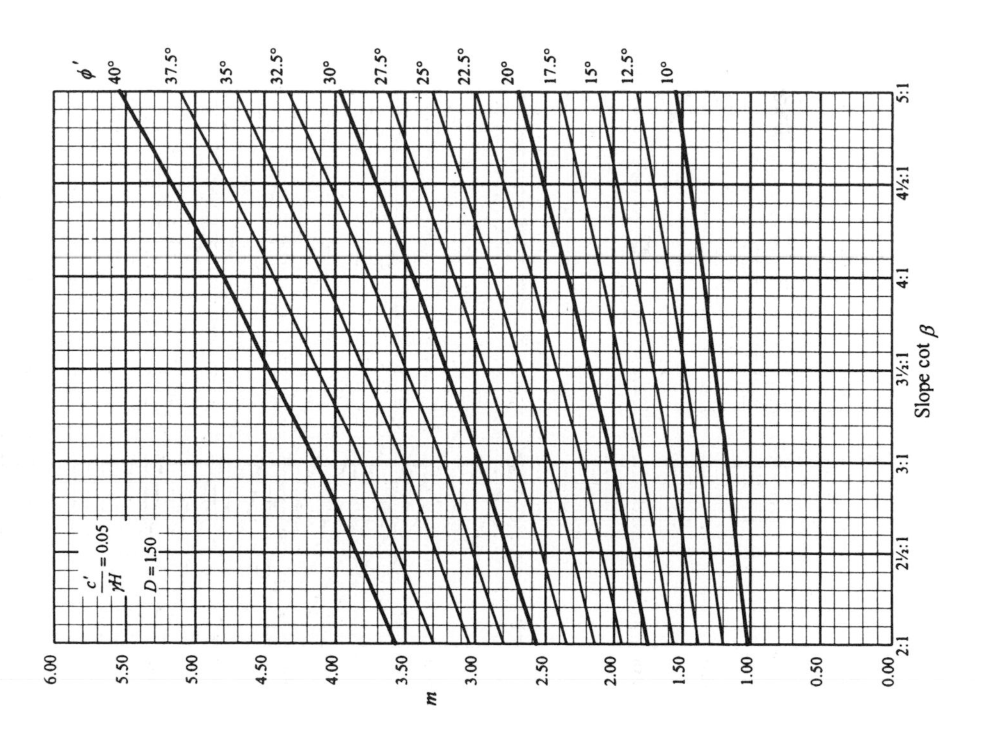
$\frac{c'}{\gamma H} = 0.05$
$D = 1.50$
ϕ'
40°
37.5°
35°
32.5°
30°
27.5°
25°
22.5°
20°
17.5°
15°
12.5°
10°
m
6.00
5.50
5.00
4.50
4.00
3.50
3.00
2.50
2.00
1.50
1.00
0.50
0.00
2:1
2½:1
3:1
3½:1
4:1
4½:1
5:1
Slope cot β

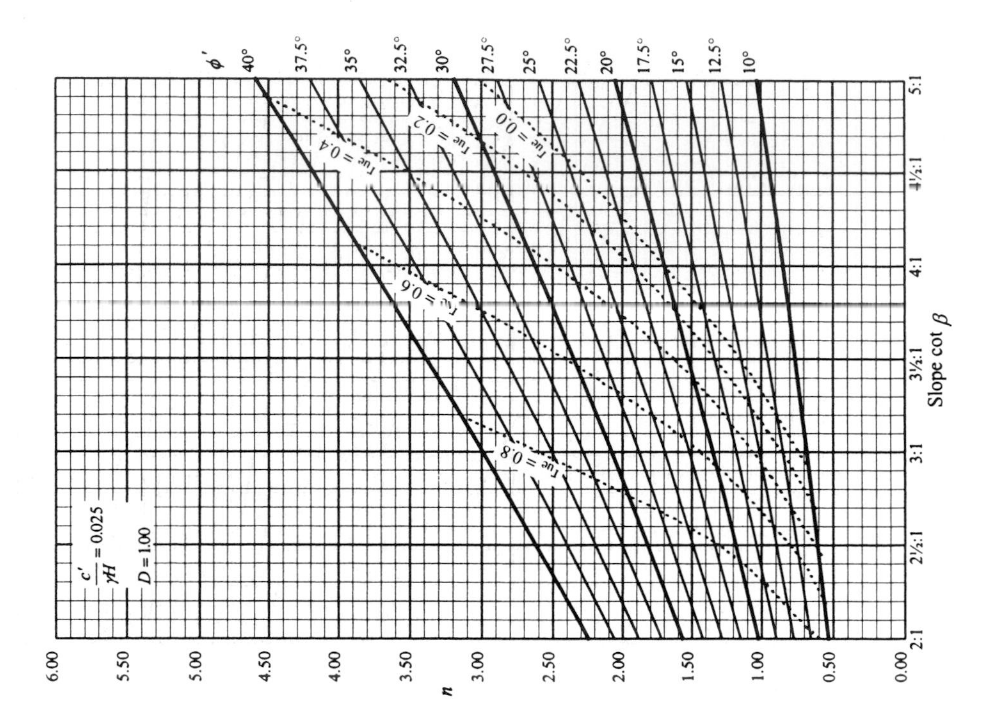
c'/γH = 0.025
D = 1.00
φ'
40°
37.5°
35°
32.5°
30°
27.5°
25°
22.5°
20°
17.5°
15°
12.5°
10°
r_ue = 0.0
r_ue = 0.2
r_ue = 0.4
r_ue = 0.6
r_ue = 0.8
n
6.00
5.50
5.00
4.50
4.00
3.50
3.00
2.50
2.00
1.50
1.00
0.50
0.00
2:1
2½:1
3:1
3½:1
4:1
4½:1
5:1
Slope cot β

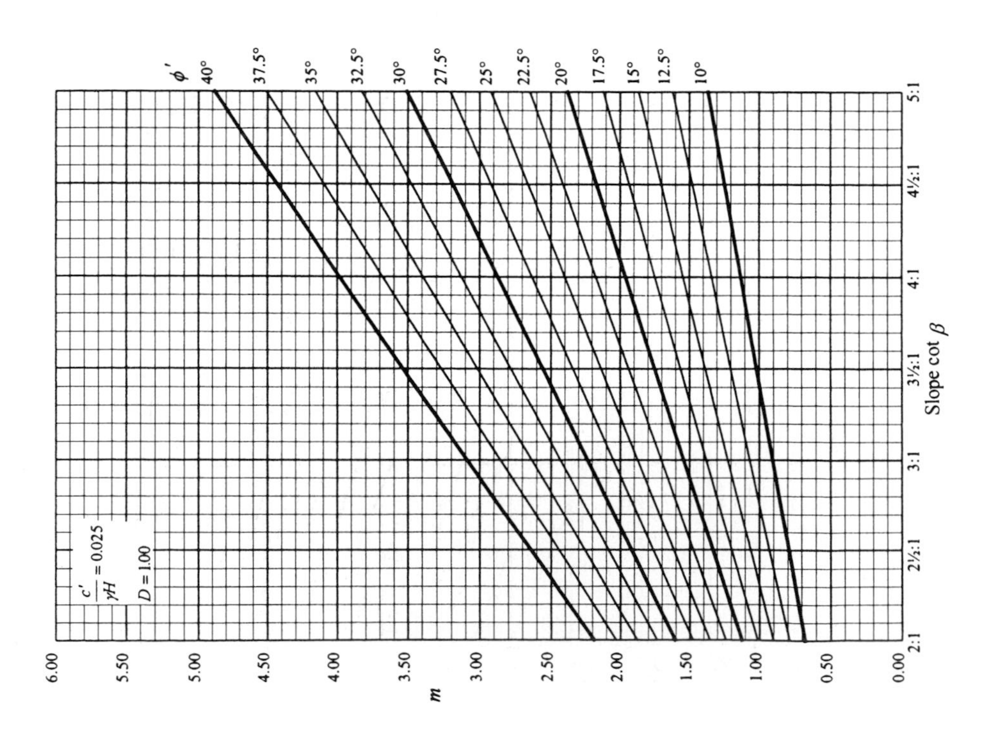
c'/γH = 0.025
D = 1.00
φ'
40°
37.5°
35°
32.5°
30°
27.5°
25°
22.5°
20°
17.5°
15°
12.5°
10°
m
6.00
5.50
5.00
4.50
4.00
3.50
3.00
2.50
2.00
1.50
1.00
0.50
0.00
2:1
2½:1
3:1
3½:1
4:1
4½:1
5:1
Slope cot β

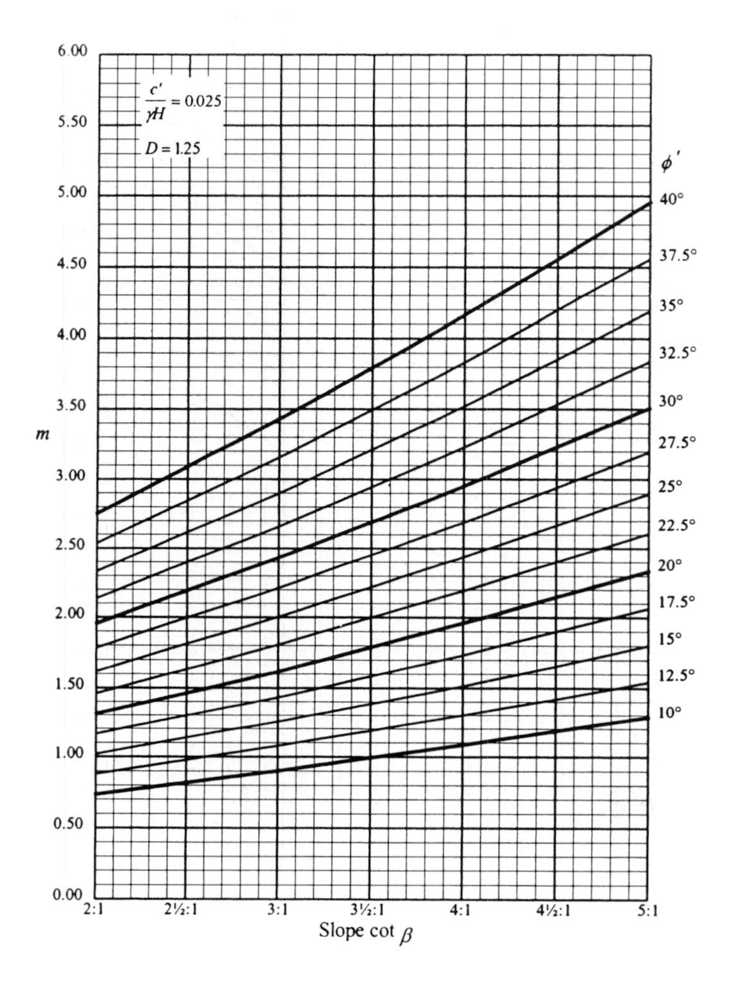

$\frac{c'}{\gamma H} = 0.025$
$D = 1.25$
φ'
40°
37.5°
35°
32.5°
30°
27.5°
25°
22.5°
20°
17.5°
15°
12.5°
10°
m
6.00
5.50
5.00
4.50
4.00
3.50
3.00
2.50
2.00
1.50
1.00
0.50
0.00
2:1
2½:1
3:1
3½:1
4:1
4½:1
5:1
Slope cot β

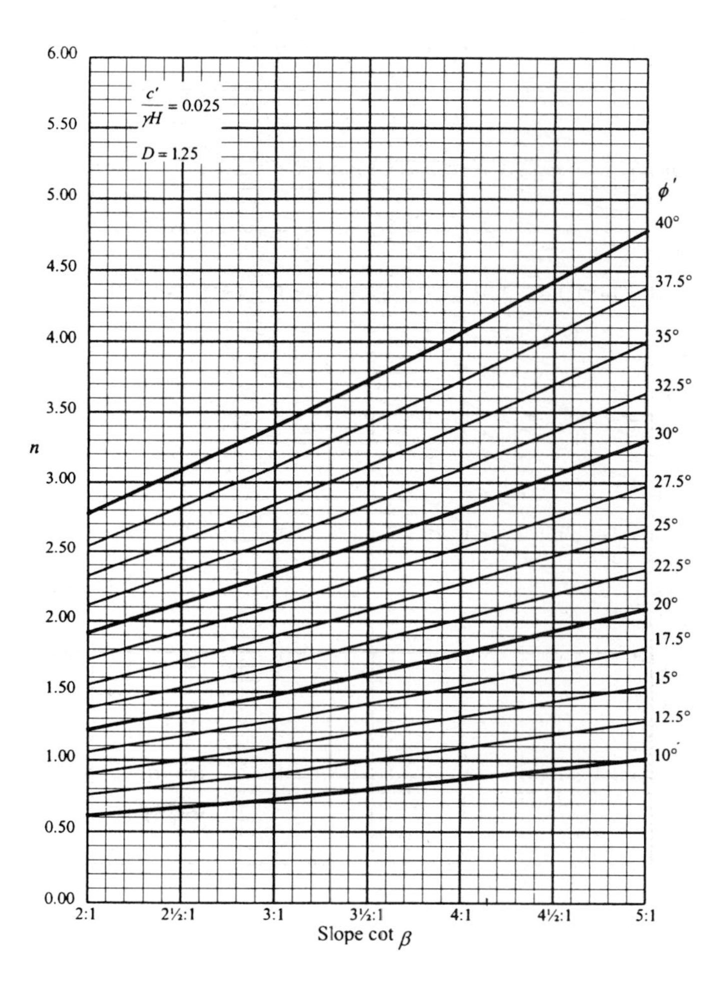

$\frac{c'}{\gamma H} = 0.025$
$D = 1.25$
φ'
40°
37.5°
35°
32.5°
30°
27.5°
25°
22.5°
20°
17.5°
15°
12.5°
10°
n
6.00
5.50
5.00
4.50
4.00
3.50
3.00
2.50
2.00
1.50
1.00
0.50
0.00
2:1
2½:1
3:1
3½:1
4:1
4½:1
5:1
Slope cot β

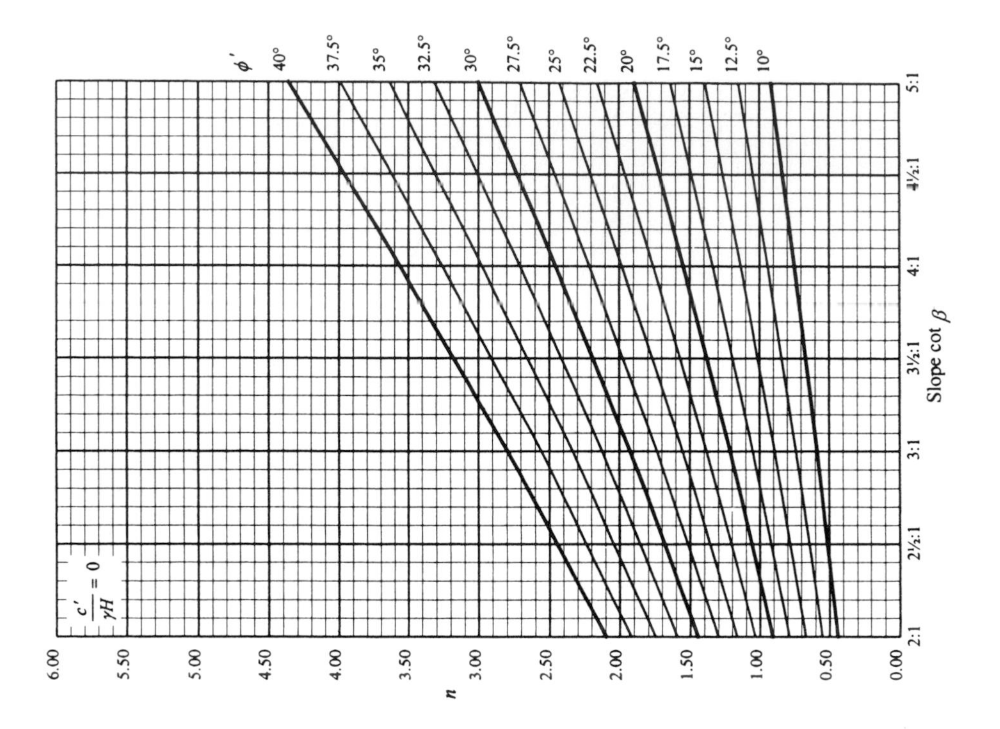
c'/γH = 0
n
6.00
5.50
5.00
4.50
4.00
3.50
3.00
2.50
2.00
1.50
1.00
0.50
0.00
Slope cot β
2:1
2½:1
3:1
3½:1
4:1
4½:1
5:1
φ'
40°
37.5°
35°
32.5°
30°
27.5°
25°
22.5°
20°
17.5°
15°
12.5°
10°

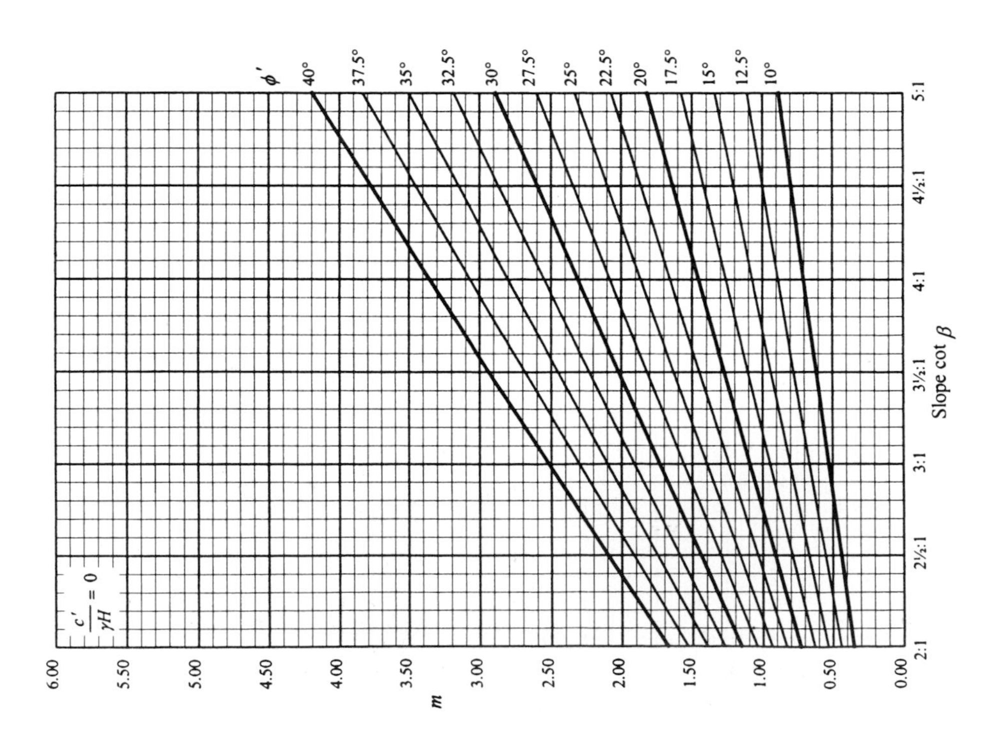
c'/γH = 0
m
6.00
5.50
5.00
4.50
4.00
3.50
3.00
2.50
2.00
1.50
1.00
0.50
0.00
Slope cot β
2:1
2½:1
3:1
3½:1
4:1
4½:1
5:1
φ'
40°
37.5°
35°
32.5°
30°
27.5°
25°
22.5°
20°
17.5°
15°
12.5°
10°

Chapter 6

Lateral Earth Pressure

6.1 Introduction

The variation in the values of a soil's strength parameters with drainage conditions has been discussed in Chapter 3 and it is important that the reader has an understanding of this phenomenon. A soil can exhibit shear resistance in one of three ways:

(i) Due entirely to friction, its cohesive intercept = 0. The soil acts as a cohesionless soil,
(i.e. as a 'ϕ soil' or as a 'ϕ' soil').

(ii) Due entirely to cohesion, its angle of shearing resistance equals 0°. The soil acts as a cohesive soil,
(i.e. as a 'c soil' or as a 'c′ soil').

(iii) A mixture of cohesive and frictional strength, with both the cohesive intercept and the angle of shearing resistance having values above 0.0. The soil acts as a cohesive–frictional soil,
(i.e. as a 'c–ϕ soil' or as a 'c′–ϕ' soil').

It can be seen therefore that the calculated value for the lateral pressure generated by the weight of a soil mass can be considerably in error if wrong values are assumed for the operative values of the cohesion and angle of shearing resistance of the soil. This aspect is considered later in this chapter but in the discussions in the early parts of this chapter the general symbols c and ϕ are used. The reader should appreciate that these symbols can be exchanged for c′ and ϕ' when necessary.

6.2 Active and passive earth pressure

Let us consider the simple case of a retaining wall with a vertical back (details of wall design and construction are given in Chapter 7) supporting a cohesionless soil with a horizontal surface (Fig. 6.1). Let the angle of shearing resistance of the soil be ϕ and let its unit weight, γ, be of a constant value. Then the vertical stress acting at a point at depth h below the top of the wall will be equal to γh.

If the wall is allowed to yield, i.e. to move forward slightly, the soil is able to expand and there will be an immediate reduction in the value of lateral pressure at depth h, but if the wall is pushed slightly into the soil then the soil will tend to be compressed and there will be an increase in the value of the lateral pressure.

The above indicates that there are two possible modes of failure that can occur within the soil mass. If we assume that the value of the vertical pressure at depth h

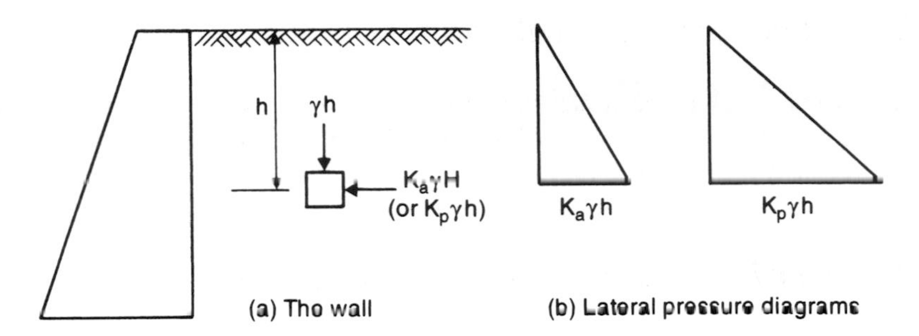

Fig. 6.1 Active and passive pressure.

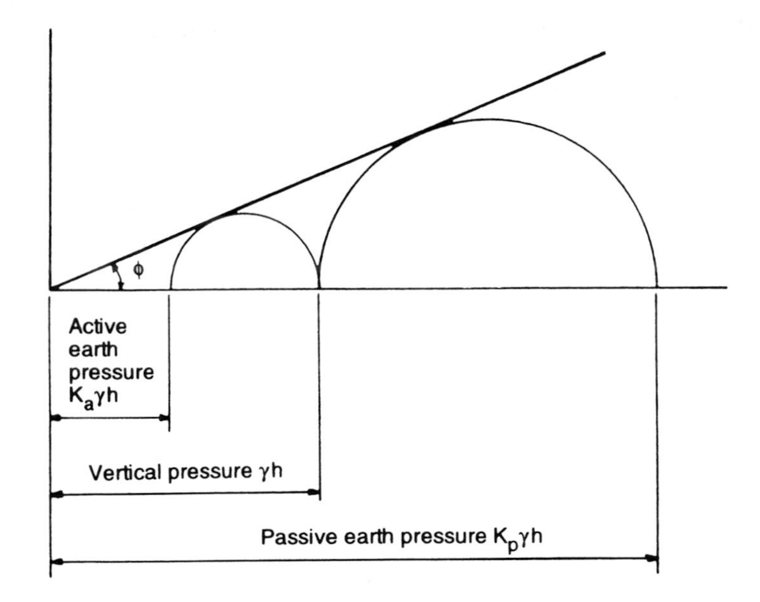

Fig. 6.2 Active and passive earth pressures.

remains unchanged at γh during these operations, then the minimum and maximum values of lateral earth pressure that will be achieved can be obtained from the Mohr circle diagram (Fig. 6.2).

The lateral pressure can reduce to a minimum value at which the stress circle is tangential to the strength envelope of the soil; this minimum value is known as the active earth pressure and equals $K_a\gamma h$ where K_a = the coefficient of active earth pressure. The lateral pressure can rise to a maximum value (with the stress circle again tangential to the strength envelope) known as the passive earth pressure, which equals $K_p\gamma h$ where K_p = coefficient of passive earth pressure.

It can be seen from Fig. 6.2 that when considering active pressure the vertical pressure due to the soil weight, γh, is a major principal stress and that when considering passive pressure the vertical pressure due to the soil weight, γh, is a minor principal stress.

6.3 Active pressure in cohesionless soils

The two major theories to estimate active and passive pressure values are those by Coulomb (1776) and by Rankine (1857). Both theories are very much in use today and both are described below.

6.3.1 *Rankine's theory (soil surface horizontal)*

Imagine a smooth, vertical retaining wall holding back a cohesionless soil with an angle of internal friction ϕ. The top of the soil is horizontal and level with the top of the wall. Consider a point in the soil at a depth h below the top of the wall (Fig. 6.3), assuming that the wall has yielded sufficiently to satisfy active earth pressure conditions.

In the Mohr diagram:

$$K_a = \frac{K_a \gamma h}{\gamma h} = \frac{OA}{OB} = \frac{OC - AC}{OC + CB} = \frac{OC - DC}{OC + DC} = \frac{1 - \frac{DC}{OC}}{1 + \frac{DC}{OC}} = \frac{1 - \sin\phi}{1 + \sin\phi}$$

It can be shown by trigonometry that

$$\frac{1 - \sin\phi}{1 + \sin\phi} = \tan^2\left(45° - \frac{\phi}{2}\right)$$

hence

$$K_a = \frac{1 - \sin\phi}{1 + \sin\phi} = \tan^2\left(45° - \frac{\phi}{2}\right)$$

The lateral pressure acting on the wall at any depth, σ_3, is equal to $K_a\sigma_1$. As $\sigma_1 = \gamma h$ and K_a is a constant, the lateral pressure increases linearly with depth (Fig. 6.3c). The magnitude of the resultant thrust, P_a, acting on the back of the wall is the area of the pressure distribution diagram. This force is a line load which acts through the centroid of the pressure distribution. In the case of a triangular distribution, the thrust acts at a third of the height of the triangle.

6.3.2 *Rankine's theory (soil surface sloping at angle β)*

This problem is illustrated in Fig. 6.4. The evaluation of K_a may be carried out in a similar manner to the previous case, but the vertical pressure will no longer be a principal stress. The pressure on the wall is assumed to act parallel to the surface of the soil, i.e. at angle β to the horizontal.

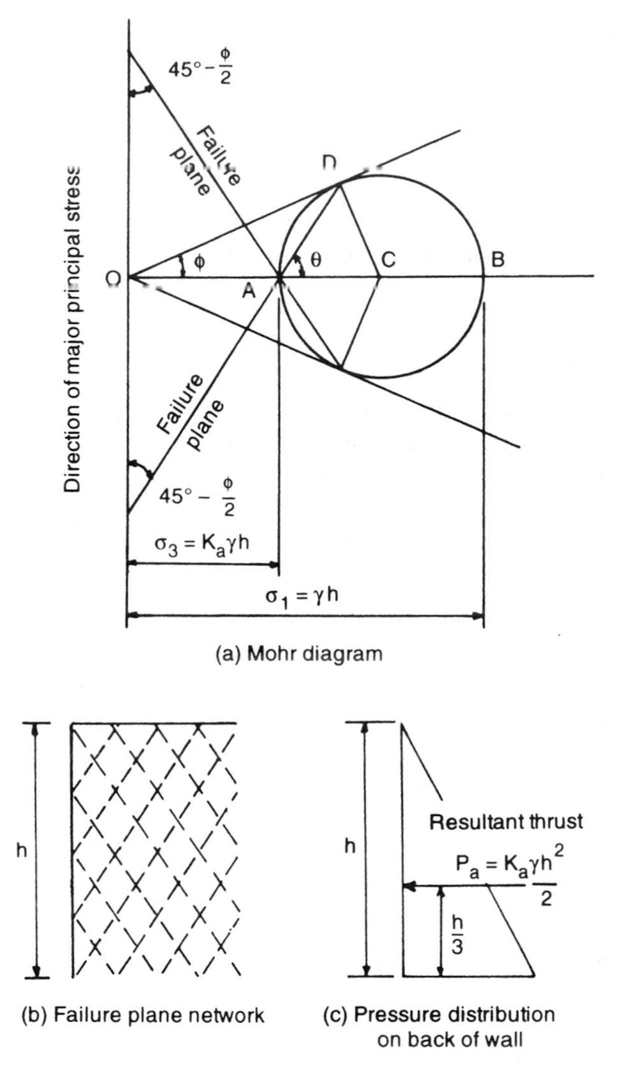

Fig. 6.3 Active pressure for a cohesionless soil with a horizontal upper surface.

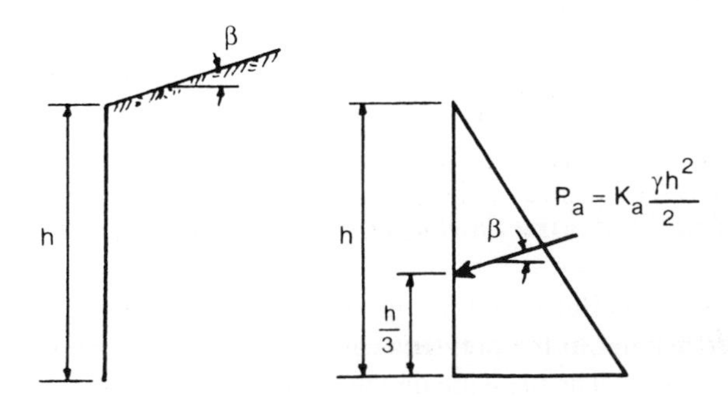

Fig. 6.4 Active pressure for a cohesionless soil with its surface sloping upwards at angle β to the horizontal.

The active pressure, p_a, is still given by the expression:

$$p_a = K_a \gamma h$$

where

$$K_a = \cos\beta \cdot \frac{\cos\beta - \sqrt{\cos^2\beta - \cos^2\phi}}{\cos\beta + \sqrt{\cos^2\beta - \cos^2\phi}}$$

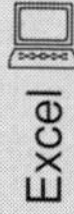

Example 6.1

Using the Rankine theory, determine the total active thrust on a vertical retaining wall 5 m high if the soil retained has a horizontal surface level with the top of the wall and has the following properties: $\phi' = 35°$; $\gamma = 19$ kN/m^3.

What is the increase in horizontal thrust if the soil slopes up from the top of the wall at an angle of 35° to the horizontal?

Solution A: Soil surface horizontal

$$K_a = \frac{1 - \sin 35°}{1 + \sin 35°} = 0.271$$

Maximum $p_a = 19 \times 5 \times 0.271 = 25.75$ kPa

Thrust = area of pressure diagram

$$= \frac{25.75 \times 5}{2} = 64 \text{ kN}$$

Solution B: Soil sloping at 35°

In this case, $\beta = \phi'$. When this happens the formula for K_a reduces to $K_a = \cos\phi'$. Hence

$$K_a = \cos 35° = 0.819$$

$$\text{Thrust} = \gamma K_a \frac{h^2}{2} = 0.819 \times 19 \times \frac{5^2}{2} = 194.5 \text{ kN}$$

This thrust is assumed to be parallel to the slope, i.e. at 35° to the horizontal.

Horizontal component = 194.5 × cos 35° = 159 kN
Increase in horizontal thrust = 95 kN/m length of wall

6.3.3 Coulomb's wedge theory

Instead of considering the equilibrium of an element in a stressed mass, Coulomb's theory considers the soil as a whole. If a wall supporting a cohesionless acting soil is

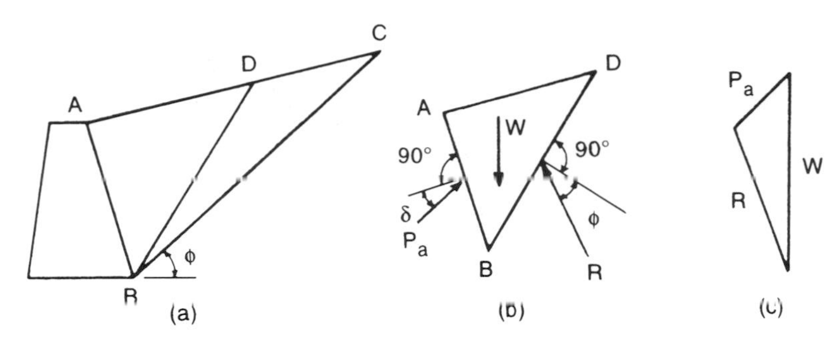

Fig. 6.5 Wedge theory for cohesionless soils.

suddenly removed the soil will slump down to its angle of shearing resistance, ϕ, on the plane BC in Fig. 6.5a. It is therefore reasonable to assume that if the wall only moved forward slightly a rupture plane BD would develop somewhere between AB and BC: the wedge of soil ABD would then move down the back of the wall AB and along the rupture plane BD. These wedges do in fact exist and have failure surfaces approximating to planes.

Coulomb analysed this problem analytically in 1776 on the assumption that the surface of the retained soil was a plane. He derived this expression for K_a:

$$K_a = \left\{ \frac{\operatorname{cosec} \psi \sin(\psi - \phi)}{\sqrt{\sin(\psi + \sigma)} + \sqrt{\dfrac{\sin(\phi + \delta)\sin(\phi - \beta)}{\sin(\psi - \beta)}}} \right\}^2$$

where

ψ = angle of back of wall to the horizontal
δ = angle of wall friction
β = angle of inclination of surface of retained soil to the horizontal
ϕ = angle of friction of retained soil (see Fig. 6.6).

Total active thrust = $\frac{1}{2}K_a\gamma H^2$, where H = total height of the wall. This thrust is assumed to act at angle δ to the normal to the wall (see Fig. 6.6).

It is of interest to note that Coulomb's expression for K_a reduces to the Rankine formula when $\psi = 90°$ and when $\delta = \beta$, i.e.

$$K_a = \cos\beta \times \frac{\cos\beta - \sqrt{(\cos^2\beta - \cos^2\phi)}}{\cos\beta + \sqrt{(\cos^2\beta - \cos^2\phi)}}$$

and further reduces to

$$K_a = \frac{1 - \sin\phi}{1 + \sin\phi}$$

when $\psi = 90°$ and when $\delta = 0°$.

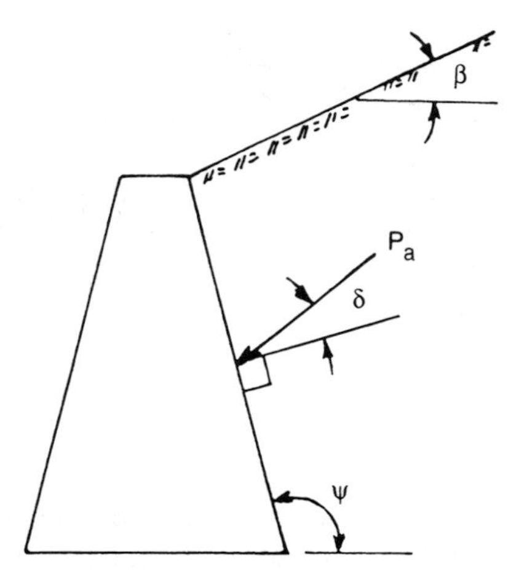

Fig. 6.6 Symbols used in Coulomb's formula.

Excel

Example 6.2

Solve Example 6.1 using the Coulomb formula. Assume that $\delta = \frac{1}{2}\phi'$.

Solution

Coulomb's formula for K_a is:

$$K_a = \left\{ \frac{\text{cosec}\,\psi \sin(\psi - \phi)}{\sqrt{\sin(\psi + \delta)} + \sqrt{\dfrac{\sin(\phi + \delta)\sin(\phi - \beta)}{\sin(\psi - \beta)}}} \right\}^2$$

Solution A: Soil surface horizontal

$\delta = 17.5°$; $\psi = 90°$; $\beta = 0°$; $\phi' = 35°$.

$$K_a = \left\{ \frac{\sin 55°/\sin 90°}{\sqrt{\sin 107.5°} + \sqrt{\sin 52.5° \sin 35°/\sin 90°}} \right\}^2$$

$$= \left\{ \frac{0.819}{0.976 + 0.675} \right\}^2 = 0.246$$

$$P_a = 0.5K_a\gamma H^2 = 0.5 \times 0.246 \times 19 \times 5^2 = 58.43 \text{ kN}$$

This value is inclined at 17.5° to the normal to the back of the wall so that the total horizontal active thrust, according to Coulomb, is 58.43 × cos 17.5° = 55.7 kN.

Note If δ had been assumed equal to 0° the calculated value of total horizontal thrust would have been the same as that obtained by the Rankine theory of Example 6.1.

Solution B: Soil surface sloping at 35°
Substituting $\phi' = 35°$, $\beta = 35°$, $\delta = 17.5°$ and $\psi = 90°$ into the formula gives $K_a = 0.704$. Hence

Total active thrust = $0.5 \times 0.704 \times 19 \times 5^2 = 167.2$ kN
Total horizontal thrust = $167.2 \times \cos 17.5° = 159.5$ kN
Increase in horizontal thrust = $159.5 - 55.7 = 104$ kN

The Culmann line construction

When the surface of the retained soil is irregular Coulomb's analytical solution becomes difficult to apply and it is generally simpler to make use of a graphical method proposed by Culmann in 1866, known as the Culmann line construction. Besides being able to cope with irregular soil surfaces the method can also deal with irregular combinations of uniform and line loads.

The procedure is to select a series of trial wedges and find the one that exerts the greatest thrust on the wall. A wedge is acted upon by three forces:

W, the weight of the wedge;
P_a, the reaction from the wall;
R, the reaction on the plane of failure.

At failure, the reaction on the failure plane will be inclined at maximum obliquity, ϕ, to the normal to the plane. If the angle of wall friction is δ then the reaction from the wall will be inclined at δ to the normal to the wall (δ cannot be greater than ϕ). As active pressures are being developed the wedge is tending to move downwards, and both R and P_a will consequently be on the downward sides of the normals (Fig. 6.5b). W is of known magnitude (area ABD $\times$ unit weight) and direction (vertical) and R and P_a are both of known direction, so the triangle of forces can be completed and the magnitude of P_a found (Fig. 6.5c). The value of the angle of wall friction, δ, can be obtained from tests, but if test values are not available δ is usually assumed as 0.5 to 0.75ϕ.

In Fig. 6.7 the total thrust on the wall due to earth pressure is to be evaluated, four trial wedges having been selected with failure surfaces BC, BD, BE and BF. At some point along each failure surface a line normal to it is drawn, after which a second line is constructed at ϕ to the normal. The resulting four lines give the lines of action of the reactions on each of the trial planes of failure. The direction of the wall reaction is similarly obtained by drawing a line normal to the wall and then another line at angle δ to it.

The weight of each trial slice is next obtained, and starting at a point X these weights are set off vertically upwards as points d_1, d_2, etc. such that Xd_1 represents the weight of slice 1 to some scale, Xd_2 represents the weight of slice 2 + slice 1, and so on.

A separate triangle of forces is now completed for each of the four wedges, the directions of the corresponding reaction on the failure plane and of P_a being obtained

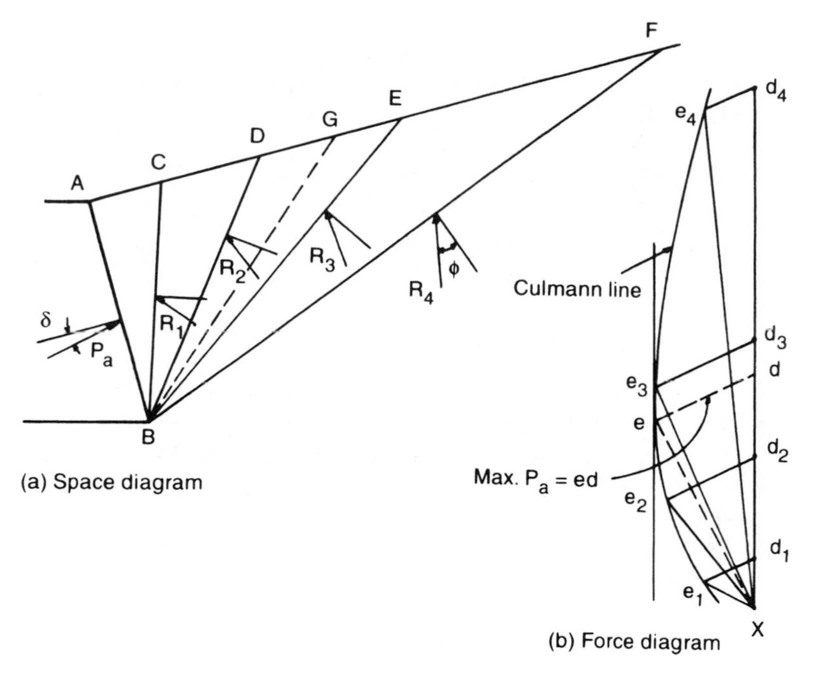

Fig. 6.7 Culmann line construction for a cohesionless soil.

from the space diagram. The point of intersection of R and P_a is given the symbol e with a suffix that tallies with the wedge analysed, e.g. the point e_1 represents the intersection of P_{a1} and R_1.

The maximum thrust on the wall is obviously represented by the maximum value of the length ed. To obtain this length a smooth curve (the Culmann line) is drawn through the points e_1, e_2, e_3 and e_4. A tangent to the Culmann line which is parallel to Xd_4 will cut the line at point e: hence the line ed can be drawn on the force diagram, and the length ed represents the thrust on the back of the wall due to the soil.

If required, the position of the actual failure plane can be plotted on the space diagram, the angle e_3Xe_2 on the force diagram equalling the angle EBD on the space diagram whilst the angle eXe_2 similarly equals the angle GBD where BG = failure plane.

6.3.4 *Point of application of the total active thrust*

With either the Rankine or the Coulomb analytical methods the total active thrust, P_a, is given by the expression:

$$P_a = \frac{1}{2}\gamma H^2 K_a$$

where K_a is the respective value of the coefficient of active earth pressure, H = height of wall and γ = unit weight of retained soil.

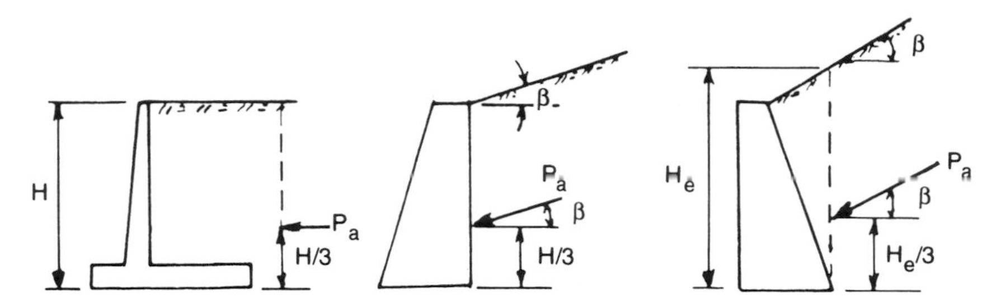

Fig. 6.8 Point of application of total active thrust (Rankine theory).

The position of the centre of pressure on the back of the wall, i.e. the point of application of P_a, is largely indeterminate. Locations suitable for design purposes are given in Fig. 6.8 and are based on the Rankine theory (with its assumption of a triangular distribution of pressure). For most practical purposes these locations of P_a can also be used in conjunction with P_a values obtained from a Coulomb analytical solution.

When using the Culmann line construction, the magnitude of P_a is obtained directly from the force diagram. Its point of application may be assumed to be where a line drawn through the centroid of the failure wedge, and parallel to the failure plane, intersects the back of the wall. (See Fig. 6.19.)

6.4 Surcharges

The extra loading carried by a retaining wall is known as a surcharge and can be a uniform load (roadway, stacked goods, etc.), a line load (trains running parallel to a wall), an isolated load (column footing), or a dynamic load (traffic).

Uniform load

Soil surface horizontal

When the surface of the soil behind the wall is horizontal, the pressure acting on the back of the wall due to the surcharge, q, is uniform with depth and has magnitude equal to K_aq (Fig. 6.9).

Soil surface sloping at angle β to horizontal

When the surface of the soil is not horizontal, the surcharge can be considered as equivalent to an extra height of soil, h_e placed on top of the soil.

$$h_e = \frac{q \sin \psi}{\gamma \sin(\psi + \beta)}$$

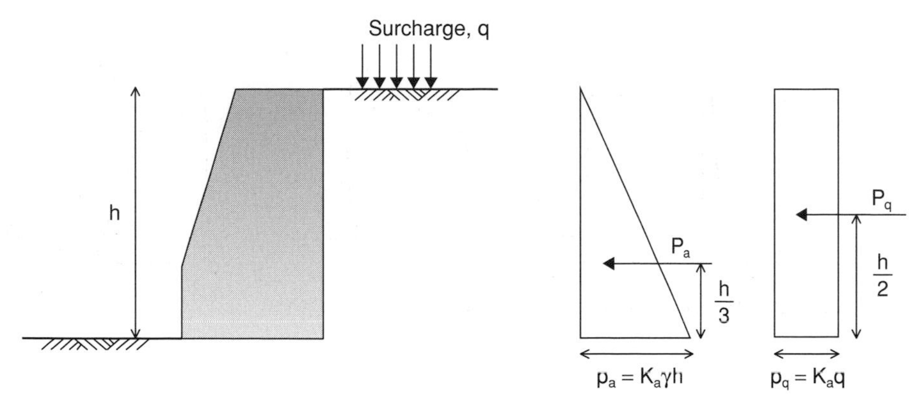

Fig. 6.9 Effect of uniform surcharge on a retaining wall.

where

γ = unit weight of soil (kN/m^3)
q = magnitude of surcharge (kPa)
ψ = angle of back of wall to horizontal
β = angle of inclination of retained soil.

Once again the pressure acting on the back of the wall due to the surcharge is considered uniform, but this time is of magnitude $K_a \gamma h_e$.

With the Culmann line construction the weight of surcharge on each slice is added to the weight of the slice. The weight of each wedge plus its surcharge is plotted as Xd_1, Xd_2, etc. and the procedure is as described earlier.

Even when a retaining wall is not intended to support a uniform surcharge it should be remembered that it may be subject to surface loadings due to plant movement during construction. It is at this time that the wall will be at its weakest state and BS 8002: 1994 *Code of practice for earth retaining structures* recommends that walls greater than 3 m in height be designed to carry a minimum uniform surcharge of 10 kPa (see Chapter 7).

Line load

The lateral thrust acting on the back of the wall as a result of a line load surcharge is best estimated by plastic analysis, as described in BS 8002: 1994, *Code of practice for earth retaining structures*.

It is possible to use a Boussinesq analysis (see Chapter 4) to determine the vertical stress increments due to the surface load and then to use these values in the plastic analysis combined with the design value of K_a (see Chapter 7).

With the Culmann line construction the weight of the line load, W_L is simply added to the trial wedges affected by it (Fig. 6.10). The Culmann line is first constructed as before, ignoring the line load. On this basis the failure plane would be BC and P_a would have a value 'ed' to some force scale.

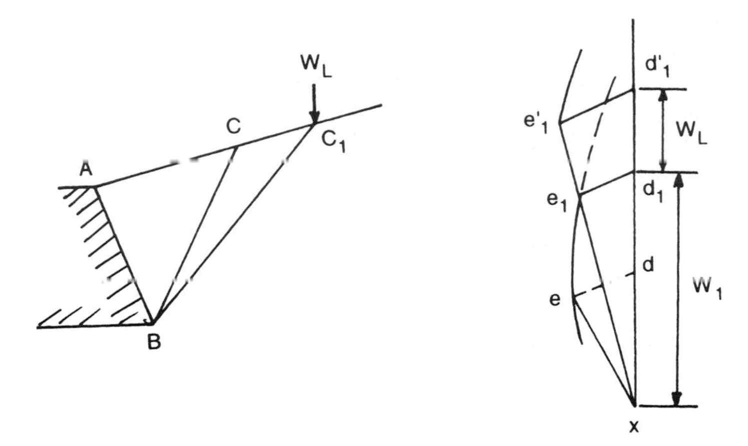

Fig. 6.10 Culmann line construction for a line load.

Slip occurring on BC_1 and all planes further from the wall will be due to the wedge weight plus W_L. For plane BC_1, set off $(W_1 + W_L)$ from X to d_1' and continue the construction of the Culmann line as before (i.e. for every trial wedge to the right of plane BC_1, add W_L to its weight). The Culmann line jumps from e_1 to e_1' and then continues to follow a similar curve.

The wall thrust is again determined from the maximum ed value by drawing a tangent, the maximum value of ed being in this case $e_1'd_1'$. If W_L is located far enough back from the wall it may be that ed is still greater than $e_1'd_1'$; in this case W_L is taken as having no effect on the wall.

Compaction effects

During the construction of gravity retaining walls layers of fill are compacted behind the wall, and this compaction process induces lateral stresses within the fill which can act against the back of the wall. If the stresses are high enough they can lead to movement or deformation of the wall, and so the effect of the compaction is taken into account during the design of the wall. Guidance on the effects of the compaction of backfill is given by Broms (1971) and by Clayton and Symons (1992).

Example 6.3

A smooth-backed vertical wall is 6 m high and retains a soil with a bulk unit weight of 20 kN/m^3 and $\phi' = 20°$. The top of the soil is level with the top of the wall and its surface is horizontal and carries a uniformly distributed load of 50 kPa. Using the Rankine theory determine the total active thrust on the wall/linear metre of wall and its point of application.

Solution

Figure 6.11a shows the problem and Fig. 6.11b shows the resultant pressure diagram.

Using the Rankine theory:

$$K_a = \frac{1 - \sin 20°}{1 + \sin 20°} = 0.49$$

$$p_a = K_a \gamma h = 0.49 \times 20 \times 6 = 58.8 \text{ kPa}$$

Since soil surface behind wall is horizontal,

$$p_q = K_a q = 0.49 \times 50 = 24.5 \text{ kPa}$$

The pressure diagram is now plotted (Fig. 6.11b).

$$\begin{aligned}\text{Total thrust } &= \text{Area of pressure diagram}\\ &= P_a + P_q\\ &= (\tfrac{1}{2} \times 58.8 \times 6) + (24.5 \times 6)\\ &= 176.4 + 147 = 323.4 \text{ kN}\end{aligned}$$

The point of application of this thrust is obtained by taking moments of forces about the base of the wall, i.e.

$$323.4 \times h = 147 \times 3 + 176.4 \times \frac{6}{3}$$

$$\Rightarrow \quad h = \frac{793.8}{323.4} = 2.45 \text{ m}$$

Resultant thrust acts at 2.45 m above base of wall.

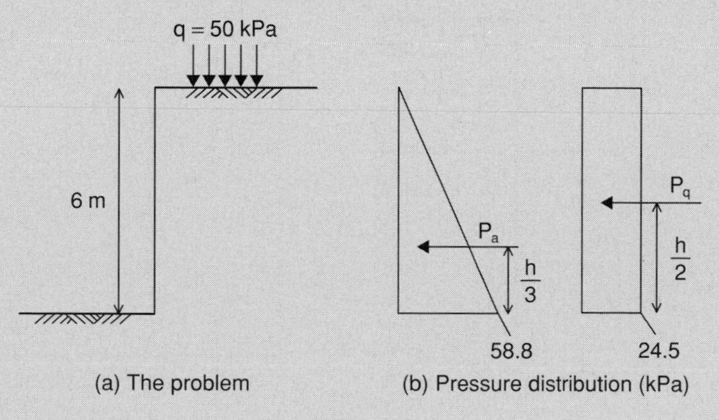

Fig. 6.11 Example 6.3.

Example 6.4

Details of the soil retained behind a smooth wall are given in Fig. 6.12. Draw the diagram of the pressure distribution on the back of the wall and determine the total horizontal active thrust acting on the back of the wall: (a) by the Rankine theory, (b) by the Coulomb theory. Take $\delta = \phi'/2$.

Solution

With either theory the active pressure at the top of the wall, $p_{a_0} = 0$.

(a) Rankine theory

p_{a_3}

Consider the upper soil layer:

$$K_a = \frac{1 - \sin 30°}{1 + \sin 30°} = 0.33$$

$$P_{a_3} = 0.33 \times 16 \times 3 = 16 \text{ kPa}$$

Consider the lower soil layer:

$$K_a = \frac{1 - \sin 20°}{1 + \sin 20°} = 0.49$$

$$P_{a_3} = 0.49 \times 16 \times 3 = 23.5 \text{ kPa}$$

The active pressure jumps from 16 to 23.5 kPa at a depth of 3 m.

For $P_{a_{7.5}}$:

$$P_{a_{7.5}} = 0.49 \times 24 \times 4.5 + 23.5 = 76.4 \text{ kPa}$$

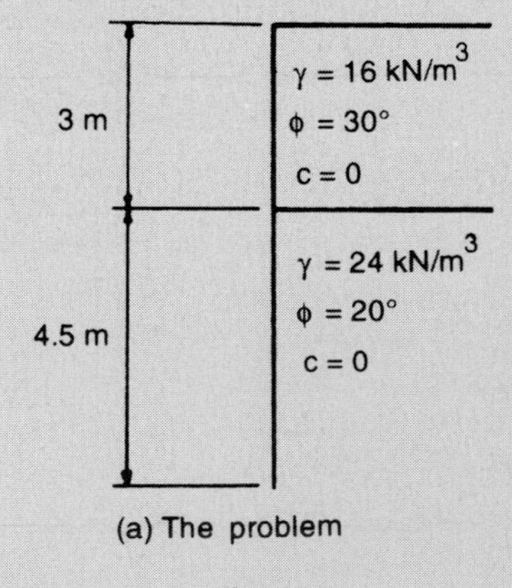

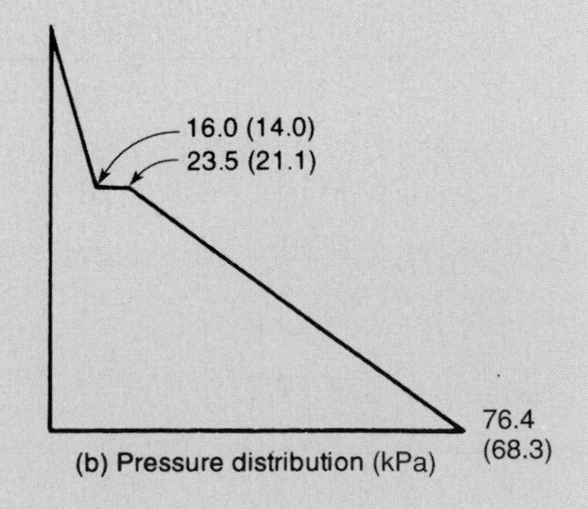

Fig. 6.12 Example 6.4.

The active pressure diagram is shown in Fig. 6.12b and the value of the total active thrust is simply the area of this diagram:

$$16 \times \frac{3}{2} + 23.5 \times 4.5 + 52.9 \times \frac{4.5}{2} = 248.8 \text{ kN}$$

(b) Coulomb theory

Consider the upper soil layer:

$$\text{For } \phi' = 30°, \delta = \phi'/2 = 15°, \beta = 0° \text{ and } \psi = 90°, K_a = 0.301$$

Hence active pressure at a depth of 3 m = 0.301 × 16 × 3 = 14.5 kPa.
But this pressure acts at 15° to the horizontal (as $\delta = 15°$).
Horizontal pressure at depth = 3 m = p_{a_3} = 14.5 cos 15° = 14.0 kPa.

Consider the lower soil layer:

$$\text{For } \phi' = 20°, \delta = \phi'/2 = 10°, \beta = 0° \text{ and } \psi = 90°, K_a = 0.447$$

$$p_{a_3} = 0.447 \times 16 \times 3 \times \cos 10° = 21.1 \text{ kPa}$$

$$p_{a_{7.5}} = [(0.447 \times 24 \times 4.5) + 21.1] \times \cos 10° = 68.3 \text{ kPa}$$

These values are shown in brackets on the pressure diagram in Fig. 6.12b.

Example 6.5

A vertical retaining wall 6 m high is supporting soil which is saturated and has a unit weight of 22.5 kN/m^3. The angle of friction of the soil, ϕ', is 35° and the surface of the soil is horizontal and level with the top of the wall. A groundwater level has been established within the soil and occurs at a level of 2 m from the top of the wall.

Using the Rankine theory calculate the significant pressure values and draw the diagram of pressure distribution that will occur on the back of the wall.

Solution

Figure 6.13a illustrates the problem and Figs 6.13b and 6.13c show the pressure distribution due to the soil and the water.

$$K_a = \frac{1 - \sin 35°}{1 + \sin 35°} = 0.27$$

Although there is the same soil throughout, there is a change in unit weight at a depth of 2 m as the unit weight of the soil below the GWL is equal to the submerged unit weight. The problem can therefore be regarded as two layers of different soil, the upper having a unit weight of 22.5 kN/m^3 and the lower (22.5 – 9.81) = 12.7 kN/m^3.

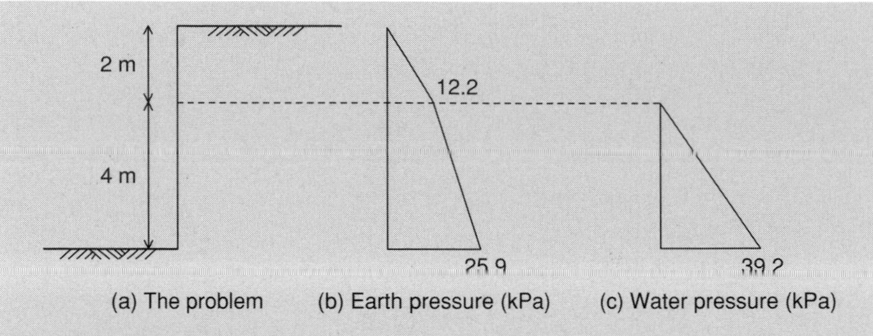

Fig. 6.13 Example 6.5.

Consider the upper soil:

At depth = 2 m: $p_{a_2} = K_a \gamma h = 0.27 \times 22.5 \times 2 = 12.2$ kPa

Consider the lower soil:

At depth = 6 m: $p_{a_6} = 12.2 + (0.27 \times 12.7 \times 4) = 25.9$ kPa

(Note that at the interface of two cohesionless soil layers, the pressure values are the same if the ϕ values are equal.)

Water pressure

At depth = 2 m, the water pressure = 0
At depth = 6 m, the water pressure = $9.81 \times 4 = 39.2$ kPa

The two pressure diagrams are shown in Figs 6.13b and 6.13c; the resultant pressure diagram is the addition of these two drawings:

$$= (\tfrac{1}{2} \times 12.2 \times 2) + (12.2 \times 4) + (\tfrac{1}{2} \times 13.7 \times 4) + (\tfrac{1}{2} \times 39.2 \times 4) = 167 \text{ kPa}$$

Had no groundwater table been present and the soil remained saturated throughout, the active thrust would have been:

$$= (\tfrac{1}{2} \times 0.27 \times 22.5 \times 6^2) = 109.4 \text{ kPa} \quad \text{(i.e. a lower value)}$$

Example 6.5 illustrates the significant increase in lateral pressure that the presence of a water table causes on a retaining wall. Except in the case of quay walls, a situation in which there is a water table immediately behind a retaining wall should not be allowed to arise. Where such a possibility is likely an adequate drainage system should be provided.

6.5 The effect of cohesion on active pressure

6.5.1 The Rankine theory

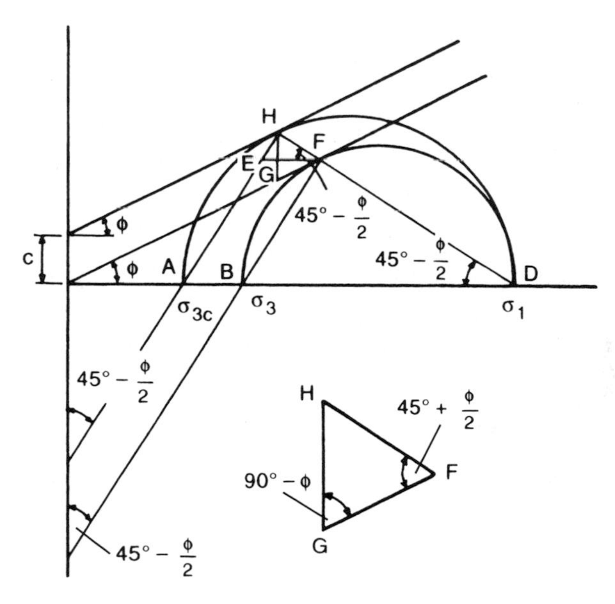

Fig. 6.14 The effect of cohesion on active pressure.

Consider two soils of the same unit weight, one acting as a purely frictional soil with an angle of shearing resistance, ϕ, and the other acting as a cohesive–frictional soil with the same angle of shearing resistance, ϕ, and a unit cohesion = c. The Mohr circle diagrams for the two soils are shown in Fig. 6.14.

At depth, h, both soils are subjected to the same major principal stress $\sigma_1 = \gamma h$. The minor principal stress for the cohesionless soil is σ_3 but for the cohesive soil it is only σ_{3c}, the difference being due to the cohesive strength, c, that is represented by the lengths AB or EF.

Consider triangle HGF:

$$\frac{HF}{GH} = \frac{HF}{c} = \frac{\sin(90^\circ - \phi)}{\sin\left(45^\circ + \frac{\phi}{2}\right)} = 2\frac{\sin\left(45^\circ - \frac{\phi}{2}\right)\cos\left(45^\circ - \frac{\phi}{2}\right)}{\cos\left(45^\circ - \frac{\phi}{2}\right)}$$

or

$$HF = 2c\sin\left(45^\circ - \frac{\phi}{2}\right)$$

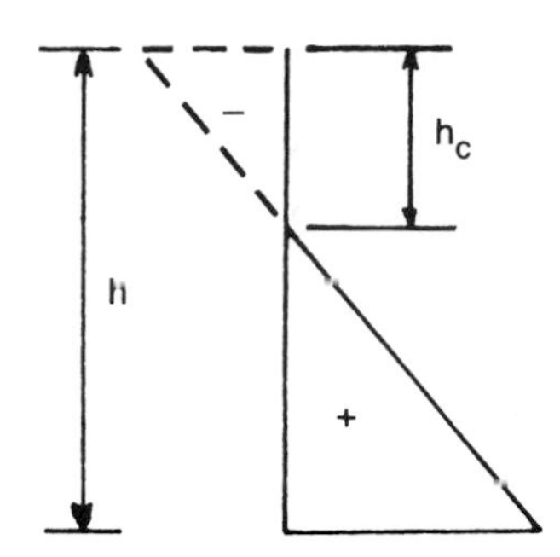

Fig. 6.15 Active pressure diagram for a soil with both cohesive and frictional strength.

Difference between σ_3 and σ_{3c}

$$= EF = \frac{HF}{\cos\left(45° - \frac{\phi}{2}\right)}$$

$$= 2c\,\frac{\sin\left(45° - \frac{\phi}{2}\right)}{\cos\left(45° - \frac{\phi}{2}\right)} = 2c\,\tan\left(45° - \frac{\phi}{2}\right)$$

Hence the active pressure, p_a, at depth h in a soil exhibiting both frictional and cohesive strength and having a horizontal upper surface is given by the expression:

$$p_a = K_a\gamma h - 2c\,\tan\left(45° - \frac{\phi}{2}\right) = K_a\gamma h - 2c\sqrt{K_a}$$

This expression was formulated by Bell (1915) and is often referred to as Bell's solution.

The active pressure diagram for such a soil is shown in Fig. 6.15. The negative values of p_a extending down from the top of the wall to a depth of h_c indicate that this zone of soil is in a state of suction. However, soils cannot really withstand tensile stress, and cracks may form within the soil. It is therefore unwise to assume that any negative active pressures exist within the depth h_c. For design purposes the active pressure value over the depth h_c should be taken as equal to zero.

6.5.2 Depth of the tension zone

In Fig. 6.15 the depth of the tension zone was given the symbol h_c. It is possible for cracks to develop over this depth, and a value for h_c is often required.

If p_a in the expression

$$p_a = K_a \gamma h_c - 2c \tan\left(45° - \frac{\phi}{2}\right)$$

is put equal to zero we can obtain an expression for h_c:

$$h_c = \frac{2c}{\gamma K_a} \tan\left(45° - \frac{\phi}{2}\right)$$

$$= \frac{2c}{\gamma} \tan\left(45° + \frac{\phi}{2}\right)$$

h_c may also be expressed:

$$h_c = \frac{2c}{\gamma\sqrt{K_a}}$$

When

$$\phi = 0°, \qquad h_c = \frac{2c}{\gamma}$$

6.5.3 *The occurrence of tensile cracks*

A tension zone, and therefore tensile cracking, can only occur when the soil exhibits cohesive strength. Gravels, sands and most silts generally operate in a drained state and, having no cohesion, do not experience tensile cracking.

Clays, when undrained, can have substantial values of c_u but, when fully drained, almost invariably have effective cohesive intercepts that are either zero or, have a small enough value to be considered negligible.

It is therefore apparent that tensile cracks can only occur in clays and are only important in undrained conditions. The value of h_c, as determined from the formula derived above, is seen to become smaller as the value of c becomes smaller. This illustrates that, as a clay wets up and its cohesive intercept reduces from c_u to c', any tensile cracks within it tend to close.

The Rankine formula for h_c must therefore be expressed in terms of total stress and is:

$$h_c = \frac{2c_u}{\gamma}\left(\tan 45° + \frac{\phi}{2}\right) \qquad \text{(for compacted silts and clays with both cohesive and frictional strength)}$$

and

$$h_c = \frac{2c_u}{\gamma} \qquad \text{(for clays)} \qquad (\text{since } \phi_u = 0)$$

If there is a uniform surcharge acting on the surface of the retained soil such that its equivalent height is h_e then the depth of the tension zone becomes equal to z_0

where $z_0 = h_c - h_e$. If, of course, the surcharge value is such that h_e is greater than h_c then no tension zone will exist.

6.5.4 The Coulomb theory

The theory assumes that at the top of the wall there is a zone of soil within which there are no friction or cohesive effects along both the back of the wall and the plane of rupture (Fig. 6.16). The depth of the zone is taken as z_0 and, as before, $z_0 - h_c$ or $z_0 = h_c - h_e$.

Graphical solution

There are now five forces acting on the wedge:

R, the reaction on plane of failure;
W, the weight of whole wedge ABED;
P_a, the resultant thrust on wall;
C_w, the adhesive force along length BF of wall ($C_w = c_w BF$);
C, the cohesive force along rupture plane BE ($C = cBE$).

The unit wall adhesion, c_w, cannot be greater than the unit cohesion, c_u. In the absence of tests that indicate higher values may be used, c_w can be taken as equal to c_u for soils up to $c_u = 50$ kPa. For soils with a cohesion value greater than 50 kPa, c_w should be taken as 50 kPa.

The value of W is obtained as before, so there are only two unknown forces: R and P_a.

In order to draw the Culmann line a polygon of forces must be constructed. The weights of the various wedges are set off as before, vertically up from the point X.

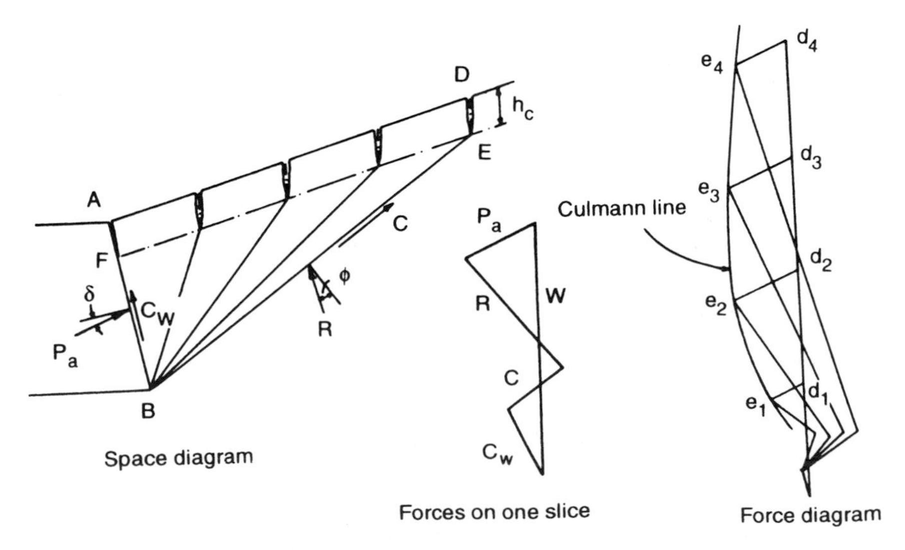

Fig. 6.16 Culmann line construction adapted to allow for cohesion.

As the force C_w is common to all polygons it is drawn next, and the C force is then plotted. The direction of P_a is drawn from point d and the direction of R is drawn from the end of force C; these two lines cross at the point e on the Culmann line.

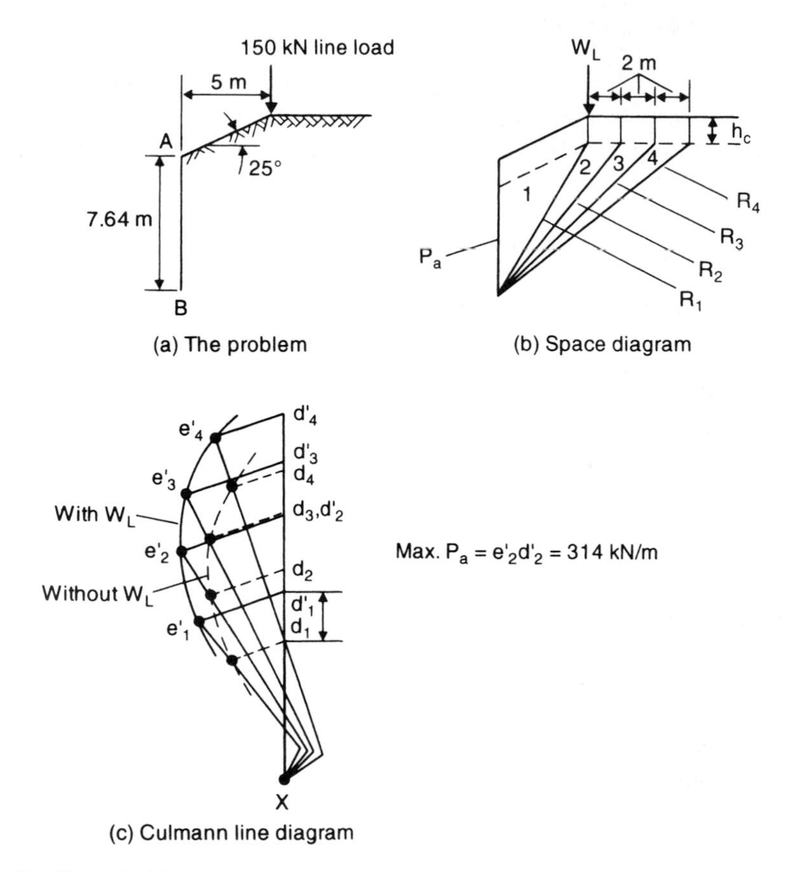

Fig. 6.17 Example 6.6.

Example 6.6

Determine the maximum thrust on the wall shown in Fig. 6.17a. The properties of the soil are: $\gamma = 17.4$ kN/m^3, $c_u = 9.55$ kPa, $\delta = \phi = 19°$.

Solution

$$h_c = \frac{2c_u}{\gamma}\tan\left(45° + \frac{\phi}{2}\right)$$

$$= \frac{2 \times 9.55}{17.4} \times \tan 54.5°$$

$$= 1.52 \text{ m}$$

Wall adhesion = (7.64 – 1.52)9.55 = 58.5 kN

Cohesion on failure planes:

1: $9.93 \times 9.55 = 94.7$ kN
2: $11.11 \times 9.55 = 106.0$ kN
3: $12.20 \times 9.55 = 116.4$ kN
4: $13.25 \times 9.55 = 126.5$ kN

Weight of wedges:

1: $22.6 \times 17.4 = 393$ kN
2: $35.1 \times 17.4 = 611$ kN
3: $43.8 \times 17.4 = 762$ kN
4: $51.9 \times 17.4 = 903$ kN

Space and force diagrams are given in Figs 6.17b and 6.17c.

$$\text{Maximum } P_a = e_2'd_2' = 314 \text{ kN/m}$$

Excel

Kerisel and Absi (1990) published values of the horizontal components of K_a and K_p for a range of values of ϕ, β, δ and ψ to ease calculation when using the Coulomb theory. In addition, the spreadsheet *earth pressure coefficients.xls* is available for download, which can be used to determine the horizontal components of K_a and K_p too. In this section we are concerned with the horizontal component of K_a (i.e. $K_a \cos \delta$) only. The horizontal component of K_p is given in Section 6.13.2.

The active pressure acting normally to the wall at a depth h can be defined by:

$$p_{ah} = K_a \gamma h - cK_{ac}$$

where

c = operating value of cohesion
K_{ac} = coefficient of active earth pressure.

Various values of K_a and K_{ac} are given in Table 6.1 for the straightforward case of $\beta = 0$ and $\psi = 90°$. Note that, where appropriate, ϕ is the operating value of the angle of shearing resistance of the soil.

Intermediate values of K_a and K_{ac} can be obtained from the spreadsheet. It should be noted that the values in both the spreadsheet and in Table 6.1 are for pressure components acting in the horizontal direction, not at an angle δ to the horizontal as in the original Coulomb theory. The values in the spreadsheet are derived from the following widely recognised formulae, which are sufficiently accurate for most purposes:

$$K_a = \text{Coulomb's value} \times \cos \delta$$

$$K_{ac} = 2\sqrt{K_a\left(1 + \frac{c_w}{c}\right)}$$

Table 6.1 Values of K_a and K_{ac} for a cohesive soil for $\beta = 0$, $\psi = 90°$.

Coefficient	Values of δ	Values of c_w/c	Values of ϕ					
			0°	5°	10°	15°	20°	25°
K_a	0	All	1.00	0.85	0.70	0.59	0.48	0.40
	ϕ	values	1.00	0.78	0.64	0.50	0.40	0.32
K_{ac}	0	0	2.00	1.83	1.68	1.54	1.40	1.29
	0	1	2.83	2.60	2.38	2.16	1.96	1.76
	ϕ	$\frac{1}{2}$	2.45	2.10	1.82	1.55	1.32	1.15
	ϕ	1	2.83	2.47	2.13	1.85	1.59	1.41

Example 6.7

Determine Coulomb's K_a value for $\phi = 20°$, $\delta = 10°$, $\beta = 0°$, $\psi = 90°$, $c_u = 10$ kPa, $c_w = 10$ kPa.

Solution

$$K_a = \left\{ \frac{\operatorname{cosec} 90° \sin 70°}{\sqrt{\sin 100°} + \sqrt{\sin 30° \sin 20°/\sin 90°}} \right\}^2$$

$$= 0.4467$$

Hence the K_a value for horizontal pressure $= 0.4467 \times \cos 10° = 0.44$.

$$K_{ac} = 2\sqrt{0.44\left(1 + \frac{10}{10}\right)} = 1.88$$

In part (b) of the next example (6.8) the values of K_a and K_{ac} are obtained via Table 6.1. It is interesting to compare the answers with those found in this example (6.7).

Example 6.8

A vertical retaining wall is 5 m high and supports a soil whose surface is horizontal and level with the top of the wall and is carrying a uniform surcharge of 75 kPa.

The properties of the soil are: $\phi = 20°$; $c_u = 10$ kPa; $\gamma = 20$ kN/m^3; $\delta = \phi/2$.

Determine the value of the maximum horizontal thrust on the back of the wall:

(a) by the Culmann line construction;
(b) by the K_a and K_{ac} coefficients of Table 6.1.

Solution

$$c_u < 50 \text{ kPa} \quad \Rightarrow \quad c_w = c_u = 10 \text{ kPa}$$

$$h_c = \frac{2c_u}{\gamma}\tan(45° + \phi/2) = 1.43 \text{ m}$$

$$h_e = \frac{q}{\gamma} = \frac{75}{20} = 3.75 \text{ m}$$

$$\Rightarrow \quad z_o = h_c - h_e = -2.32 \text{ m}, \qquad \text{i.e. take } z_o = 0$$

(a) The space and force diagrams for the Culmann line construction are shown in Figs 6.18a and 6.18b respectively. Three slices have been chosen and the calculations are best tabulated.

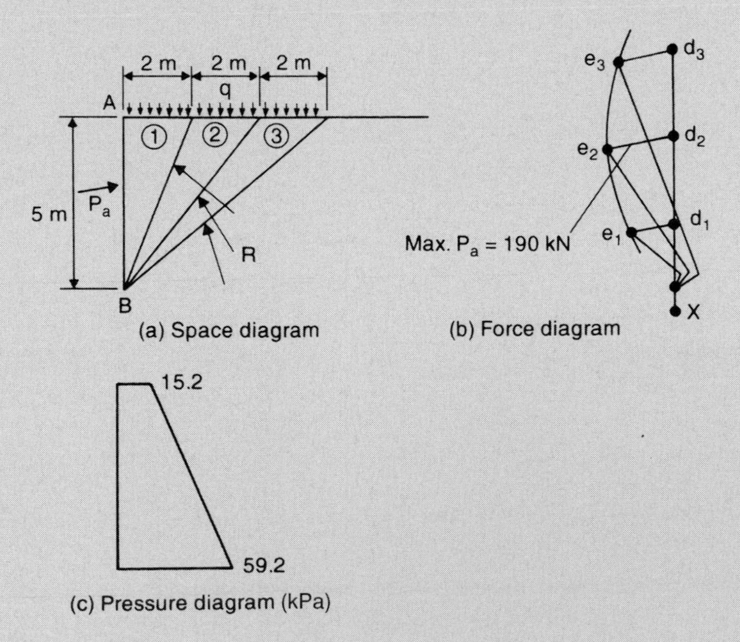

Fig. 6.18 Example 6.8.

Slice	Area (m^2)	$\gamma \times$ area (kN)	q (kN)	ΣW (kN)	C (= c × base) (kN)
1	5	100	150	250	53
2	10	200	300	500	64
3	15	300	450	750	79

Cohesive force on back of wall, $C_w = c_w \times AB = 10 \times 5 = 50$ kN.

From the force diagram, maximum $P_a = 190$ kN/m run of wall, acting at δ to the normal to the wall.

$\Rightarrow$ Maximum horizontal thrust on back of wall = 190 cos 10° = 187 kN/m run of wall.

(b) Coefficients K_a and K_{ac} (Table 6.1) can be obtained by linear interpolation:

For $c_w/c_u = 1.0$ and $\phi = 20°$; $K_a = 0.48$ for $\delta = 0°$

For $c_w/c_u = 1.0$ and $\phi = 20°$; $K_a = 0.40$ for $\delta = \phi$

$\Rightarrow \quad K_a = 0.44$

For $c_w/c_u = 1.0$ and $\phi = 20°$; $K_{ac} = 1.96$ for $\delta = 0°$

For $c_w/c_u = 1.0$ and $\phi = 20°$; $K_{ac} = 1.59$ for $\delta = \phi$

$$\Rightarrow \quad K_{ac} = \frac{1.96 + 1.59}{2} = 1.78$$

Active pressure at top of wall,

$$P_{a_s} = \gamma h_e K_a - cK_{ac}$$
$$= (20 \times 3.75 \times 0.44) - (10 \times 1.78) = 15.2 \text{ kPa}$$

Active pressure at base of wall,

$$P_{a_5} = \gamma(H + h_e)K_a - cK_{ac}$$
$$= 20(5 + 3.75)0.44 - 17.8 = 59.2 \text{ kPa}$$

The pressure diagram on the back of the wall is shown in Fig. 6.18c. Remembering that these are the values of pressure acting normal to the wall, the maximum horizontal thrust will be the area of the diagram.

$$\text{Maximum horizontal thrust} = \frac{15.2 + 59.2}{2} \times 5 = 186 \text{ kN/m run of wall.}$$

6.5.5 Point of application of total active thrust

For both of the preceding analytical solutions the area of the resulting active pressure diagram will give the magnitude of the total active thrust, P_a. If required, its point of application can be obtained by taking moments of forces about some convenient point on the space diagram. If this approach is not practical then the assumptions of Fig. 6.8 should generally be sufficiently accurate.

As mentioned earlier, for the Culmann line construction the point of application of P_a can be taken as the point where a line drawn through the centroid of the failure wedge, and parallel to the failure plane, cuts the back of the wall (Fig. 6.19).

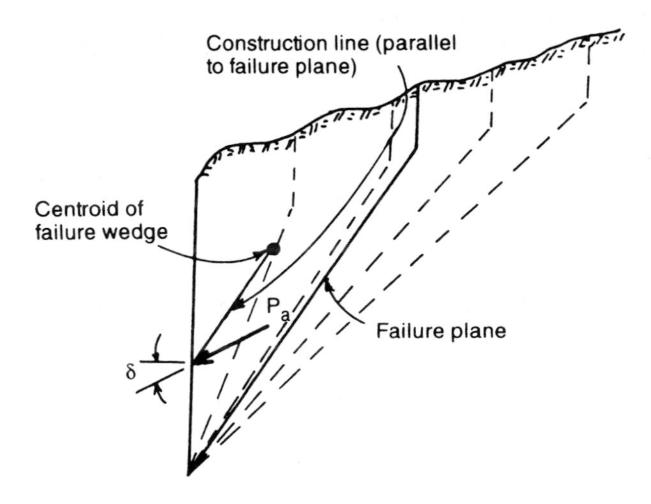

Fig. 6.19 Determination of line of action p_a.

6.6 Choice of method for prediction of active pressure

The main criticism of the Rankine theory is that it assumes conditions that are unrealistic in soils. There will invariably be friction and/or adhesion developed between the soil and the wall as it will have some degree of roughness and will never be perfectly smooth. Hence, in many cases, the Rankine assumption that no shear forces develop on the back of the wall is simply not true and it may be appropriate to use the Coulomb theory.

As noted earlier it is not easy to obtain measured values of the value of wall friction, δ, and the value of the wall adhesion, c_w, which are usually estimated. δ is obviously a function of the angle of friction, ϕ, of the retained soil immediately adjacent to the wall and can have any value from virtually zero up to some maximum value, which cannot be greater than ϕ. Similarly the operative value of c_w is related to the value of unit cohesion, c, of the soil immediately adjacent to the wall.

Just what will be the actual operating value of δ depends upon the amount of relative movement between the soil and the wall. A significant downward movement of the soil relative to the wall will result in the development of the maximum δ value.

Cases of significant relative downward movement of the soil are not necessarily all that common. Often there are cases in which there is some accompanying downward movement of the wall resulting in the smaller relative displacement. Examples of such cases can be gravity and sheet piled walls and a value of δ less than the maximum should obviously be used. (Descriptions of different wall types are given in Chapter 7.)

When the retained soil is supported on a foundation slab, as with a reinforced concrete cantilever or counterfort wall, there will be virtually no movement of the soil relative to the back of the wall. In this case the adoption of a 'virtual face' in the design procedure, as illustrated in Example 7.2, justifies the use of the Rankine approach.

The use of the Rankine method affords a quick means for determining a conservative value of active pressure, which can be useful in preliminary design work.

Full explanations of the design procedures advocated in BS 8002: 1994 and Eurocode 7 are given in Chapter 7.

6.7 Design parameters for different soil types

Owing to various self-compensating factors, the operative values of the strength parameters that determine the value of the active earth pressure are close to the peak values obtained from the triaxial test, even although a retaining wall operates in a state of plane strain. As has been discussed in Chapter 3, the values of these strength parameters vary with both the soil type and the drainage conditions. For earth pressure calculations, attention should be paid to the following.

Sands and gravels

For all stages of construction and for the period after construction the appropriate strength parameter is ϕ'. Take c′ as being equal to zero.

Clays

The manner in which a clay soil behaves during its transition from an undrained to a drained state depends upon the previous stress history of the soil and has been described in Chapter 3.

Soft or normally consolidated clay

During and immediately after construction of a wall supporting this type of soil the vertical effective stress is small, the strength of the soil is at a minimum and the value of the active earth pressure exerted on to the back of the wall is at a maximum. After construction and after sufficient time has elapsed, the soil will achieve a drained condition. The effective vertical stress will then be equal to the total vertical stress and the soil will have achieved its greatest strength. At this stage therefore the back of the wall will be subjected to the smallest possible values of active earth pressure (if other factors do not alter).

Obviously it is possible to use effective stress analyses to estimate the value of pressure on the back of the wall for any stage of the wall's life. A designer is interested chiefly in the maximum pressure values, which occur during and immediately after construction. As it is not easy to predict accurate values of pore water pressures for this stage, an effective stress analysis can be difficult and it is simplest to use the undrained strength parameters in any earth pressure calculations, i.e. assume that $\phi = 0°$ and that the undrained strength of the clay is c_u.

As mentioned in Chapter 3, the sensitivity of a normally consolidated clay can vary from 5 to 10. If it is considered that the soil will be severely disturbed during construction then the c_u value used in the design calculations should be the undrained strength of the clay remoulded to the same density and at the same moisture content as the *in situ* values.

If required, the final pressure values on the back of the wall, which apply when the clay is fully drained, can be evaluated in terms of effective stresses using the effective stress parameters ϕ' ($c' = 0$ for a normally consolidated clay). Soft clays usually have to be supported by a sheet pile type of wall, as a constructional or permanent feature, or by a slurry trench form of construction. In neither case is it easy to provide drainage behind the wall and, therefore, water pressures acting on the wall must be considered in the design.

Overconsolidated clay

In the undrained state negative pore water pressures are generated during shear. This simply means that this type of clay is at its strongest and the pressure on the wall is at its minimum value during and immediately after construction. The maximum value of active earth pressure will occur when the clay has reached a fully drained condition and the retaining wall should be designed to withstand this value, obtained from the effective stress parameters ϕ' and c'.

With an overconsolidated clay, c' has a finite value (Fig. 3.31) but, for retaining wall design, this value cannot be regarded as dependable as it could well decrease. It is therefore safest to assume that $c' = 0$ and to work with ϕ' only in any earth pressure calculations involving overconsolidated clay. The assumption also helps to allow for any possible increase in lateral pressure due to swelling in an expansive clay as its pore water pressures change from negative (in the undrained state) to zero (when fully drained).

Silts

In many cases a silt can be assumed to be either purely granular, with the characteristics of a fine sand, or purely cohesive, with the characteristics of a soft clay. When such a classification is not possible then the silt must be regarded as a c–ϕ soil. The total stress parameters ϕ and c should be used for the evaluation of active earth pressures which will be applicable to the period of during and immediately after construction.

The final active earth pressure to which the wall will be subjected can be determined from an effective stress analysis using the parameters ϕ' and c'.

Rain water in tension cracks

If tension cracks develop within a retained soil and if the surface of the soil is not rendered impervious then rain water can penetrate into them. If the cracks become full of water we can consider that we have a triangular distribution of water pressure acting on the back of the wall over the depth of the cracks, z_0. The value of this pressure will vary from zero at the top of the wall to approximately $10z_0$ kPa at the base of the cracks. This water pressure should be allowed for in design calculations, see Exercise 6.3.

The ingress of water, if prolonged, can lead eventually to softening and swelling of the soil. Swelling could partially close the cracks but would then cause swelling pressures that could act on the back of the wall. The prediction of values of lateral pressure due to soil swelling is quite difficult.

Shrinkage cracks may also occur and, in Britain, can extend downwards to depths of about 1.5 m below the surface of the soil. If water can penetrate these shrinkage cracks then the resulting water pressures should be allowed for as for tension cracks.

6.8 The choice of backfill material

The ideal backfill material is granular, such as suitably graded stone, gravel, or clean sand with a small percentage of fines. Such a soil is free draining and of good durability and strength but, unfortunately, it can be expensive, even when obtained locally.

Economies can sometimes be achieved by using granular material in retaining wall construction in the form of a wedge as shown in Fig. 6.20. The wedge separates

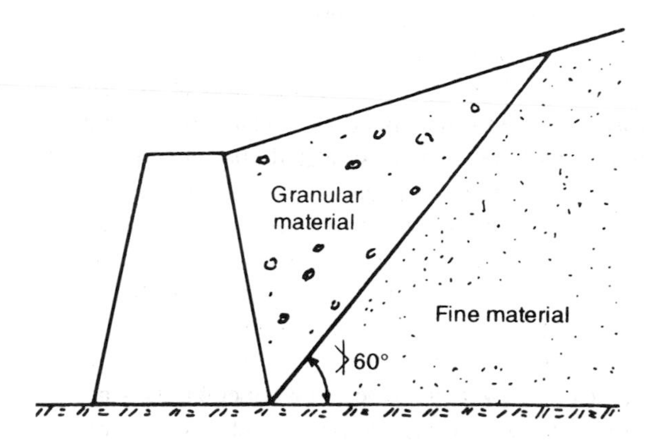

Fig. 6.20 Use of granular material in retaining wall construction.

the finer material making up the bulk of the backfill from the back of the wall. With such a wedge lateral pressures exerted on to the back of the wall can be evaluated with the assumption that the backfill is made up entirely of the granular material.

Slag, clinker, burnt colliery shale and other manufactured materials that approximate to a granular soil will generally prove satisfactory as backfill material provided that they do not contain harmful chemicals. Inorganic silts and clays can be used as backfills but require special drainage arrangements and can give rise to swelling and shrinkage problems that are not encountered in granular material. Peat, organic soil, chalk, unburnt colliery shale, pulverised fuel ash and other unsuitable material should not be used as backfill if at all possible.

Backfill drainage

No matter what material is used as a backfill its drainage is of great importance. A retaining wall is designed generally to withstand only lateral pressures exerted by the soil that it is supporting. In any design the possibility of a groundwater level occurring in the material behind a retaining wall must be examined and an appropriate drainage system decided upon.

For a granular backfill the only drainage often necessary is the provision of weep holes that go through the wall and are spaced at some 3 m centres, both horizontally and vertically. The holes can vary in diameter from 75 to about 150 mm and are protected against clogging by the provision of gravel pockets placed in the backfill immediately behind each weep hole (Fig. 6.21a).

Generally weep holes can only be provided in outside walls and an alternative arrangement for granular backfill is illustrated in Fig. 6.21b. It consists of a continuous longitudinal back drain, placed at the foot of the wall and consisting of open jointed pipes packed around with gravel or some other suitable filter material. The design of filters is discussed in Chapter 2. Provision for rodding out should be provided.

If the backfill material is granular but has more than 5 per cent fine sand, silt or clay particles mixed within it then it is only semipervious. For such a material the provision of weep holes on their own will provide inefficient drainage, with the further complication of there being a much greater tendency for clogging to occur. The answer is to provide additional drainage, in the form of vertical strips of filter material (about 0.33×0.33 m^2 in cross-section) placed midway between the weep holes and led down to a continuous longitudinal strip of the same filter material of the same cross-section as shown in Fig. 6.21c.

For clayey materials blanket drains of suitable filter material are necessary. These blankets should be about 0.33 m thick and typical arrangements are shown in Figs 6.21d and 6.21e. Generally the vertical drainage blanket of Fig. 6.21d will prove satisfactory, especially if the surface of the retained soil can be protected with some form of impervious covering. If this protection cannot be given then there is the chance of high seepage pressure being created during heavy rain (see Example 6.9). In such a situation the alternative arrangement of the inclined filter blanket of Fig. 6.21e can substantially reduce such seepage pressures.

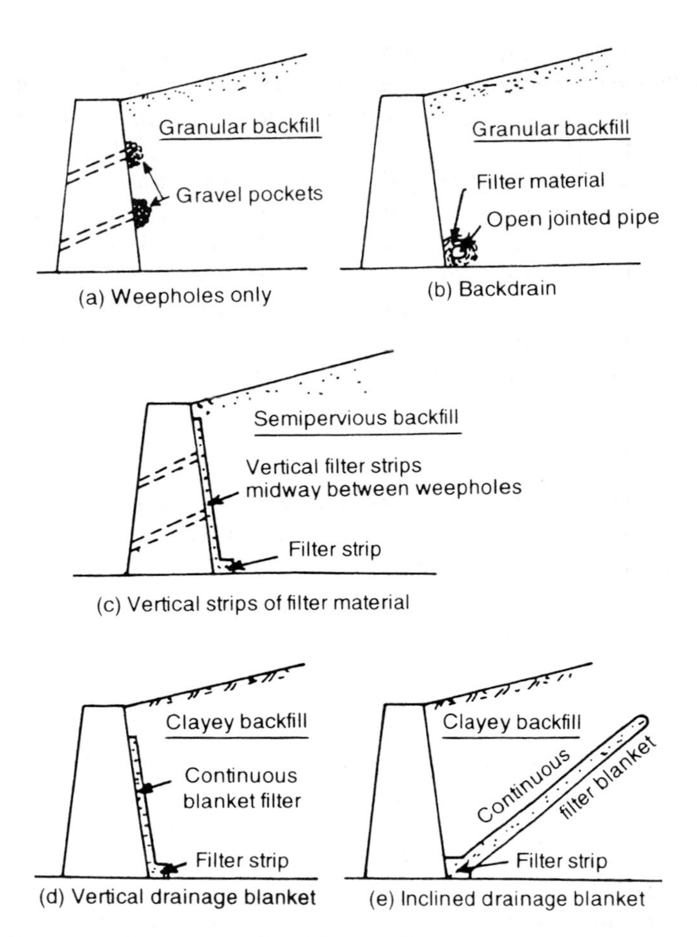

Fig. 6.21 Common drainage systems for retaining walls.

The reason for the different effects of the two drainage systems can be seen when we consider the respective seepage flow nets that are generated during flooded conditions.

The flow net for the vertical drain is shown in Fig. 6.22a. It must be appreciated that the drain is neither an equipotential nor a flow line. It is a drained surface and therefore the only head of water that can exist along it is that due to elevation. Hence, if a square flow net has been drawn, the vertical distances between adjacent equipotentials entering the drain will be equal to each other (in a manner similar to the upstream slope of an earth dam).

Owing to the seepage forces, an additional force, P_w, now acts upwards and at right-angles to the failure plane. From the flow net it is possible to determine values of excess hydrostatic pressure, h_w, at selected points along the failure plane (see Fig. 6.22a). If a smooth curve is drawn through these h_w values (when plotted along the failure plane), it becomes possible to evaluate P_w (see Example 6.9).

The resulting force diagram is shown in Fig. 6.22b. In theory the polygon of forces for a $c-\phi$ soil will be as shown in Fig. 6.22c but, as seepage will only occur once

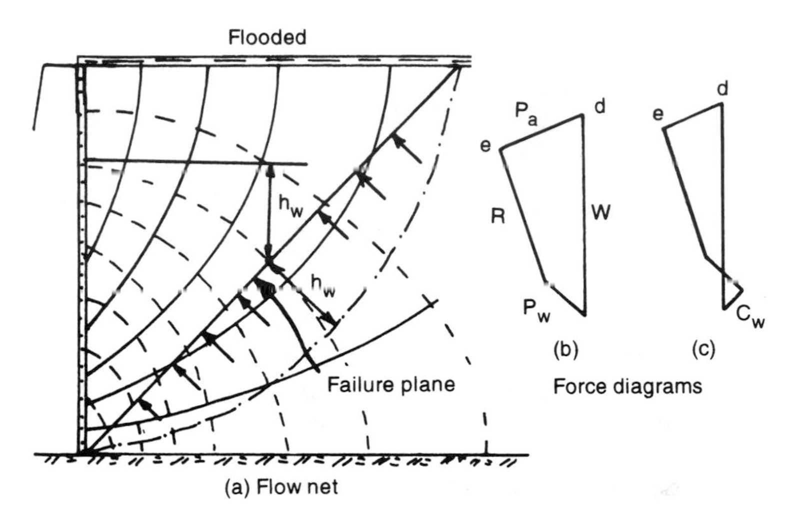

Fig. 6.22 Seepage forces behind a retaining wall with a vertical drain during heavy rain.

the soil has achieved a drained state, the operative strength parameter is ϕ' with c′ generally being assumed to be zero.

The seepage flow net for the inclined drain in Fig. 6.21e is shown in Fig. 6.23. Such a drain induces vertical drainage of the rain water and it is seen that the portion of the flow net above the drain is absolutely regular and, more important, that the equipotentials are horizontal. This latter fact means that, within the soil above the drain, the value of excess hydrostatic head at any point must be zero. The failure plane will not be subjected to the upward force P_w and the pressure exerted on the back of the wall can only be from the saturated soil.

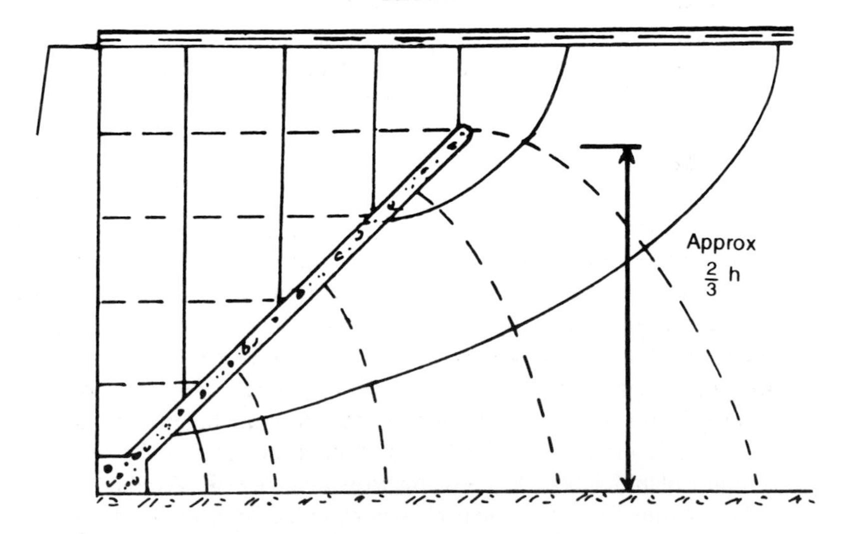

Fig. 6.23 Effect of an inclined drain on seepage forces.

Differential hydrostatic head

When there is a risk of a groundwater level developing behind the wall then the possible increase in lateral pressure due to submergence must be allowed for. This problem will occur in tidal areas, and quay walls must be designed to withstand the most adverse difference created by tidal lag between the water level in front of and the groundwater level behind the wall. As there is no real time for steady seepage conditions to develop between the two head levels, the effect of possible seepage forces can safely be ignored.

Example 6.9

A vertical 4 m high wall is founded on a relatively impervious soil and is supporting soil with the properties: $\phi' = 40°$, $c' = 0$, $\delta = 20°$, $\gamma_{sat} = 20$ kN/m^3.

The surface of the retained soil is horizontal and is level with the top of the wall. If the wall is subjected to heavy and prolonged rain such that the retained soil becomes saturated and its surface flooded, determine the maximum horizontal thrust that will be exerted on to the wall:

(i) if there is no drainage system;
(ii) if there is the drainage system of Fig. 6.21d;
(iii) if there is the drainage system of Fig. 6.21e.

Solution

(i) No drainage

As we have been given a value for the angle of wall friction it is more realistic to use the Culmann line construction. The total pressure on the back of the wall will be the summation of the pressure from the submerged soil and the pressure from the water.

Four trial wedges have been chosen and are shown in Fig. 6.24a and the corresponding force diagram in Fig. 6.24b.

Maximum P_a due to submerged soil = 16 kN
Horizontal component of $P_a = 16.0 \times \cos 20° = 15$ kN

$$\text{Horizontal thrust from water pressure} = 9.81 \times \frac{4^2}{2} = 78.5 \text{ kN}$$

Total horizontal thrust = 93.5 kN/m run of wall

(ii) With vertical drain on back of wall

The flow net for steady seepage from the flooded surface of the soil into the drain is shown in Fig. 6.24c. From this diagram it is possible to determine the distribution of the excess hydrostatic head, h_w, along the length of the failure surface of each of the four trial wedges. These distributions are shown in Fig. 6.24d and the area of each diagram times the unit weight of water gives the upward force, P_w, acting at right-angles to each failure plane.

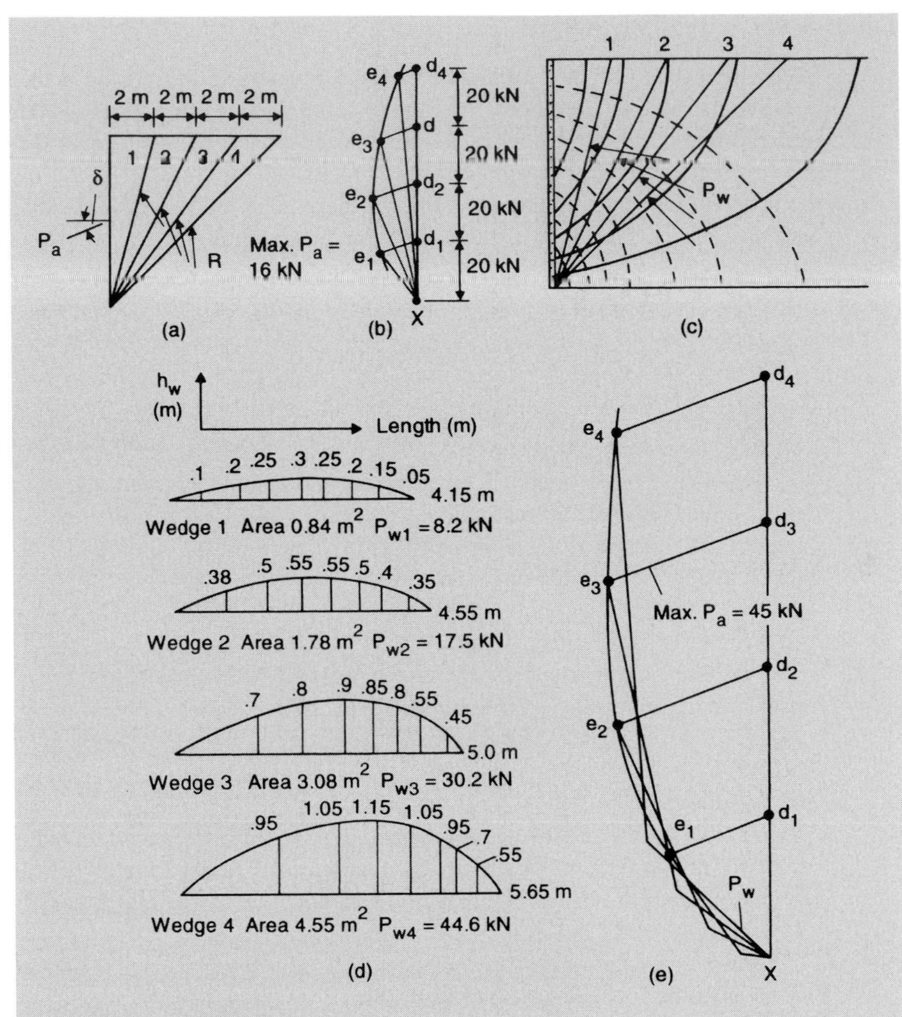

Fig. 6.24 Example 6.9.

The tabulated calculations are:

Wedge	Saturated weight (kN)	P_w (kN)
1	40	8
2	80	18
3	120	30
4	160	45

The force diagrams and the Culmann line construction are shown in Fig. 6.24e. From the force diagram, maximum $P_a = 45$ kN.

$\Rightarrow$ Maximum horizontal thrust on wall = $45 \times \cos 20° = 42$ kN/m

(iii) With inclined drain
As has been shown earlier, for all points in the soil above the drain there can be no excess hydrostatic heads. The force diagram is therefore identical with Fig. 6.24e except that, as P_w is zero for all wedges, it is removed from each polygon of forces. When this is done it is found that the maximum value of P_a is 30 kN.

$\Rightarrow$ Maximum horizontal thrust on back of wall = $30 \times \cos 20°$
= 28 kN/m

6.9 Earth pressure at rest

Consider a mass of soil with a horizontal upper surface and let this soil be completely at rest, undisturbed by any forces other than its own weight. If the unit weight of the soil is γ then an element at a depth h below the surface will be subjected to a vertical stress γh. This stress is a major principal effective stress, i.e. $\sigma'_1 = \gamma h$, and it will induce a horizontal minor principal effective stress, σ'_3.

The effective stress ratio σ'_1/σ'_3 for a soil at rest is given the symbol K_0, which is called the coefficient of earth pressure at rest, i.e. the lateral pressure in a soil at rest = $K_0\gamma h$ (Fig. 6.25).

It has been shown experimentally that, for granular soils and normally consolidated clays, $K_0 \approx 1 - \sin \phi'$ (Jaky, 1944).

Eurocode 7 actually relates K_0 to the overconsolidation ratio and states that $K_0 = (1 - \sin \phi) \times \sqrt{OCR}$ for soils with not very high values of overconsolidation ratio.

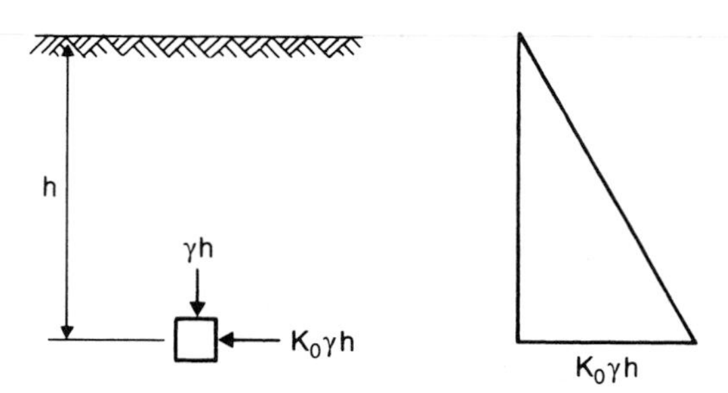

Fig. 6.25 Earth pressure at rest.

6.10 Influence of wall yield on design

A wall can yield in one of two ways: either by rotation about its lower edge (Fig. 6.26b) or by sliding forward (Fig. 6.26c). Provided that the wall yields sufficiently, a state of active earth pressure is reached and the thrust on the back of the wall is in both cases about the same (P_a).

The pressure distribution that gives this total thrust value can be very different in each instance, however. For example, consider a wall that is unable to yield (Fig. 6.26a). The pressure distribution is triangular and is represented by the line AC.

Consider that the wall now yields by rotation about its lower edge until the total thrust = P_a (Fig. 6.26b). This results in conditions that approximate to the Rankine theory and is known as the totally active case.

Suppose, however, that the wall yields by sliding forward until active thrust conditions are achieved (Fig. 6.26c). This hardly disturbs the upper layers of soil so that the top of the pressure diagram is similar to the earth pressure at rest diagram. As the total thrust on the wall is the same as in rotational yield, it means that the pressure distribution must be roughly similar to the line AE in Fig. 6.26c.

This type of yield gives conditions that approximate to the wedge theory, the centre of pressure moving up to between 0.45 and 0.55 h above the wall base, and is referred to as the arching–active case.

The differences between the various pressure diagrams can be seen in Fig. 6.26d where the three diagrams have been superimposed. It has been found that if the top of a wall moves 0.1 per cent of its height, i.e. a movement of 10 mm in a 10 m high wall, an arching–active case is attained. This applies whether the wall rotates or slides. In order to achieve the totally active case the top of the wall must move about 0.5 per cent, or 50 mm in a 10 m wall.

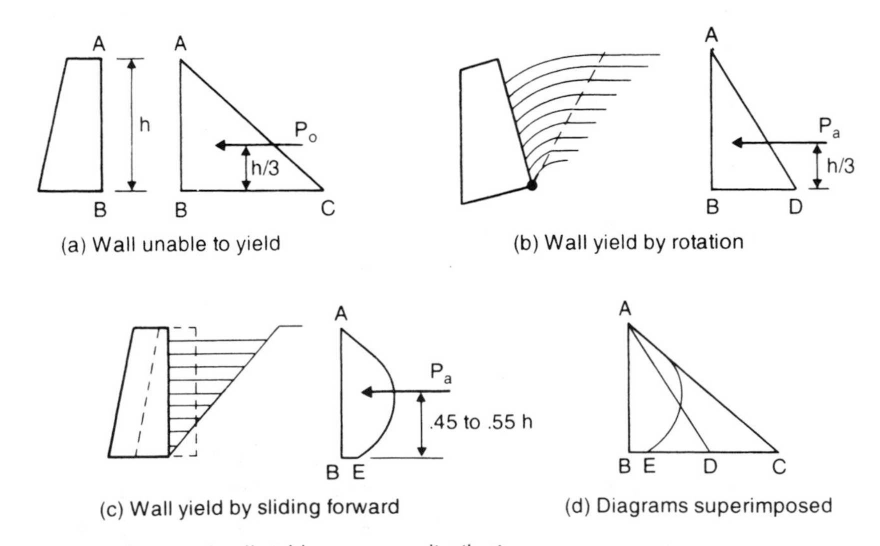

Fig. 6.26 Influence of wall yield on pressure distribution.

It can therefore be seen that if a retaining wall with a cohesionless backfill is held so rigidly that little yield is possible (e.g. if it is joined to an adjacent structure) it must be designed to withstand earth pressure values much larger than active pressure values.

If such a wall is completely restrained it must be designed to take earth pressure at rest values, although this condition does not often occur; if a wall is so restrained that only a small amount of yielding can take place arching–active conditions may be achieved, as in the strutting of trench timbers. In this case the assumption of triangular pressure distribution is incorrect, the actual pressure distribution being indeterminate but roughly parabolic.

If the wall yields 0.5 per cent of its height then the totally active case is attained and the assumption of triangular pressure distribution is satisfactory. Almost all retaining walls, unless propped at the top, can yield a considerable amount with no detrimental effects and attain this totally active state.

In the case of a wall with a cohesive backfill, the totally active case is reached as soon as the wall yields but due to plastic flow within the clay there is a slow build-up of pressure on the back of the wall, which will eventually yield again to re-acquire the totally active pressure conditions. This process is repetitive and over a number of years the resulting movement of the wall may be large. For such soils one can either design for higher pressure or, if the wall is relatively unimportant, design for the totally active case bearing in mind that the useful life of the wall may be short.

6.11 Strutted excavations

When excavating a deep trench the insertion of shuttering to hold up the sides becomes necessary. The excavation is carried down first to some point, X, and rigidly strutted timbering is inserted between the levels D to X (Fig. 6.27a).

As further excavation is carried out, timbering and strutting are inserted in stages, but before the timbering is inserted the soil yields by an amount that tends to increase with depth (it is relatively small at the top of the trench).

In Fig. 6.27b the shape A′B′C′D′ represents, to an enlarged scale, the original form of the surface that has yielded to the position ABCD of Fig. 6.27a; the resulting pressure on the back of the wall is roughly parabolic and is indicated in Fig. 6.27c.

For design purposes a trapezoidal distribution is assumed, of the form recommended in BS 8002: 1994 after the work of Terzaghi and Peck (1967), since revised (Terzaghi, Peck and Mesri, 1996). The design procedure for the struts is semi-empirical. For sands the pressure distribution is assumed to be uniform over the full depth of the excavation (Fig. 6.27d). For clays, the pressure distribution depends on the stability number, N:

$$N = \frac{\gamma H}{c_u}$$

If N is greater than 4, the distribution in Fig. 6.27e is used, provided that K_a is greater than 0.4. If N is less than 4, or if $0.2 < K_a < 0.4$, the distribution in Fig. 6.27f is used. With respect to Fig. 6.27e, m is generally taken as 1.0. For soft clays, however, m can reduce to ~0.4.

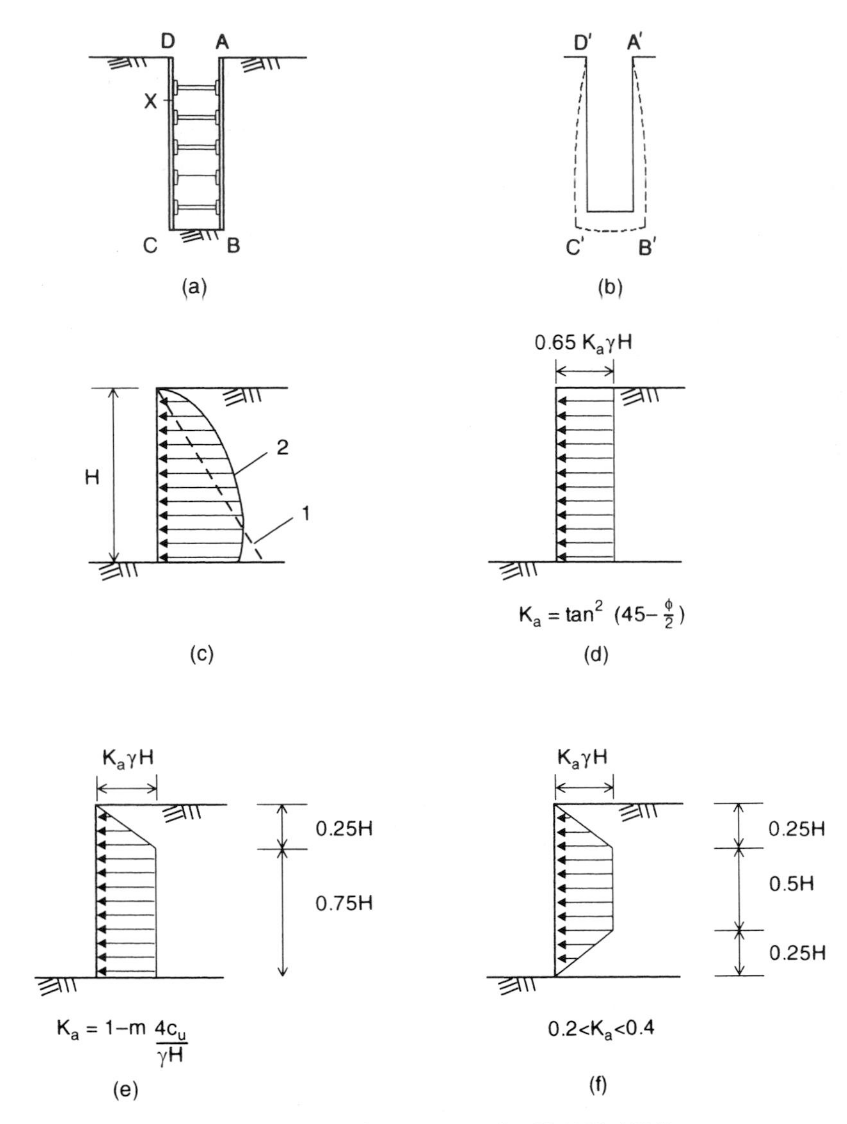

Fig. 6.27 Pressure distribution in strutted excavation (after BS 8002: 1994).

6.12 Passive pressure in cohesionless soils

6.12.1 Rankine's theory (soil surface horizontal)

In this case the vertical pressure due to the weight of the soil, γh, is acting as a minor principal stress. Figure 6.28a shows the Mohr circle diagram representing these stress conditions and drawn in the usual position, i.e. with the axis OX (the direction of the major principal plane) horizontal. Figure 6.28b shows the same diagram correctly orientated with the major principal stress, $K_p\gamma h$, horizontal and the major principal plane vertical. The Mohr diagram, it will be seen, must be rotated through 90°.

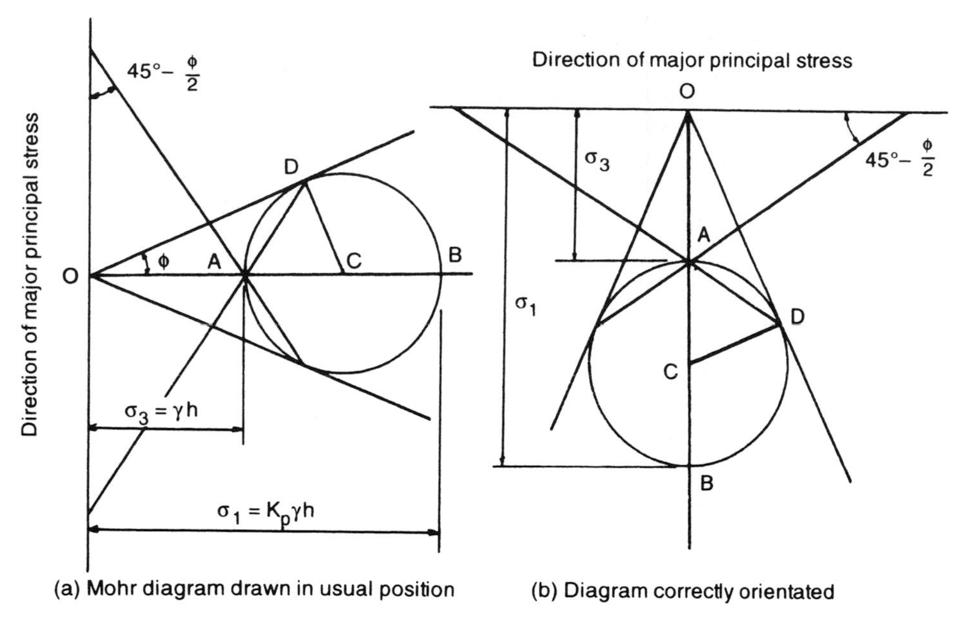

Fig. 6.28 Passive earth pressure for a cohesionless soil with a horizontal upper surface.

In the Mohr diagram:

$$\frac{\sigma_1}{\sigma_3} = \frac{\text{OB}}{\text{OA}} = \frac{\text{OC} + \text{DC}}{\text{OC} - \text{DB}} = \frac{1 + \sin\phi}{1 - \sin\phi} = \tan^2\left(45° + \frac{\phi}{2}\right)$$

hence

$$K_p = \frac{1 + \sin\phi}{1 - \sin\phi} = \tan^2\left(45° + \frac{\phi}{2}\right)$$

As with active pressure, there is a network of shear planes inclined at $(45° - \phi/2)$ to the direction of the major principal stress, but this time the soil is being compressed as opposed to expanded.

6.12.2 Rankine's theory (soil surface sloping at angle β)

The directions of the principal stresses are not known, but we assume that the passive pressure acts parallel to the surface of the slope. The analysis gives:

$$K_p = \cos\beta \cdot \frac{\cos\beta + \sqrt{(\cos^2\beta - \cos^2\phi)}}{\cos\beta - \sqrt{(\cos^2\beta - \cos^2\phi)}}$$

Note: The amount of friction developed between a retaining wall and the soil can be of a high magnitude (particularly in the case of passive pressure). The Rankine theory's assumption of a smooth wall with no frictional effects can therefore lead to a significant underestimation (up to about a half) of the true K_p value. The theory can obviously lead to conservative design which, although safe, might at times be over-safe and lead to an uneconomic structure.

6.12.3 The Coulomb theory

With the assumption of a plane failure surface leading to a wedge failure, Coulomb's expression for K_p for a granular soil is:

$$K_p = \left\{ \frac{\operatorname{cosec}\psi \sin(\psi - \phi)}{\sqrt{\sin(\psi - \delta)} - \sqrt{\left[\frac{\sin(\phi + \delta)\sin(\phi + \beta)}{\sin(\psi - \beta)}\right]}} \right\}^2$$

the symbols having the same meanings as previously.

The expression reduces to:

$$K_p = \frac{1 + \sin\phi}{1 - \sin\phi}$$

when $\psi = 90°$, $\delta = 0°$ and $\beta = 0°$.

With passive pressure, unfortunately, the failure surface only approximates to a plane surface when the angle of wall friction is small.

The situation arises because the behaviour of the soil is not only governed by its weight but also by the compression forces induced by the wall tending to push into the soil. These forces, unlike the active case, do not act on only one plane within the soil, resulting in a non-uniform strain pattern and the development of a curved failure surface (Fig. 6.29).

It is apparent that in most cases the assumption of a Coulomb wedge for a passive failure can lead to a serious overestimation of the resistance available. Terzaghi (1943) first analysed this problem and concluded that, provided the angle of friction developed between the soil and the wall is not more than $\phi/3$, where ϕ is the operative value of the angle of friction of the soil, the assumption of a plane failure surface

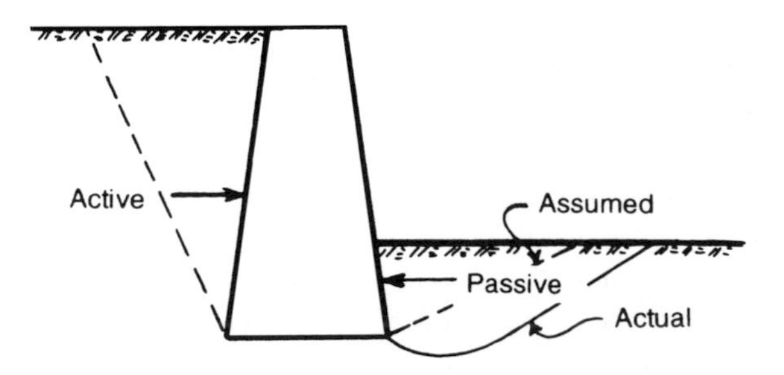

Fig. 6.29 Departure of passive failure surface from a plane.

Table 6.2 Values of K_p for cohesionless soils (Kerisel and Absi, 1990).

Values of δ	Values of ϕ			
	25°	30°	35°	40°
	Values of K_p			
0°	2.5	3.0	3.7	4.6
10°	3.1	4.0	4.8	6.5
20°	3.7	4.9	6.0	8.8
30°	–	5.8	7.3	11.4

generally gives reasonable results. For values of δ greater than $\phi/3$, the errors involved can be very large.

Adjusted values for K_p that allow for a curved failure surface are given in Table 6.2. These values apply to a vertical wall and a horizontal soil surface and include the multiplier cos δ as the values in the table give the components of pressure that will act normally to the wall.

It is seen therefore that for a smooth wall where $\delta = 0°$ the Rankine theory can be used for the evaluation of passive pressure. If wall friction is mobilised then $\delta \neq 0°$ and the coefficients of Table 6.2 should be used (unless $\delta \leq \phi/3$ in which case the Coulomb equation can be used directly).

6.13 The effect of cohesion on passive pressure

6.13.1 *The Rankine theory*

Rankine's theory was developed by Bell (1915) for the case of a frictional/cohesive soil. His solution for a soil with a horizontal surface is:

$$p_p = \gamma h \tan^2\left(45° + \frac{\phi}{2}\right) + 2c \tan\left(45° + \frac{\phi}{2}\right) = K_p \gamma h + 2c\sqrt{K_p}$$

6.13.2 *The Coulomb theory*

As has been discussed, a clay has a non-linear stress–strain relationship and its shear strength depends upon its previous stress history. Add to this the complications of non-uniform strain patterns within a passive resistance zone and it is obvious that any design approach must be an empirical approach based on experimental work.

A similar equation to that of Bell can be used for passive pressure values when the effect of wall friction and adhesion are taken into account.

Table 6.3 Values of K_p and K_{pc} for a cohesive soil for $\beta = 0$; $\psi = 90°$.

Coefficient	Values of δ	Values of c_w/c	Values of ϕ					
			0°	5°	10°	15°	20°	25°
K_p	0	All	1.0	1.2	1.4	1.7	2.1	2.5
	ϕ	values	1.0	1.3	1.6	2.2	2.9	3.9
K_{pc}	0	0	2.0	2.2	2.4	2.6	2.8	3.1
	0	$\frac{1}{2}$	2.4	2.6	2.9	3.2	3.5	3.8
	0	1	2.6	2.9	3.2	3.6	4.0	4.4
	ϕ	$\frac{1}{2}$	2.4	2.8	3.3	3.8	4.5	5.5
	ϕ	1	2.6	2.9	3.4	3.9	4.7	5.7

The passive pressure acting normally to the wall at a depth h can be defined as:

$$p_{ph} = K_p \gamma h + cK_{pc}$$

where

c = operating value of cohesion
K_{pc} = coefficient of passive earth pressure.

Various values of K_p and K_{pc} are given in Table 6.3 for the straightforward case of $\beta = 0$, $\psi = 90°$. As with the active pressure coefficients given in Table 6.1, they give the value of the pressure acting normally to the wall.

An alternative to using the values set out in Table 6.3 is that of the work of Sokolovski (1960), part of which is presented in Table 6.4. This offers a more realistic set of values than those listed in Tables 6.2 and 6.3.

The K_{pc} values were obtained from the approximate relationship:

$$K_{pc} = 2\sqrt{K_p}\left\{1 + \frac{c_w}{c}\right\}$$

Excel

This relationship has been used in the *earth pressure coefficients.xls* spreadsheet which can be used to determine intermediate values of K_p and K_{pc}.

It should be noted that, in the case of passive earth pressure, the amount of wall movement necessary to achieve the ultimate value of ϕ can be large, particularly in the case of a loose sand where one cannot reasonably expect that more than one half the value of the ultimate passive pressures will be developed.

The following design parameters are recommended for δ and c_w:

For timber, steel and precast concrete: $= \phi/2$
For cast *in situ* concrete: $= 2\phi/3$

where ϕ is the operative condition of the angle of friction of the soil.

Table 6.4 Values of K_p and K_{pc} (after Sokolovski, 1960).

Values of δ	Values of c_w/c	Values of ϕ (degrees)			
		10°	20°	30°	40°
		Values of K_p			
0	All values	1.42	2.04	3.00	4.60
$\phi/2$		1.55	2.51	4.46	9.10
ϕ		1.63	2.86	5.67	14.10
		Values of K_{pc}			
0	0	2.38	2.86	3.46	4.29
$\phi/2$	0	2.49	3.17	4.22	6.03
ϕ	0	2.55	3.38	4.76	7.51
0	0.5	2.92	3.50	4.24	5.25
$\phi/2$	0.5	3.05	3.88	5.17	7.39
ϕ	0.5	3.13	4.14	5.83	9.20
0	1.0	3.37	4.04	4.90	6.07
$\phi/2$	1.0	3.52	4.48	5.97	8.53
ϕ	1.0	3.61	4.78	6.73	10.62

Generally, c_w should be assumed to be half of the value for the active pressure conditions.

6.14 Operative values for ϕ and c for passive pressure

Granular soils

It is generally agreed that, for passive pressures in a granular soil, the operative value of ϕ is lower than ϕ_t, the peak triaxial angle obtained from drained tests, particularly for high values of ϕ_t.

With a granular soil ϕ_t is most often estimated from the results of some *in situ* test such as the standard penetration test. It is suggested therefore that values for ϕ, to be used in the determination of passive pressure values, can be obtained from Fig. 6.30 (which is a modified form of Fig. 3.34). The corrected N′ value can be used in place of the direct blow count N.

BS 8002: 1994 states that the design value of soil strength be taken as the lower value of the peak soil strength reduced by a mobilisation factor, and the critical state strength (see Section 7.4.1). For granular soils this refers to the tangent of the angle of shearing resistance, ϕ. Eurocode 7 does not give particular guidance on the operative value of ϕ to adopt but does recommend that the wall ground interface parameter, δ,

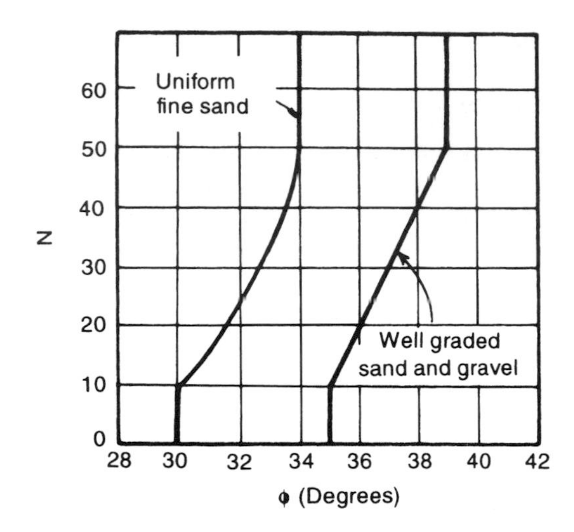

Fig. 6.30 Relationship between N and ϕ.

be determined from the design critical state angle of shearing resistance, $\phi_{cv;d}$ for concrete or steel sheet pile walls.

Normally consolidated clays

As with the active state, this type of clay is at its weakest when in its undrained state, i.e. during and immediately after construction. For a normally consolidated clay the operative strength parameters are $c = c_u$ and $\phi = 0°$.

Overconsolidated clays

With this soil its weakest strength occurs once the soil has reached its drained state. The operative parameters are therefore $c = 0$ and $\phi = \phi'$, although this is an over-simplification for the case when the level of soil in front of the wall has been reduced by excavation. In this instance there will be a relief of overburden pressure which could result in softening occurring within the soil. When this happens some estimation of the strength reduction of the soil must be made, possibly by shear tests on samples of the softened soil.

Silts

As for active pressure, the passive resistance of a silt can be estimated either from the results of *in situ* penetration tests or from a drained triaxial test. For passive pressure it is best to take the parameters to be $c = 0$ and $\phi = \phi_t$.

Exercises

Exercise 6.1

A 6 m high retaining wall with a smooth vertical back retains a mass of dry cohesionless soil that has a horizontal surface level with the top of the wall and carries a uniformly distributed load of 10 kPa. The soil weighs 20 kN/m^3 and has an angle of internal friction of 36°.

Determine the active thrust on the back of the wall per metre length of wall (i) without the uniform surcharge and (ii) with the surcharge.

Answers (i) 93.6 kN; (ii) 109.2 kN

Exercise 6.2

The back of a 10.7 m high wall slopes away from the soil it retains at an angle of 10° to the vertical. The surface of the soil slopes up from the top of the wall at a surcharge angle of 20°. The soil is cohesionless with a density of 17.6 kN/m^3 and $\phi' = 33°$.

If the angle of wall friction, $\delta' = 19°$ determine the maximum thrust on the wall: (a) graphically and (b) analytically using the Coulomb theory.

Answer (a) and (b) $P_a = 476.3$ kN

Exercise 6.3

The soil profile acting against the back of a retaining wall is shown in Fig. 6.31. Assuming that Rankine's conditions apply, determine both the theoretical lateral earth pressure distribution and the distribution that would be used in design, and sketch the two pressure distribution diagrams.

From the pressure distribution that would be used in design, determine the magnitude of the total thrust that acts on the wall.

Determine the point of application of the total thrust, and express it as the distance from the base of the wall.

Answer P = 183.6 kN; x = 2.56 m above base.

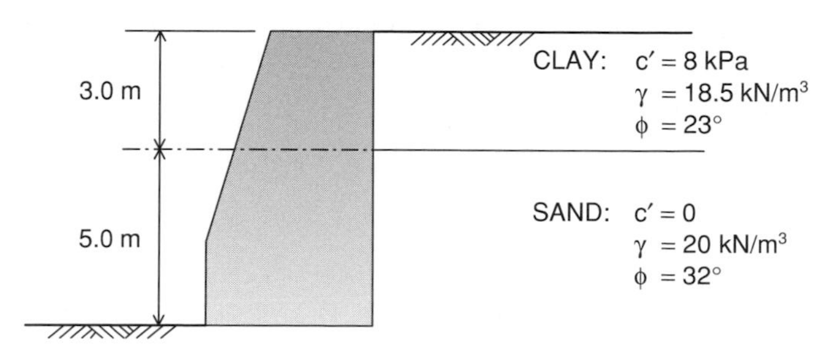

Fig. 6.31 Exercise 6.3.

Exercise 6.4

A soil has the following properties: $\gamma = 18$ kN/m^3, $\phi' = 30°$, $c' = 5$ kPa. The soil is retained behind a 6 m high vertical wall and has a horizontal surface level with the top of the wall.

If $c'_w = 5$ kPa and $\delta' = 15°$ determine the total active horizontal thrust acting on the back of the wall:

(i) with no surcharge acting on the retained soil;
(ii) when the surface of the soil is subjected to a vertical uniformly distributed pressure of 30 kPa.

(Use the values of K_a and K_{ac} from Table 6.1)

Answer (i) 59.5 kN; (ii) 100.8 kN ($K_a = 0.29$ and $K_{ac} = 1.53$)

Chapter 7
Earth Retaining Structures

7.1 Main types of earth retaining structures

Various types of earth retaining structures are used in civil engineering, the main ones being:

- mass construction gravity walls;
- reinforced concrete walls;
- crib walls;
- gabion walls;
- sheet pile walls;
- diaphragm walls;
- reinforced soil walls;
- anchored earth walls.

The last two structures are different from the rest in that the soil itself forms part of these structures. Because of this fundamental difference reinforced soil and anchored earth walls are discussed separately at the end of this chapter.

Earth retaining structures are commonly used to support soils and structures to maintain a difference in elevation of the ground surface and are normally grouped into gravity walls or embedded walls.

7.2 Gravity walls

7.2.1 Mass construction gravity walls

This type of wall depends upon its weight for its stability and is built of such a thickness that the overturning effect of the lateral earth pressure that it is subjected to does not induce tensile stresses within it.

The walls are built in mass concrete or cemented precast concrete blocks, brick, stone, etc. and are generally used for low walls; they become uneconomic for high walls.

The cross-section of the wall is trapezoidal with a base width between 0.3 and 0.5 h, where h = the height of the wall. This base width includes any projections of the heel or toe of the wall, which are usually not more than 0.25 m each and are intended to reduce the bearing pressure between the base of the wall and the supporting soil. If the wall is built of concrete then its width at the top should be not less than 0.2 m, and preferably 0.3 m, to allow for the proper placement of the concrete.

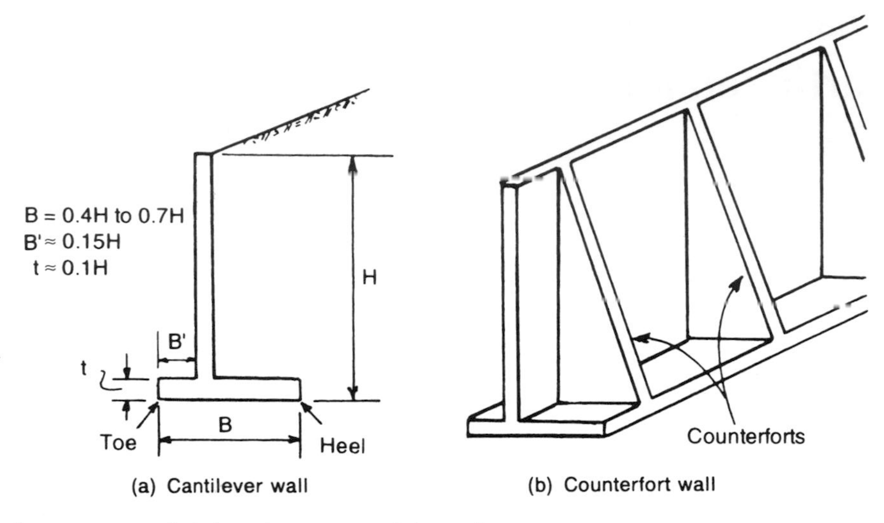

Fig. 7.1 Types of reinforced concrete retaining walls.

7.2.2 Reinforced concrete walls

Cantilever wall

This wall has a vertical, or inclined, stem monolithic with a base slab and is suitable for heights up to about 7 m. Typical dimensions for the wall are given in Fig. 7.1. Its slenderness is possible as the tensile stresses within its stem and base are resisted by steel reinforcement. If the face of the wall is to be exposed then general practice is to provide it with a small backward batter of about 1 in 50 in order to compensate for any slight forward tilting of the wall.

Relieving platforms

A retaining wall is subjected to both shear and bending stresses caused by the lateral pressures induced from the soil that it is supporting. A mass construction gravity wall can take such stresses in its stride but this is not so for the vertical stem of a reinforced concrete retaining wall. If structural failure of the stem is to be avoided then it must be provided with enough steel reinforcement to resist the bending moment and to have a sufficient thickness to withstand the shear stresses, for all sections throughout its height.

It is this situation that imposes a practical height limitation of about 7 m on the wall stem of a conventional retaining wall. As a wall is increased in dimensions it becomes less flexible and the lateral pressures exerted on it by the soil will tend to be higher than the active values assumed in the design. It is possible therefore to enter a sort of upwards spiral – if a wall is strengthened to withstand increased lateral pressures then its rigidity is increased and the lateral pressures are increased – and so on.

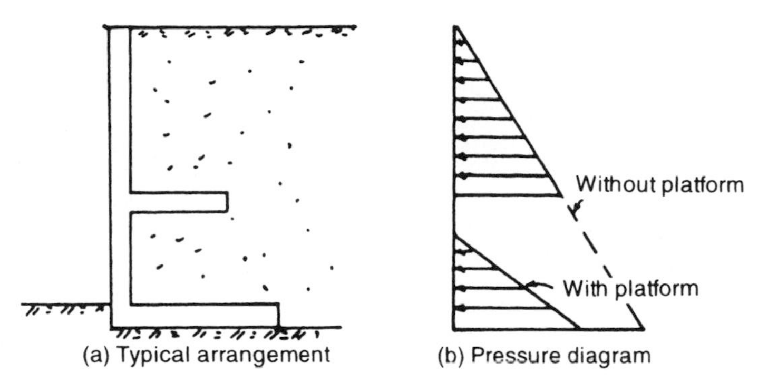

Fig. 7.2 Moment relief platforms (after Tsagareli, 1967).

A way out of the problem was suggested by Tsagareli (1967) who proposed the provision of one or more horizontal concrete slabs, or platforms, placed within the backfill and rigidly connected to the wall stem. A platform carries the weight of the material above it (up as far as the next platform if there are more than one). This vertical force exerts a cantilever moment on to the back of the wall in the opposite direction to the bending moment caused by the lateral soil pressure. The resulting bending moment diagram becomes a series of steps and the wall is subjected to a maximum bending moment value that is considerably less than the value when there are no platforms (Fig. 7.2).

With the reduction of bending moment values to a manageable level the wall stem can be kept slim enough for the assumption of active pressure values to be realistic, with a consequential more economical construction.

Counterfort wall

This wall can be used for heights greater than about 6 m. Its wall stem acts as a slab spanning between the counterfort supports which are usually spaced at about 0.67 H but not less than 2.5 m, because of construction considerations. Details of the wall are given in Fig. 7.1b.

A form of the counterfort wall is the buttressed wall where the counterforts are built on the face of the wall and not within the backfill. There can be occasions when such a wall is useful but, because of the exposed buttresses, it can become unsightly and is not very popular.

7.2.3 Crib walls

Details of the wall are shown in Fig. 7.3a. It consists of a series of pens made up from prefabricated timber, precast concrete or steel members which are filled with granular soil. It acts like a mass construction gravity wall with the advantage of quick erection and, due to its flexible nature, the ability to withstand relatively large differential settlements. A crib wall is usually tilted so that its face has a batter of

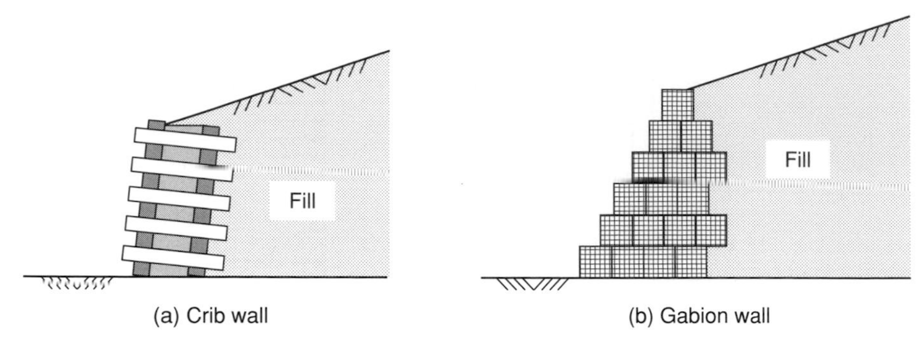

Fig. 7.3 Crib and gabion walls.

about 1 in 6. The width of the wall can vary from 0.5 H to 1.0 H and the wall is suitable for heights up to about 6.5 m. It is important to note that, apart from earth fill, a crib wall should not be subjected to surcharge loadings.

7.2.4 Gabion walls

A gabion wall is built of cuboid metal cages or baskets made up from a square grid of steel fabric, usually 5 mm in diameter and spaced 75 mm apart. These baskets are usually 2 m long and 1 m^2 in cross-section, filled with stone particles. A central diaphragm fitted in each metal basket divides it into two equal 1 m^3 sections and adds stability. During construction the stone-filled baskets are secured together with steel wire some 2.5 mm in diameter. The base of a gabion wall is usually about 0.5 H, and a typical wall is illustrated in Fig. 7.3b. It is seen that a front face batter can be provided by slightly stepping back each succeeding layer.

7.3 Embedded walls

Embedded walls rely on the passive resistance of the soil in front of the lower part of the wall to provide stability. Anchors or props, where incorporated, provide additional support.

7.3.1 Sheet pile walls

These walls are made up from a series of interlocking piles individually driven into the foundation soil. Most modern sheet pile walls are made of steel but earlier walls were also made from timber or precast concrete sections and may still be encountered. There are two main types of sheet pile walls: cantilever and anchored.

Cantilever wall

This wall is held in the ground by the active and passive pressures that act on its lower part (Fig. 7.8).

Anchored wall

This wall is fixed at its base, as is the cantilever wall, but it is also supported by a row, or two rows, of ties or struts placed near its top (Fig. 7.11).

7.3.2 Diaphragm walls

A diaphragm wall could be classed either as a reinforced concrete wall or as a sheet pile wall but it really merits its own classification. It consists of a vertical reinforced concrete slab fixed in position in the same manner as a sheet pile in that the lower section is held in place by the active and passive soil pressures that act upon it.

A diaphragm wall is constructed by a machine digging a trench in panels of limited length, filled with bentonite slurry as the digging proceeds to the required depth. This slurry has thixotropic properties, i.e. it forms into a gel when left undisturbed but becomes a liquid when disturbed. There is no penetration of the slurry into clays, and in sands and silts water from the bentonite slurry initially penetrates into the soil and creates a virtually impervious skin of bentonite particles, only a few millimetres thick, on the sides of the trench. The reason for the slurry is that it creates lateral pressures which act on the sides of the short trench panel and thus prevent collapse. When excavation is complete the required steel reinforcement is lowered into position. The trench is then filled with concrete by means of a tremie pipe, the displaced slurry being collected for cleaning and further use.

The wall is constructed in alternating short panel lengths. When the concrete has developed sufficient strength, the remaining intermediate panels are excavated and constructed to complete the wall. The length of each panel is limited to the amount that the soil will arch, in a horizontal direction, to support the ground until the concrete has been placed.

The various construction stages are shown in a simplified form in Fig. 7.4.

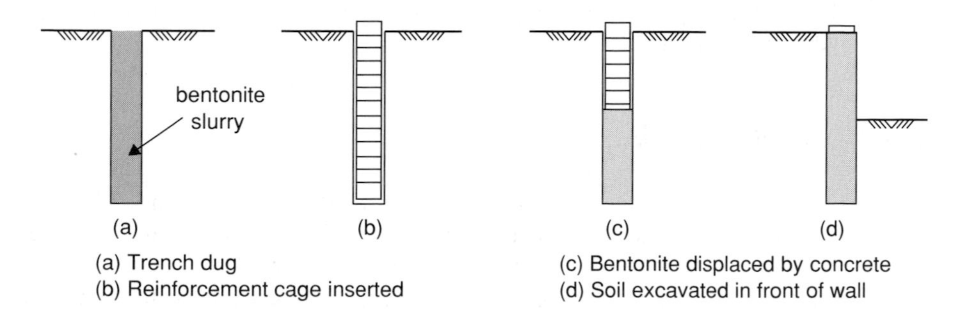

Fig. 7.4 The construction stages of a diaphragm wall.

7.3.3 Contiguous and secant bored pile walls

Contiguous bored pile walls

This type of wall is constructed from a single or double row of piles placed beside each other. Alternate piles are cast first and the intermediate piles are then installed. The construction technique allows gaps to be left between piles which can permit inflow of water in granular conditions. The secant bored pile wall offers a watertight alternative.

Secant bored pile walls

The construction technique is similar to that of the contiguous bored pile wall except that the alternate piles are drilled at a closer spacing. Then, while the concrete is still green, the intermediate holes are drilled along a slightly offset line so that the holes cut into the first piles. These holes are then concreted to create a watertight continuous wall.

7.4 Design of earth retaining structures

The traditional approach for the design of earth retaining structures involved establishing the ratio of the restoring moment (or force) to the disturbing moment (or force) and declaring this ratio as a factor of safety. This factor had to be high enough to allow for any uncertainties in the soil parameters used in the analysis, and the approach was generally referred to as the *factor of safety* or *gross pressure* approach. An alternative approach now becoming widely adopted is the *limit state* design approach. This method is advocated in both BS 8002: 1994 *Code of practice for earth retaining structures* and in Eurocode 7.

7.4.1 Design to BS 8002: 1994

In BS 8002: 1994 the design values of loads are intended to be the most pessimistic and unfavourable, whether derived by factoring or otherwise. To satisfy the requirements of both the ultimate and serviceability limit states, the design soil strength values are obtained from consideration of the representative values for peak and ultimate strength. The design values are taken as the lower of:

(a) the soil strength mobilised at a strain acceptable for serviceability: this can be expressed as the peak strength reduced by a mobilisation factor, M;
(b) the soil strength which would be mobilised at collapse, following significant ground movements: this can generally be taken as the critical state strength.

The value of the mobilisation factor, M, depends on whether the design is concerned with undrained or drained conditions. For undrained conditions, the design clay strength (design c_u) is taken as the representative undrained strength divided by a value of M not less than 1.5, if the wall displacement is not to exceed 0.5 per cent of the wall height.

For drained conditions, the lesser of two values of soil strength should be used:

(a) the representative peak strength divided by a value of M = 1.2, i.e.

$$\text{Design } \tan \phi' = \frac{\text{Representative } \tan \phi'_{max}}{M}$$

$$\text{Design } c' = \frac{\text{Representative } c'}{M}$$

(b) the representative critical state strength.

Again, this should ensure a maximum wall displacement of 0.5 per cent wall height, for non-soft and non-loose soils.

When considering the design values of wall friction, δ, and undrained wall adhesion, c_w, BS 8002 recommends that the design value be the lesser of the representative value determined by test, or 75 per cent of the design shear strength to be actually mobilised in the soil.

$$\text{Design } \tan \delta = 0.75 \times \text{Design } \tan \phi'$$
$$\text{Design } c_w = 0.75 \times \text{Design } c_u$$

BS 8002 also provides a design recommendation for unplanned future excavation in front of the wall. This really only applies to embedded walls and is considered as 10 per cent of the clear height retained, up to a maximum depth of 0.5 m. For cantilever walls the clear height is equal to the height of the excavation, and for propped walls is equal to the height below the bottom prop. The recommendation provides for unforeseen and accidental events after construction. The code also recommends a minimum design surcharge of 10 kPa to be applied to the design of walls retaining 3 m or more of soil. For walls of lower height, the designer may adopt a lower value of surcharge.

The design values of actions and ground properties are adopted in the analysis and the conformity of a particular limit state is checked for, by ensuring that the magnitude of the restoring moment (or force) is greater than the disturbing moment (or force), see Example 7.2. Examples of the limit states to be checked are listed in Section 7.5.1.

7.4.2 Geotechnical design to Eurocode 7

The Eurocode Programme was initiated to establish a set of harmonised technical rules for the design of building and civil engineering works across Europe. The rules are known collectively as the Structural Eurocodes which comprise a series of ten design documents. Eurocode 7 is the document that concerns geotechnical design and its introduction has been anticipated across Europe for a number of years. Now, with the publication in 2004 of Part 1 of the Standard: *Eurocode 7: Geotechnical design – Part 1: General rules* (BSI, 2004), the period of anticipation is over and routine geotechnical design in the UK and other EU member states can accordingly follow the procedures set out in this new design standard. Full implementation of the Standard is to be adopted across Europe in 2010. Until then, national design codes may continue to be used alongside Eurocode 7. In the UK this means that the British Standard codes for geotechnical design will continue to be used until 2010, and thereafter Eurocode 7 will be used. That said, it is likely that these British Standards will continue to be used beyond 2010 to aid designers already used to these documents. However, to ensure compatibility with Eurocode 7, revisions to the British Standards may be necessary to remove any conflicting clauses.

Eurocode 7 is published in two parts, and Part 2, *Eurocode 7: Geotechnical design – Part 2: Design assisted by field and laboratory testing*, is due for publication in 2006. Only Part 1 of the Standard is covered in this book.

Using Eurocode 7

The design philosophy adopted in Eurocode 7 is the same as that adopted in all the Eurocodes and advocates the use of limit state design to ensure that the serviceability limit states are not exceeded. *Serviceability limit states* are those states that, if exceeded, render the structure unsafe even though no collapse situation is reached, such as excessive deflection, settlement or rotation. In contrast to the traditional method of the use of lumped factors of safety, the Standard promotes the use of *partial factors of safety* and thus reflects a significant shift from current UK geotechnical design practice. Where once the design methods treated material properties and loads in an unmodified state and applied a factor of safety at the end of the design process to allow for the uncertainty in the unmodified values, Eurocode 7 guides the designer to modify each parameter early in the design by use of the partial factor of safety. This approach sees the *representative* or *characteristic* value of the parameters (e.g. loads, soil strength parameters) converted to the *design* value by combining it with the particular partial factor of safety for that parameter. Worked examples in this chapter, and also in Chapters 5 and 8, will help the student to follow this new approach to design. Additional explanation of the use of Eurocode 7 is given by Frank *et al.* (2004) and Bright *et al.* (2004).

The clauses throughout Eurocode 7 are considered as either *Principles* (identified by the letter P immediately preceding the clause) or *Application Rules*. Principles are unique statements or definitions that must be adopted. Application Rules offer examples of how to ensure that the Principles are adhered to and thus offer guidance to the designer in following the Principles.

The Standard states that the limit states should be verified by one of four means: by (1) *calculation*, (2) *prescriptive measures*, (3) *experimental models and load tests*, or (4) *an observational method*. In this book we shall concentrate solely on geotechnical design by calculation.

To facilitate an appropriate design, projects are considered as falling into one of three *Geotechnical Categories*, based on the complexity of the geotechnical design together with the associated risks. *Category 1* is for small projects with negligible risk, *Category 2* is for conventional structures (e.g. foundations, retaining walls, embankments) and *Category 3* is for structures not covered by Categories 1 and 2. It is obvious that most routine geotechnical design work will fall into Geotechnical Category 2.

To enable the limit states to be checked, the *design* values of the geotechnical parameters, the ground resistance and the actions (e.g. forces or loads), must be determined. The design values of actions (F_d) are derived by multiplying the *representative* values (F_{rep}) by the appropriate partial factor of safety, γ_F. The design values of geotechnical parameters (X_d) are derived by dividing the *characteristic* values (X_k) by the appropriate partial factor of safety, γ_M. The resistance (R) is derived from the design values of actions and ground parameters. The design resistance (R_d) can either be taken as equal to R or as equal to a reduced value of R, which is derived by dividing by an additional partial factor of safety, γ_R. The choice of which partial factor of safety to use is governed by the nature of the action and by the *design approach* being used. Actions are classified as permanent (G) (either *favourable* or *unfavourable*), variable (Q), accidental (A) or seismic (AE). The 'effects' of actions are also considered in the design.

Eurocode 7 lists five limit states to be considered in the design process:

EQU: the loss of equilibrium of the structure or the supporting ground when considered as a rigid body and where the internal strengths of the structure and the ground do not provide resistance (e.g. Fig. 7.5a). This limit state is satisfied if the sum of the

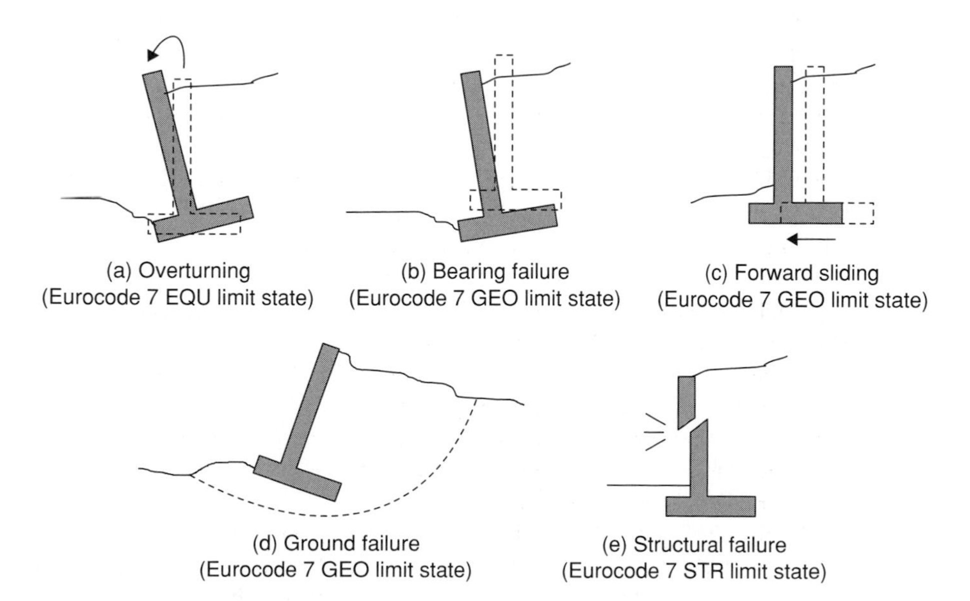

Fig. 7.5 Limit states for earth retaining structures (for both BS 8002 and Eurocode 7).

design values of the effects of destabilising actions ($E_{dst;d}$) is less than or equal to the sum of the design values of the effects of the stabilising actions ($E_{stb;d}$) together with any contribution through the resistance of the ground around the structure (T_d), i.e. $E_{dst;d} \leq E_{stb;d} + T_d$. In most cases, the contribution to stability from the resistance of the ground around the structure will be minimal so T_d will be taken as zero.

GEO: failure or excessive deformation of the ground, where the soil or rock is significant in providing resistance (e.g. Figs 7.5b, 7.5c and 7.5d). This limit state is satisfied if the design effect of the actions (E_d) is less than or equal to the design resistance (R_d), i.e. $E_d \leq R_d$.

STR: failure or excessive deformation of the structure, where the strength of the structural material is significant in providing resistance (e.g. Fig. 7.5e). As with the GEO limit state, the STR is satisfied if the design effect of the actions (E_d) is less than or equal to the design resistance (R_d), i.e. $E_d \leq R_d$.

UPL: the loss of equilibrium of the structure or the supporting ground by vertical uplift due to water pressures (buoyancy) or other actions. This limit state is verified by checking that the sum of the design permanent and variable destabilising vertical actions ($V_{dst;d}$) is less than or equal to the sum of the design stabilising permanent vertical action ($G_{stb;d}$) and any additional resistance to uplift (R_d), i.e. $V_{dst;d} \leq G_{stb;d} + R_d$.

HYD: hydraulic heave, internal erosion and piping in the ground as might be experienced, for example, at the base of a braced excavation. This limit state is verified by checking that the design total pore water pressure ($u_{dst;d}$) or seepage force ($S_{dst;d}$) at the base of the soil column under investigation is less than or equal to the total vertical stress ($\sigma_{stb;d}$) at the bottom of the column, or the submerged unit weight ($G'_{stb;d}$) of the same column, i.e. $u_{dst;d} \leq \sigma_{stb;d}$ or $S_{dst;d} \leq G'_{stb;d}$ (see Example 2.6).

The EQU, GEO and STR limit states are the most likely ones to be considered for routine design. Furthermore, in the design of retaining walls and foundations it is likely that limit state GEO will be the prevalent state for determining the size of the structural elements.

When checking the GEO and STR limit state requirements, one of three design approaches is used: *Design Approach 1*, *Design Approach 2* or *Design Approach 3*. This choice of three approaches reflects the Europe-wide adoption of the Standard and offers designers in different nations an approach most relevant to their needs. Design Approach 1 is the most likely method to be adopted in the UK and Examples 7.1–7.4 illustrate the use of this method in the design of retaining structures.

As mentioned earlier, the choice of partial factors to be used is dependent on the design approach being followed (for the GEO and STR limit states). For each design approach, a different combination of partial factor sets is used to verify the limit state. For Design Approach 1 (for retaining walls and shallow footings), two combinations are available and the designer would normally check the limit state using each combination, except on occasions where it is obvious that one combination will govern the design. (The combination of partial factor sets is different for pile foundations – see Chapter 8.)

Design Approach 1:	Combination 1: A1 + M1 + R1
	Combination 2: A2 + M2 + R1
Design Approach 2:	A1 + M1 + R2
Design Approach 3:	A* + M2 + R3

(*Note*. A*: use set A1 on structural actions, set A2 on geotechnical actions.)

The sets for actions (denoted by A), material properties (denoted by M) and ground resistance (denoted by R) for each design approach are given in Table 7.1. Also given in the table are the partial factors for the EQU limit state.

Earth pressure coefficients

During the design of retaining walls it is often appropriate to use Rankine's K_a and K_p, such as in the case of cantilever gravity walls (see Example 7.2). However, when Rankine's conditions do not apply, Eurocode 7 provides a set of charts that may be used to determine K_a and K_p for a given δ/ϕ' ratio. Charts for both horizontal and inclined retained surfaces are given, and the chart to determine K_a for a horizontal ground surface behind the wall is redrawn in Fig. 7.6.

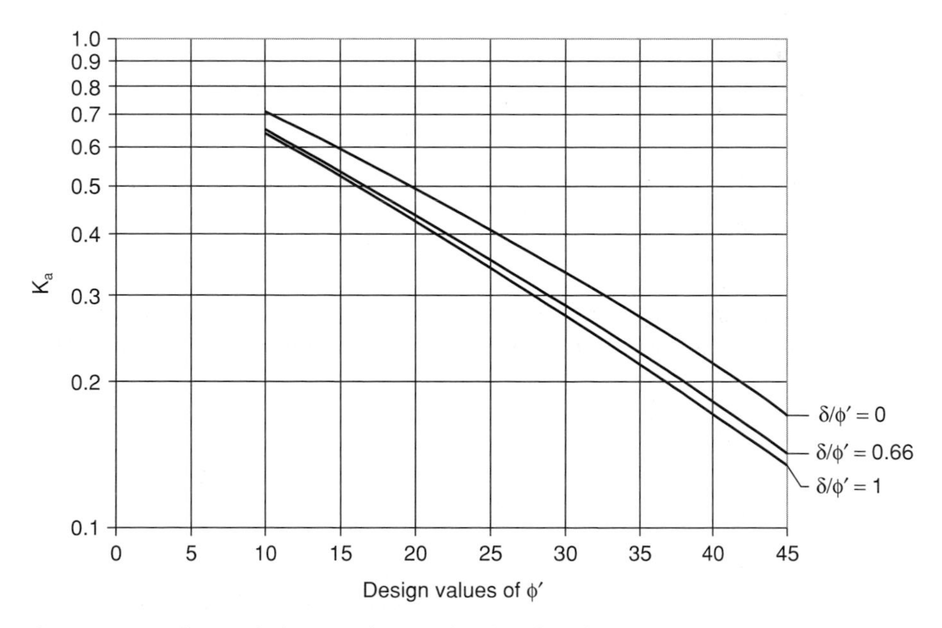

Fig. 7.6 K_a coefficients for horizontal retained surface (based on Fig. C.1.1 in Eurocode 7).

Table 7.1 Partial factor sets for EQU, GEO and STR limit states.

Parameter		Symbol	EQU	GEO/STR – Partial factor sets						
				A1	A2	M1	M2	R1	R2	R3
Permanent action (G)	Unfavourable	$\gamma_{G;dst}$	1.1	1.35	1.0					
	Favourable	$\gamma_{G;stb}$	0.9	1.0	1.0					
Variable action (Q)	Unfavourable	$\gamma_{Q;dst}$	1.5	1.5	1.3					
	Favourable	–	–	–	–					
Accidental action (A)	Unfavourable	$\gamma_{A;dst}$	1.0	1.0	1.0					
	Favourable	–	–	–	–					
Coefficient of shearing resistance (tan ϕ')		$\gamma_{\phi'}$	1.25			1.0	1.25			
Effective cohesion (c')		$\gamma_{c'}$	1.25			1.0	1.25			
Undrained shear strength (c_u)		γ_{cu}	1.4			1.0	1.4			
Unconfined compressive strength (q_u)		γ_{qu}	1.4			1.0	1.4			
Weight density (γ)		γ_{γ}	1.0			1.0	1.0			
Bearing resistance (R_v)		γ_{Rv}						1.0	1.4	1.0
Sliding resistance (R_h)		γ_{Rh}						1.0	1.1	1.0
Earth resistance (R_e)		γ_{Re}						1.0	1.4	1.0

Note: weight density ≡ unit weight.

7.5 Design of gravity walls

7.5.1 Limit states

Whether following BS 8002 or Eurocode 7, a number of limit states should be considered:

(1) Overturning (Fig. 7.5a). For a wall to be stable the resultant thrust must be within the base. Most walls are so designed that the thrust is within the middle third of the base.
(2) Bearing failure of the soil beneath the structure (Fig. 7.5b). The overturning moment from the earth's thrust causes high bearing pressures at the toe of the wall. These values must be kept within safe limits – usually not more than one-third of the supporting soil's ultimate bearing capacity.
(3) Forward sliding (Fig. 7.5c). Caused by insufficient base friction or lack of passive resistance in front of the wall.
(4) Slip of the surrounding soil (Fig. 7.5d). This effect can occur in cohesive soils and can be analysed as for a slope stability problem.
(5) Structural failure caused by faulty design, poor workmanship, deterioration of materials, etc. (Fig. 7.5e).
(6) Excessive deformation of the wall or ground such that adjacent structures or services reach their ultimate limit state.
(7) Unfavourable seepage effects and the adequacy of any drainage system provided.

7.5.2 Bearing pressures on soil

The resultant of the forces due to the pressure of the soil retained and the weight of the wall subject the foundation to both direct and bending effects.

Let R be the resultant force on the foundation, per unit length, and let R_v be its vertical component (Fig. 7.7a). Considering unit length of wall:

$$\text{Section modulus of foundation} = \frac{B^2}{6}$$

$$\begin{aligned}\text{Maximum pressure on base} &= \text{Direct pressure} + \text{pressure due to bending}\\ &= \frac{R_v}{B} + \frac{6R_v e}{B^2}\\ &= \frac{R_v}{B}\left(1 + \frac{6e}{B}\right)\end{aligned}$$

$$\text{Minimum pressure on base} = \frac{R_v}{B}\left(1 - \frac{6e}{B}\right)$$

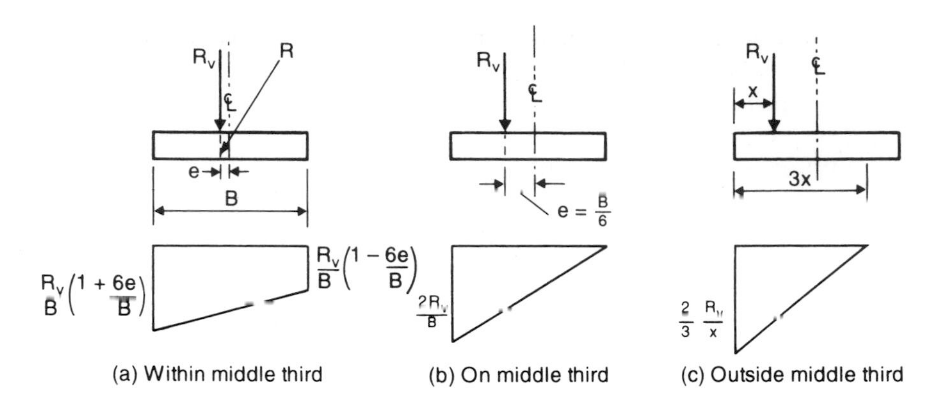

Fig. 7.7 Bearing pressures due to a retaining wall foundation.

The formulae only apply when R_v is within the middle third; when R_v is on the middle third (Fig. 7.7b), then

$$e = \frac{B}{6}$$

$$\Rightarrow \quad \text{Maximum pressure} = \frac{2R_v}{B}, \quad \text{Minimum pressure} = 0$$

If the resultant R lies outside the middle third (Fig. 7.7c) the formulae become:

$$\text{Maximum pressure} = \frac{2R_v}{3x}; \quad \text{Minimum pressure} = 0$$

7.5.3 Base resistance to sliding

Granular soils and drained clays

The base resistance to sliding is equal to $R_v \tan \delta$ where δ is the angle of friction between the base of the wall and its supporting soil, and R_v is the vertical reaction on the wall base. In limit state design, the sliding limit state will be satisfied if the base resistance to sliding is greater than, or equal to, R_h, the horizontal component of the resultant force acting on the base. In the factor of safety approach, the ratio $(R_v \tan \delta)/R_h$ is determined to establish the factor of safety against sliding. It is common practice to take the passive resistance from any soil in front of a gravity wall as equal to zero, since this soil will be small in depth and in a disturbed state following construction of the wall.

(In the case of a drained clay any value of effective cohesion, c'_w will be so small that it is best ignored.)

Undrained clays

The adhesion between the supporting soil and the base of a gravity or reinforced concrete wall can be taken as equal to the value c_w used in the determination of the active pressure values and based on the value of c_u.

Resistance to sliding = $c_w \times$ Area of base of wall

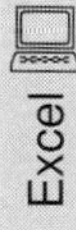

Example 7.1

Check the proposed design of the mass concrete retaining wall shown below using Eurocode 7. The wall is to be cast into the foundation soil to a depth of 1.0 m and will retain granular fill to a height of 4 m as shown. Take the unit weight of concrete as $\gamma_c = 24$ kN/m^3 and ignore any passive resistance from the soil in front of the wall. Check the overturning (EQU) and sliding (GEO) limit states (using Design Approach 1).

Solution

EQU limit state:
From Table 7.1, we obtain the partial factors:

$$\gamma_{G;dst} = 1.1;\ \gamma_{G;stb} = 0.9;\ \gamma_{Q;dst} = 1.5;\ \gamma_{\phi'} = 1.25$$

First, we determine the design material properties and the design actions:

(i) Design material properties

Retained fill:

$$\phi'_d = \tan^{-1}\left(\frac{\tan \phi'}{\gamma_{\phi'}}\right) = \tan^{-1}\left(\frac{\tan 32°}{1.25}\right) = 26.6°$$

Eurocode 7 states that for concrete walls cast into the soil, δ should be taken as equal to the design value of ϕ, i.e. $\delta/\phi'_d = 1$. From Fig. 7.6, $K_a = 0.31$.

Foundation soil:

$$\phi'_d = \tan^{-1}\left(\frac{\tan \phi'}{\gamma_{\phi'}}\right) = \tan^{-1}\left(\frac{\tan 28°}{1.25}\right) = 23°$$

From Fig. 7.6, $K_a = 0.37$.

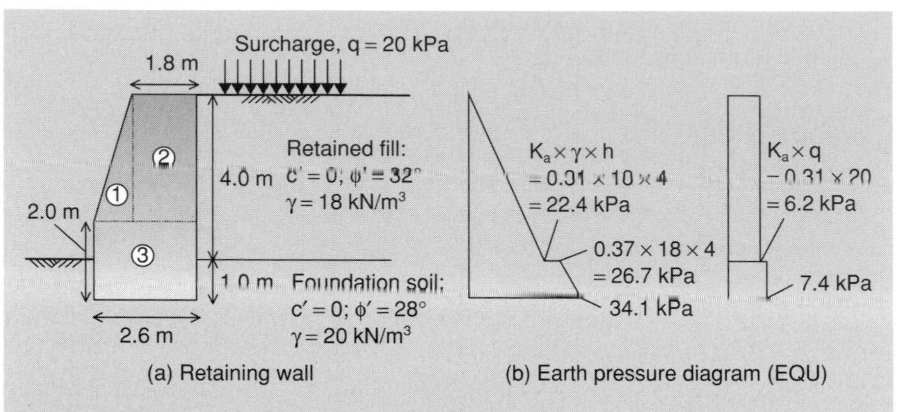

Fig. 7.8 Example 7.1.

(ii) Design actions

The self-weight of the wall is a permanent, favourable action. Consider the wall as comprising three areas as indicated in Fig. 7.8a. The design weight of each area is determined as follows:

Area 1: $G_{W1;d} = \frac{1}{2} \times 0.8 \times 3 \times \gamma_{concrete} \times \gamma_{G;stb} = 1.2 \times 24 \times 0.9 = 25.9$ kN
Area 2: $G_{W2;d} = 1.8 \times 3 \times \gamma_{concrete} \times \gamma_{G;stb} = 5.4 \times 24 \times 0.9 = 116.6$ kN
Area 3: $G_{W3;d} = 2.6 \times 2 \times \gamma_{concrete} \times \gamma_{G;stb} = 5.2 \times 24 \times 0.9 = 112.3$ kN

The thrust from the active earth pressure behind the wall is a permanent, unfavourable action. Values of the active earth pressure are indicated in Fig. 7.8b.

$P_{a;d}$ (fill) $= \frac{1}{2} \times 22.4 \times 4 \times \gamma_{G;dst} = 49.3$ kN
$P_{a;d}$ (foundation soil) $= \frac{1}{2} \times (26.7 + 34.1) \times 1.0 \times \gamma_{G;dst} = 33.4$ kN

The lateral thrust from the surcharge is a variable, unfavourable action.

$P_{q;d}$ (fill) $= 6.2 \times 4 \times \gamma_{Q;dst} = 37.2$ kN
$P_{q;d}$ (foundation soil) $= 7.4 \times 1.0 \times \gamma_{Q;dst} = 11.1$ kN

(iii) Design effect of actions and design resistance

The effect of the actions is to cause the overturning moment about the toe of the wall. This is resisted by the stabilising moment from the self-weight of the wall.

Action	Magnitude of action (kN)	Lever arm (m)	Moment (kNm)
Stabilising			
Area 1	25.9	$\frac{2}{3} \times 0.8 = 0.53$	13.7
Area 2	116.6	$0.8 + \frac{1.8}{2} = 1.7$	198.2
Area 3	112.3	$\frac{2.6}{2} = 1.3$	146.0
		Total:	357.9
Destabilising			
P_a (fill)	49.3	$1 + \frac{4}{3} = 2.33$	115.0
P_a (foundation soil)	33.4	$\frac{1.0(2 \times 26.7 + 34.1)}{3(26.7 + 34.1)} = 0.48$	16.0
P_q (fill)	37.2	$1.0 + \frac{4}{2} = 3.0$	111.6
P_q (foundation soil)	11.1	$\frac{1.0}{2} = 0.5$	5.6
		Total:	248.2

From the results it is seen that the EQU limit state requirement is satisfied since the sum of the effects of the design destabilising actions (248.2 kNm) is less than the sum of the effects of the design stabilising actions (357.9 kNm). This result may be presented as the ratio of the resistance to the effects and be termed the *over-design factor*, Γ:

$$\Gamma = \frac{357.9}{248.2} = 1.44$$

GEO Limit state:
For Design Approach 1 we must check both partial factor sets combinations.

1. Combination 1 (partial factor sets A1 + M1 + R1)
From Table 7.1: $\gamma_{G;dst} = 1.35$; $\gamma_{G;stb} = 1.0$; $\gamma_{Q;dst} = 1.5$; $\gamma_{\phi'} = 1.0$.

(a) Design material properties

Retained fill:

$$\phi'_d = \tan^{-1}\left(\frac{\tan \phi'}{\gamma_{\phi'}}\right) = \tan^{-1}\left(\frac{\tan 32°}{1.0}\right) = 32°$$

From Fig. 7.6, $K_a = 0.25$.

Foundation soil:

$$\phi'_d = \phi' = 28°$$

From Fig. 7.6, $K_a = 0.30$.

(b) Design actions

The design weight of each area of the wall is determined as before, this time taking $\gamma_{G;stb} = 1.0$.

Area 1: $G_{W1;d} = 1.2 \times 24 \times 1.0 = 28.8$ kN
Area 2: $G_{W2;d} = 5.4 \times 24 \times 1.0 = 129.6$ kN
Area 3: $G_{W3;d} = 5.2 \times 24 \times 1.0 = 124.8$ kN
Total, $R_{v;d}$: 283.2 kN

The thrust from the active earth pressure is a permanent, unfavourable action.

$P_{a;d}$ (fill) $= \frac{1}{2} \times 0.25 \times 18 \times 4^2 \times \gamma_{G;dst} = 48.6$ kN
p_a (fill/foundation soil interface) $= 0.30 \times 18 \times 4 = 21.6$ kPa
p_a (base) $= 21.6 + 0.30 \times 20 \times 1.0 = 27.6$ kPa
$P_{a;d}$ (foundation soil) $= \frac{1}{2} \times (21.6 + 27.6) \times 1.0 \times \gamma_{G;dst} = 33.2$ kN

The lateral thrust from the surcharge is a variable, unfavourable action.

$P_{q;d}$ (fill) $= 20 \times 0.25 \times 4 \times \gamma_{Q;dst} = 30.0$ kN
$P_{q;d}$ (foundation soil) $= 20 \times 0.30 \times 1.0 \times \gamma_{Q;dst} = 9.0$ kN

(c) Design effect of actions and design resistance

The effect of the actions is to cause forward sliding of the wall. This is resisted by the friction on the underside of the wall.

Total horizontal thrust, $R_{h;d} = 48.6 + 33.2 + 30.0 + 9.0 = 120.8$ kN
Design resistance $= R_{v;d} \tan \delta = 283.2 \times \tan 28° = 150.6$ kN (NB $\delta = \phi$)

Thus the GEO limit state requirement is satisfied and the over-design factor

$$\Gamma = \frac{150.6}{120.8} = 1.25$$

2. Combination 2 (partial factor sets A2 + M2 + R1)
The partial factors are: $\gamma_{G;dst} = 1.0$; $\gamma_{G;stb} = 1.0$; $\gamma_{Q;dst} = 1.3$; $\gamma_{\phi'} = 1.25$; $\gamma_{c'} = 1.25$. The calculations are the same as for Combination 1 except that this time these partial factors are used.

$$R_{v;d} = 283.2 \text{ kN}$$
$$R_{h;d} = 44.9 + 30.4 + 32.5 + 9.7 = 117.5 \text{ kN}$$
$$R_{v;d} \tan \delta = 283.2 \times \tan 23° = 120.2 \text{ kN}$$

Thus the GEO limit state requirement is satisfied and the over-design factor

$$\Gamma = \frac{120.2}{117.5} = 1.03$$

Overview
The GEO limit state requirement is satisfied for both combinations and thus the proposed design of the wall is satisfactory. The lower value of Γ obtained (in this case 1.03) governs the design.

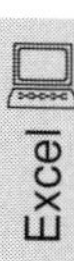

Example 7.2

The proposed design of a cantilever retaining wall is shown in Fig. 7.9. The unit weight of the concrete is 24 kN/m^3 and the soil has unit weight 18 kN/m^3. The soil peak strength parameters are $\phi' = 38°$, $c' = 0$ and the safe bearing capacity of the soil is 250 kPa. The soil behind the wall carries a uniform surcharge of intensity 10 kPa. Ignore any passive resistance from the soil in front of the wall.

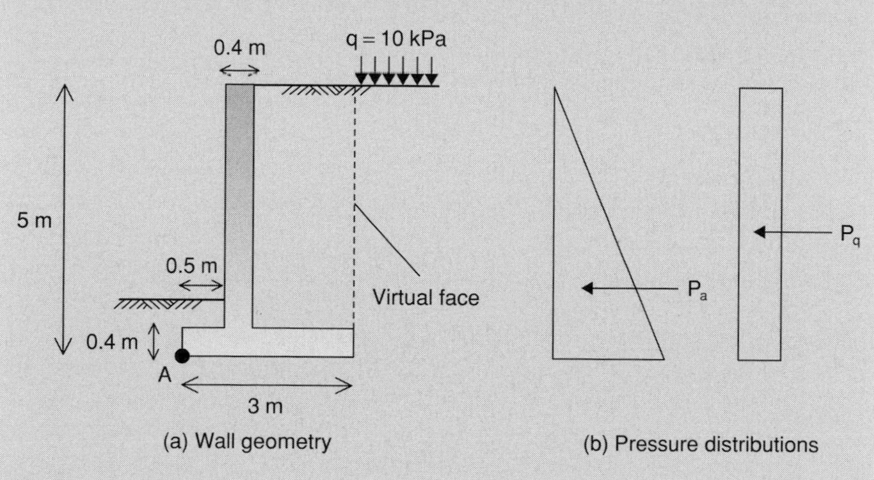

Fig. 7.9 Example 7.2.

Check the safety of the proposed design:

(a) by the traditional (gross pressure) method (assume coefficient of friction between base of wall and soil to equal tan ϕ'_{peak});
(b) in accordance with BS 8002;
(c) against the EQU and GEO (Design Approach 1) limit states of Eurocode 7.

Solution

When the retained soil is supported by a heel, the design assumes a virtual face as shown in Fig. 7.9a. Rankine's conditions apply along this face and the earth pressures acting here are established in the design.

(a) Gross pressure method

Sliding:
Using Rankine's theory (with $\phi' = 38°$) $K_a = 0.238$

Active thrust from soil, $P_a = \frac{1}{2}K_a\gamma h^2 = \frac{1}{2} \times 0.238 \times 18 \times 5^2 = 53.6$ kN
Active thrust due to surcharge, $P_q = K_a qh = 0.238 \times 10 \times 5 = 11.9$ kN
$\Sigma H = 65.5$ kN

Vertical reaction, R_v = weight of base + weight of stem + soil on heel
$= 24(0.4 \times 3.0) + 24(0.4 \times 4.6) + 18(2.1 \times 4.6)$
$= 28.8 + 44.2 + 173.9$
$= 246.9$ kN

Total force causing sliding, $R_h = 65.5$ kN
Force resisting sliding $= R_v \tan \delta = 246.9 \times \tan 38° = 192.9$ kN

$$\text{Factor of safety against sliding, } F_s = \frac{192.9}{65.5} = 2.95$$

Overturning:
Taking moments about point A, the toe of the wall.

Disturbing moment, M_S:

$$M_S = P_a \times \left(\frac{5}{3}\right) + P_q \times \left(\frac{5}{2}\right)$$

$$= 89.3 + 29.8$$

$$= 119.1 \text{ kNm}$$

Resisting moment, M_R

Due to base = $28.8 \times 1.5 = 43.2$ kN m
Due to stem = $44.2 \times 0.7 = 30.9$ kN m
Due to soil on heel = $173.9 \times 1.95 = 339.1$ kN m
$M_R = 43.2 + 30.9 + 339.1 = 413.2$ kN m

Factor of safety against overturning, $F_o = \dfrac{413.2}{119.1} = 3.47$

Bearing capacity:
Consider moments about point A.

If R_v acts at a distance x from A, then

$$R_v x = 413.2 - 119.1 = 294.1 \text{ kN m}$$

that is

$$x = \frac{294.1}{246.9} = 1.19 \text{ m} \quad \text{(within middle third of base)}$$

Eccentricity of R_v, $e = \frac{3}{2} - 1.19 = 0.31$ m

$$\text{Maximum bearing pressure} = \frac{R_v}{B}\left(1 + \frac{6e}{B}\right)$$

$$= \frac{246.9}{3}\left(1 + \frac{6 \times 0.31}{3}\right)$$

$$= 133 \text{ kPa}$$

Factor of safety against bearing capacity failure, $F_b = \dfrac{250}{133} = 1.88$

(b) BS 8002

Sliding:

Design $\phi' = \tan^{-1}\left(\dfrac{\tan 38^\circ}{1.2}\right) = 33^\circ$

Using Rankine's theory, $K_a = 0.295$

Active thrust from soil, $P_a = \frac{1}{2}K_a\gamma h^2 = \frac{1}{2} \times 0.295 \times 18 \times 25 = 66.4$ kN
Active thrust due to surcharge, $P_q = K_a qh = 0.295 \times 10 \times 5 = 14.8$ kN
$\Sigma H = 81.2$ kN

Total force causing sliding, $R_h = 81.2$ kN
Design $\tan\delta = 0.75(\text{design} \tan\phi') = 0.75 \tan 33^\circ = 0.49$
Force resisting sliding = $R_v \tan\delta = 246.9 \times 0.49 = 121.0$ kN

In terms of the sliding limit state, the design is satisfactory since $R_v \tan \delta$ (121 kN) is greater than R_h (81.2 kN).

Overturning:
Taking moments about point A, the toe of the wall.

Disturbing moment, M_S

$$M_S = P_a \times \left(\frac{5}{3}\right) + P_q \times \left(\frac{5}{2}\right)$$

$$= 110.7 + 37$$
$$= 147.7 \text{ kN m}$$

From before, resisting moment, $M_R = 413.2$ kN m

In terms of the overturning limit state, the design is satisfactory since the resisting moment (413.2 kN m) is greater than the disturbing moment (147.7 kN m).

Bearing:
Again consider moments about point A.

If R_v acts at a distance x from A, then

$$R_v x = 413.2 - 147.7 = 265.5 \text{ kN m}$$

that is,

$$x = \frac{265.5}{246.9} = 1.08 \text{ m} \qquad \text{(within middle third of base)}$$

Eccentricity of R_v, $e = \frac{3}{2} - 1.08 = 0.42$ m

$$\text{Maximum bearing pressure} = \frac{R_v}{B}\left(1 + \frac{6e}{B}\right)$$

$$= \frac{246.9}{3}\left(1 + \frac{6 \times 0.42}{3}\right)$$

$$= 151.4 \text{ kPa}$$

In terms of the bearing resistance limit state, the design is satisfactory since the maximum bearing pressure (151.4 kPa) is less than the allowable bearing capacity (250 kPa).

(c) Eurocode 7

EQU Limit state: (i.e. overturning)
From Table 7.1: $\gamma_{G;dst} = 1.1$; $\gamma_{G;stb} = 0.9$; $\gamma_{Q;dst} = 1.5$; $\gamma_{\phi'} = 1.25$.

(i) Design material properties

$$\phi'_d = \tan^{-1}\left(\frac{\tan \phi'}{\gamma_{\phi'}}\right) = \tan^{-1}\left(\frac{\tan 38^\circ}{1.25}\right) = 32.0^\circ : K_a = \frac{1 - \sin \phi'_d}{1 + \sin \phi'_d} = 0.307$$

(ii) Design actions

The weight of the wall is a permanent, favourable action:

$$\text{Stem: } G_{stem;d} = 0.4 \times 4.6 \times \gamma_{concrete} \times \gamma_{G;stb} = 1.84 \times 24 \times 0.9 = 39.7 \text{ kN}$$
$$\text{Base: } G_{base;d} = 0.4 \times 3.0 \times \gamma_{concrete} \times \gamma_{G;stb} = 1.2 \times 24 \times 0.9 = 25.9 \text{ kN}$$
$$\text{Soil on heel: } G_{heel;d} = 2.1 \times 4.6 \times \gamma \times \gamma_{G;stb} = 9.66 \times 18 \times 0.9 = 156.5 \text{ kN}$$

The thrust from the active earth pressure is a permanent, unfavourable action.

$$P_{a;d} = \tfrac{1}{2} \times 0.307 \times 18 \times 5^2 \times \gamma_{G;dst} = 76.0 \text{ kN}$$

The lateral thrust from the surcharge is a variable, unfavourable action.

$$P_{q;d} = 0.307 \times 10 \times 5 \times \gamma_{Q;dst} = 23.0 \text{ kN}$$

(iii) Design effect of actions and design resistance

$$\text{Destabilising moment, } M_{dst} = 76.0 \times \left(\frac{5}{3}\right) + 23.0 \times \left(\frac{5}{2}\right) = 184.2 \text{ kN m}$$

$$\text{Stabilising moment, } M_{stb} = 39.7 \times 0.7 + 25.9 \times 1.5 + 156.5 \times 1.95 = 371.8 \text{ kN m}$$

The EQU limit state requirement is satisfied since $M_{dst} < M_{stb}$, and the over-design factor

$$\Gamma = \frac{371.8}{184.2} = 2.02.$$

GEO Limit state, Design Approach 1: (i.e. sliding and bearing)

1. Combination 1 (partial factor sets A1 + M1 + R1)
From Table 7.1: $\gamma_{G;dst} = 1.35$; $\gamma_{G;stb} = 1.0$; $\gamma_{Q;dst} = 1.5$; $\gamma_{\phi'} = 1.0$.

(i) Design material properties

$\phi'_d = 38°$: Using Rankine's theory, $K_a = 0.238$.

(ii) Design actions

In sliding, the weight of the wall is a permanent, favourable action:

$$\text{Stem: } G_{stem;d} = 0.4 \times 4.6 \times \gamma_{concrete} \times \gamma_{G;stb} = 1.84 \times 24 \times 1.0 = 44.2 \text{ kN}$$
$$\text{Base: } G_{base;d} = 0.4 \times 3.0 \times \gamma_{concrete} \times \gamma_{G;stb} = 1.2 \times 24 \times 1.0 = 28.8 \text{ kN}$$
$$\text{Soil on heel: } G_{heel;d} = 2.1 \times 4.6 \times \gamma \times \gamma_{G;stb} = 9.66 \times 18 \times 1.0 = 173.9 \text{ kN}$$
$$\text{Total, } R_{v;d}\text{: } 246.9 \text{ kN}$$

The thrust from the active earth pressure is a permanent, unfavourable action.

$$P_{a;d} = \tfrac{1}{2} \times 0.238 \times 18 \times 5^2 \times \gamma_{G;dst} = 72.3 \text{ kN}$$

The lateral thrust from the surcharge is a variable, unfavourable action.

$$P_{q;d} = 0.238 \times 10 \times 5 \times \gamma_{Q;dst} = 17.8 \text{ kN}$$

(iii) Design effect of actions and design resistance

Sliding:

$$\text{Total horizontal thrust, } R_{h;d} = 72.3 + 17.9 = 90.2 \text{ kN}$$
$$\text{Design resistance} = R_{v;d} \tan \delta = 246.9 \times \tan 38° = 192.9 \text{ kN} \quad (\text{since } \delta = \phi)$$

The GEO limit state requirement for sliding is satisfied and the over-design factor

$$\Gamma = \frac{192.9}{90.2} = 2.14$$

Bearing:
Destabilising moment

$$M_{dst} = M_{Pa} + M_{Pq} = 72.3 \times 1.67 + 17.8 \times 2.5 = 165.2 \text{ kN m}$$

Stabilising moment

$$M_{stb} = M_{stem} + M_{base} + M_{heel} = 44.2 \times 0.7 + 28.8 \times 1.5 + 173.9 \times 1.95 = 413.2 \text{ kN m}$$

$$\text{Lever arm of } R_{v;d},\ x = \frac{413.2 - 165.2}{246.9} = 1.02 \text{ m}$$
(within middle third of base)

$$\text{Eccentricity, } e = 1.5 - 1.02 = 0.48 \text{ m}$$

R_v is now considered as a permanent, unfavourable action (= 246.9 × $\gamma_{G;dst}$):

$$\text{Maximum bearing pressure} = \frac{R_v}{B}\left(1 + \frac{6e}{B}\right) = 221.0 \text{ kPa}$$

The GEO limit state requirement for bearing is satisfied and the over-design factor

$$\Gamma = \frac{250}{221.0} = 1.13$$

2. Combination 2 (partial factor sets A2 + M2 + R1)
The calculations are the same as for Combination 1 except that this time the following partial factors (from Table 7.1) are used: $\gamma_{G;dst} = 1.0$; $\gamma_{G;stb} = 1.0$; $\gamma_{Q;dst} = 1.3$; $\gamma_{\phi'} = 1.25$; $\gamma_{c'} = 1.25$.

Sliding:

$$R_{v;d} = 246.9 \text{ kN}$$
$$R_{h;d} = 69.1 + 20.0 = 89.1 \text{ kN}$$
$$R_{v;d} \tan\delta = 246.9 \times \tan 32° = 154.3 \text{ kN}$$

The GEO limit state requirement for sliding is satisfied and the over-design factor

$$\Gamma = \frac{154.3}{89.1} = 1.73$$

Bearing:
Destabilising moment

$$M_{dst} = M_{Pa} + M_{Pq} = 69.1 \times 1.67 + 20.0 \times 2.5 = 165.4 \text{ kNm}$$

Stabilising moment

$$M_{stb} = M_{stem} + M_{base} + M_{heel} = 44.2 \times 0.7 + 28.8 \times 1.5 + 173.9 \times 1.95 = 413.2 \text{ kNm}$$

$$\text{Lever arm of } R_v,\ x = \frac{413.2 - 165.4}{246.9} = 1.01 \text{ m} \quad \text{(within middle third of base)}$$

$$\text{Eccentricity, } e = 1.5 - 1.01 = 0.49 \text{ m}$$

$$\text{Maximum bearing pressure} = \frac{R_v}{B}\left(1 + \frac{6e}{B}\right) = 163.0 \text{ kPa}$$

The GEO limit state requirement for bearing is satisfied and the over-design factor

$$\Gamma = \frac{250}{163.0} = 1.53$$

7.6 Design of sheet pile walls

A sheet pile wall is a flexible structure which depends for stability upon the passive resistance of the soil in front of and behind the lower part of the wall. Stability also depends on the anchors when incorporated.

Retaining walls of this type differ from other walls in that their weight is negligible compared with the remaining forces involved. Design methods usually neglect the effect of friction between the soil and the wall, but this omission is fairly satisfactory when determining active pressure values; it should be remembered, however, that the effect of wall friction can almost double the Rankine value of K_p. In Eurocode 7 charts are provided from which values of K_a and K_p can be determined (see Earth pressure coefficients in Section 7.4.2).

7.6.1 Cantilever walls

Sheet pile walls are flexible and sufficient yield will occur in a cantilever wall to give totally active earth pressure conditions (Fig. 7.10).

Let the height of the wall be h, and suppose it is required to find the depth of penetration, d, that will make the wall stable. For equilibrium the active pressure on the back of the wall must be balanced by the passive pressure both in front of and behind the wall. If an arbitrary point C is chosen and it is assumed that the wall will rotate outwards about this point, the theoretical pressure distribution on the wall is as shown in Fig. 7.10c. The toe of the wall (point D, Fig. 7.10a) is deep enough such that the conditions that prevail are known as *fixed earth* conditions.

Limit state design method

In BS 8002: 1994 the design procedure is to reduce the shear strength by the mobilisation factor, M (see Section 7.4.1). The depth, d, is obtained by balancing the disturbing and restoring moments about C, together with the horizontal forces established using the pressure distribution shown in Fig. 7.10c. The method generates two equations containing the unknowns d and d_0, which are solved by repeated iteration until the correct values are obtained.

In Eurocode 7 the GEO limit state is applied to assess the rotational stability using a theoretical pressure distribution similar to that advocated in BS 8002: 1994. Example 7.3 illustrates the use of both the BS 8002 and Eurocode 7 methods for the design of cantilever sheet pile walls.

Traditional methods

Various traditional methods of design exist. Each involves the determination of an overall factor of safety for passive resistance, F_p, based on different lateral earth pressure distributions. The methods are described in detail by Padfield and Mair (1984).

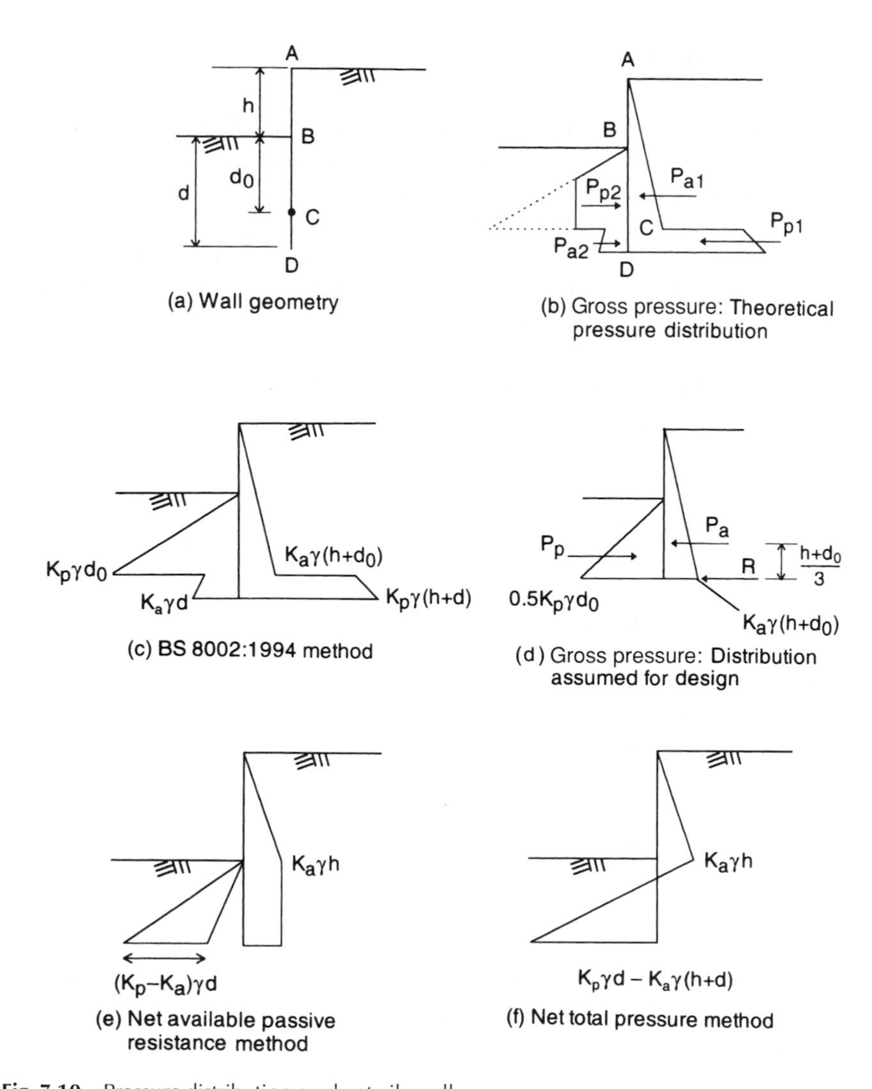

Fig. 7.10 Pressure distribution on sheet pile wall.

(1) Gross pressure method

The method is also referred to as the CP2 method, after the Institution of Structural Engineers' original Code of Practice published in 1951. It is very unlikely that the full passive resistance for the soil in front of the wall will be developed. Common practice is to divide the total theoretical value of thrust $K_p\gamma d^2/2$ by a factor of safety, traditionally taken as $F_p = 2.0$. The effective passive resistance in front of the wall is therefore assumed to have a magnitude of $K_p\gamma d^2/4$ and is of trapezoidal distribution, the centre of pressure of this trapezium lying between d/2 and d/3 above the base of the pile (for ease of calculation the value is generally taken as d/3). It is common to use lower values of F_p for low values of ϕ'. Padfield and Mair (1984) recommend the following values:

ϕ' (degrees)	F_p
> 30	2.0
20–30	1.5–2.0
< 20	1.5

Calculations are considerably simplified if it is assumed that the passive resistance on the back of the wall, P_{p1}, acts as a concentrated load, R, on the foot of the pile, leading to the pressure distribution shown in Fig. 7.10d, from which d can be obtained by taking moments of thrusts about the base of the pile. The value of d obtained by this method is more nearly the value of d_0 in Fig. 7.10a, the customary practice being to increase the value of d by 20 per cent to allow for this effect.

(2) Net available passive resistance method

The method is also referred to as the Burland, Potts and Walsh method after Burland *et al.* (1981). They advocate a modified pressure distribution (Fig. 7.10e) with the effect that the factor of safety is applied to the net available passive resistance.

(3) Strength factor method

This is based on the gross pressure method distribution but with a factor of safety applied to the shear strength of the soil, i.e.

$$\tan \phi'_m = \frac{\tan \phi'}{F_s}$$

$$c'_m = \frac{c'}{F_s}$$

By factoring the strength parameters, K_a is increased and K_p is decreased, leading to modified pressure distributions relative to those obtained using the gross pressure method.

(4) Net total pressure method

This was advocated by British Steel in the *British Steel piling handbook* (1997), where the net horizontal pressure distribution is used (Fig. 7.10f). The pressure distribution is derived by subtracting the active earth and water pressures from the passive earth and water pressures.

Excel

Example 7.3

Calculate the minimum depth of embedment, d, to provide stability to a cantilever sheet pile wall, retaining an excavated depth of 5 m using:

(a) BS 8002 method;
(b) Eurocode 7 GEO limit state, Design Approach 1;
(c) Gross pressure method.

The soil properties are $\phi'_{peak} = 30°$, $c' = 0$, $\gamma = 20$ kN/m^3.

Solution

(a) BS 8002 method

In accordance with the methods set out in BS 8002, we apply a surcharge of 10 kPa, and allow for a future unplanned excavation of 10 per cent of the clear height (= 0.5 m) in front of the wall (Fig. 7.11a). The pressure distribution is then as shown in Fig. 7.11b.

$$\text{Design } \phi' = \tan^{-1}\left(\frac{\tan 30°}{1.2}\right) = 25.7°$$

Using Rankine's theory, $K_a = 0.395$, $K_p = 2.53$

The earth pressures acting at the salient points in the distribution are established.

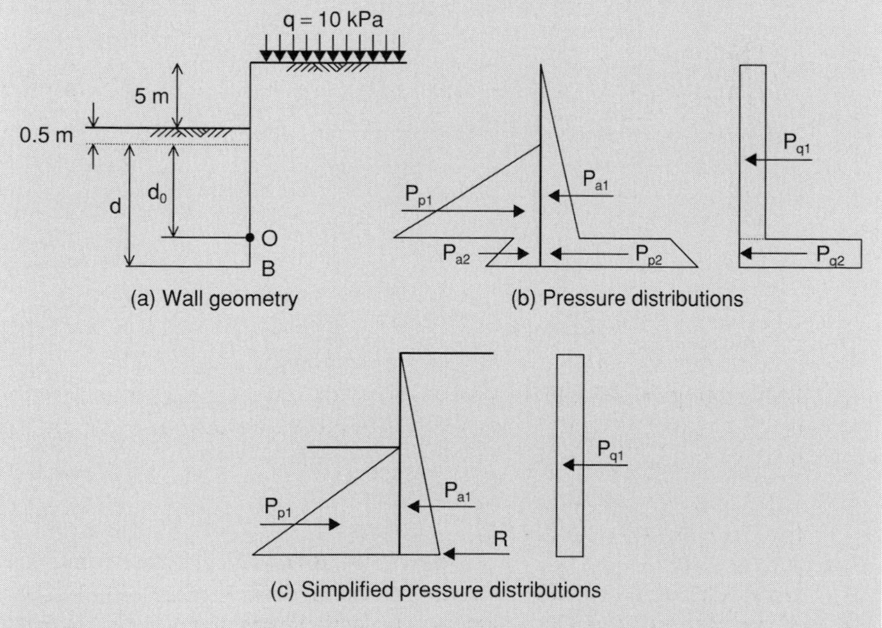

Fig. 7.11 Example 7.3.

Behind wall:

$$p_{a,O} = 0.395 \times 20 \times (d_0 + 5.5) = 7.9(d_0 + 5.5)\ \text{kPa}$$
$$p_{p,O} = 2.53 \times 20 \times (d_0 + 5.5) = 50.6(d_0 + 5.5)\ \text{kPa}$$
$$p_{p,B} = 2.53 \times 20 \times (d + 5.5) = 50.6(d + 5.5)\ \text{kPa}$$

In front of wall:

$$p_{p,O} = 2.53 \times 20 \times d_0 = 50.6d_0\ \text{kPa}$$
$$p_{a,O} = 0.395 \times 20 \times d_0 = 7.9d_0\ \text{kPa}$$
$$p_{a,B} = 0.395 \times 20 \times d = 7.9d\ \text{kPa}$$

From the earth pressures, the forces acting on the wall are now established.

$$P_{a1} = \tfrac{1}{2} \times 7.9(d_0 + 5.5) \times (d_0 + 5.5) = 3.95(d_0 + 5.5)^2$$
$$P_{a2} = \tfrac{1}{2} \times (7.9d_0 + 7.9d) \times (d - d_0)$$
$$= 3.95(d_0 + d) \times (d - d_0) = 3.95(d^2 - d_0^2)$$

$$P_{p1} = \tfrac{1}{2} \times 50.6d_0 \times d_0 = 25.3d_0^2$$
$$P_{p2} = \tfrac{1}{2} \times [50.6(d + 5.5) + 50.6(d_0 + 5.5)] \times (d - d_0)$$
$$= 25.3(d - d_0)[(d + 5.5) + (d_0 + 5.5)] = 25.3(d - d_0)[(d + d_0) + 11.0]$$
$$= 25.3(d^2 - d_0^2) + 278.3(d - d_0)$$

$$P_{q1} = 0.395 \times 10 \times (d_0 + 5.5) = 3.95(d_0 + 5.5)$$
$$P_{q2} = 2.53 \times 10 \times (d - d_0) = 25.3(d - d_0)$$

Consider moments about point O.

	Force (kN)	Lever arm (m)	Moment (kN m)
P_{a1}	$3.95(d_0 + 5.5)^2$	$\frac{(d_0 + 5.5)}{3}$	$1.32(d_0 + 5.5)^3$
P_{a2}	$3.95(d^2 - d_0^2)$	$\frac{(d - d_0)(7.9d_0 + 15.8d)}{23.7(d_0 + d)}$	$0.167(d - d_0)^2(7.9d_0 + 15.8d)$
P_{p1}	$25.3d_0^2$	$\frac{d_0}{3}$	$8.43d_0^3$
P_{p2}	$25.3(d^2 - d_0^2) + 278.3(d - d_0)$	$\frac{(d - d_0)(2p_{p,B} + p_{p,O})}{3(p_{p,O} + p_{p,B})}$	$8.43(d - d_0)^2[(d_0 + 5.5) + 2(d + 5.5)]$
P_{q1}	$3.95(d_0 + 5.5)$	$\frac{(d_0 + 5.5)}{2}$	$1.98(d_0 + 5.5)^2$
P_{q2}	$25.3(d - d_0)$	$\frac{(d - d_0)}{2}$	$12.65(d - d_0)^2$

$\Sigma M_O = 0$

i.e. clockwise moments – anticlockwise moments = 0

$$M_{Pp1} + M_{Pp2} + M_{Pq2} - M_{Pa1} - M_{Pa2} - M_{Pq1} = 0$$
$$8.43d_0^3 + 8.43(d - d_0)^2[(d_0 + 5.5) + 2(d + 5.5)] + 12.65(d - d_0)^2 - 1.32(d_0 + 5.5)^3 - 0.167(d - d_0)^2(7.9d_0 + 15.8d) - 1.98(d_0 + 5.5)^2 = 0 \quad (1)$$

$\Sigma H = 0$

i.e.

$$P_{p1} + P_{a2} - P_{a1} - P_{p2} - P_{q1} - P_{q2} = 0$$
$$25.3d_0^2 + 3.95(d^2 - d_0^2) - 3.95(d_0 + 5.5)^2 - [25.3(d^2 - d_0^2) + 278.3(d - d_0)] - 3.95(d_0 + 5.5) - 25.3(d - d_0) = 0 \quad (2)$$

Excel

Equations (1) and (2) are best solved using a programmable calculator or computer program such as the spreadsheet *Example 7.3.xls*, available for download. For the case above

$d_0 = 6.4$ m
$d = 7.3$ m

Note: Few engineers would perform the detailed calculations set out in Equations (1) and (2). Instead, most would simplify the pressure distributions of Fig. 7.11b to the approximate distributions shown in Fig. 7.11c, and determine d_0 using an approach similar to the gross pressure method.

In this example, the moment equilibrium equation would then become:

$\Sigma M_0 = 0$

i.e. $M_{Pp1} - M_{Pa1} - M_{q1} = 0$

$$25.3d_0^2 \times \frac{d_0}{3} - 3.95(d_0 + 5.5)^2 \frac{(d_0 + 5.5)}{3} - 3.95(d_0 + 5.5) \frac{(d_0 + 5.5)}{2} = 0$$

which solves for $d_0 = 6.98$ m

To obtain the design depth, d, d_0 is increased by an amount equal to the extent required to generate a net passive resistance force below the point of rotation at least as large as R. (R is obtained from simple horizontal force equilibrium.) This demands additional calculations and it is common practice to avoid this by simply increasing d_0 by 20 per cent to give d.

i.e., $d = d_0 \times 1.2 = 6.98 \times 1.2 = 8.4$ m.

(b) Eurocode 7, GEO Limit State, Design Approach 1

Allowance is made for a future unplanned excavation Δa equal to 10 per cent of the clear height (= 0.5 m). The pressure distribution is shown in the left-hand part of Fig. 7.11b.

1. Combination 1 (partial factor sets A1 + M1 + R1)
From Table 7.1: $\gamma_{G;dst} = 1.35$; $\gamma_{\phi'} = 1.0$.

$$\phi'_d = \tan^{-1}\left(\frac{\tan\phi'}{\gamma_{\phi'}}\right) = \tan^{-1}\left(\frac{\tan 30°}{1.0}\right) = 30°$$

Using Rankine's theory, $K_a = 0.333$, $K_p = 3.0$

Earth pressures:
Behind wall:

$$p_{a,O} = 0.333 \times 20 \times (d_0 + 5.5) = 6.67(d_0 + 5.5) \text{ kPa}$$
$$p_{p,O} = 3.0 \times 20 \times (d_0 + 5.5) = 60(d_0 + 5.5) \text{ kPa}$$
$$p_{p,B} = 3.0 \times 20 \times (d + 5.5) = 60(d + 5.5) \text{ kPa}$$

In front of wall:

$$p_{p,O} = 3.0 \times 20 \times d_0 = 60d_0 \text{ kPa}$$
$$p_{a,O} = 0.333 \times 20 \times d_0 = 6.67d_0 \text{ kPa}$$
$$p_{a,B} = 0.333 \times 20 \times d = 6.67d \text{ kPa}$$

Design actions:
The active thrust due to the earth pressure is a permanent, unfavourable action:

$$P_{a1;d} = \tfrac{1}{2} \times 6.67(d_0 + 5.5) \times (d_0 + 5.5) \times \gamma_{G;dst} = 3.33(d_0 + 5.5)^2 \times 1.35$$
$$= 4.5(d_0 + 5.5)^2$$
$$P_{a2;d} = \tfrac{1}{2} \times (6.67d_0 + 6.67d) \times (d - d_0) \times \gamma_{G;dst} = 4.5(d_0 + d) \times (d - d_0)$$
$$= 4.5(d^2 - d_0^2)$$

There is debate on how passive earth pressures should be treated in Eurocode 7. One view is that the thrust produced by earth pressures should be treated as an action and therefore multiplied by γ_F to obtain the design value. Another is to treat active earth pressures as an action and passive earth pressures as a resistance, in which case the latter should be divided by γ_{Re} to obtain the design value. In the UK National Annex, values of $\gamma_{Re} = 1.0$ and the prevailing view is to treat passive pressures as a permanent, unfavourable action, since the thrust derives from the same sources as the active pressure, and the level of uncertainty in its value should be the same.

$$P_{p1;d} = \tfrac{1}{2} \times 60d_0 \times d_0 \times \gamma_{G;dst} = 40.5d_0^2$$

$$P_{p2;d} = \tfrac{1}{2} \times [60(d + 5.5) + 60(d_0 + 5.5)] \times (d - d_0) \times \gamma_{G;dst}$$

$$= 40.5(d - d_0)[(d + 5.5) + (d_0 + 5.5)]$$

$$= 40.5(d^2 - d_0^2) + 330(d - d_0)$$

Effect of actions:
As with the BS 8002 approach, d and d_0 are obtained by resolving the moment and force equilibrium equations:

$$\Sigma M_O = 0$$

i.e.

$$M_{Pp1} + M_{Pp2} - M_{Pa1} - M_{Pa2} = 0$$

$$13.5d_0^3 + 13.5(d - d_0)^2[(d_0 + 5.5) + 2(d + 5.5)] - 1.5(d_0 + 5.5)^3 - 0.19(d - d_0)^2(7.9d_0 + 15.8d) = 0 \quad (1)$$

$$\Sigma H = 0$$

i.e.

$$P_{p1;d} + P_{a2;d} - P_{a1;d} - P_{p2;d} = 0$$

$$40.5d_0^2 + 4.5(d^2 - d_0^2) - 4.5(d_0 + 5.5)^2 - [40.5(d^2 - d_0^2) + 330(d - d_0)] = 0 \quad (2)$$

Excel

Using *Example 7.3.xls:*

$d_0 = 5.2$ m
$d = 5.7$ m

2. Combination 2 (partial factor sets A2 + M2 + R1)
The calculations are the same as for Combination 1 except that this time the following partial factors are used: $\gamma_{G;dst} = 1.0$; $\gamma_{\phi'} = 1.25$.

The following expressions are then derived ($K_a = 0.409$; $K_p = 2.444$):

$$\Sigma M_O = 0$$

$$8.13d_0^3 + 8.13(d - d_0)^2[(d_0 + 5.5) + 2(d + 5.5)] - 1.36(d_0 + 5.5)^3 - 0.172(d - d_0)^2(7.9d_0 + 15.8d) = 0 \quad (1)$$

$$\Sigma H = 0$$

$$24.4d_0^2 + 4.09(d^2 - d_0^2) \times (d - d_0) - 4.09(d_0 + 5.5)^2 - [24.4(d^2 - d_0^2) + 268.4(d - d_0)] = 0 \quad (2)$$

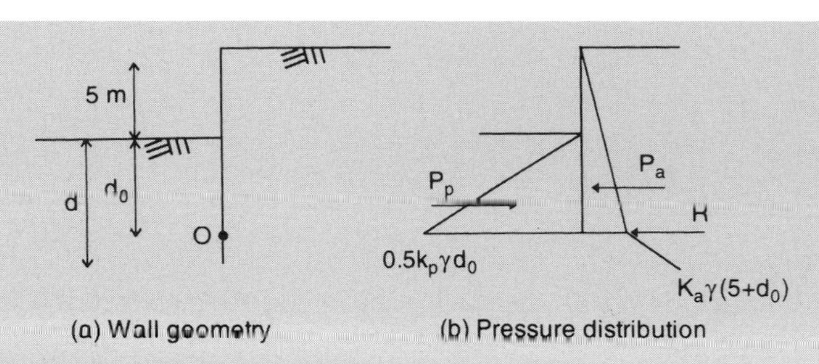

Fig. 7.12 Example 7.3 Part (b) gross pressure method.

Excel

Using *Example 7.3.xls:*

$$d_0 = 5.7 \text{ m}$$
$$d = 6.6 \text{ m}$$

(c) Gross pressure method

In the gross pressure method, the net passive resistance below the point of rotation is replaced by the horizontal force R, as shown in Fig. 7.12.

Using Rankine's theory (with $\phi' = 30°$) $K_a = \frac{1}{3}$; $K_p = 3.0$.

	Force (kN)	Lever arm (m)	Moment (kN m)
P_a	$\frac{20}{2 \times 3}(5 + d_0)^2$	$\frac{(5 + d_0)}{3}$	$\frac{10}{9}(5 + d_0)^3$
P_p	$15d_0^2$	$\frac{d_0}{3}$	$5d_0^3$

Minimum depth is required, and since $F_p = 2.0$ has already been applied to the pressure distribution,

$$\frac{5d_0^3}{\frac{10}{9}(5 + d_0)^3} = \frac{9d_0^3}{2(5 + d_0)^3} = 1$$

$$d_0 = 7.7 \text{ m}$$
$$d = 1.2 \times 7.7 = 9.24 \text{ m}$$

7.6.2 *Anchored and propped walls*

When the top of a sheet pile wall is anchored, a considerable reduction in the penetration depth can be obtained. Due to this anchorage the lateral yield in the upper part of the wall is similar to the yield in a timbered trench, whereas in the lower part the yield is similar to that of a retaining wall yielding by rotation. As a result the pressure distribution on the back of an anchored sheet pile is a combination of the totally

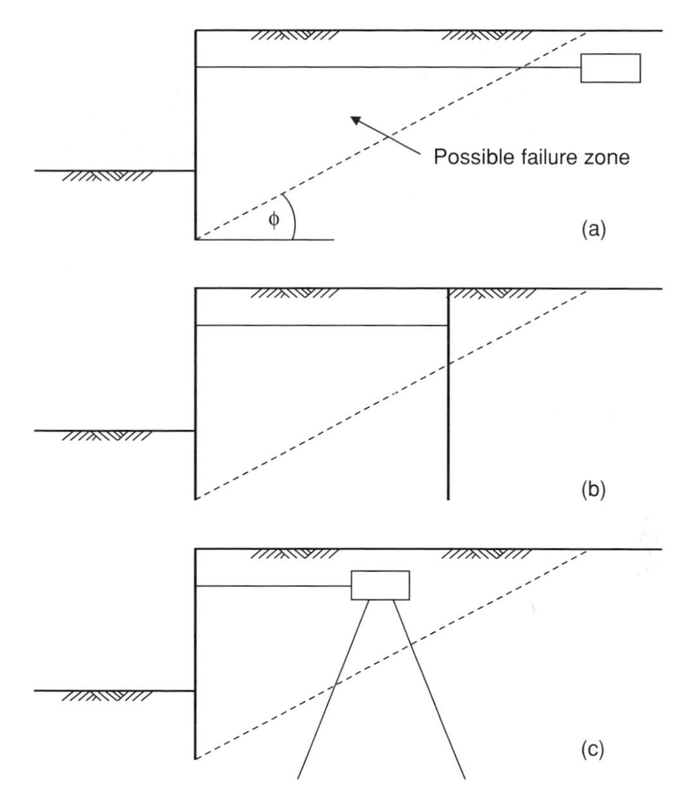

Fig. 7.13 Anchorage systems for sheet pile walls.

active and the arching-active cases, the probable pressure distribution being indicated in Fig. 7.14b. In practice the pressure distribution behind the wall is assumed to be totally active.

The anchor or prop force required can be obtained by equating horizontal forces: $T = P_a - P_p$, from which a value is obtained per metre run of wall. The resulting value of T is increased by 25 per cent to allow for flexibility in the piling and arching in the soil. Anchors are usually spaced at 2–3 m intervals and secured to stiffening wales.

Anchorage can be obtained by the use of additional piling or by anchor blocks (large concrete blocks in which the tie is embedded). Any anchorage block must be outside the possible failure plane (Fig. 7.13a), and when space is limited piling becomes necessary (Fig. 7.13b). If bending is to be avoided in the anchorage pile, then a pair of raking piles can be used (Fig. 7.13c).

7.6.3 Depth of embedment for anchored walls

Free earth support method

It is assumed that rotation occurs about the anchor point and that sufficient yielding occurs for the development of active and passive pressures. The pressure distribution

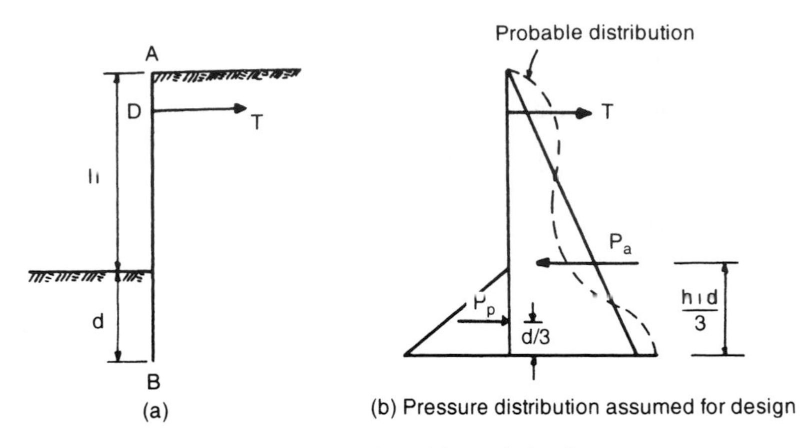

Fig. 7.14 Free earth support method for anchored sheet piled walls.

assumed in design is shown in Fig. 7.14, the wall being considered free to move at its base (the point B). By taking moments about the tie rod at D an expression for the penetration depth, d, can be obtained, the actual penetration depth being taken as equal to 1.2d. The four methods of assessing the ratio of restoring moments to overturning moments described for cantilever walls are also used for anchored walls. Design to Eurocode 7 involves the use of the GEO limit state to assess the rotational stability, as illustrated in Example 7.4.

Example 7.4

If an anchor is placed 1 m below the ground level behind the sheet pile wall described in Example 7.3, calculate the minimum depth of embedment, d, to provide stability using:

(a) BS 8002 method;
(b) Eurocode 7 GEO Limit State, Design Approach 1;
(c) gross pressure method, taking F_p, the factor of safety on passive resistance, as equal to 2.0.

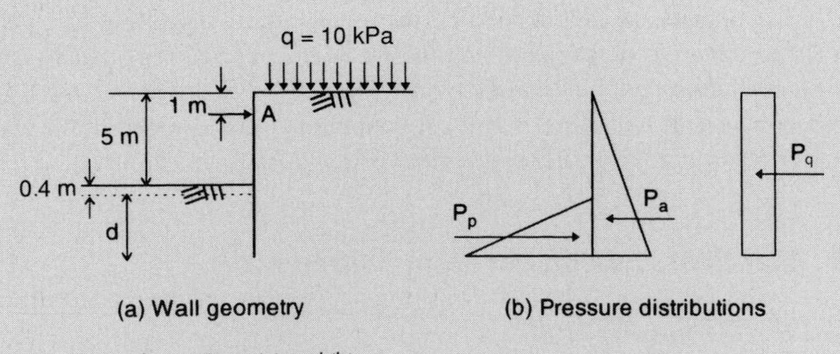

Fig. 7.15 Example 7.4 Parts (a) and (b).

Solution

(a) BS 8002 method

As before, we apply a surcharge of 10 kPa, and allow for a future unplanned excavation of 0.4 m in front of the wall (Fig. 7.15a). The pressure distribution is then as shown in Fig. 7.15b.

From before, $K_a = 0.395$, $K_p = 2.53$.

	Force (kN)	Lever arm about A (m)	Moment (kN m)
P_a	$3.95(d + 5.4)^2$	$\frac{2}{3}(d + 5.4) - 1$	$3.95(d + 5.4)^2\left[\frac{2}{3}(d + 5.4) - 1\right]$
P_q	$3.95(d + 5.4)$	$\frac{(d + 5.4)}{2} - 1$	$3.95(d + 5.4)\left[\frac{(d + 5.4)}{2} - 1\right]$
P_p	$25.3d^2$	$\frac{2}{3}d + 4.4$	$25.3d^2\left[\frac{2}{3}d + 4.4\right]$

$\sum M = 0$

i.e.

$$M_{Pa} + M_{Pq} - M_{Pp} = 0$$

$$3.95(d + 5.4)^2\left[\frac{2}{3}(d + 5.4) - 1\right] + 3.95(d + 5.4)\left[\frac{(d + 5.4)}{2} - 1\right] - 25.3d^2\left[\frac{2}{3}d + 4.4\right] = 0$$

$d = 2.9$ m

Design depth $= 2.9 + 0.4 = 3.3$ m

(b) Eurocode 7, GEO Limit State, Design Approach 1

The pressure distribution is shown in the left hand part of Fig. 7.15b.

1. Combination 1 (partial factor sets A1 + M1 + R1)

From Table 7.1: $\gamma_{G;dst} = 1.35$; $\gamma_{\phi'} = 1.0$.

From before, $K_a = 0.333$, $K_p = 3.0$.

Design actions:

The active thrust due to the earth pressure is a permanent, unfavourable action:

$$P_{a;d} = \tfrac{1}{2} \times K_a \times \gamma \times (d + 5.4)^2 \times \gamma_{G;dst} = 4.5(d + 5.4)^2$$

The passive resistance is also considered a permanent, unfavourable action:

$$P_{p;d} = \tfrac{1}{2} \times 60 \times d \times d \times \gamma_{G;dst} = 40.5d^2$$

Effect of actions:
As with the BS 8002 approach, d is obtained by resolving the moment equilibrium equation. The lever arms are as before.

$$\Sigma M = 0$$

i.e.

$$M_{Pa} - M_{Pp} = 0$$

$$4.5(d + 5.4)^2 \times [\tfrac{2}{3}(d + 5.4) - 1] - 40.5d^2 \times (\tfrac{2}{3}d + 4.4) = 0$$

Using *Example 7.4.xls*, d = 2.1 m
Design depth = 2.1 + 0.4 = 2.5 m

2. Combination 2 (partial factor sets A2 + M2 + R1)
The calculations are the same as for Combination 1 except that this time the following partial factors are used: $\gamma_{G;dst} = 1.0$; $\gamma_{\phi'} = 1.25$.
The following expressions are then derived ($K_a = 0.409$; $K_p = 2.444$):

$$\Sigma M = 0$$

i.e.

$$4.09(d + 5.4)^2 \times [\tfrac{2}{3}(d + 5.4) - 1] - 24.4d^2 \times (\tfrac{2}{3}d + 4.4) = 0$$

Using *Example 7.4.xls*, d = 2.9 m
Design depth = 2.9 + 0.4 = 3.3 m

(c) Gross pressure method

The pressure distribution is shown in Fig. 7.16.
Using Rankine's theory (with $\phi' = 30°$) $K_a = 0.33$; $K_p = 3.0$:

	Force (kN)	Lever arm about A (m)	Moment (kN m)
P_a	$\frac{10}{3}(5 + d)^2$	$\frac{2}{3}(d + 5) - 1$	$\frac{10}{3}(5 + d)^2\left[\frac{2}{3}(d + 5) - 1\right]$
P_p	$15d^2$	$\frac{2}{3}d + 4$	$15d^2\left[\frac{2}{3}d + 4\right]$

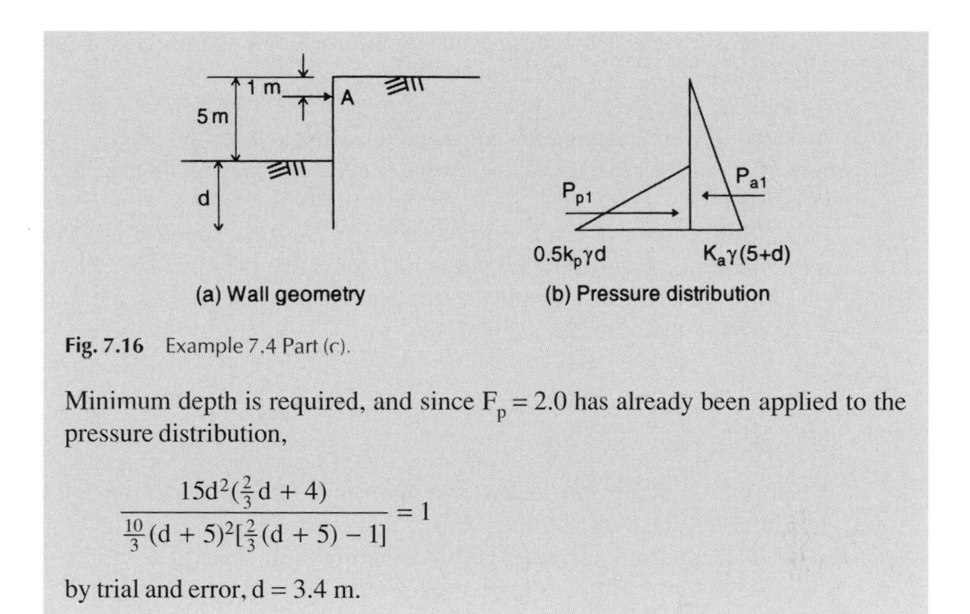

Fig. 7.16 Example 7.4 Part (c).

Minimum depth is required, and since $F_p = 2.0$ has already been applied to the pressure distribution,

$$\frac{15d^2(\frac{2}{3}d + 4)}{\frac{10}{3}(d + 5)^2[\frac{2}{3}(d + 5) - 1]} = 1$$

by trial and error, d = 3.4 m.

7.6.4 Reduction of design moments in anchored sheet pile walls

Rowe (1952) conducted a series of model tests in which he showed that the bending moments that actually occur in an anchored sheet pile wall are less than the values computed by the free earth support method. This difference in values is due mainly to arching effects within the soil which create a passive pressure distribution in front of the wall that is considerably different from the theoretical triangular distribution assumed for the analysis. Because of this phenomenon the point of application of the passive resistive force occurs at a much shallower depth than the generally assumed value of d/3 (where d = depth of penetration of the pile). Soil arching is discussed later in this chapter.

Rowe later extended his work to cover clay soils (1957, 1958) and suggested a semi-empirical approach, covering the main soil types, whereby the values computed by the free earth support method for both the moments in the pile and the tension in the tie can be realistically reduced. The method involves the use of two coefficients, r_d and r_t, and worked examples illustrating the use of the method have been prepared by Barden (1974). Numerical studies performed by Potts and Fourie (1984) have confirmed Rowe's findings for normally consolidated clays. However, their studies showed that Rowe's results do not stand for overconsolidated clays, and they have produced separated design charts for this case.

7.6.5 Treatment of groundwater conditions

In order to carry out the stability analysis of a retaining wall involving groundwater it is necessary to know the values of the water pressures acting on both sides of the wall.

If there is a water level on one side of the wall only the problem is simple to analyse and is illustrated in Example 6.5.

If there are water levels on both sides of the wall but at the same elevation then the two water pressure diagrams are equal and therefore balance out. Hence, apart from allowing for the fact that the soil below the water is submerged, no special treatment is necessary.

With different water levels on both sides of the pile, seepage can occur. An approximate method to allow for this is to assume that the excess head causing the flow is distributed linearly around the length of the pile that is within the water zone, i.e. (2d + h − i − j), as shown in Fig. 7.17a.

Example 7.5

Determine an approximation for the water pressure distribution on each side of the sheet pile wall shown in Fig. 7.17a, if h = 8 m, d = 6 m and i = j = 0 (i.e. GWL at ground surface on both sides of the wall). Take $\gamma_w = 9.81\ \text{kN/m}^3$.

Solution

With the assumption that the excess head is linearly distributed around the length of the pile within the water zone, the formula for u, the water pressure on both sides at the pile toe, is

$$u = \frac{2(h + d - j)(d - i)\gamma_w}{(2d + h - i - j)} = 82.4\ \text{kPa}$$

The assumed diagrams for water pressure on each side of the wall are shown in Fig. 7.17b and the net water pressure diagram is shown in Fig. 7.17c.

During design to Eurocode 7, the water pressures are considered as geotechnical actions, and the appropriate partial factors of safety are selected and applied to the net water pressures.

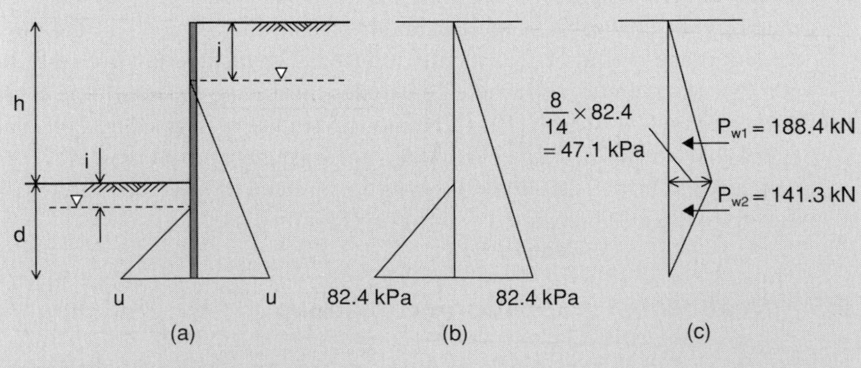

Fig. 7.17 Example 7.5.

7.7 Reinforced soil

The principle of reinforced soil is that a mass of soil can be given tensile strength in a specific direction if lengths of a material capable of carrying tension are embedded within it in the required direction.

This idea has been known for centuries. The Bible quotes the use of straw to strengthen unburnt clay bricks, and, from ancient times, fascine mattresses have been used to strengthen soft soil deposits prior to road construction. Ziggurats, built in Iraq, consisted of dried earth blocks, reinforced across the width of the structure with tarred ropes. However the full potential of reinforced soil was never realised until Vidal, who coined the term 'reinforced earth', demonstrated its wide potential and produced a rational design approach in his paper of 1966. There is no doubt that the present day use of reinforced soil structures stems directly from the pioneer work of Vidal.

Reinforced soil can be used in many geotechnical applications but, in this chapter, we are only concerned with earth retaining structures.

A reinforced soil retaining wall is a gravity structure and a simple form of such a wall is illustrated in Fig. 7.18. Brief descriptions of the components listed in the figure are set out below.

Soil fill

The soil should be granular and free draining with not more than 10 per cent passing the 63 μm sieve.

Reinforcing elements

(1) Metals

Originally many reinforced soil structures used thin metallic strips usually 50–100 mm wide and some 3–5 mm thick. Metals used were aluminium alloy, copper, stainless steel and galvanised steel, the latter being the most common. The common property of these materials is that they all have high moduli of elasticity so that negligible strains are created within the soil mass.

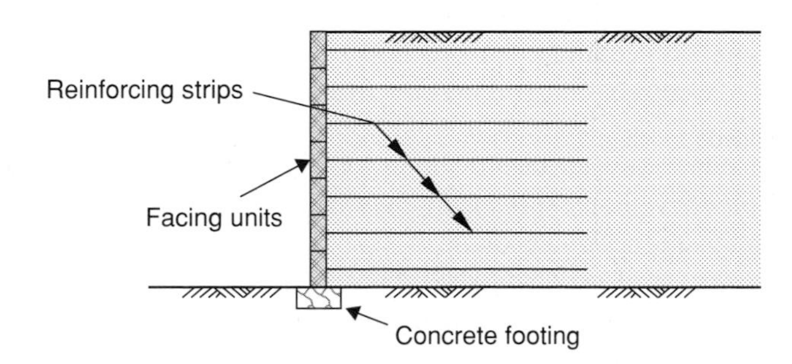

Fig. 7.18 Typical reinforced soil retaining wall.

(2) Plastics
Since the mid-1970s there has been an increasing use of geosynthetics as reinforcement in reinforced soil, either in strip form or in grid form, such as Tensar geogrid. Geosynthetics have the advantage of greater durability than metal in corrosive soil, and their tensile strength can approach that of steel. In grid form plastic reinforcement can achieve high frictional properties between itself and the surrounding soil. The main disadvantage of plastic reinforcement is that it experiences plastic deformation when subjected to tensile forces, which can lead to relatively large strains within the soil mass.

Another type of polymer reinforcement material is when it is reinforced with glass fibres. Known as glass fibre reinforced plastic, GRP, this material has a tensile strength similar to mild steel with the advantage that it does not experience plastic deformation.

Facing units

At the free boundary of a reinforced soil structure it is necessary to provide a barrier in order that the fill is contained. This is provided by a thin weatherproof facing which in no way contributes structural strength to the wall. The facing is usually built up from prefabricated units small and light enough to be manhandled. The units are generally made of precast concrete although steel, aluminium and plastic units are sometimes encountered. In order to form a platform from which the facing units can be built up a small mass concrete foundation is required.

Design of reinforced soil retaining structures

In 1995, BS 8006 *Code of practice for strengthened/reinforced soils and other fills* was published. This code adopts a limit state design approach using partial factors. However, it is not compatible with Eurocode 7, and revisions to BS 8006 seem necessary to remove conflicting clauses. BS 8006 describes which design methods are acceptable for reinforced soil slopes and walls but it does not explain how to actually design these geotechnical structures. Therefore, the reader may wish to refer to the book by Jones (1996), where the subject of reinforced soil is described in more comprehensive detail.

Reinforced soil can provide a method for retaining soil when existing ground conditions do not allow construction by other, more conventional, methods. For example a compressible soil may be perfectly capable of supporting a reinforced soil retaining structure whereas it would probably require some form of piled foundation if a more conventional retaining wall were to be constructed. The technique can also be used when there is insufficient land space to construct the sloping side of a conventional earth embankment.

However, reinforced soil should not be thought of as only a form of alternative construction as it is often the first choice of design engineers when considering an earth retaining structure.

7.8 Soil nailing

Soil nailing is an *in situ* reinforcement technique used to stabilise slopes and retain excavations but, in this chapter, we are concerned only with earth retaining structures. The technique uses steel bars fully bonded into the soil mass. The bars are inserted into the soil either by direct driving or by drilling a borehole, inserting the bar and then filling the annulus around the bar with grout. The face of the exposed soil is sprayed with concrete to produce a zone of reinforced soil. The zone then acts as a homogeneous unit supporting the soil behind in a similar manner to a conventional retaining wall. The construction phases of a soil nailed wall are shown in Fig. 7.19.

Although the completed soil structure may be expected to behave similarly to a conventional reinforced soil structure, there are notable differences between the two construction methods:

- natural soil properties may be greatly inferior to those permitted in a reinforced soil structure where selected fill is used;
- soil nails are installed by driving or by drilling and grouting rather than by placement within compacted fill;
- the construction process for nailing follows a 'top-down' sequence rather than a 'bottom-up' sequence for reinforced soils;
- the facing to a nailed structure is usually formed from sprayed concrete (shotcrete) or geosynthetics rather than precast units;
- nails are commonly installed at an inclination to the horizontal in contrast to reinforced soil where the reinforcements are placed horizontally.

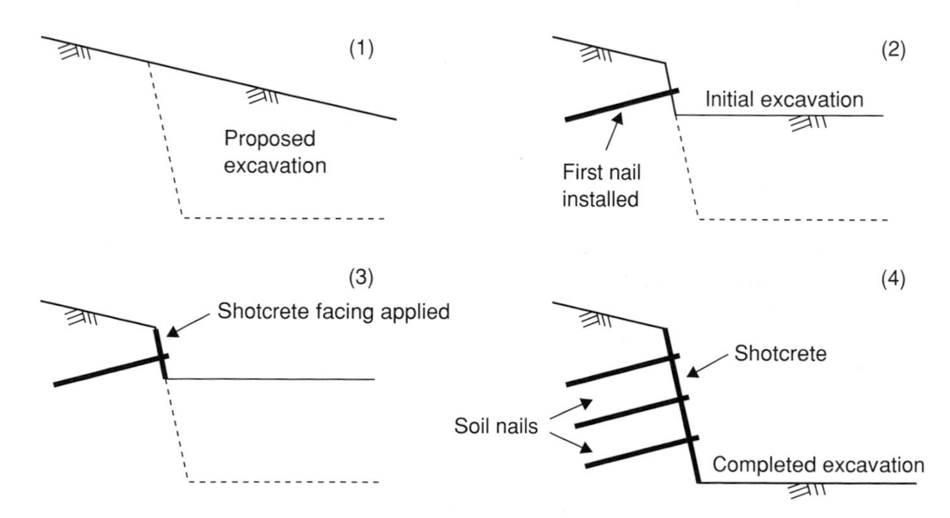

Fig. 7.19 Construction stages of a soil nailed wall (adapted from BS 8006: 1995).

There are two methods of forming the nail: *drill and gout* and *driving*. With the drill and grout method, steel bars are installed into pre-drilled holes and grout injected around them to bond them fully into the soil mass. This generates a reasonably large contact area between the grout and the soil thereby providing a high pull-out resistance. With the driving method, nails are either driven into the soil using a hydraulic or pneumatic hammer, or fired into the soil from a nail launcher which uses an explosive release of compressed air. This method of installation requires the nails to be relatively robust and to have a reasonably small cross-sectional area. Details of the driving technique are given by Myles and Bridle (1991).

To date, soil nailing has not been widely used in the UK. This may have been partly due to the previous lack of published design procedures and specifications for soil nailing, which was rectified in 1995 with the publication of BS 8006 *Code of practice for strengthened/reinforced soils and other fills*. Full details of soil nail techniques and design methods are given by Gassler (1990), Schlosser (1982), Schlosser and de Buhan (1990) and RDGC (1991) and a recent extension of the technique is described by Pokharel and Ochiai (1997).

Exercises

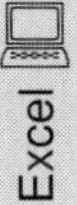

Exercise 7.1

A reinforced concrete cantilever retaining wall, supporting a granular soil, has dimensions shown in Fig. 7.20.

Using a gross factor of safety approach, calculate the factors of safety against sliding and overturning and check the bearing pressure on the soil beneath the wall if the allowable bearing pressure is 300 kPa. Take the unit weight of concrete as 23.5 kN/m^3 and assume that the friction between the

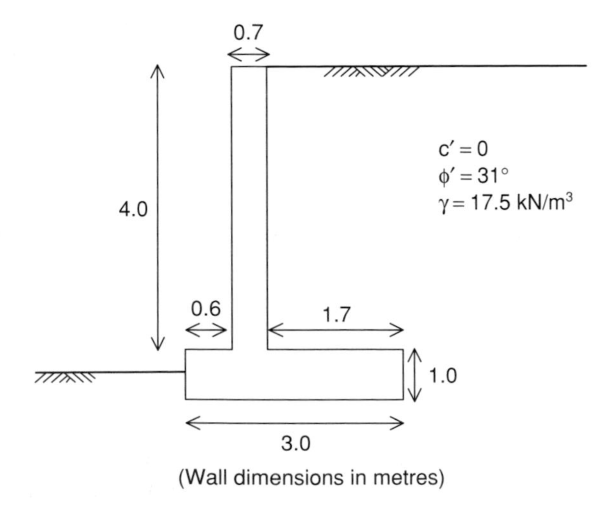

Fig. 7.20 Exercise 7.1.

base of the wall and the soil is equal to ϕ'. Ignore any passive resistance from the soil in front of the wall.

Answer $F_s = 2.19$; $F_o = 3.63$; $p_{max} = 135.5$ kPa $\{< 300 \Rightarrow \text{OK}\}$

Exercise 7.2

Check the safety of the mass concrete retaining wall shown in Fig. 7.21 in terms of sliding and overturning using:

(a) a gross factor of safety approach;
(b) the BS 8002 approach, using a mobilisation factor, M, of 1.2.

You may assume that Rankine's conditions apply to the soil behind the wall, and that the friction between the base of the wall and the soil, $\delta = \frac{2}{3}\phi'$. Ignore any passive resistance from the soil in front of the wall.

Answer $F_o = 3.53$; $F_s = 1.85$; overturning and sliding limit states satisfied.

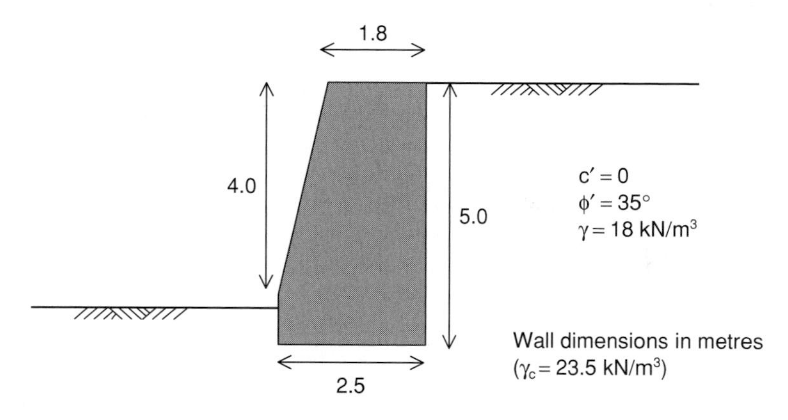

Fig. 7.21 Exercise 7.2.

Exercise 7.3

A cantilever sheet pile wall is to be constructed in a granular soil with the following properties:

$\gamma = 19.2$ kN/m^3
$\phi' = 29°$
$c' = 0$

The depth of the excavation is to be 4 m. Determine the minimum depth of embedment, d, to provide stability in accordance with BS 8002, taking the mobilisation factor, M = 1.2.

If the wall is now to be designed to carry a surcharge of 20 kPa, what would be the required depth of embedment?

Answer $d_o = 6.0$ m, d = 7.2 m; d = 7.9 m

Exercise 7.4

Consider again the situation in Exercise 7.3. If groundwater is present and the GWL is coincident with the ground surface on both sides of the wall, determine the new required depth of embedment to provide stability.

Answer 13.6 m

Exercise 7.5

Check each situation of Exercises 7.1–7.3 in accordance with Eurocode 7, using Design Approach 1 in all cases.

Answer

Example 7.1: Overturning (EQU), $\Gamma = 2.41$
Sliding (GEO) DA1-1, $\Gamma = 1.62$;
DA1-2, $\Gamma = 1.42$

Example 7.2: Overturning (EQU), $\Gamma = 2.74$
Sliding (GEO) DA1-1, $\Gamma = 2.76$;
DA1-2, $\Gamma = 2.27$

Example 7.3: Surcharge = 0: DA1-1, $d_0 = 4.4$ m, d = 4.8 m;
DA1-2, $d_0 = 4.4$ m, d = 5.2 m
Surcharge = 20 kPa: DA1-1, $d_0 = 4.1$ m, d = 4.8 m;
DA1-2, $d_0 = 4.6$ m, d = 5.7 m

Chapter 8
Bearing Capacity of Soils

8.1 Bearing capacity terms

The following terms are used in bearing capacity problems.

Ultimate bearing capacity

The value of the average contact pressure between the foundation and the soil which will produce shear failure in the soil.

Safe bearing capacity

The maximum value of contact pressure to which the soil can be subjected without risk of shear failure. This is based solely on the strength of the soil and is simply the ultimate bearing capacity divided by a suitable factor of safety.

Allowable bearing pressure

The maximum allowable net loading intensity on the soil allowing for both shear and settlement effects.

8.2 Types of foundation

Strip foundation

Often termed a *continuous footing* this foundation has a length significantly greater than its width. It is generally used to support a series of columns or a wall.

Pad footing

Generally an individual foundation designed to carry a single column load although there are occasions when a pad foundation supports two or more columns.

Raft foundation

This is a generic term for all types of foundations that cover large areas. A raft foundation is also called a *mat foundation* and can vary from a fascine mattress supporting a farm road to a large reinforced concrete basement supporting a high rise block.

Pile foundation

Piles are used to transfer structural loads to either the foundation soil or the bedrock underlying the site. They are usually designed to work in groups, with the column loads they support transferred into them via a capping slab.

Pier foundation

This is a large column built up either from the bedrock or from a slab supported by piles. Its purpose is to support a large load, such as that from a bridge. A pier operates in the same manner as a pile but it is essentially a short squat column whereas a pile is relatively longer and more slender.

Shallow foundation

A foundation whose depth below the surface, z, is equal to or less than its least dimension, B. Most strip and pad footings fall into this category.

Deep foundation

A foundation whose depth below the surface is greater than its least dimension. Piles and piers fall into this category.

8.3 Analytical methods for the determination of the ultimate bearing capacity of a foundation

The ultimate bearing capacity of a foundation is given the symbol q_u and there are various analytical methods by which it can be evaluated. As will be seen, some of these approaches are not all that suitable but they still form a very useful introduction to the study of the bearing capacity of a foundation.

8.3.1 Earth pressure theory

Consider an element of soil under a foundation (Fig. 8.1). The vertical downward pressure of the footing, q_u, is a major principal stress causing a corresponding Rankine

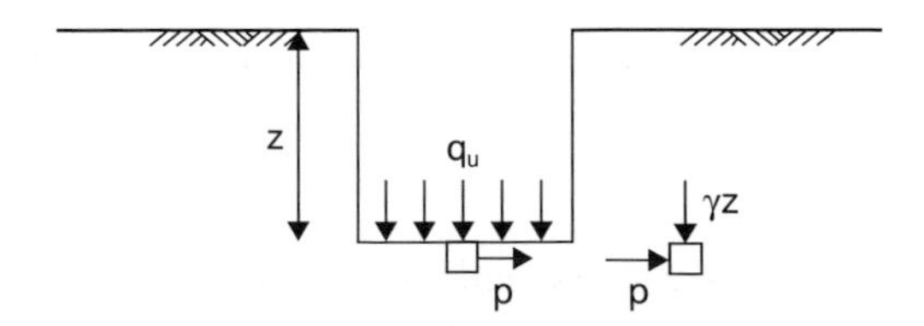

Fig. 8.1 Earth pressure conditions immediately below a foundation.

active pressure, p. For particles beyond the edge of the foundation this lateral stress can be considered as a major principal stress (i.e. passive resistance) with its corresponding vertical minor principal stress γz (the weight of the soil).

Now

$$p = q_u \frac{1 - \sin\phi}{1 + \sin\phi}$$

also

$$p = \gamma z \frac{1 + \sin\phi}{1 - \sin\phi}$$

$$\Rightarrow \quad q_u = \gamma z \left(\frac{1 + \sin\phi}{1 - \sin\phi}\right)^2$$

This is the formula for the ultimate bearing capacity, q_u. It will be seen that it is not satisfactory for shallow footings because when $z = 0$ then, according to the formula, q_u also $= 0$.

Bell's development of the Rankine solution for c–ϕ soils gives the following equation:

$$q_u = \gamma z \left(\frac{1 + \sin\phi}{1 - \sin\phi}\right)^2 + 2c\sqrt{\left(\frac{1 + \sin\phi}{1 - \sin\phi}\right)^3} + 2c\sqrt{\frac{1 + \sin\phi}{1 - \sin\phi}}$$

For $\phi = 0°$,

$$q_u = \gamma z + 4c$$

or $q_u = 4c$ for a surface footing.

8.3.2 *Slip circle methods*

With slip circle methods the foundation is assumed to fail by rotation about some slip surface, usually taken as the arc of a circle. Almost all foundation failures exhibit rotational effects, and Fellenius (1927) showed that the centre of rotation is slightly above the base of the foundation and to one side of it. He found that in a saturated cohesive soil the ultimate bearing capacity for a surface footing is

$$q_u = 5.52c_u$$

To illustrate the method we will consider a foundation failing by rotation about one edge and founded at a depth z below the surface of a saturated clay of unit weight γ and undrained strength c_u (Fig. 8.2).

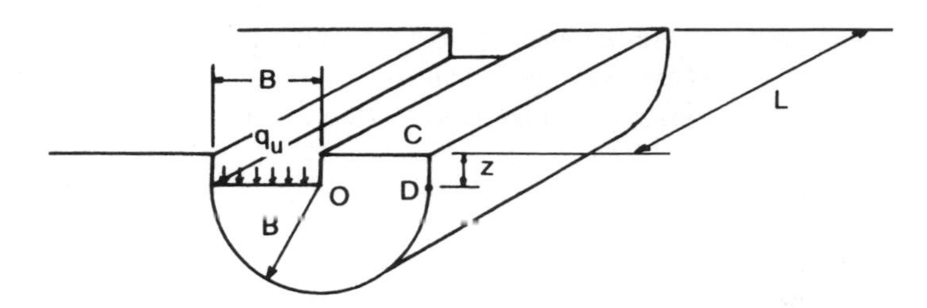

Fig. 8.2 Foundation failure rotation about one edge.

Disturbing moment about O:

$$q_u \times LB \times \frac{B}{2} = \frac{q_u LB^2}{2} \tag{1}$$

Resisting moments about O

Cohesion along cylindrical sliding surface $= c_u \pi LB$

$$\Rightarrow \quad \text{Moment} = \pi c_u LB^2 \tag{2}$$

Cohesion along CD $= c_u zL$

$$\Rightarrow \quad \text{Moment} = c_u zLB \tag{3}$$

Weight of soil above foundation level $= \gamma zLB$

$$\Rightarrow \quad \text{Moment} = \frac{\gamma zLB^2}{2} \tag{4}$$

For limit equilibrium (1) = (2) + (3) + (4)

i.e.

$$\frac{q_u LB^2}{2} = \pi c_u LB^2 + c_u zLB + \frac{\gamma zLB^2}{2}$$

$$\Rightarrow \quad q_u = 2\pi c_u + \frac{2c_u z}{B} + \gamma z$$

$$= 2\pi c_u \left(1 + \frac{1}{\pi}\frac{z}{B} + \frac{1}{2\pi}\frac{\gamma z}{c_u}\right)$$

$$= 6.28 c_u \left(1 + 0.32\frac{z}{B} + 0.16\frac{\gamma z}{c_u}\right)$$

Cohesion of end sectors

The above formula only applies to a strip footing, and if the foundation is of finite dimensions then the effect of the ends must be included.

To obtain this it is assumed that when the cohesion along the perimeter of the sector has reached its maximum value, c_u, the value of cohesion at some point on the sector at distance r from O is $c_r = c_u r/B$, as shown in Fig. 8.3.

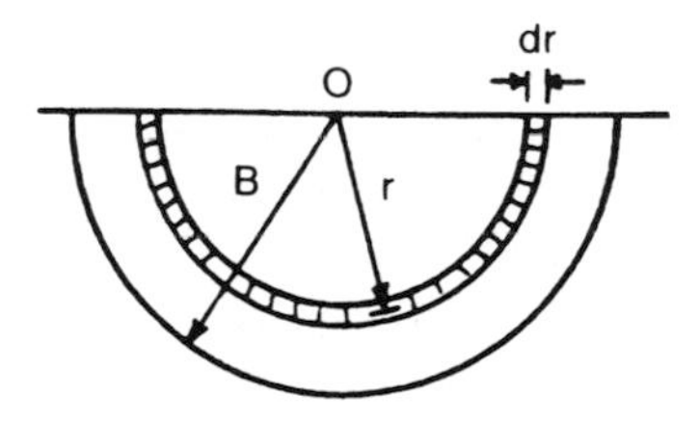

Fig. 8.3 Cohesion of end sectors.

Rotational resistance of an elemental ring, dr thick

$$= \frac{c_u r}{B} \times \pi r \, dr$$

$$\text{Moment about O} = \frac{c_u r}{B} \times \pi r \, dr \times r = \pi \frac{c_u}{B} r^3 \, dr$$

$$\text{Total moment of both ends} = 2 \int_0^B \pi \frac{c_u}{B} r^3 \, dr$$

$$= 2\pi \frac{c_u}{B} \times \frac{B^4}{4} = \frac{\pi c_u B^3}{2} \qquad (5)$$

This analysis ignores the cohesion of the soil above the base of the foundation at the two ends, but unless the foundation is very deep this will have little effect on the value of q_u. The term (5) should be added into the original equation.

For a surface footing the formula for q_u is:

$$q_u = 6.28c_u$$

This value is high because the centre of rotation is actually above the base, but in practice a series of rotational centres are chosen and each circle is analysed (as for a slope stability problem) until the lowest q_u value has been obtained. The method can be extended to allow for frictional effects but is considered most satisfactory when used for cohesive soils; it was extended by Wilson (1941), who prepared a chart (Fig. 8.4) which gives the centre of the most critical circle for cohesive soils (his technique is not applicable to other categories of soil or to surface footings).

The slip circle method is useful when the soil properties beneath the foundation vary, since an approximate position of the critical circle can be obtained from

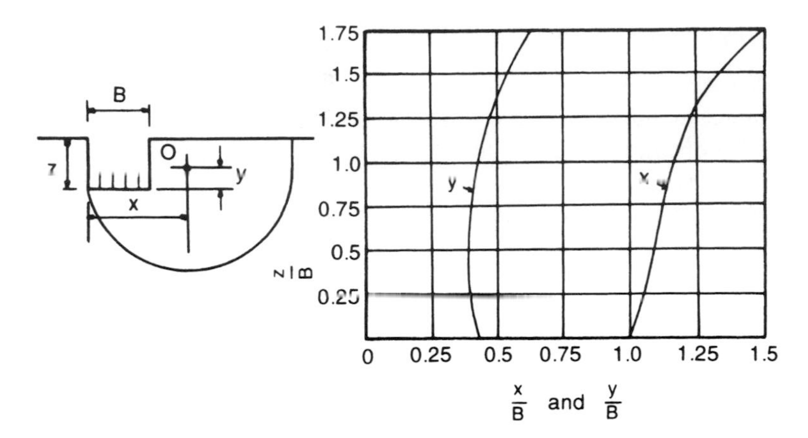

Fig. 8.4 Location of centre of critical circle for use with Fellenius' method (after Wilson, 1941).

Fig. 8.4 and then other circles near to it can be analysed. When the soil conditions are uniform Wilson's critical circle gives

$$q_u = 5.52c_u$$

for a surface footing.

8.3.3 *Plastic failure theory*

Forms of bearing capacity failure

Terzaghi (1943) stated that the bearing capacity failure of a foundation is caused by either a general soil shear failure or a local soil shear failure. Vesic (1963) listed punching shear failure as a further form of bearing capacity failure.

(1) General shear failure
The form of this failure is illustrated in Fig. 8.5, which shows a strip footing. The failure pattern is clearly defined and it can be seen that definite failure surfaces develop within the soil. A wedge of compressed soil (I) goes down with the footing, creating slip surfaces and areas of plastic flow (II). These areas are initially prevented from moving outwards by the passive resistance of the soil wedges (III). Once this passive resistance is overcome, movement takes place and bulging of the soil surface around the foundation occurs. With general shear failure collapse is sudden and is accompanied by a tilting of the foundation.

(2) Local shear failure
The failure pattern developed is of the same form as for general shear failure but only the slip surfaces immediately below the foundation are well defined. Shear failure is local and does not create the large zones of plastic failure which develop with

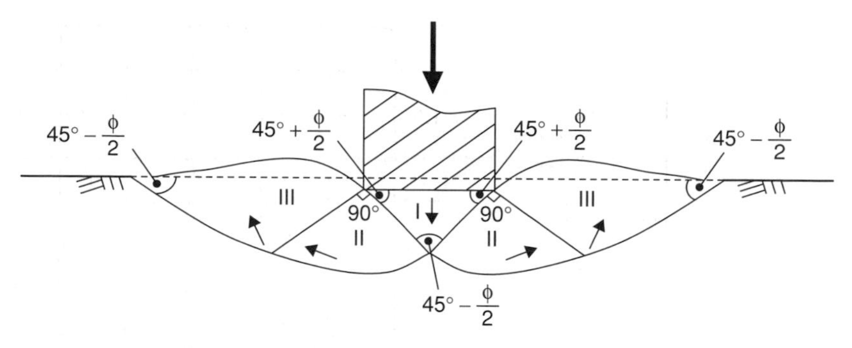

Fig. 8.5 General shear failure.

general shear failure. Some heaving of the soil around the foundation may occur but the actual slip surfaces do not penetrate the surface of the soil and there is no tilting of the foundation.

(3) Punching shear failure
This is a downward movement of the foundation caused by soil shear failure only occurring along the boundaries of the wedge of soil immediately below the foundation. There is little bulging of the surface of the soil and no slip surfaces can be seen.

For both punching and local shear failure, settlement considerations are invariably more critical than those of bearing capacity so that the evaluation of the ultimate bearing capacity of a foundation is usually obtained from an analysis of general shear failure.

Prandtl's analysis

Prandtl (1921) was interested in the plastic failure of metals and one of his solutions (for the penetration of a punch into metal) can be applied to the case of a foundation penetrating downwards into a soil with no attendant rotation.

The analysis gives solutions for various values of ϕ, and for a surface footing with $\phi = 0$, Prandtl obtained:

$$q_u = 5.14c$$

Terzaghi's analysis

Working on similar lines to Prandtl's analysis, Terzaghi (1943) produced a formula for q_u which allows for the effects of cohesion and friction between the base of the footing and the soil and is also applicable to shallow ($z/B \leq 1$) and surface foundations. His solution for a strip footing is:

$$q_u = cN_c + \gamma z N_q + 0.5\gamma B N_\gamma \qquad (6)$$

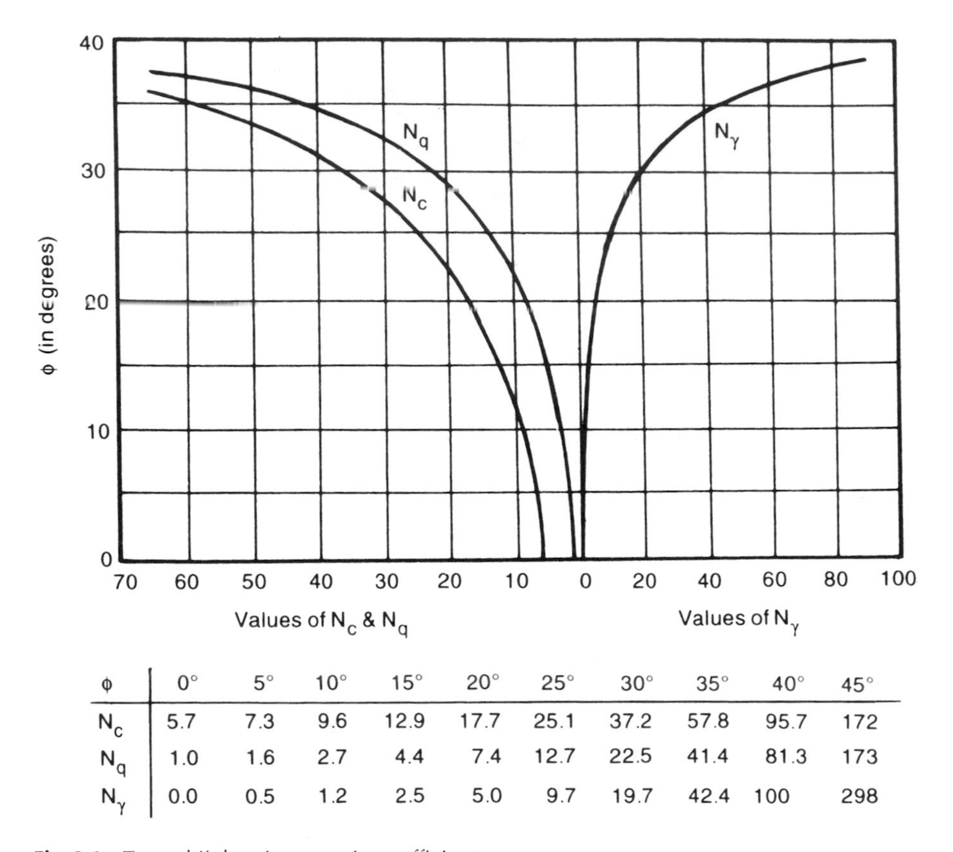

φ	0°	5°	10°	15°	20°	25°	30°	35°	40°	45°
N_c	5.7	7.3	9.6	12.9	17.7	25.1	37.2	57.8	95.7	172
N_q	1.0	1.6	2.7	4.4	7.4	12.7	22.5	41.4	81.3	173
N_γ	0.0	0.5	1.2	2.5	5.0	9.7	19.7	42.4	100	298

Fig. 8.6 Terzaghi's bearing capacity coefficients.

The coefficients N_c, N_q and N_γ depend upon the soil's angle of shearing resistance and can be obtained from Fig. 8.6. When $\phi = 0°$, $N_c = 5.7$; $N_q = 1.0$; $N_\gamma = 0$.

$$\Rightarrow \quad q_u = 5.7c + \gamma z$$

or $q_u = 5.7c$ for a surface footing.

The increase in the value of N_c from 5.14 to 5.7 is due to the fact that Terzaghi allowed for frictional effects between the foundation and its supporting soil.

The coefficient N_q allows for the surcharge effects due to the soil above the foundation level, and N_γ allows for the size of the footing, B. The effect of N_γ is of little consequence with clays, where the angle of shearing resistance is usually assumed to be the undrained value, ϕ_u, and assumed equal to 0°, but it can become significant with wide foundations supported on cohesionless soil.

Terzaghi's solution for a circular footing is:

$$q_u = 1.3cN_c + \gamma z N_q + 0.3\gamma B N_\gamma \quad \text{(where B = diameter)} \qquad (7)$$

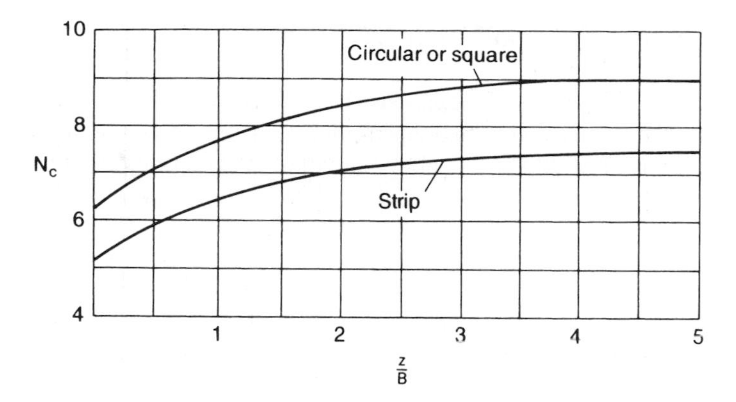

Fig. 8.7 Variation of the coefficient N_c with depth (after Skempton, 1951).

For a square footing:

$$q_u = 1.3cN_c + \gamma z N_q + 0.4\gamma B N_\gamma \tag{8}$$

and for a rectangular footing:

$$q_u = cN_c\left(1 + 0.3\,\frac{B}{L}\right) + \gamma z N_q + 0.5\gamma B N_\gamma\left(1 - 0.2\,\frac{B}{L}\right) \tag{9}$$

Skempton (1951) showed that for a cohesive soil ($\phi = 0°$) the value of the coefficient N_c increases with the value of the foundation depth, z. His suggested values for N_c, applicable to circular, square and strip footings, are given in Fig. 8.7. In the case of a rectangular footing on a cohesive soil a value for N_c can either be estimated from Fig. 8.7 or obtained from the formula:

$$N_c = 5\left(1 + 0.2\,\frac{B}{L}\right)\left(1 + 0.2\,\frac{z}{B}\right)$$

with a limiting value for N_c of $N_c = 7.5(1 + 0.2B/L)$, which corresponds to a z/B ratio greater than 2.5 (Skempton, 1951).

8.3.4 *Summary of bearing capacity formula*

It can be seen that Rankine's theory does not give satisfactory results and that, for variable subsoil conditions, the slip surface analysis of Fellenius provides the best solution. For normal soil conditions, Equations (6)–(9) can generally be used and may be applied to foundations at any depth in c–ϕ soils and to shallow foundations in cohesive soils. For deep footings in cohesive soil the values of N_c suggested by Skempton may be used in place of the Terzaghi values.

8.3.5 Choice of soil parameters

As with earth pressure equations, bearing capacity equations can be used with either the undrained or the drained soil parameters. As granular soils operate in the drained state at all stages during and after construction, the relevant soil strength parameter is ϕ'.

Saturated cohesive soils operate in the undrained state during and immediately after construction and the relevant parameters are c_u and ϕ_u (with ϕ_u generally assumed equal to zero). If required, the long-term stability can be checked with the assumption that the soil will be drained and the relevant parameters are c′ and ϕ' (with c′ generally taken as equal to zero) but this procedure is not often carried out.

Example 8.1

A rectangular foundation, 2 m × 4 m, is to be founded at a depth of 1 m below the surface of a deep stratum of soft saturated clay (unit weight = 20 kN/m^3).

Undrained and consolidated undrained triaxial tests established the following soil parameters: $\phi_u = 0°$, $c_u = 24$ kPa; $\phi' = 25°$, $c' = 0$.

Determine the ultimate bearing capacity of the foundation, (i) immediately after construction and, (ii) some years after construction.

Solution

(i) It may be assumed that immediately after construction the clay will be in an undrained state. The relevant soil parameters are therefore $\phi_u = 0°$ and $c_u = 24$ kPa.

From Fig. 8.6: $N_c = 5.7$, $N_q = 1.0$, $N_\gamma = 0.0$.

$$
\begin{aligned}
q_u &= cN_c(1 + 0.3B/L) + \gamma z N_q \\
&= 24 \times 5.7(1 + 0.3 \times 2/4) + 20 \times 1 \times 1 \\
&= 177.3 \text{ kPa}
\end{aligned}
$$

(ii) It can be assumed that, after some years, the clay will be fully drained so that the relevant soil parameters are $\phi' = 25°$ and $c' = 0$.

From Fig. 8.6: $N_c = 25.1$, $N_q = 12.7$. $N_\gamma = 9.7$.

$$
\begin{aligned}
q_u &= \gamma z N_q + 0.5\gamma B N_\gamma(1 - 0.2B/L) \\
&= 20 \times 1 \times 12.7 + 0.5 \times 20 \times 2 \times 9.7(1 - 0.2 \times 2/4) \\
&= 428.6 \text{ kPa}
\end{aligned}
$$

Example 8.2

A continuous foundation is 1.5 m wide and is founded at a depth of 1.5 m in a deep layer of sand of unit weight 18.5 kN/m^3.

Determine the ultimate bearing capacity of the foundation if the soil strength parameters are $c' = 0$ and $\phi' =$ (i) 35°, (ii) 30°.

Solution

(i) From Fig. 8.6: for $\phi' = 35°$, $N_c = 57.8$, $N_q = 41.4$, $N_\gamma = 42.4$. For a continuous footing:

$$q_u = c'N_c + \gamma z N_q + 0.5\gamma B N_\gamma$$
$$= 18.5 \times 1.5 \times 41.4 + 0.5 \times 18.5 \times 1.5 \times 42.4$$
$$= 1737 \text{ kPa}$$

(ii) From Fig. 8.6: for $\phi' = 30°$, $N_c = 37.2$, $N_q = 22.5$, $N_\gamma = 19.7$.

$$q_u = 18.5 \times 1.5 \times 22.5 + 0.5 \times 18.5 \times 1.5 \times 19.7$$
$$= 898 \text{ kPa}$$

The ultimate bearing capacity is reduced by some 48 per cent when the value of ϕ' is reduced by some 15 per cent.

8.4 Determination of the safe bearing capacity

Lumped factor of safety approach

The value of the safe bearing capacity is simply the value of the net ultimate bearing capacity divided by a suitable factor of safety, F. The value of F is usually not less than 3.0, except for a relatively unimportant structure, and sometimes can be as much as 5.0. At first glance these values for F appear high but the necessity for them is illustrated in Example 8.2, which demonstrates the effect on q_u of a small variation in the value of ϕ.

The net ultimate bearing capacity is the increase in vertical pressure, above that of the original overburden pressure, that the soil can just carry before shear failure occurs.

The original overburden pressure is γz and this term should be subtracted from the bearing capacity equations, i.e. for a strip footing:

$$q_{u\,net} = cN_c + \gamma z(N_q - 1) + 0.5\gamma B N_\gamma$$

The safe bearing capacity is therefore the above expression divided by F plus the term γz:

$$\text{Safe bearing capacity} = \frac{cN_c + \gamma z(N_q - 1) + 0.5\gamma B N_\gamma}{F} + \gamma z$$

In the case of a footing founded in undrained clay, where $\phi_u = 0°$, the net ultimate bearing capacity is, of course, $c_u N_c$.

The safe bearing capacity notion is not used during design to Eurocode 7 where, as will be demonstrated in Section 8.7, conformity of the bearing resistance limit state is achieved by ensuring that the design effect of the actions does not exceed the design bearing resistance.

8.5 The effect of groundwater on bearing capacity

Water table below the foundation level

If the water table is at a depth of not less than B below the foundation, the expression for net ultimate bearing capacity is the one given above, but when the water table rises to a depth of less than B below the foundation the expression becomes:

$$q_{u\,net} = cN_c + \gamma z(N_q - 1) + 0.5\gamma' BN_\gamma$$

where

γ = unit weight of soil above groundwater level
γ' = effective unit weight.

For cohesive soils ϕ_u is small and the term $0.5\gamma' BN_\gamma$ is of little account, the value of the bearing capacity being virtually unaffected by groundwater. With sands, however, the term cN_c is zero and the term $0.5\gamma' BN_\gamma$ is about one half of $0.5\gamma BN_\gamma$, so that groundwater has a significant effect.

Water table above the foundation level

For this case Terzaghi's expressions are best written in the form:

$$q_{u\,net} = cN_c + \sigma'_v(N_q - 1) + 0.5\gamma' BN_\gamma$$

where σ'_v = effective overburden pressure removed.

From the expression it will be seen that, in these circumstances, the bearing capacity of a cohesive soil can be affected by groundwater.

8.6 Developments in bearing capacity equations

Terzaghi's bearing capacity equations have been successfully used in the design of numerous shallow foundations throughout the world and are still in use. However, they are viewed by many to be conservative as they do not consider factors that affect bearing capacity such as inclined loading, foundation depth and the shear resistance of the soil above the foundation. This section describes developments that have been made to the original equations.

8.6.1 General form of the bearing capacity equation

Meyerhof (1963) proposed the following general equation for q_u:

$$q_u = cN_c s_c i_c d_c + \gamma z N_q s_q i_q d_q + 0.5\gamma B N_\gamma s_\gamma i_\gamma d_\gamma \quad (10)$$

where

s_c, s_q and s_γ are shape factors
i_c, i_q and i_γ are inclination factors
d_c, d_q and d_γ are depth factors.

Other factors, G_c, G_q and G_γ to allow for a sloping ground surface, and B_c, B_q and B_γ to allow for any inclination of the base, can also be included when required.

It must be noted that the values of N_c, N_q and N_γ used in the general bearing capacity equation are not the Terzaghi values. The values of N_c and N_q are now obtained from Meyerhof's equations (1963), as they are recognised as probably being the most satisfactory.

$$N_c = (N_q - 1)\cot\phi, \qquad N_q = \tan^2\left(45° + \frac{\phi}{2}\right)e^{\pi\tan\phi}$$

Unfortunately there is not the same agreement about the remaining factor N_γ and the following expressions all have their supporters:

$N_\gamma = (N_q - 1)\tan 1.4\phi$ Meyerhof (1963)
$N_\gamma = 1.5(N_q - 1)\tan\phi$ Hansen (1970)
$N_\gamma = 2(N_q + 1)\tan\phi$ Vesic (1973)

It should be noted that Hansen suggested that the operating value of ϕ should be that corresponding to plane strain, which is some 10 per cent greater than the value of ϕ obtained from the triaxial test and normally used. With this approach Hansen's expression for $N_\gamma = 1.5(N_q - 1)\tan 1.1\phi$, which applies to a continuous footing but is probably not so relevant to other shapes of footings.

In order to give the reader some guidance it can be said that the expression suggested by Vesic is being increasingly used. Further examples in this chapter will therefore use the following expressions for the bearing capacity coefficients:

$$N_c = (N_q - 1)\cot\phi$$

$$N_q = \tan^2\left(45° + \frac{\phi}{2}\right)e^{\pi\tan\phi}$$

$$N_\gamma = 2(N_q + 1)\tan\phi$$

Typical values are shown in Table 8.1.

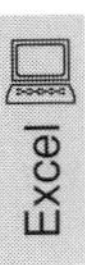

Table 8.1 Bearing capacity factors in common use.

ϕ (°)	N_c	N_q	N_γ
0	5.14	1.00	0.00
5	6.49	1.57	0.45
10	8.34	2.47	1.22
15	10.98	3.94	2.65
20	14.83	6.40	5.39
25	20.72	10.66	10.88
30	30.14	18.40	22.40
35	46.12	33.30	48.03
40	75.31	64.20	109.41
45	133.87	134.87	271.75
50	266.88	319.06	762.86

8.6.2 Shape factors

These factors are intended to allow for the effect of the shape of the foundation on its bearing capacity. The factors have largely been evaluated from laboratory tests and the values in present use are those proposed by De Beer (1970):

$$s_c = 1 + \frac{B}{L} \cdot \frac{N_q}{N_c}$$

$$s_q = 1 + \frac{B}{L} \tan \phi$$

$$s_\gamma = 1 - 0.4 \frac{B}{L}$$

8.6.3 Depth factors

These factors are intended to allow for the shear strength of the soil above the foundation. Hansen (1970) proposed the following values:

	$z/B \leq 1.0$	$z/B > 1.0$
d_c	$1 + 0.4(z/B)$	$1 + 0.4 \arctan(z/B)$
d_q	$1 + 2 \tan \phi(1 - \sin \phi)^2(z/B)$	$1 + 2 \tan \phi(1 - \sin \phi)^2 \arctan(z/B)$
d_γ	1.0	1.0

Note The arctan values must be expressed in radians, e.g. if z = 1.5 and B = 1.0 m then arctan(z/B) = arctan(1.5) = 56.3° = 0.983 radians.

Excel

Example 8.3

Recalculate Example 8.1 using Meyerhof's general bearing capacity formula.

Solution

(i) From Table 8.1, for $\phi_u = 0°$, $N_c = 5.14$, $N_q = 1.0$ and $N_\gamma = 0.0$.

Shape factors:

$$s_c = 1 + (2/4)(1.0/5.14) = 1.1$$
$$s_q = 1 + (2/4)\tan 0° = 1.0$$
$$s_\gamma = 1 - 0.4(2/4) = 0.8$$

Depth factors:

$z/B = 1/2 = 0.5$. Using Hansen's values for $z/B \le 1.0$:

$$d_c = 1 + 0.4(1/2) = 1.2, \qquad d_q = 1.0 \text{ (as } \phi_u = 0°), \qquad d_\gamma = 1.0$$

$$q_u = cN_c s_c d_c + \gamma z N_q s_q d_q$$
$$= 24 \times 5.14 \times 1.1 \times 1.2 + 20 \times 1.0 \times 1.0 \times 1.0$$
$$= 182.8 \text{ kPa}$$

(ii) From Table 8.1, for $\phi' = 25°$, $N_q = 10.66$ and $N_\gamma = 10.88$.

The expressions for s_q and d_q involve ϕ. These two factors will therefore have different values from those in case (i):

$$s_q = 1 + (2/4)\tan 25° = 1.23$$
$$d_q = 1 + 2\tan 25°(1 - \sin 25°)^2(1/2) = 1.16$$
$$q_u = \gamma z N_q s_q d_q + 0.5\gamma B N_\gamma s_\gamma d_\gamma$$
$$= 20 \times 1 \times 10.66 \times 1.23 \times 1.16 + 0.5 \times 20 \times 2 \times 10.88 \times 0.8 \times 1.0$$
$$= 478.3 \text{ kPa}$$

Example 8.4

Using a factor of safety = 3.0 determine the values of safe bearing capacity for cases (i) and (ii) in Example 8.3.

Solution

Case (i):

$$q_{u\,net} = q_u - \gamma z = 162.8 \text{ kPa}$$

$$\text{Safe bearing capacity} = \frac{162.8}{3} + 20 \times 1$$
$$= 74.3 \text{ kPa}$$

For case (ii):

$$q_{u\,net} = \gamma z(N_q s_q d_q - 1) + 0.5\gamma B N_\gamma s_\gamma d_\gamma$$
$$= 458.3 \text{ kPa}$$

$$\text{Safe bearing capacity} = \frac{458.3}{3} + 20 \times 1$$
$$= 172.8 \text{ kPa}$$

8.6.4 Effect of eccentric and inclined loading on foundations

A foundation can be subjected to eccentric loads and/or to inclined loads, eccentric or concentric.

Eccentric loads

Let us consider first the relatively simple case of a vertical load acting on a rectangular foundation of width B and length L such that the load has eccentricities e_B and e_L (Fig. 8.8). To solve the problem we must think in terms of the rather artificial concept of effective foundation width and length. That part of the foundation that is symmetrical about the point of application of the load is considered to be useful, or effective, and is the area of the rectangle of effective length $L' = L - 2e_L$ and of effective width $B' = B - 2e_B$.

In the case of a strip footing of width B, subjected to a line load with an eccentricity e, then $B' = B - 2e$ and the ultimate bearing capacity of the foundation is found from either equation (6) or the general equation (10) with the term B replaced by B′.

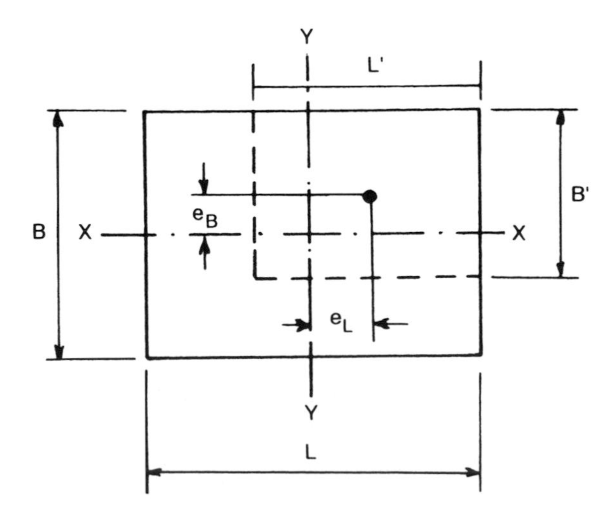

Fig. 8.8 Effective widths and area.

The overall eccentricity of the bearing pressure, e, must consider the self-weight of the foundation and is equal to:

$$e = \frac{P \times e_P}{P + W}$$

where

P = magnitude of the eccentric load
W = self-weight of the foundation
e_p = eccentricity of P.

Inclined loads

The usual method of dealing with an inclined line load, such as P in Fig. 8.9, is to first determine its horizontal and vertical components P_H and P_V and then, by taking moments, determine its eccentricity, e, in order that the effective width of the foundation B′ can be determined from the formula B′ = B − 2e.

The ultimate bearing capacity of the strip foundation (of width B) is then taken to be equal to that of a strip foundation of width B′ subjected to a concentric load, P, inclined at α to the vertical.

Various methods of solution have been proposed for this problem, e.g. Janbu (1957), Hansen (1957), but possibly the simplest approach is that proposed by Meyerhof (1953) in which the bearing capacity coefficients N_c, N_q and N_y are reduced by multiplying them by the factors i_c, i_q and i_γ in his general equation (10). Meyerhof's expressions for these factors are:

$$i_c = i_q = (1 - \alpha/90°)^2$$
$$i_\gamma = (1 - \alpha/\phi)^2$$

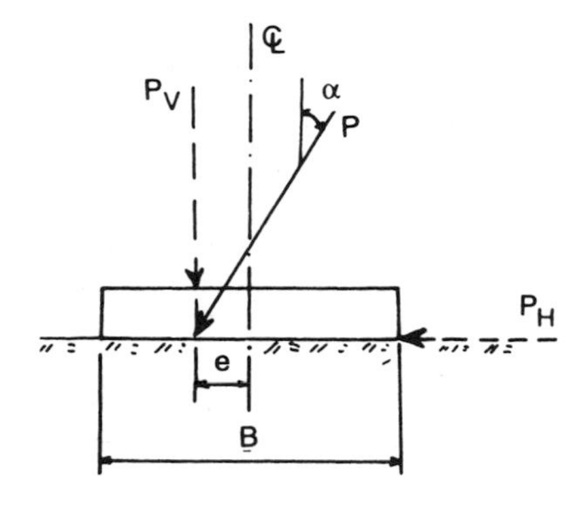

Fig. 8.9 Strip foundation with inclined load.

8.7 Designing spread foundations to Eurocode 7

The design of spread foundations is covered in Section 6 of Eurocode 7. The limit states to be checked and the partial factors to be used in the design are the same as we saw when we looked at the design of retaining walls in Section 7.4.2.

In terms of establishing the bearing resistance, the code states that a commonly recognised method should be used, and Annex D of the Standard gives a sample calculation. Interestingly the depth factors are excluded in Eurocode 7 (without explanation) and for this reason they are excluded too from the solutions to Examples 8.5 and 8.6 in this chapter. Spreadsheets *Example 8.5.xls* and *Example 8.6.xls*, however, offer the choice whether to include the depth factors or not.

While the design procedure required to satisfy the conditions of Eurocode 7 involves essentially the same methods as we have seen so far in this chapter, there are a few differences listed in Annex D which can be considered for drained conditions. These concern the shape and inclination factors as well as the bearing resistance factor, N_γ, and are listed below:

$N_\gamma = 2(N_q - 1)\tan\phi'$ (for a rough base, such as a typical foundation)
$s_q = 1 + (B'/L')\sin\phi'$ (for a rectangular foundation)
$s_q = 1 + \sin\phi'$ (for a square or circular foundation)
$s_\gamma = 1 - 0.3(B'/L')$ (for a rectangular foundation)
$s_\gamma = 0.7$ (for a square or circular foundation)

$$s_c = \frac{s_q N_q - 1}{N_q - 1}$$ (rectangular, square and circle foundation)

$$i_c = i_q - \left(\frac{1 - i_q}{N_c \tan\phi'}\right); \quad i_q = \left[1 - \frac{H}{V + A'c'\cot\phi'}\right]^m; \quad i_\gamma = i_q^{\left(\frac{m+1}{m}\right)}$$

where

V = vertical load acting on foundation
H = horizontal load (or component of inclined load) acting on foundation
A′ = design effective area of foundation

$$m = m_B = \frac{2 + \frac{B'}{L'}}{1 + \frac{B'}{L'}}$$ when H acts in the direction of B′;

$$m = m_L = \frac{2 + \frac{L'}{B'}}{1 + \frac{L'}{B'}}$$ when H acts in the direction of L′.

Eurocode 7 also states that the vertical total action should include the weight of any backfill acting on top of the foundation in addition to the weight of the foundation itself plus the applied load it is carrying.

Excel

Example 8.5

A continuous footing is 1.8 m wide by 0.5 m deep and is founded at a depth of 0.75 m in a clay soil of unit weight 20 kN/m^3 with $\phi_u = 0°$ and $c_u = 30$ kPa. The foundation is to carry a vertical line load of magnitude 50 kN/m run, which will act at a distance of 0.4 m from the centre-line. Take the unit weight of concrete as 24 kN/m^3.

(i) Determine the safe bearing capacity for the footing, taking F = 3.0.
(ii) Check the Eurocode 7 GEO limit state (Design Approach 1) by establishing the magnitude of the over-design factor.

Solution

(i) Safe bearing capacity

Self-weight of foundation, $W_f = 0.5 \times 24 \times 1.8 = 21.6$ kN/m run
Weight of soil on top of foundation, $W_s = 0.25 \times 20 \times 1.8 = 9.0$ kN/m run
Total weight of foundation + soil, $W = 21.6 + 9.0 = 30.6$ kN/m run

$$\text{Eccentricity of bearing pressure, } e = \frac{P \times e_P}{P + W} = \frac{50 \times 0.4}{50 + 30.6} = 0.25 \text{ m}$$

Since $e \le \frac{B}{6}$, the total force acts within the middle third of the foundation.
Effective width of footing, $B' = 1.8 - 2 \times 0.25 = 1.3$ m
From Table 8.1, for $\phi_u = 0°$, $N_c = 5.14$, $N_q = 1.0$, $N_\gamma = 0$.
Footing is continuous, i.e. $L \to \infty$; $s_c = 1.0$.

$$d_c = 1 + 0.4\left(\frac{0.75}{1.8}\right) = 1.17$$

$$\text{Safe bearing capacity (per metre run)} = \frac{q_{u\,net}}{3} + \gamma z = \frac{cN_cs_cd_c}{3} + \gamma z$$

$$= \frac{30 \times 5.14 \times 1.0 \times 1.17}{3} + 20 \times 0.75$$

$$= 75 \text{ kPa}$$

Safe bearing capacity $= 75 \times B' = 97.5$ kN/m run

(ii) Eurocode 7 GEO limit state

1. Combination 1 (partial factor sets A1 + M1 + R1)
From Table 7.1: $\gamma_{G;dst} = 1.35$; $\gamma_{Q;dst} = 1.5$; $\gamma_{cu} = 1.0$; $\gamma_{Rv} = 1.0$.

$$\text{Design material property: } c_{u;d} = \frac{c_u}{\gamma_{cu}} = \frac{30}{1} = 30 \text{ kPa}$$

Design actions:

Weight of foundation, $W_d = W \times \gamma_{G;dst} = 30.6 \times 1.35 = 41.3$ kN/m run
Applied line load, $P_d = P \times \gamma_{G;dst} = 50 \times 1.35 = 67.5$ kN/m run

Effect of design actions:

Total vertical force, $F_d = 41.3 + 67.5 = 108.8$ kN/m run

$$\text{Eccentricity, } e = \frac{P_d \times e_P}{P_d + W_d} = \frac{67.5 \times 0.4}{67.5 + 41.3} = 0.248 \text{ m}$$

Since $e \leq \frac{B}{6}$, the total force acts within the middle-third of the foundation.

Effective width of footing, $B' = 1.8 - 2 \times 0.248 = 1.3$ m

Design resistance:

From before, $N_c = 5.14$, $N_q = 1.0$, $N_\gamma = 0$, $s_c = 1.0$.

$$\begin{aligned}\text{Ultimate bearing capacity, } q_u &= c_{u;d}N_c s_c + \gamma z N_q \\ &= 30 \times 5.14 \times 1 + 20 \times 0.75 \times 1.0 \\ &= 169.2 \text{ kPa}\end{aligned}$$

Ultimate bearing capacity per metre run, $Q_u = 169.2 \times 1.3 = 220$ kN/m run

$$\text{Bearing resistance, } R_d = \frac{Q_u}{\gamma_{Rv}} = \frac{220}{1} = 220 \text{ kN/m run}$$

$$\text{Over-design factor, } \Gamma = \frac{R_d}{F_d} = \frac{220}{108.8} = 2.03$$

Since $\Gamma > 1$, the GEO limit state requirement is satisfied.

2. Combination 2 (partial factor sets A2 + M2 + R1)

The calculations are the same as for Combination 1 except that this time the following partial factors (from Table 7.1) are used: $\gamma_{G;dst} = 1.0$; $\gamma_{Q;dst} = 1.3$; $\gamma_{cu} = 1.40$; $\gamma_{Rv} = 1.0$.

$c_{u;d} = 21.4$ kPa
$W_d = 30.6 \times \gamma_{G;dst} = 30.6$ kN/m run
$P_d = 50.0 \times \gamma_{G;dst} = 50.0$ kN/m run
$F_d = 30.6 + 50.0 = 80.6$ kN/m run
$e = 0.248$ m; $B' = 1.3$ m
$Q_u = (c_{u;d}N_c s_c + \gamma z N_q) \times B' = 125.1 \times 1.3 = 163.1$ kN/m run

$$R_d = \frac{Q_u}{\gamma_{Rv}} = \frac{163.1}{1} = 163.1 \text{ kN/m run}$$

$$\Gamma = \frac{R_d}{F_d} = \frac{163.1}{80.6} = 2.02$$

Since $\Gamma > 1$, the GEO limit state requirement is satisfied.

Excel

Example 8.6

A concrete foundation 3 m wide, 9 m long and 0.75 m deep is to be founded at a depth of 1.5 m in a deep deposit of dense sand. The angle of shearing resistance of the sand is 35° and its unit weight is 19 kN/m³. The unit weight of concrete is 24 kN/m³.

(a) Using a lumped factor of safety approach (take F = 3.0):

(i) Determine the safe bearing capacity for the foundation.
(ii) Determine the safe bearing capacity of the foundation if it is subjected to a vertical line load of 220 kN/m at an eccentricity of 0.3 m, together with a horizontal line load of 50 kN/m acting at the base of the foundation as illustrated in Figure 8.10.

(b) For the situation described in (ii) above, establish the magnitude of the over-design factor for the Eurocode 7 GEO limit state, using Design Approach 1.

Solution

(a) Lumped factor of safety

(i)

Safe bearing capacity

$$= \frac{q_{u\,net}}{3} + \gamma z$$

$$= \frac{\gamma z(N_q s_q d_q - 1) + 0.5\gamma B N_\gamma s_\gamma d_\gamma}{3} + \gamma z$$

From Table 8.1, for $\phi' = 35°$, $N_q = 33.3$, $N_\gamma = 48.03$:

$$s_q = 1 + \left(\frac{3}{9}\right)\tan 35° = 1.23; \qquad s_\gamma = 1 - 0.4\left(\frac{3}{9}\right) = 0.87$$

$$d_q = 1 + 2\tan 35°(1 - \sin 35°)^2\left(\frac{1.5}{3}\right) = 1.13; \qquad d_\gamma = 1$$

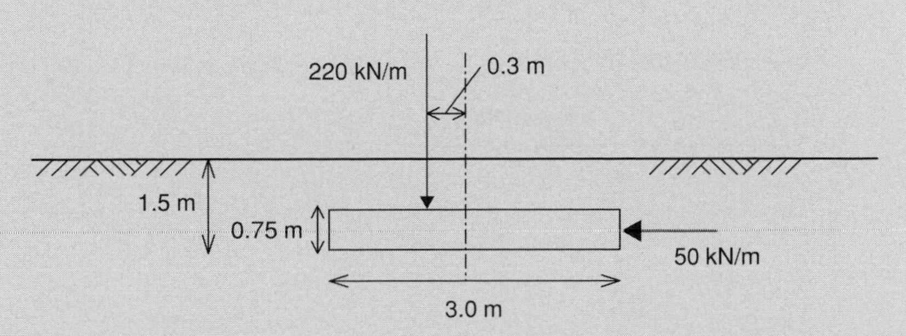

Fig. 8.10 Example 8.6.

Safe bearing capacity

$$= \frac{19 \times 1.5(33.3 \times 1.23 \times 1.13 - 1) + 0.5 \times 19 \times 3 \times 48.03 \times 0.87 \times 1.0}{3} + 19 \times 1.5$$

$= 855.7$ kPa

(ii)
Self-weight of foundation, $W = 0.75 \times 9 \times 3 \times 24 = 486$ kN
Total applied vertical load, $P = 220 \times 9 = 1980$ kN
Total applied horizontal load, $H = 50 \times 9 = 450$ kN
Total vertical load acting on soil, $V = 486 + 1980 = 2466$ kN

Eccentricity of bearing pressure

$$e = \frac{P \times e_P}{P + W} = \frac{1980 \times 0.3}{2466} = 0.24 \text{ m}$$

Since $e \leq \frac{B}{6}$, the total force acts within the middle-third of the foundation.

Effective width of footing, $B' = 3.0 - 2 \times 0.24 = 2.52$ m

The foundation is effectively acted upon by a load of magnitude, F inclined at an angle to the vertical, α:

$$F = \sqrt{V^2 + H^2} = \sqrt{2466^2 + 450^2} = 2506.7 \text{ kN}$$

$$\alpha = \tan^{-1}\left(\frac{450}{2466}\right) = 10.3°$$

$$i_q = \left(1 - \frac{10.3}{90}\right)^2 = 0.78; \qquad i_\gamma = \left(1 - \frac{10.3}{35}\right)^2 = 0.50$$

$$s_q = 1 + \left(\frac{2.52}{9}\right)\tan 35° = 1.2; \qquad s_\gamma = 1 \times 0.4\left(\frac{2.52}{9}\right) = 0.89$$

$$d_q = 1 + 2\tan 35°(1 - \sin 35°)^2\left(\frac{1.5}{2.52}\right) = 1.15; \qquad d_\gamma = 1$$

Safe bearing capacity

$$= \frac{\gamma z(N_q s_q d_q i_q - 1) + 0.5\gamma B' N_\gamma s_\gamma d_\gamma i_\gamma}{3} + \gamma z$$

$$= \frac{19 \times 1.5(33.3 \times 1.2 \times 1.15 \times 0.78 - 1) + 0.5 \times 19 \times 2.52 \times 48.03 \times 0.89 \times 1.0 \times 0.5}{3}$$

$+ 19 \times 1.5$

$= 530$ kPa

(b) Eurocode 7

Weight of soil on top of foundation, $W_s = 0.75 \times 9 \times 3 \times 19 = 384.8$ kN
Total weight of foundation + soil, $W = 486 + 384.8 = 870.8$ kN

1. Combination 1 (partial factor sets A1 + M1 + R1)
From Table 7.1: $\gamma_{G;dst} = 1.35$; $\gamma_{Q;dst} = 1.5$; $\gamma_{\phi'} = 1.0$; $\gamma_{Rv} = 1.0$.

Design material property: $\phi'_d = \tan^{-1}\left(\dfrac{\tan\phi'}{\gamma_{\phi'}}\right) = 35°$

Design actions:

Weight of foundation, $W_d = W \times \gamma_{G;dst} = 870.8 \times 1.35 = 1175.6$ kN
Applied vertical line load, $P_d = P \times \gamma_{G;dst} = 1980 \times 1.35 = 2673$ kN
Applied horizontal line load, $H_d = H \times \gamma_{G;dst} = 450 \times 1.35 = 607.5$ kN

Effect of design actions:

Total vertical force, $F_d = W_d + P_d = 1175.6 + 2673 = 3848.6$ kN

$$\text{Eccentricity, } e = \frac{P_d \times e_P}{P_d + W_d} = \frac{2673 \times 0.3}{3848.6} = 0.208 \text{ m}$$

Since $e \leq \dfrac{B}{6}$, the total force acts within the middle-third of the foundation.
Effective width of footing, $B' = 3.0 - 2 \times 0.208 = 2.58$ m
Effective area of footing, $A' = 2.58 \times 9 = 23.2\ m^2$

Design resistance:

From Table 8.1, $N_c = 46.1$, $N_q = 33.3$.

From Eurocode 7, Annex D,

$$N_\gamma = 2\,(N_q - 1)\tan\phi' = 45.2$$

$$s_q = 1 + \frac{B'}{L}\sin\phi' = 1 + \left(\frac{2.58}{9}\right)\sin 35° = 1.16$$

$$s_c = \frac{s_q N_q - 1}{N_q - 1} = 1.17$$

$$s_\gamma = 1 - 0.3\left(\frac{B'}{L}\right) = 0.91$$

$$m = \frac{2 + \dfrac{B'}{L}}{1 + \dfrac{B'}{L}} = 1.78$$

$$i_q = \left[1 - \frac{H}{V + A'c' \cot \phi'}\right]^m = \left[1 - \frac{607.5}{3848.6 + 0}\right]^{1.78} = 0.74 \qquad (V = F_d)$$

$$i_c = i_q - \left(\frac{1 - i_q}{N_c \tan \phi'}\right) = 0.74 - \left(\frac{1 - 0.74}{46.1 \tan 35^\circ}\right) = 0.72$$

$$i_\gamma = i_q^{\left(\frac{m+1}{m}\right)} = 0.74^{\frac{2.78}{1.78}} = 0.62$$

Ultimate bearing capacity, per m^2,

$$\begin{aligned} q_u &= c'_d N_c s_c i_c + \gamma_d z N_q s_q i_q + 0.5B'\gamma_d N_\gamma s_\gamma i_\gamma \\ &= 0 + (19 \times 1.5 \times 33.3 \times 1.16 \times 0.74) \\ &\quad + (0.5 \times 2.58 \times 19 \times 45.2 \times 0.91 \times 0.62) \\ &= 1439 \text{ kPa} \end{aligned}$$

Ultimate bearing capacity, $Q_u = q_u \times L \times B = 1439 \times 9 \times 3 = 38\,853$ kN

Bearing resistance, $R_d = \dfrac{Q_u}{\gamma_{Rv}} = \dfrac{38\,853}{1} = 38\,853$ kN

Over-design factor, $\Gamma = \dfrac{R_d}{F_d} = \dfrac{38\,853}{3848.6} = 10.1$

Since $\Gamma > 1$, the GEO limit state requirement is satisfied.

2. Combination 2 (partial factor sets A2 + M2 + R1)

The calculations are the same as for Combination 1 except that this time the following partial factors (from Table 7.1) are used: $\gamma_{G;dst} = 1.0$; $\gamma_{Q;dst} = 1.3$; $\gamma_{\phi'} = 1.25$; $\gamma_{Rv} = 1.0$.

$$\phi'_d = \tan^{-1}\left(\frac{\tan \phi'}{\gamma_{\phi'}}\right) = \tan^{-1}\left(\frac{\tan 35^\circ}{1.25}\right) = 29.3^\circ$$

$$W_d = 870.8 \times \gamma_G = 870.8 \text{ kN}$$
$$P_d = 1980 \times \gamma_G = 1980 \text{ kN}$$
$$H_d = 450 \times \gamma_G = 450 \text{ kN}$$

$$e = \frac{P_d \times e_P}{P_d + W_d} = \frac{1980 \times 0.3}{1980 + 870.8} = 0.208 \text{ m} \qquad \text{(within the middle-third)}$$

$$B' = 3.0 - 2 \times 0.208 = 2.58 \text{ m}$$

$N_q = 16.9$, $N_\gamma = 17.8$, $s_q = 1.14$, $s_\gamma = 0.91$, $i_q = 0.74$, $i_\gamma = 0.62$.

Ultimate bearing capacity, per m^2,

$$\begin{aligned} q_u &= c'_d N_c s_c i_c + \gamma z N_q s_q i_q + 0.5B'\gamma N_\gamma s_\gamma i_\gamma \\ &= 653.5 \text{ kPa} \end{aligned}$$

Ultimate bearing capacity, $Q_u = 653.5 \times L \times B = 17\,644$ kN

Bearing resistance, $R_d = \dfrac{Q_u}{\gamma_{Rv}} = \dfrac{17\,644}{1} = 17\,644$ kN

Over-design factor, $\Gamma = \dfrac{R_d}{F_d} = \dfrac{17\,644}{2850.1} = 6.19$

Since $\Gamma > 1$, the GEO limit state requirement is satisfied.

8.8 Non-homogeneous soil conditions

The bearing capacity equations (6)–(10) are based on the assumption that the foundation soil is homogeneous and isotropic.

In the case of variable soil conditions the analysis of bearing capacity can be carried out using some form of slip circle method, as described earlier in this chapter. This procedure can take time and designs based on one of the bearing capacity formulae are consequently quite often used.

For the case of a foundation resting on thin layers of soil, of thicknesses H_1, H_2, H_3, . . . H_n and of total depth H, Bowles (1982) suggests that these layers can be treated as one layer with an average c value c_{av} and an average ϕ value ϕ_{av}, where

$$c_{av} = \frac{c_1H_1 + c_2H_2 + c_3H_3 + \cdots + c_nH_n}{H}$$

$$\phi_{av} = \arctan \frac{H_1 \tan\phi_1 + H_2 \tan\phi_2 + H_3 \tan\phi_3 + \cdots + H_n \tan\phi_n}{H}$$

Vesic (1975) suggested that, for the case of a foundation founded in a layer of soft clay which overlies a stiff clay, the ultimate bearing capacity of the foundation can be expressed as:

$$q_u = c_uN_{cm} + \gamma_z$$

where c_u = the undrained strength of the soft clay and N_{cm} = a modified form of N_c, the value of which depends upon the ratio of the c_u values of both clays, the thickness of the upper layer, the foundation depth and the shape and width of the foundation. Values of N_{cm} are quoted in Vesic's paper.

The converse situation, i.e. that of a foundation founded in a layer of stiff clay which overlies a soft clay, has been studied by Brown and Meyerhof (1969), who quoted a formula for N_{cm} based on a punching shear failure analysis.

For other cases of more heterogeneous soil conditions there is at present no recognised method by which the bearing capacity equations can be realistically applied.

At first glance a safe way of determining the bearing capacity of a foundation might be to base it on the shear strength of the weakest soil below it, but such a procedure can be uneconomical, particularly if the weak soil is overlain by much stronger soil. A more suitable method is to calculate the safe bearing capacity using the shear strength of the stronger material and then to check the amount of overstressing that this will cause in the weaker layers. The method is shown in Example 8.7, which illustrates a typical problem that may arise during the selection of a site for a new spoil heap.

For structural foundations the factor of safety against bearing capacity failure is generally not less than 3.0, but for spoil heaps this factor can often be reduced to 2.0.

Example 8.7

The effective width of a proposed spoil heap will be about 61 m. The subsoil conditions on which the tip is to be built are shown in Fig. 8.11a.

Determine a value for the maximum safe pressure that may be exerted by the tip on to the soil.

Solution

The average undrained cohesion of the stiff clay is about 165 kPa.

Using this value with Terzaghi's formula:

$$q_u = cN_c = 165 \times 5.7 = 940 \text{ kPa}$$

Assign safe bearing capacity = 430 kPa; $F = \dfrac{940}{430} = 2.19$

Various vertical sections through the soil must now be selected (A, B, C, D and E in Fig. 8.11a). Using a contact pressure value of 430 kPa, the induced shear stresses are obtained from Fig. 4.11b, and for each section the variation in soil strength with depth is plotted along with the corresponding values of shear stress increments (Fig. 8.11b). From these plots the areas of overstressing (shown hatched) are apparent and it is possible to plot this area on a cross-section (Fig. 8.11c).

A considerable portion of the silt is overstressed and if this were applied to the design of a raft foundation carrying a normal structure it would not be acceptable. With a spoil heap, however, the amount of settlement induced would hardly be detrimental. Also, as the load will be applied gradually, there will be a chance for the silt to consolidate partially and obtain some increase in strength before the full load is applied.

Owing to the thickness of boulder clay there is little chance of a heave of the ground surface around the tip. For interest, the overstressed zone corresponding to a contact pressure of 320 kPa is also shown in Fig. 8.11c.

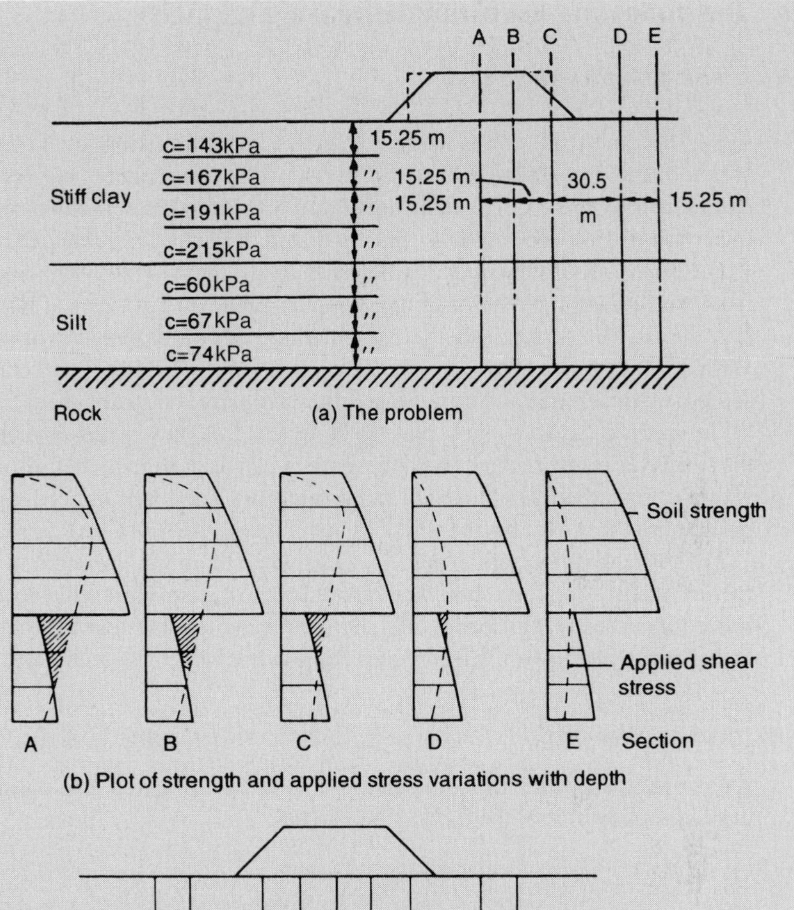

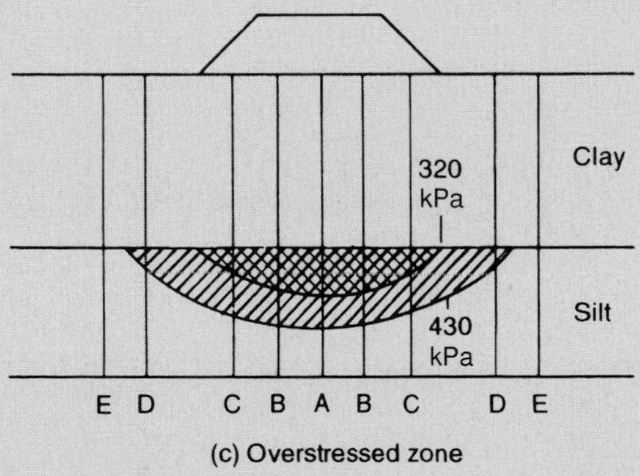

Fig. 8.11 Example 8.7.

If the contact pressure had been determined by considering the strength of the silt (average c = 67 kPa):

$q_u = 5.7 \times 67$ kPa = 382 kPa
Safe bearing capacity (F = 2) = 191 kPa

8.9 *In situ* testing for ultimate bearing capacity

8.9.1 The plate loading test

In this test an excavation is made to the expected foundation level of the proposed structure and a steel plate, usually from 300 to 750 mm square, is placed in position and loaded by means of a kentledge. During loading the settlement of the plate is measured and a curve similar to that illustrated in Fig. 8.12 is obtained.

On dense sands and gravels and stiff clays there is a pronounced departure from the straight line relationship that applies in the initial stages of loading, and the q_u value is then determined by extrapolating backwards (as shown in the figure). With a soft clay or a loose sand the plate experiences a more or less constant rate of settlement under load and no definite failure point can be established.

In spite of the fact that a plate loading test can only assess a metre or two of the soil layer below the test level, the method can be extremely helpful in stony soils where undisturbed sampling is not possible provided it is preceded by a boring programme, to prove that the soil does not exhibit significant variations.

The test can give erratic results in sands when there is a variation in density over the site, and several tests should be carrried out to determine a sensible average. This procedure is costly, particularly if the groundwater level is near the foundation level and groundwater lowering techniques consequently become necessary.

Estimation of allowable bearing pressure from the plate loading test

As would be expected, the settlement of a square footing kept at a constant pressure increases as the size of the footing increases.

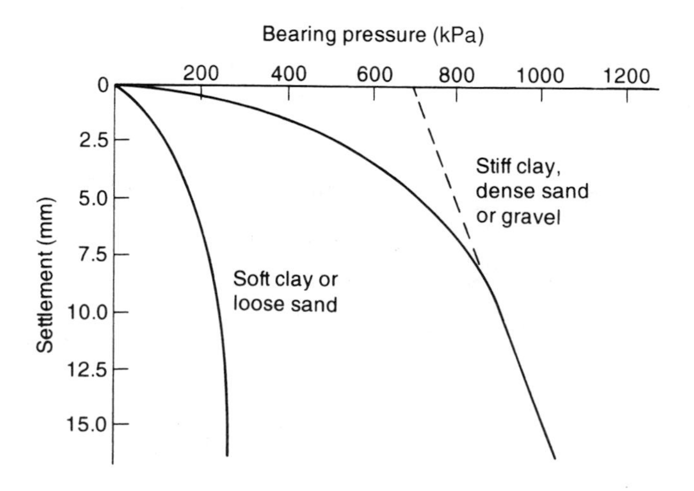

Fig. 8.12 Typical plate loading test results.

Terzaghi and Peck (1948) investigated this effect and produced the relationship:

$$S = S_1\left(\frac{2B}{B + 0.3}\right)^2$$

where

S_1 = settlement of a loaded area 0.305 m square under a given loading intensity p
S = settlement of a square or rectangular footing of width B (in metres) under the same pressure p.

In order to use plate loading test results the designer must first decide upon an acceptable value for the maximum allowable settlement. Unless there are other conditions to be taken into account it is generally accepted that maximum allowable settlement is 25 mm.

The method for determining the allowable bearing pressure for a foundation of width B m is apparent from the formula. If S is put equal to 25 mm and the numerical value of B is inserted in the formula, S_1 will be obtained. From the plate loading test results we have the relationship between S_1 and p (Fig. 8.12), so the value of p corresponding to the calculated value of S_1 is the allowable bearing pressure of the foundation subject to any adjustment that may be necessary for certain groundwater conditions. The adjustment procedure is the same as that employed to obtain the allowable bearing pressure from the standard penetration test.

8.9.2 *Standard penetration test*

This test is generally used to determine the bearing capacity of sands or gravels and is conducted with a split spoon sampler (a sample tube which can be split open longitudinally after sampling) with internal and external diameters of 35 and 50 mm respectively. The sampler is sometimes referred to as the Raymond spoon sampler after the piling firm that evolved the test (Fig. 8.13). A full guide on the methods and use of the SPT is given by Clayton (1995).

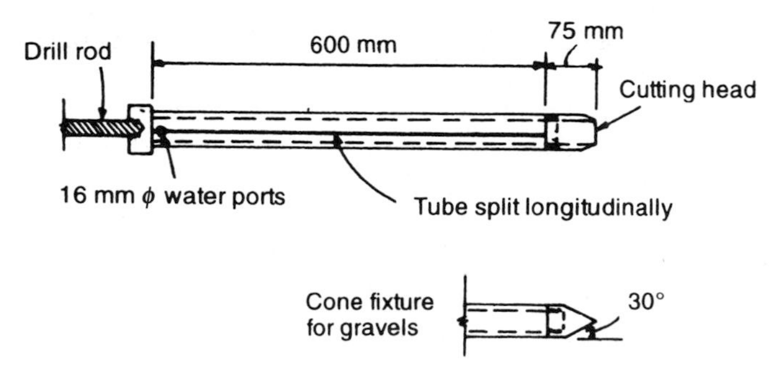

Fig. 8.13 Standard penetration test.

Table 8.2 Relative density of sands.

N	Relative density	
	Terzaghi and Peck	Gibbs and Holtz
0–4	very loose	0–15%
4–10	loose	15–35
10–30	medium	35–65
30–50	dense	65–85
over 50	very dense	85–100

The sampler is lowered down the borehole until it rests on the layer of cohesionless soil to be tested. It is then driven into the soil for a length of 450 mm by means of a 65 kg hammer free-falling 760 mm for each blow. The number of blows required to drive the last 300 mm is recorded and this figure is designated as the N value of the soil (the first 150 mm of driving is ignored because of possible loose soil in the bottom of the borehole from the boring operations). After the tube has been removed from the borehole it can opened and its contents examined.

In gravelly soils damage can occur to the cutting head, and a solid cone, evolved by Palmer and Stuart (1957), is fitted in its place. The N value derived from such soils appears to be of the same order as that obtained when the cutting head is used in finer soils.

Terzaghi and Peck (1948) evolved a qualitative relationship between the relative density of the soil tested and the number of blows from the standard penetration test, N. Gibbs and Holtz (1957) put figures to this relationship, which are given in Table 8.2.

Corrections to the measured N value

An important feature of the standard penetration test is the influence of the effective overburden pressure on the N count. Sand can exhibit different N values at different depths even though its relative density is constant. Terzaghi and Peck make no reference to the effects that this can have, but Gibbs and Holtz examined the effects of most of the variables involved and concluded that the significant factors affecting the N value are the relative density of the soil and the value of the effective overburden pressure removed.

Various workers have investigated this problem (Coffmann, 1960; Bazaraa, 1967), but the method proposed by Thorburn (1963) now seems to have gained general acceptance, at least in the UK.

Thorburn assumed that the original Terzaghi and Peck relationships between N and the relative density corresponded to an effective overburden pressure of 138 kPa. His correction chart therefore dealt with a range of effective overburden pressure for 0 to 138 kPa, it being tacitly assumed that for values of effective overburden pressure greater than 138 kPa, N′ can be taken as equal to N.

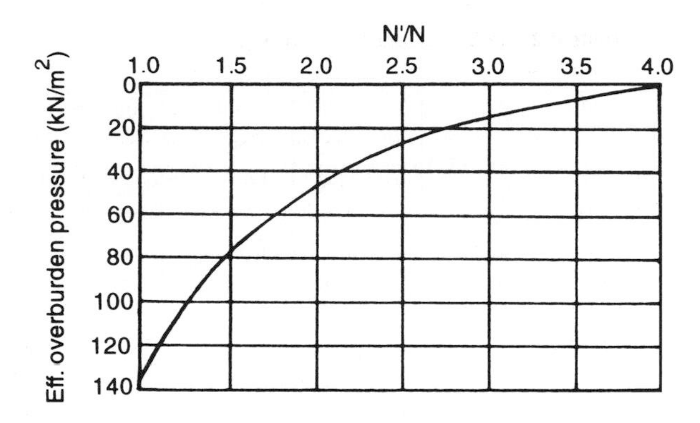

Fig. 8.14 Estimation of N′ from the test value N (after Thorburn, 1963).

It is possible, by the use of Thorburn's chart, to prepare the plot of the N′/N ratio relationship to effective overburden pressure, over the range 0 to 138 kPa (roughly from 0 to 7 m depth of overburden).

This relationship is reproduced in Fig. 8.14 and can be used directly in design.

Terzaghi and Peck point out that in saturated (i.e. below the water table) fine and silty sands the N value can be altered by the low permeability of the soil. If the void ratio of the soil is higher than that corresponding to its critical density, the penetration resistance is less than in a large-grained soil of the same relative density. Conversely, if the void ratio is less than that corresponding to critical density the penetration resistance is increased.

The value of N corresponding to the critical density appears to be about 15, and Terzaghi and Peck suggest that if the number of measured blows, N, is greater than 15 it should be assumed that the density of the tested soil is equal to that of a sand for which the number of blows is equal to 15 + 0.5 (N – 15), i.e.:

$$\text{True } N = 15 + 0.5\,(N - 15)$$

where

N = actual number of blows recorded in the test
True N = number of blows from which N′ should be evaluated

Estimation of allowable bearing pressure from the standard penetration test

Having obtained N′, the determination of the allowable bearing pressure is generally based upon an empirical relationship evolved by Terzaghi and Peck (1948) that is based on the measured settlements of various foundations on sand (Fig. 8.15). The allowable bearing pressure for these curves (which are applicable to both square and rectangular foundations) was defined by Terzaghi and Peck as the pressure that will not cause a settlement greater than 25 mm.

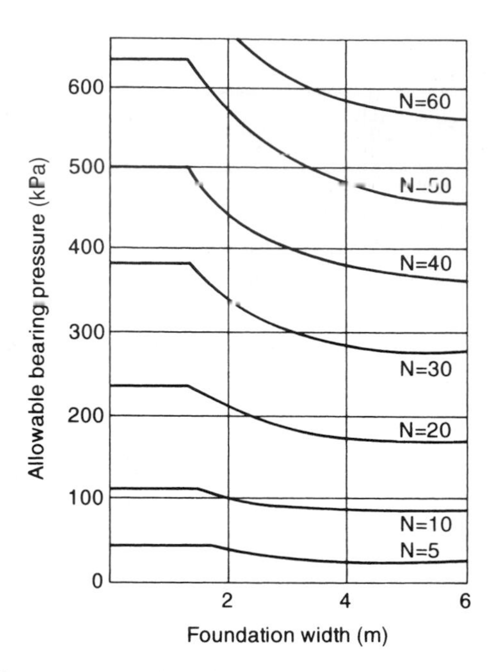

Fig. 8.15 Allowable bearing pressure from the standard penetration test (after Terzaghi and Peck, 1948).

When several foundations are involved the normal design procedure is to determine an average value for N′ from all the boreholes. The allowable bearing pressure for the widest foundation is then obtained with this figure and this bearing pressure is used for the design of all the foundations. The procedure generally leads to only small differential settlements, but even in extreme cases the differential settlement between any two foundations will not exceed 20 mm.

The curves of Fig. 8.15 apply to unsaturated soils, i.e. when the water table is at a depth of at least 1.0 B below the foundation. When the soil is submerged the value of allowable bearing pressure obtained from the curves should be reduced. Originally the values were reduced to 50 per cent but this is now considered excessively conservative as the influence of the groundwater will have already been included in the observed penetration resistance. General practice is now to apply the 50 per cent reduction if the groundwater level is at or above the foundation level, and to apply no reduction if the groundwater level occurs at a depth of at least B below the foundation level. Between these two limits the amount of reduction can be estimated by linear interpolation.

If settlement is of no consequence it is possible to think in terms of ultimate bearing capacity using the approximate relationship between ϕ' and N′ given in Fig. 3.34. Knowing N′, a ϕ' value, from which bearing capacity coefficients are evaluated, can be obtained. This procedure is not generally adopted.

Example 8.8

A granular soil was subjected to standard penetration tests at depths of 3 m. Groundwater level occurred at a depth of 1.5 m below the surface of the soil which was saturated and had a unit weight of 19.3 kN/m^3. The average N count was 15.

(i) Determine the corrected value N′.
(ii) A strip footing, 3 m wide, is to be founded at a depth of 3 m. Assuming that the sand's strength characteristics are constant with depth, determine the allowable bearing pressure.

Solution

(i) Effective overburden pressure = $3 \times 19.3 - 1.5 \times 10 = 43$ kPa
From Fig. 8.14, for $\sigma'_v = 43$ kPa, N′/N = 2.1.
Therefore N′ = $15 \times 2.1 = 31$.

(ii) From Fig. 8.15, for N′ = 31 and B = 3 m:

Allowable bearing pressure = 300 kPa

But this value is for *dry* soil and the sand below the foundation is also below groundwater level and is therefore submerged.

$$\text{It seems that allowable bearing pressure} = \frac{300}{2} = 150 \text{ kPa}$$

8.9.3 *Correlation between the plate loading and the standard penetration tests*

Meigh and Nixon (1961) compared the results of plate loading tests with those of standard penetration tests carried out at the same sites by determining from both sets of results the allowable bearing pressure, p (defined as the pressure causing 25 mm settlement of the foundation) for a 3.05 m square foundation. The differences were quite marked: for fine and silty sands the plate loading test led to values of p about 1.5 times the value obtained from the standard penetration test results, whilst for gravels the plate loading test gave values of p that were from 4 to 6 times greater.

It should be pointed out that Meigh and Nixon used the uncorrected N test values in their calculations, and when Sutherland (1963) examined Meigh and Nixon's results he showed that the disparity between the allowable bearing pressures calculated from the two tests became much less when the corrected N′ value (in which overburden pressure is allowed for) was used.

8.9.4 The static cone penetration test

This penetrometer, often called the Dutch cone penetrometer, is headed by a cone of overall diameter 35.7 mm, giving an end area of 1000 mm^2, and having an apex angle of 60°. The cone is forced downwards at a steady rate (15–20 mm/s) through the soil by means of a load from a hydraulic cylinder transmitted to solid 15 mm diameter rods. These solid rods are centrally placed within 36 mm diameter outer rods. The load acting at the top of the inner rods can be determined from pressure gauge readings and the cone resistance, C_r, is taken to be this load divided by the end area.

Improved forms of the Dutch cone, such as that introduced by Begemann (1965), make it possible to measure cone and side resistances separately, an advantage if the test results are to be used in pile design.

A further development has been the electrical friction–cone penetrometer, described by Lousberg *et al.* (1974), in which the cone penetration resistance is measured and recorded continuously by means of a load cell within the instrument. The penetrometer also has a frictional sleeve connected to a second and independent load cell so that frictional resistance can also be recorded.

A full description of cone penetration testing and its application in geotechnical and geo-environmental engineering is given by Lunne *et al.* (1997).

8.9.5 Presumed bearing capacity

The British Standard BS 8004: 1986 gives a list of safe bearing capacity values and this is reproduced in Table 8.3. The values are based on the following assumptions:

(i) The site and adjoining sites are reasonably level.
(ii) The ground strata are reasonably level.
(iii) There is no softer layer below the foundation stratum.
(iv) The site is protected from deterioration.

Foundations designed to these values will normally have an adequate factor of safety against bearing capacity failure, provided that they are not subjected to inclined loading, but it must be remembered that settlement effects have not been considered.

For cohesive soils the consistency is related to the undrained strength, c_u. Such a relationship is suggested in BS 5930 and is reproduced in Table 8.4.

Table 8.3 Presumed safe bearing capacity, q_s, values (based on BS 8004: 1986).

	q_s (kPa)
Rocks	
(Values based on assumption that foundation is carried down to unweathered rock)	
Hard igneous and gneissic	10 000
Hard sandstones and limestones	4 000
Schists and slates	3 000
Hard shale and mudstones, soft sandstone	2 000
Soft shales and mudstones	1000–600
Hard chalk, soft limestone	600
Cohesionless soils	
(Values to be halved if soil submerged)	
Compact gravel, sand and gravel	>600
Medium dense gravel, or sand and gravel	600–200
Loose gravel, or sand and gravel	<200
Compact sand	>300
Medium dense sand	300–100
Loose sand	<100
Cohesive soils	
(Susceptible to long-term consolidation settlement)	
Very stiff boulder clays and hard clays	600–300
Stiff clays	300–150
Firm clays	150–75
Soft clays and silts	<75
Very soft clays and silts	Not applicable

Table 8.4 Undrained shear strength of cohesive soils.

Consistency	c_u (kPa)	Field behaviour
Hard	>300	Brittle
Very stiff	300–150	Brittle or very tough
Stiff	150–75	Cannot be moulded in fingers
Firm	75–40	Can just be moulded in fingers
Soft	40–20	Easily moulded in fingers
Very soft	<20	Exudes between fingers if squeezed

8.10 Pile foundations

The use of sheet piling, which can be of timber, concrete or steel, for earth retaining structures has been described in Chapter 7. Piled foundations form a separate category and are generally used:

(i) to transmit a foundation load to a solid soil stratum;
(ii) to support a foundation by friction of the piles against the soil;
(iii) to resist a horizontal or uplift load;
(iv) to compact a loose layer of granular soil.

8.10.1 Classification of piles

Piles can be classified by different criteria, such as their material (e.g. concrete, steel, timber), their method of installation (e.g. driven or bored), the degree of soil displacement during installation, or their size (e.g. large diameter, small diameter). However, in terms of pile design, the most appropriate classification criteria is the behaviour of the pile once installed (e.g. end bearing pile, friction pile, combination pile).

End bearing (Fig. 8.16a)

Derive most of their carrying capacity from the penetration resistance of the soil at the toe of the pile. The pile behaves as an ordinary column and should be designed as such except that, even in weak soil, a pile will not fail by buckling and this effect need only be considered if part of the pile is unsupported, i.e. it is in either air or water.

Friction (Fig. 8.16b)

Carrying capacity is derived mainly from the adhesion or friction of the soil in contact with the shaft of the pile.

Combination (Fig. 8.16c)

Really an extension of the end bearing pile when the bearing stratum is not hard, such as a firm clay. The pile is driven far enough into the lower material to develop adequate frictional resistance. A further variation of the end bearing pile is piles with enlarged bearing areas. This is achieved by forcing a bulb of concrete into the soft stratum immediately above the firm layer to give an enlarged base. A similar effect is

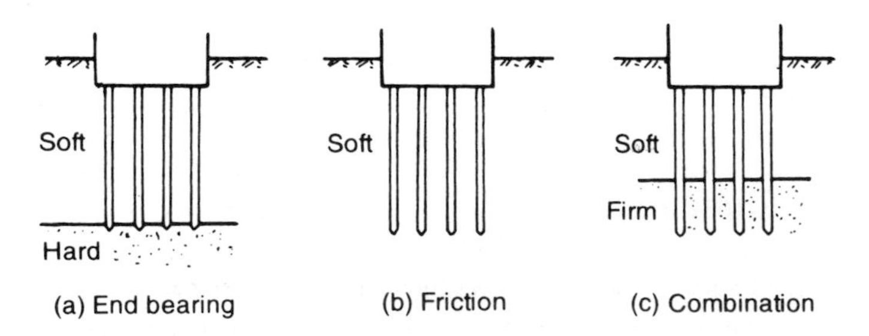

Fig. 8.16 Classification of piles.

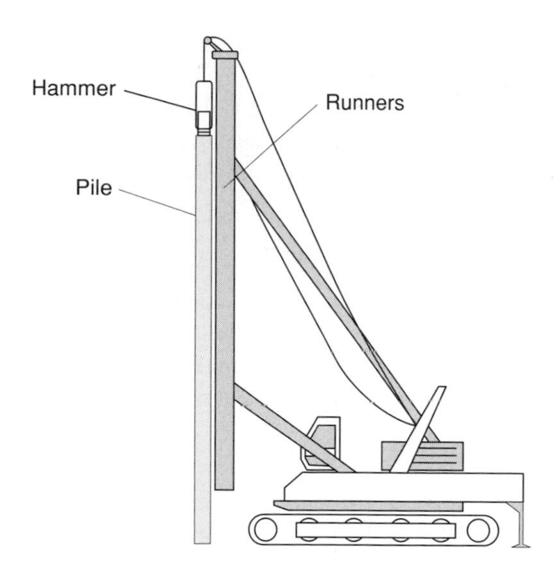

Fig. 8.17 Pile driving rig.

produced with bored piles by forming a large cone or bell at the bottom with a special reaming tool.

8.10.2 Driven piles

These are prefabricated piles that are installed into the ground through the use of a pile driver as illustrated in Fig. 8.17. The pile is hoisted into position on the pile driver and aligned against the runners so that the pile is driven into the ground at exactly the required angle, to exactly the required depth. The pile is driven into the soil by striking the top of the pile repeatedly with a pneumatic or percussive hammer or by driving the pile down using a hydraulic ram. Most commonly the piles are made from precast concrete although timber and steel piles are also available.

Precast concrete

These are usually of square or octagonal section. Reinforcement is necessary within the pile to help withstand both handling and driving stresses. Prestressed concrete piles are also used and are becoming more popular than ordinary precast as less reinforcement is required.

Timber

Timber piles have been used from earliest recorded times and are still used for permanent work where timber is plentiful. In the UK, timber piles are used mainly in temporary works, due to their lightness and shock resistance, but they are also used for piers and fenders and can have a useful life of some 25 years or more if kept completely below the water table. However, they can deteriorate rapidly if used in ground in which the water level varies and allows the upper part to come above the water surface. Pressure creosoting is the usual method of protection. In tropical climes timber piles above groundwater level are liable to be destroyed by wood-eating insects, sometimes in a matter of weeks.

Steel piles: tubular, box or H-section

These are suitable for handling and driving in long lengths. They have a relatively small cross-sectional area and penetration is easier than with other types. The risk from corrosion is not as great as one might think although tar coating or cathodic protection can be employed in permanent work.

Jetted pile

When driving piles in non-cohesive soils the penetration resistance can often be considerably reduced by jetting a stream of high-pressured water into the soil just below the pile. There have been cases where piles have been installed by jetting alone. The method requires considerable experience, particularly when near to existing foundations.

Vibrated pile

As an alternative to jetting, vibration techniques can be used to place piles in granular soils. Vibrators are not efficient in clays but can be used if piles are to be extracted.

Jacked pile

Generally built up with a series of short sections of precast concrete, this pile is jacked into the ground and progressively increased in length by the addition of a pile section whenever space becomes available. The jacking force is easily measured and the load to pile penetration relationship can be obtained as jacking proceeds. Jacked piles are often used to underpin existing structures where lack of space excludes the use of pile driving hammers.

Screw pile

A screw pile consists of a steel, or concrete, cylinder with helical blades attached to its lower end. The pile is made to screw down into the soil by rotating the cylinder with a capstan at the top of the pile. A screw pile, due to the large size of its screw blades, can offer large uplift resistance.

8.10.3 Driven and cast-in-place piles

Two of the main types of this pile, used in Britain, are described below.

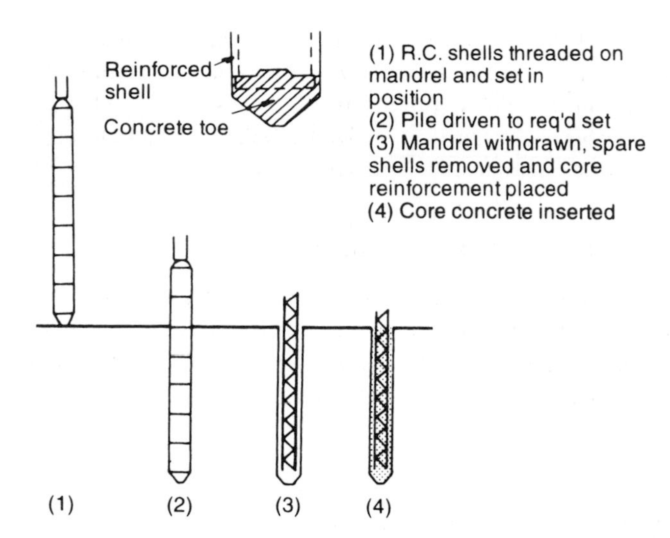

Fig. 8.18 West's shell pile.

West's shell pile

Precast, reinforced concrete tubes, about 1 m long, are threaded on to a steel mandrel and driven into the ground after a concrete shoe has been placed at the front of the shells. Once the shells have been driven to specification the mandrel is withdrawn and reinforced concrete inserted in the core. Diameters vary from 325 to 600 mm. Details of the pile and the method of installation are shown in Fig. 8.18.

Franki pile

A steel tube is erected vertically over the place where the pile is to be driven, and about a metre depth of gravel is placed at the end of the tube. A drop hammer, 1500 to 4000 kg mass, compacts the aggregate into a solid plug which then penetrates the soil and takes the steel tube down with it. When the required set has been achieved the tube is raised slightly and the aggregate broken out. Dry concrete is now added and hammered until a bulb is formed. Reinforcement is placed in position and more dry concrete is placed and rammed until the pile top comes up to ground level. The sequence of operations is illustrated in Fig. 8.19.

8.10.4 Bored and cast-in-situ piles

These piles are formed within a drilled borehole. During the drilling process the sides of the borehole are supported to prevent the soil from collapsing inwards and temporary sections of steel cylindrical casing are advanced along with the drilling process to provide this required support. As the drilling progresses, the soil is removed from within the casing and brought to the surface. Once the full depth of the borehole has been reached, the casing is gradually withdrawn, the reinforcement cage is placed

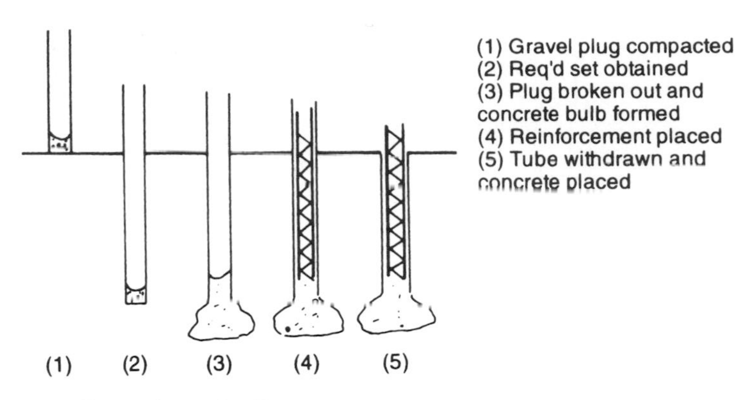

Fig. 8.19 Installation of a Franki pile.

and the concrete which forms the pile is pumped into the borehole. For very deep boreholes the installation of many sections of temporary casing can be an expensive and slow process, and an alternative means of supporting the sides is through the use of a bentonite slurry in the same manner as for a diaphragm wall (see Section 7.3.2).

An alternative technique which does not use borehole side-support is the *continuous flight auger* (CFA) pile. With this technique a continuous flight auger with a hollow stem is used to create the borehole. The sides of the borehole are supported by the soil on the flights of the auger and so no casing is required. Once the required depth has been reached, the concrete is pumped down the hollow stem and the auger is steadily withdrawn. The steel reinforcement is placed once the auger is clear of the borehole.

8.10.5 Large diameter bored piles

The driven or bored and cast-in-place piles discussed previously generally have maximum diameters in the order of 0.6 m and are capable of working loads round about 2 MN. With modern buildings column loads in the order of 20 MN are not uncommon. A column carrying such a load would need about ten conventional piles, placed in a group and capped by a concrete slab, probably some 25 m^2 in area.

A consequence of this problem has been the increasing use of the large diameter bored pile. This pile has a minimum shaft diameter of 0.75 m and may be under-reamed to give a larger bearing area if necessary. Such a pile is capable of working loads in the order of 25 MN and, if taken down through the soft to the hard material, will minimise settlement problems so that only one such pile is required to support each column of the building. Large diameter bored piles have been installed in depths down to 60 m.

8.10.6 Determination of the bearing capacity of a pile by load tests

The load test is the only really reliable means of determining a pile's load capacity, but it is expensive, particularly if the ground is variable and a large number of piles must therefore be tested.

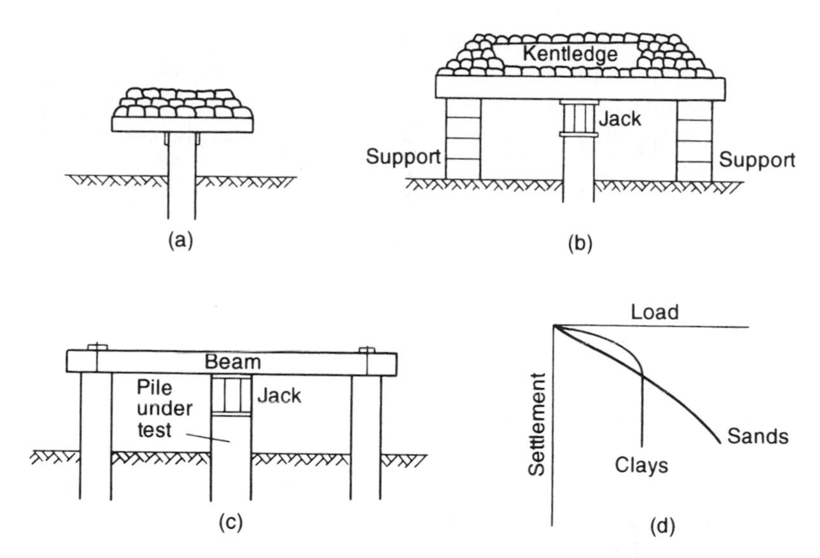

Fig. 8.20 (a)–(c) Methods for testing a pile. (d) Load to settlement relationship.

Full-scale piles should be used-and these should be driven in the same manner as those placed for the permanent work.

Figure 8.20 gives rough indications of how a test pile may be loaded. A large mass of dead weight is placed on a platform supported by the pile. The load is applied in increments and the settlement is recorded when the rate of settlement has reduced to 0.25 mm in an hour, at which stage a further increment can be applied (Fig. 8.20a). The method has the disadvantage that the platform must be balanced on top of the pile and there is always the risk of collapse. An alternative, and better, technique is to jack the pile against a kentledge using an arrangement similar to Fig. 8.20b.

Sometimes the piles to be used permanently can be used to test a pile as shown in Fig. 8.20c.

The form of load to settlement relationship obtained from a loading test is shown in Fig. 8.20d. Loading is continued until failure occurs, except for large diameter bored piles which, having a working load of some 25 MN, would require massive kentledges if failure loads were to be achieved. General practice has become to test load these piles to the working load plus 50 per cent.

Design standards offer some limited guidance on static load pile test methods. BS 8004 specifies two types of test, described below, from which the ultimate load of a pile can be obtained, and Eurocode 7 (see Section 8.11) makes reference to the ASTM suggested method for the axial pile loading test, described by Smoltczyk (1985). Furthermore, it is likely that the forthcoming European standard for pile testing will adopt the recommendations and procedures described by De Cock *et al.* (2003).

(1) The maintained load test

Here the load is applied to the pile in a series of increments, usually equal to 25 per cent of the designated working load for the pile. The ultimate pile load is taken to be the load that achieves some specified amount of settlement, usually 10 per cent of the pile's diameter.

(2) The constant rate of penetration test

In this test the pile is jacked downwards at a constant rate of penetration. The ultimate pile load is considered to be the load at which either a shear failure takes place within the soil or the penetration of the pile equals 10 per cent of its diameter.

The figure of one tenth is intended for normal sized piles and, if applied to large diameter bored piles, could lead to excessive settlements if a factor of safety of 2.5 were adopted. This, of course, only applies to large diameter piles resting on soft rocks. In the case of a large diameter bored pile resting on hard rock the ultimate load depends upon the ultimate stress in the concrete.

8.10.7 *Determination of the bearing capacity of a pile by soil mechanics*

A pile is supported in the soil by the resistance of the toe to futher penetration plus the frictional or adhesive forces along its embedded length.

Ultimate bearing capacity = Ultimate base resistance + Ultimate skin friction:

$$Q_u = Q_b + Q_s$$

Cohesive soils

Q_b for piles in cohesive soils is based on Meyerhof's equation (1951):

$$Q_b = N_c \times c_b \times A_b$$

where

N_c = bearing capacity factor, widely accepted as equal to 9.0
c_b = undisturbed undrained shear strength of the soil at base of pile.

Q_s is given by the equation:

$$Q_s = \alpha \times \bar{c}_u \times A_s$$

where

α = adhesion factor
$\bar{c}_u$ = average undisturbed undrained shear strength of soil adjoining pile
A_s = surface area of embedded length of pile.

Hence

$$Q_u = c_b N_c A_b + \alpha \bar{c}_u A_s$$

The adhesion factor α

Most of the bearing capacity of a pile in cohesive soil is derived from its shaft resistance, and the problem of determining the ultimate load resolves into determining a value for α. For soft clays α can be equal to or greater than 1.0 as, after driving, soft clays tend to increase in strength. In overconsolidated clays α has been found to vary from 0.3 to 0.6. The usual value assumed for London clay was, for many years, taken as 0.45 but more recently a value of 0.6 for this type of soil has become more accepted.

Cohesionless soils

The ultimate load of a pile installed in cohesionless soil is estimated using only the value of the drained parameter, ϕ', and assuming that any contribution due to c′ is zero.

$$Q_b = q_b A_b = \sigma'_v N_q A_b$$

where

σ'_v = the effective overburden pressure at the base of the pile
N_q = the bearing capacity coefficient
A_b = the area of the pile base.

The selection of a suitable value for N_q is obviously a crucial part of the design of the pile. The values suggested by Berezantzev *et al.* (1961) are often used and are reproduced in Fig. 8.21. Note that the full value of N_q is used as it is assumed that the weight of soil removed or displaced is equal to the weight of the pile that replaced it.

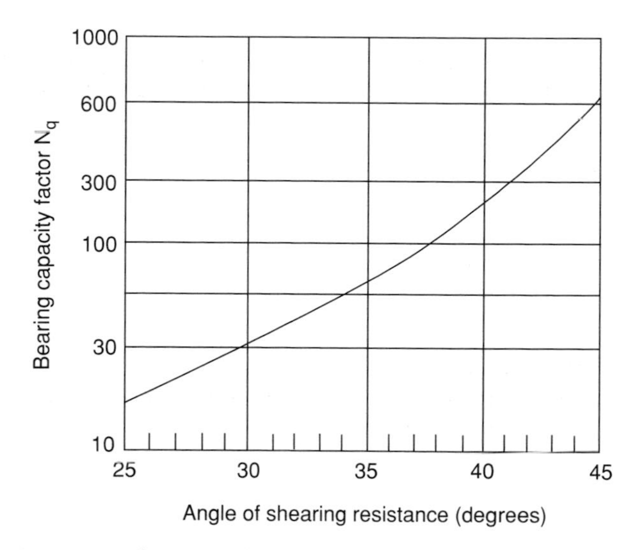

Fig. 8.21 Bearing capacity factor N_q (after Berezantzev *et al.*, 1961).

$$Q_s = f_s A_s$$

where

f_s = average value of the ultimate skin friction over the embedded length of the pile
A_s = surface area of embedded length of pile.

Meyerhof (1959) suggested that for the average value of the ultimate skin friction:

$$f_s = K_s \overline{\sigma'_v} \tan \delta$$

where

K_s = the coefficient of lateral earth pressure
$\overline{\sigma'_v}$ = average effective overburden pressure acting along the embedded length of the pile shaft
δ = angle of friction between the pile and the soil.

Hence

$$Q_s = A_s K_s \overline{\sigma'_v} \tan \delta$$

and

$$Q_u = \sigma'_v N_q A_b + A_s K_s \overline{\sigma'_v} \tan \delta$$

Typical values for δ and K_s were derived by Broms (1966), and are listed in Table 8.5.

Vesic (1973) pointed out that the value of q_b, i.e. $\sigma'_v N_q$, does not increase indefinitely but has a limiting value at a depth of some 20 times the pile diameter. There is therefore a maximum value of $\sigma'_v N_q$ that can be used in the calculations for Q_b.

In a similar manner there is a limiting value that can be used for the average ultimate skin friction, f_s. This maximum value of f_s occurs when the pile has an embedded

Table 8.5 Typical values for δ and K_s suggested by Broms (1966).

Pile material	δ	K_s	
		Relative density of soil	
		Loose	Dense
Steel	20°	0.5	1.0
Concrete	$0.75\phi'$	1.0	2.0
Timber	$0.67\phi'$	1.5	4.0

length between 10 and 20 pile diameters. Vesic (1970) suggested that the maximum value of the average ultimate skin resistance should be obtained from the formula:

$$f_s = 0.08(10)^{1.5(D_r)^4}$$

where D_r = the relative density of the cohesionless soil.

In practice f_s is often taken as 100 kPa if the formula gives a greater value.

Unlike piles embedded in cohesive soils, the end resistances of piles in cohesionless soils are of considerable significance and short piles are therefore more efficient in cohesionless soils.

8.10.8 Determination of soil piling parameters from *in situ* tests

With cohesionless soils it is possible to make reasonable estimates of the values of q_b and f_s from *in situ* penetration tests. Meyerhof (1976) suggests the following formulae to be used in conjunction with the standard penetration test.

Driven piles

Sands and gravel $\quad q_b \approx \dfrac{40ND}{B} \leq 400N \text{ (kPa)}$

Non-plastic silts $\quad q_b \approx \dfrac{40ND}{B} \leq 300N \text{ (kPa)}$

Bored piles

Any type of granular soil $\quad q_b \approx \dfrac{14ND}{B} \text{ kPa}$

Large diameter driven piles $\quad f_s \approx 2\bar{N}$ kPa

Average diameter driven piles $\quad f_s \approx \bar{N}$ kPa

Bored piles $\quad f_s \approx 0.67\bar{N}$ kPa

where

N = the *uncorrected* blow count at the pile base

$\bar{N}$ = the average *uncorrected* N value over the embedded length of the pile

D = embedded length of the pile in the end bearing stratum

B = width, or diameter, of pile.

An alternative method is to use the results of the Dutch cone test. Typical results from such a test are shown in Fig. 8.22 and are given in the form of a plot showing the variation of the cone penetrations resistance with depth.

For the ultimate base resistance, C_r, the cone resistance is taken as being the average value of C_r over the depth 4d as shown, where d = diameter of shaft. Then:

$$Q_b = C_r A_b$$

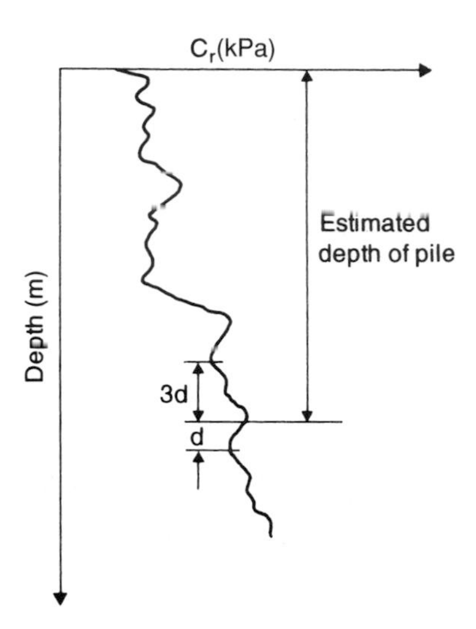

Fig. 8.22 Typical results from a Dutch cone test.

The ultimate skin friction, f_s, can be obtained from one of the following:

$$f_s \approx \frac{\overline{C}_r}{200} \text{ kPa} \qquad \text{for driven piles in dense sand}$$

$$f_s \approx \frac{\overline{C}_r}{400} \text{ kPa} \qquad \text{for driven piles in loose sand}$$

$$f_s \approx \frac{\overline{C}_r}{150} \text{ kPa} \qquad \text{for driven piles in non-plastic silts}$$

where $\overline{C}_r$ = average cone resistance along the embedded length of the pile (De Beer, 1963).

Then $Q_s = f_sA_s$ and, as before, $Q_u = Q_b + Q_s$.

Example 8.9

A 5 m thick layer of medium sand overlies a deep deposit of dense gravel. A series of standard penetration tests carried out through the depth of the sand has established that the average blow count, $\bar{N}$, is 22. Further tests show that the gravel has a standard penetration value of N = 40 in the region of the interface with the sand. A precast pile of square section 0.25×0.25 m^2 is to be driven down through the sand and to penetrate sufficiently into the gravel to give good end bearing.

Adopting a safety factor of 3.0 determine the allowable load that the pile will be able to carry.

Solution

Ultimate bearing capacity of the pile = $Q_u = Q_s + Q_b$

Q_b: All end bearing effects will occur in the gravel. Now

$$q_b \approx 40\ N\ \frac{D}{B}\ \text{kPa or } 400 \times N\ \text{kPa (whichever is the lesser)}$$

i.e.

$$q_b = 40 \times 40 \times \frac{D}{0.25} = 400 \times 40 = 16\ 000\ \text{kPa}$$

$$\text{Penetration into gravel, D,} = \frac{16\ 000 \times 0.25}{40 \times 40} = 2.5\ \text{m}$$

and

$$Q_b = 16\ 000 \times 0.25^2 = 1000\ \text{kN}$$

Q_s in sand: $Q_s = f_s A_s = 22 \times 5 \times 0.25 \times 4 = 110$ kN
Q_s in gravel: $Q_s = f_s A_s = 40 \times 2.5 \times 0.25 \times 4 = 100$ kN

i.e.

$$Q_u = 210 + 1000 = 1210\ \text{kN}$$

$$\text{Allowable load} = \frac{1210}{3} = 400\ \text{kN}$$

Example 8.9 illustrates that, as discussed earlier, the end bearing effects are much greater than those due to side friction. It can be argued that, in order to develop side friction (shaft resistance) fully, a significant downward movement of the pile is required which cannot occur in this example because of the end resistance of the gravel. As a result of this phenomenon, it is common practice to apply a different factor of safety to the shaft resistance than that applied to the end bearing resistance. Typically a factor of safety of around 1.5 is applied to shaft resistance, and a factor of safety between 2.5 and 3.0 is applied to the end bearing resistance.

Returning to Example 8.9, and adopting $F_b = 3$, $F_s = 1.5$, the allowable load now becomes:

$$\frac{1000}{3} + \frac{210}{1.5} = 473\ \text{kN}$$

Negative skin friction

If a soil subsides or consolidates around a group of piles these piles will tend to support the soil and there can be a considerable increase in the load on the piles.

The main causes for this state of affairs are that:

(i) bearing piles have been driven into recently placed fill;
(ii) fill has been placed around the piles after driving.

If negative friction effects are likely to occur then the piles must be designed to carry the additional load. In extreme cases the value of negative skin friction can equal the positive skin friction but, of course, this maximum value cannot act over the entire bedded length of the pile, being virtually zero at the top of the pile and reaching some maximum value at its base.

8.11 Designing pile foundations to Eurocode 7

The principles of Eurocode 7, as described in Section 7.4.2, apply to the design of pile foundations, and the reader is advised to refer back to that section whilst studying the following few pages.

The design of pile foundations is covered in Section 7 of Eurocode 7. There are 11 limit states listed that should be considered, though only those limit states most relevant to the particular situation would normally be considered in the design. These include the loss of overall stability, bearing resistance failure of the pile, uplift of the pile and structural failure of the pile. In this chapter we will look only at checking against ground resistance failure through the compressive loading of the pile.

Pile design methods acceptable to Eurocode 7 are in the main based on the results of static pile load tests, and the design calculations should be validated against the test results. When considering the compressive ground resistance limit state the task is to demonstrate that the design axial compression load on a pile or pile group, $F_{c;d}$, is less than or equal to the design compressive ground resistance, $R_{c;d}$, against the pile or pile group. In the case of pile groups, $R_{c;d}$ is taken as the lesser value of the design ground resistance of an individual pile and that of the whole group.

In keeping with the rules of Eurocode 7, the design value of the compressive resistance of the ground is obtained by dividing the characteristic value by a partial factor of safety. The characteristic value is obtained by one of three approaches: from static load tests, from ground tests results or from dynamic tests results.

(i) Ultimate compressive resistance from static load tests

The characteristic value of the compressive ground resistance, $R_{c;k}$, is obtained by combining the measured value from the pile load tests with a correlation factor, ξ (related to the number of piles tested). More explicitly, $R_{c;k}$ is taken as the lesser value of:

$$R_{c;k} = \frac{(R_{c;m})_{mean}}{\xi_1} \quad \text{and} \quad R_{c;k} = \frac{(R_{c;m})_{min}}{\xi_2}$$

where

$(R_{c;m})_{mean}$ = the mean measured resistance
$(R_{c;m})_{min}$ = the minimum measured resistance
ξ_1, ξ_2 = correlation factors obtained from Table 8.6.

Table 8.6 Correlation factors – static load tests results.

	Number of piles tested				
	1	2	3	4	≥ 5
ξ_1	1.4	1.3	1.2	1.1	1.0
ξ_2	1.4	1.2	1.05	1.0	1.0

It may be that the characteristic compressive resistance of the ground is more appropriately determined from the characteristic values of the base resistance, $R_{b;k}$ and the shaft resistance, $R_{s;k}$:

$$R_{c;k} = R_{b;k} + R_{s;k}$$

The design compressive resistance of the ground may be derived by either:

$$R_{c;d} = \frac{R_{c;k}}{\gamma_t}$$

or

$$R_{c;d} = \frac{R_{b;k}}{\gamma_b} + \frac{R_{s;k}}{\gamma_s}$$

where γ_b, γ_s and γ_t are partial factors on base resistance, shaft resistance and the total resistance respectively. The partial factors for piles in compression recommended in Eurocode 7 are given in Table 8.7.

Considering Design Approach 1, the following partial factor sets (see Section 7.4.2) are used for the design of axially loaded piles:

Combination 1: A1 + M1 + R1
Combination 2: A2 + (M1 or M2)* + R4

* M1 is used for calculating pile resistance; M2 is used for calculating unfavourable actions on piles.

Table 8.7 Piles in compression: partial factor sets R1, R2, R3 and R4.

Partial factor set	R1			R2	R3	R4		
	Driven	Bored	CFA	All	All	Driven	Bored	CFA
Base, γ_b	1.0	1.25	1.1	1.1	1.0	1.3	1.6	1.45
Shaft, γ_s	1.0	1.00	1.0	1.1	1.0	1.3	1.3	1.30
Total, γ_t	1.0	1.15	1.1	1.1	1.0	1.3	1.5	1.40

Example 8.10

A series of static load tests on a set of four bored piles gave the following results:

	Test no.			
	1	2	3	4
Measured load (kN)	382	425	365	412

From an understanding of the ground conditions, it is assumed that the ratio of base resistance to shaft resistance is 3:1.

Determine the design compressive resistance of the ground in accordance with Eurocode 7, Design Approach 1.

Solution

$$(R_{c;m})_{mean} = \frac{382 + 425 + 365 + 412}{4} = 396 \text{ kN}$$

$$(R_{c;m})_{min} = 365 \text{ kN}$$

From Table 8.6, $\xi_1 = 1.1$; $\xi_2 = 1.0$

$$R_{c;k} = \frac{(R_{c;m})_{mean}}{\xi_1} = \frac{396}{1.1} = 360 \text{ kN}$$

$$R_{c;k} = \frac{(R_{c;m})_{min}}{\xi_2} = \frac{365}{1.0} = 365 \text{ kN}$$

that is

$$R_{c;k} = 360 \text{ kN} \quad \text{(i.e. the minimum value)}$$

Since the ratio of base resistance to shaft resistance is 3:1, we have:

Characteristic base resistance, $R_{b;k} = 360 \times 0.75 = 270$ kN
Characteristic shaft resistance, $R_{s;k} = 360 \times 0.25 = 90$ kN

1. Design Approach 1, Combination 1

Partial factor set R1 is used:

$$R_{c;d} = \frac{R_{c;k}}{\gamma_t} = \frac{360}{1.15} = 313 \text{ kN}$$

or

$$R_{c;d} = \frac{R_{b;k}}{\gamma_b} + \frac{R_{s;k}}{\gamma_s} = \frac{270}{1.25} + \frac{90}{1.0} = 306 \text{ kN}$$

2. Design Approach 1, Combination 2

Partial factor set R4 is used:

$$R_{c;d} = \frac{R_{c;k}}{\gamma_t} = \frac{360}{1.5} = 240 \text{ kN}$$

or

$$R_{c;d} = \frac{R_{b;k}}{\gamma_b} + \frac{R_{s;k}}{\gamma_s} = \frac{270}{1.6} + \frac{90}{1.3} = 238 \text{ kN}$$

The design compressive resistance of the ground is thus determined:

$$R_{c;d} = \min(313, 306, 240, 238) = 238 \text{ kN}$$

(ii) Ultimate compressive resistance from ground tests results

The design compressive resistance can be determined from ground tests results. Here the characteristic compressive resistance, $R_{c;k}$, is taken as the lesser value of:

$$R_{c;k} = \frac{(R_{b;cal} + R_{s;cal})_{mean}}{\xi_3} \quad \text{and} \quad R_{c;k} = \frac{(R_{b;cal} + R_{s;cal})_{min}}{\xi_4}$$

where

$(R_{b;cal})_{mean}$ = the mean calculated base resistance
$(R_{s;cal})_{mean}$ = the mean calculated shaft resistance
$(R_{b;cal})_{min}$ = the minimum calculated base resistance
$(R_{s;cal})_{min}$ = the minimum calculated shaft resistance
ξ_3, ξ_4 = correlation factors obtained from Table 8.8.

The calculated base and shaft resistances are determined using the equations set out in Section 8.10.7.

Table 8.8 Correlation factors – ground tests results.

	Number of test profiles						
	1	2	3	4	5	7	10
ξ_3	1.4	1.35	1.33	1.31	1.29	1.27	1.25
ξ_4	1.4	1.27	1.23	1.20	1.15	1.12	1.08

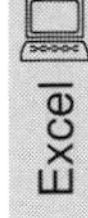

Example 8.11

A 10 m long by 0.7 m diameter CFA pile is to be founded in a uniform soft clay. The following test results were established in a geotechnical laboratory as part of a site investigation:

Borehole no.	1	2	3	4
Mean undrained strength along shaft, $c_{u;shaft}$ (kPa)	65	62	70	73
Mean undrained strength at base, $c_{u;base}$ (kPa)	90	79	96	100

The pile will carry a permanent axial load of 500 kN (includes the self-weight of the pile) and an applied transient (variable) axial load of 150 kN.

Check the bearing resistance (GEO) limit state in accordance with Eurocode 7, Design Approach 1 by establishing the magnitude of the over-design factor. Assume $N_c = 9$ and $\alpha = 0.7$.

Solution

Area of base of pile, $A_b = \dfrac{\pi D^2}{4} = \dfrac{\pi \times 0.7^2}{4} = 0.385\ \text{m}^2$

The total resistance is determined from the results from each borehole:

$$(R_{b;cal})_1 = (N_c \times c_u \times A_b) + (\pi \times D \times L \times \alpha \times c_u)$$
$$= (9 \times 90 \times 0.385) + (\pi \times 0.7 \times 10 \times 0.7 \times 65) = 1312\ \text{kN}$$
$$(R_{b;cal})_2 = (9 \times 79 \times 0.385) + (\pi \times 0.7 \times 10 \times 0.7 \times 62) = 1228\ \text{kN}$$
$$(R_{b;cal})_3 = (9 \times 96 \times 0.385) + (\pi \times 0.7 \times 10 \times 0.7 \times 70) = 1410\ \text{kN}$$
$$(R_{b;cal})_4 = (9 \times 100 \times 0.385) + (\pi \times 0.7 \times 10 \times 0.7 \times 73) = 1470\ \text{kN}$$

$$(R_{c;cal})_{mean} = \frac{1312 + 1228 + 1410 + 1470}{4} = 1355\ \text{kN}$$

$(R_{c;cal})_{min} = 1228$ kN (i.e. Borehole 2)

From Table 8.8, $\xi_3 = 1.31$; $\xi_4 = 1.2$.

$$R_{c;k} = \frac{(R_{c;cal})_{mean}}{\xi_3} = \frac{1355}{1.31} = 1034 \text{ kN}$$

$$R_{c;k} = \frac{(R_{c;cal})_{min}}{\xi_4} = \frac{1228}{1.2} = 1023 \text{ kN}$$

that is, $(R_{c;cal})_{min}$ governs and this lower value of $R_{c;k}$ is taken as the characteristic compressive resistance.

Therefore, using ξ_4:

$$\text{Characteristic base resistance, } R_{b;k} = \frac{9 \times 79 \times 0.385}{1.20} = 228 \text{ kN}$$

$$\text{Characteristic shaft resistance, } R_{s;k} = \frac{\pi \times 0.7 \times 10 \times 0.7 \times 62}{1.20} = 795 \text{ kN}$$

1. Design Approach 1, Combination 1:
Design resistance: partial factor set R1 is used (Table 8.7):

$$R_{c;d} = \frac{R_{b;k}}{\gamma_b} + \frac{R_{s;k}}{\gamma_s} = \frac{228}{1.1} + \frac{795}{1.0} = 1002 \text{ kN}$$

Design actions: partial factor set A1 is used (Table 7.1):

$$F_{c;d} = 500 \times 1.35 + 150 \times 1.5 = 900 \text{ kN}$$

$$\text{Over-design factor, } \Gamma = \frac{1002}{900} = 1.11$$

2. Design Approach 1, Combination 2:
Design resistance: partial factor set R4 is used (Table 8.7):

$$R_{c;d} = \frac{R_{b;k}}{\gamma_b} + \frac{R_{s;k}}{\gamma_s} = \frac{228}{1.45} + \frac{795}{1.3} = 769 \text{ kN}$$

Design actions: partial factor set A2 is used (Table 7.1):

$$F_{c;d} = 500 \times 1.0 + 150 \times 1.3 = 695 \text{ kN}$$

$$\text{Over-design factor, } \Gamma = \frac{769}{695} = 1.11$$

Since $\Gamma \geq 1$, the design of the pile satisfies the GEO limit state requirement.

(iii) Ultimate compressive resistance from dynamic tests results
Although static load tests and ground tests are the most common methods of determining the compressive resistance of the pile, the resistance can also be estimated from dynamic tests provided that the test procedure has been calibrated against static load tests.

8.12 Pile groups

8.12.1 Action of pile groups

Piles are usually driven in groups (see Fig. 8.23).

In the case of end bearing piles the pressure bulbs of the individual piles will overlap (if spacing < 5d – the usual condition). Provided that the bearing strata are firm throughout the affected depth of this combined bulb then the bearing capacity of the group will be equal to the summation of the individual strengths of the piles. However, if there is a compressible soil layer beneath the firm layer in which the piles are founded, care must be taken to ensure that this weaker layer is not overstressed.

Pile groups in cohesionless soils

Pile driving in sands and gravels compacts the soil between the piles. This compactive effect can make the bearing capacity of the pile group greater than the sum of the individual pile strengths. Spacing of piles is usually from two to three times the diameter, or breadth, of the piles.

Pile groups in cohesive soils

A pile group placed in a cohesive soil has a collective strength which is considerably less than the summation of the individual pile strengths which compose it.

One characteristic of pile groups in cohesive soils is the phenomenon of 'block failure'. If the piles are placed very close together (a common temptation when

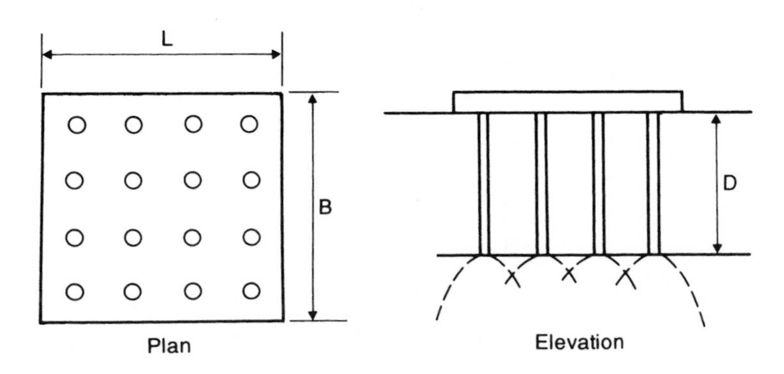

Fig. 8.23 A typical pile group.

dealing with a limited site area), the strength of the groups may be governed by its strength at block failure. This is when the soil fails along the perimeter of the group.

For block failure:

$$Q_u = 2D(B + L) \times \bar{c}_u + 1.3c_b N_c BL$$

where

D = depth of pile penetration
L = length of pile group
B = breadth of pile group
N_c = bearing capacity coefficient (taken generally as 9.0).

Whitaker (1957), in a series of model tests, showed that block failure will not occur if the piles are spaced at not less than 1.5d apart. General practice is to use 2d to 3d spacings.

In such cases:

$$Q_u = En\,Q_{up}$$

where

E = efficiency of pile group (0.7 for spacings 2d–3d)
Q_{up} = ultimate bearing capacity of single pile
n = number of piles in group.

8.12.2 Settlement effects in pile groups

Quite often it is the allowable settlement, rather than the safe bearing capacity, that decides the working load that a pile group may carry.

For bearing piles the total foundation load is assumed to act at the base of the piles on a foundation of the same size as the plan of the pile group. With this assumption it becomes a simple matter to examine settlement effects.

With friction piles it is virtually impossible to determine the level at which the foundation load is effectively transferred to the soil. An approximate method, often used in design, is to assume that the effective transfer level is at a depth of 2D/3 below the top of the piles. It is also assumed that there is a spread of the total load, one horizontal to four vertical. The settlement of this equivalent foundation (Fig. 8.24) can then be determined by the normal methods.

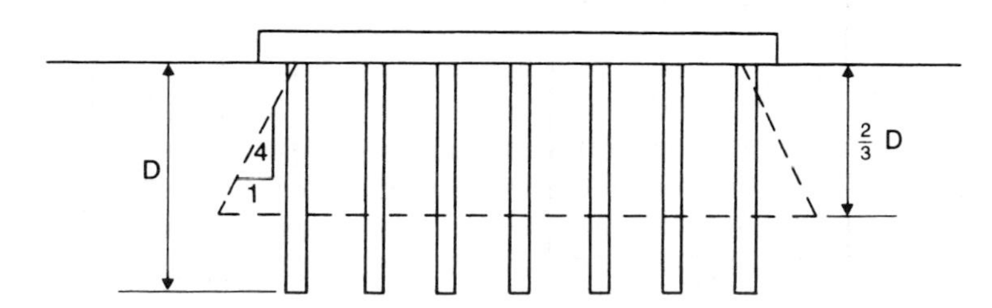

Fig. 8.24 Transference of load in friction piles.

Exercises

Note Where applicable the answers quoted incorporate a factor of safety equal to 3.0.

Exercise 8.1

A fine sand deposit is saturated throughout with a unit weight of 20 kN/m^3. Ground water level is at a depth of 1 m below the surface. A standard penetration test, carried out at a depth of 2 m, gave an N value of 18. If the settlement is to be limited to not more than 25 mm, determine an allowable bearing pressure value for a 2 m square foundation founded at a depth of 2 m.

Answers 460/2 = 230 kPa (N $\approx$ 40)

Exercise 8.2

A strip footing 3 m wide is to be founded at a depth of 2 m in a saturated soil of unit weight 19 kN/m^3. The soil has an angle of friction, ϕ, of 28° and a cohesion, c, of 5 kPa. Groundwater level is at a depth of 4 m. Determine a value for the safe bearing capacity of the foundation.

If the groundwater level was to rise to the ground surface, determine the new value of safe bearing capacity.

Answer 459 kPa; 249 kPa

Exercise 8.3

A 2.44 m wide strip footing is to be founded in a coarse sand at a depth of 3.05 m. The unit weight of the sand is 19.3 kN/m^3 and standard penetration tests at the 3.05 m depth gave an N value of 12.

(i) Determine the safe bearing capacity of the foundation if settlement is of no account.
(ii) Determine the allowable bearing pressure if settlement of the foundation is not to exceed 25 mm.

Answers (i) 1300 kPa, (ii) 300 kPa

Exercise 8.4

A single test pile, 300 mm diameter, is driven through a depth of 8 m of clay which has an undrained cohesive strength varying from 10 kPa at its surface to 50 kPa at a depth of 8 m. Estimate the safe load that the pile can carry.

Answer 60 kN

Exercise 8.5

A continuous concrete footing ($\gamma_c = 24$ kN/m^3) of breadth 2.0 m and thickness 0.5 m is to be founded in a clay soil ($\phi_u = 0°$; $c_u = 22$ kPa; $\gamma = 19$ kN/m^3) at a depth of 1.0 m. The footing will carry an applied vertical load of magnitude 85 kN per metre run. The load will act on the centre-line of the footing.

Using Eurocode 7 Design Approach 1, determine the magnitude of the over-design factor for both Combination 1 and Combination 2.

Answer 1.53 (DA1-1); 1.56 (DA1-2)

If you were to include depth factors in the design procedure, what would be the revised value of the over-design factor for each combination?

Answer 1.79 (DA1-1); 1.81 (DA1-2)

Note: Adopting depth factors in the design will invariably lead to higher values of over-design factor.

Exercise 8.6

A rectangular foundation (2.5 m × 6 m × 0.8 m deep) is to be founded at a depth of 1.2 m in a dense sand ($c' = 0$; $\phi' = 32°$; $\gamma = 19.4$ kN/m^3). The unit weight of concrete = 24 kN/m^3. The foundation will carry a vertical line load of 250 kN/m at an eccentricity of 0.4 m.

By following Eurocode 7, Design Approach 1 establish the proportion of the available resistance that will be used.

Answer 16 per cent (DA1-1); 24 per cent (DA1-2)

Note: The proportion of available resistance that will be used is determined by taking the reciprocal of the over-design factor.

Chapter 9

Foundation Settlement and Soil Compression

9.1 Settlement of a foundation

Probably the most difficult of the problems that a soils engineer is asked to solve is the accurate prediction of the settlement of a loaded foundation.

The problem is in two distinct parts: (i) the value of the total settlement that will occur, and (ii) the rate at which this value will be achieved.

When a soil is subjected to an increase in compressive stress due to a foundation load the resulting soil compression consists of elastic compression, primary compression and secondary compression.

Elastic compression

This compression is usually taken as occurring immediately after the application of the foundation load. Its vertical component causes a vertical movement of the foundation (immediate settlement) that in the case of a partially saturated soil is mainly due to the expulsion of gases and to the elastic bending reorientation of the soil particles. With saturated soils immediate settlement effects are assumed to be the result of vertical soil compression before there is any change in volume.

Primary compression

The sudden application of a foundation load, besides causing elastic compression, creates a state of excess hydrostatic pressure in saturated soil. These excess pore water pressure values can only be dissipated by the gradual expulsion of water through the voids of the soil, which results in a volume change that is time dependent. A soil experiencing such a volume change is said to be consolidating, and the vertical component of the change is called the consolidation settlement.

Secondary compression

Volume changes that are more or less independent of the excess pore water pressure values cause secondary compression. The nature of these changes is not fully understood but they are apparently due to a form of plastic flow resulting in a displacement of the soil particles. Secondary compression effects can continue over long periods of time and, in the consolidation test (see Section 9.3.2), become apparent towards the end of the primary compression stage: due to the thinness of the sample, the excess pore water pressures are soon dissipated and it may appear that the main part of secondary compression occurs after primary compression is completed. This effect is absent in the case of an *in situ* clay layer because the large dimensions involved mean that a considerable time is required before the excess pore pressures drain away. During this time the effects of secondary compression are also taking place so that, when primary compression is complete, little, if any, secondary effect is noticeable. The terms 'primary' and 'secondary' are therefore seen to be rather arbitrary divisions of the single, continuous consolidation process. The time relationships of these two factors will be entirely different if they are obtained from two test samples of different thicknesses.

9.2 Immediate settlement

9.2.1 *Cohesive soils*

If a saturated clay is loaded rapidly, the soil will be deformed during the load application and excess hydrostatic pore pressures are set up. This deformation occurs with virtually no volume change, and due to the low permeability of the clay, little water is squeezed out of the voids. Vertical deformation due to the change in shape is the immediate settlement.

This change in shape is illustrated in Fig. 9.1a, where an element of soil is subjected to a vertical major principal stress increase $\Delta\sigma_1$, which induces an excess pore water pressure, Δu. The lateral expansion causes an increase in the minor principal stress, $\Delta\sigma_3$.

The formula for immediate settlement of a flexible foundation was provided by Terzaghi (1943) and is

$$\rho_i = \frac{pB(1 - \mu^2)N_p}{E}$$

which gives the immediate settlement at the corners of a rectangular footing, length L and width B. In the case of a uniformly loaded, perfectly flexible square footing, the immediate settlement under its centre is twice that at its corners.

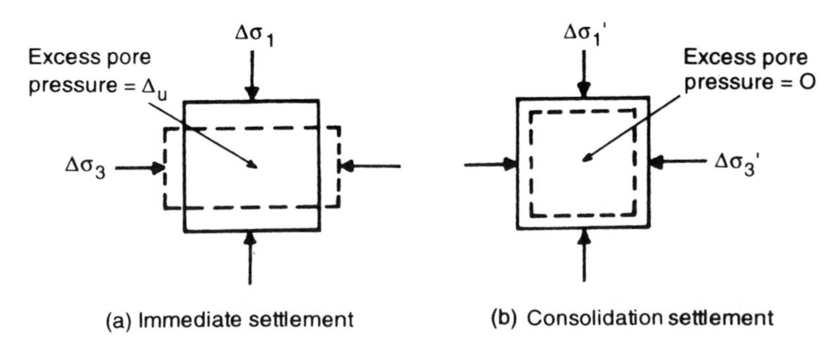

Fig. 9.1 Compressive deformation.

Table 9.1

L/B	N_p
1.0	0.56
2.0	0.76
3.0	0.88
4.0	0.96
5.0	1.00

Various values for N_p are given in Table 9.1.

By the principle of superposition it is possible to determine the immediate settlement under any point of the base of a foundation (Example 9.2). A spoil heap or earth embankment can be taken as flexible and to determine the immediate settlement of deposits below such a construction the coefficients of Table 9.1 should be used.

Foundations are generally more rigid than flexible and tend to impose a uniform settlement which is roughly the same value as the mean value of settlement under a flexible foundation. The mean value of settlement for a rectangular foundation on the surface of a semi-elastic medium is given by the expression:

$$\rho_i = \frac{pB(1 - \mu^2)I_p}{E}$$

where

B = width of foundation
p = uniform contact pressure
E = Young's modulus of elasticity for the soil
μ = Poisson's ratio for the soil (= 0.5 in saturated soil)
I_p = an influence factor depending upon the dimensions of the foundation.

Skempton (1951) suggests the values for I_p given in Table 9.2.

Table 9.2

L/B	I_p
circle	0.73
1	0.82
2	1.00
5	1.22
10	1.26

Immediate settlement of a thin clay layer

The coefficients of Tables 9.1 and 9.2 only apply to foundations on deep soil layers. Vertical stresses extend to about 4B below a strip footing and the formulae, strictly speaking, are not applicable to layers thinner than this, although little error is incurred if the coefficients are used for layers of thicknesses greater than 2.0B. A drawback of the method is that it can only be applied to a layer immediately below the foundation.

For cases when the thickness of the layer is less than 4.0B a solution is possible with the use of coefficients prepared by Steinbrenner (1934), whose procedure was to determine the immediate settlement at the top of the layer (assuming infinite depth) and to calculate the settlement at the bottom of the layer (again assuming infinite depth) below it. The difference between the two values is the actual settlement of the layer.

The total immediate settlement at the corner of a rectangular foundation on an infinite layer is

$$\rho_i = \frac{pB(1-\mu^2)I_p}{E}$$

The values of the coefficient I_p (when $\mu = 0.5$) are given in Fig. 9.2c. To determine the settlement of a point beneath the foundation the area is divided into rectangles that meet over the point (the same procedure used when determining vertical stress increments by Steinbrenner's method). The summation of the settlements of the corners of the rectangles gives the total settlement of the point considered.

This method can be extended to determine the immediate settlement of a clay layer which is at some depth below the foundation. In Fig. 9.2b the settlement of the lower layer (of thickness $H_2 - H_1$) is obtained by first determining the settlement of a layer extending from below the foundation that is of thickness H_2 (using E_2); from this value is subtracted the imaginary settlement of the layer H_1 (again using E_2).

It should be noted that the settlement values obtained by this method are for a perfectly flexible foundation. Usually the value of settlement at the centre of the

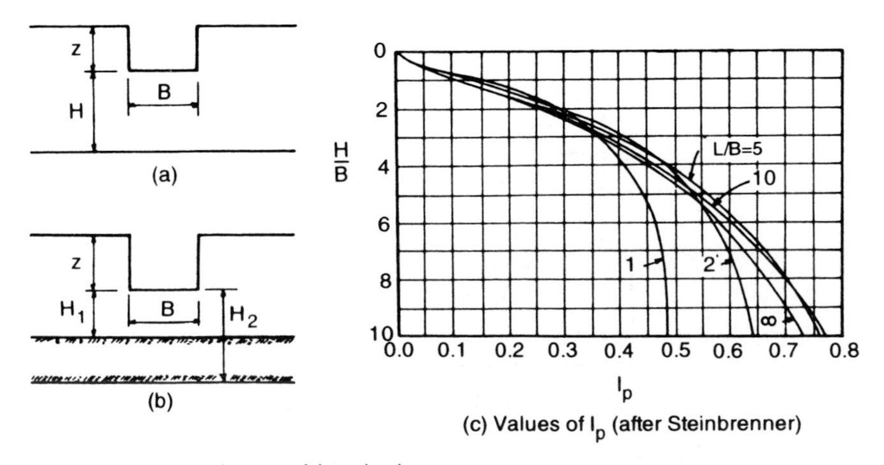

Fig. 9.2 Immediate settlement of thin clay layer.

foundation is evaluated and reduced by a rigidity factor (generally taken as 0.8) to give a mean value of settlement that applies over the whole foundation.

Example 9.1

A reinforced concrete foundation, of dimensions 20 m × 40 m exerts a uniform pressure of 200 kPa on a semi-infinite saturated soil layer (E = 50 MPa).

Determine the value of immediate settlement under the foundation using Table 9.2.

Solution

$$\frac{L}{B} = \frac{40}{20} = 2.0$$

From Table 9.2, $I_p = 1.0$.

$$\rho_i = \frac{pB(1-\mu^2)}{E} I_p = \frac{200 \times 20 \times 0.75}{50\,000} \times 1.0 = 0.06 \text{ m} = 60 \text{ mm}$$

Example 9.2

The plan of a proposed spoil heap is shown in Fig. 9.3a. The tip will be about 23 m high and will sit on a thick, soft alluvial deposit (E = 15 MPa). It is estimated that the eventual uniform bearing pressure on the soil will be about 300 kPa. Estimate the immediate settlement under the point A at the surface of the soil.

Solution

The procedure is to divide the plan area into a number of rectangles, the corners of which must meet at the point A; in Fig. 9.3b it is seen that three rectangles are required. As the structure is flexible and the soil deposit is thick, the coefficients of Table 9.1 should be used:

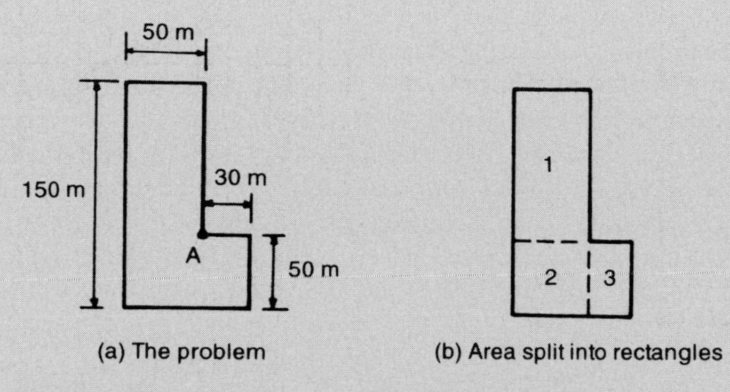

Fig. 9.3 Example 9.2.

Rectangle (1): 100 m × 50 m $\frac{L}{B} = 2.0$; $N_p = 0.76$

Rectangle (2): 50 m × 50 m $\frac{L}{B} = 1.0$; $N_p = 0.56$

Rectangle (3): 50 m × 30 m $\frac{L}{B} = 1.67$; $N_p = 0.64$

$$\rho_i = \frac{p}{E}(1 - \mu^2)(N_{p1}B_1 + N_{p2}B_2 + N_{p3}B_3)$$

$$= \frac{300 \times 0.75}{15\ 000}(0.76 \times 50 + 0.56 \times 50 + 0.64 \times 30)$$

$$= 1.28 \text{ m}$$

The effect of depth

Fox (1948) showed that for deep foundations ($z > B$) the calculated immediate settlements are more than the actual ones, and a reduction may be applied. If $z = B$ the reduction is approximately 25 per cent, increasing to about 50 per cent for infinitely deep foundations.

Most foundations are shallow, however, and although this reduction can be allowed for when a layer of soil is some depth below a foundation, the settlement effects in this case are small so it is not customary practice to reduce them further.

Determination of E

The modulus of elasticity, E, is usually obtained from the results of a consolidated undrained triaxial test carried out on a representative sample of the soil that is consolidated under a cell pressure approximating to the effective overburden pressure at the level from which the sample was taken. The soil is then sheared undrained to obtain the plot of total deviator stress against strain; this is never a straight line and to determine E a line must be drawn from the origin up to the value of deviator stress that will be experienced in the field when the foundation load is applied. In deep layers there is the problem of assessing which depth represents the average, and ideally the layer should be split into thinner layers with a value of E determined for each.

A certain amount of analysis work is necessary in order to carry out the above procedure. The increments of principal stress $\Delta\sigma_1$ and $\Delta\sigma_3$ must be obtained so that the value of $\Delta\sigma_1 - \Delta\sigma_3$ is known, and a safety factor of 3.0 is generally applied against bearing capacity failure. Skempton (1951) points out that when the factor of safety is 3.0 the maximum shear stress induced in the soil is not greater than 65 per cent of the ultimate shear strength, so that a value of E can be obtained directly from the triaxial test results by simply determining the strain corresponding to 65 per cent of the maximum deviator stress and dividing this value into its corresponding stress. The method produces results that are well within the range of accuracy possible with other techniques.

9.2.2 Cohesionless soils

Owing to the high permeabilities of cohesionless soils, both the elastic and the primary effects occur more or less together. The resulting settlement from these factors is termed the immediate settlement.

The chance of bearing capacity failure in a foundation supported on a cohesionless soil is remote, and for these soils it has become standard practice to use settlement as the design criterion. The allowable bearing pressure, p, is generally defined as the pressure that will cause an average settlement of 25 mm in the foundation.

The determination of p from the results of the standard penetration test has been discussed in Chapter 8. If the actual bearing pressure is not equal to the value of p then the value of settlement is not known and, since it is difficult to obtain this value from laboratory tests, resort must be made to *in situ* test results. Most methods used required the value of C_r, the penetration resistance of the Dutch cone, which is usually expressed in MPa or kPa.

Meyerhof's method

A quick estimate of the settlement, ρ, of a footing on sand has been proposed by Meyerhof (1974):

$$\rho = \frac{\Delta pB}{2\overline{C}_r}$$

where

B = the least dimension of the footing
$\overline{C}_r$ = the average value of C_r over a depth below the footing equal to B
Δp = the net foundation pressure increase, which is simply the foundation loading less the value of vertical effective stress at foundation level, σ'_{v0}.

The two other methods commonly in use were proposed by De Beer and Martens (1957) and by Schmertmann (1970). Both methods require a value for C_r and, if either is to be used with standard penetration test results, it is necessary to have the correlation between C_r and N.

Obviously the value of C_r obtained from the Dutch cone penetration test must be related to the number of recorded blows, N, obtained from the standard penetration test. Various workers have attempted to find this relationship but, so far, the results have not been encouraging. Meigh and Nixon (1961) showed that, over a number of sites, C_r varied from $(430 \times N)$ to $(1930 \times N)$ kPa.

The relationship most commonly used at the present time is that proposed by Meyerhof (1956):

$$C_r = 400 \times N \text{ kPa}$$

where N = actual number of blows recorded in the standard penetration test.

It goes without saying that, whenever possible, C_r values obtained from actual cone tests should be used in preference to values estimated from N values.

The relationships between N and C_r determined by various workers and the implications involved have been discussed by Meigh (1987).

De Beer and Martens' method

From the results of the *in situ* tests carried out, a plot of C_r (or N) values against depth is prepared, similar to that shown in Fig. 8.22. With the aid of this plot the profile of the compressible soil beneath the proposed foundation can be divided into a suitable number of layers, preferably of the same thickness, although this is not essential.

In the case of a deep soil deposit the depth of soil considered as affected by the foundation should not be less than 2.0B, ideally 4.0B, where B = foundation width.

The method proposes the use of a constant of compressibility, C_S, where

$$C_S = 1.5\,\frac{C_r}{p_{o1}}$$

where

C_r = static cone resistance (kPa)
p_{o1} = effective overburden pressure at the point tested:

Total immediate settlement is

$$\rho_i = \frac{H}{C_S}\ln\frac{p_{o2} + \Delta\sigma_z}{p_{o2}}$$

where

$\Delta\sigma_z$ = vertical stress increase at the centre of the consolidating layer of thickness H
p_{o2} = effective overburden pressure at the centre of the layer before any excavation or load application.

Note Meyerhof (1956) suggests that a more realistic value for C_S is

$$C_S = 1.9\,\frac{C_r}{p_{o1}}$$

Such a refinement may be an advantage if the calculations use C_r values which have been determined from Dutch cone penetration tests, but if the C_r values used have been obtained from the relationship $C_r = 400$ kPa, such a refinement seems naïve.

Schmertmann's method

Originally proposed by Schmertmann in 1970 and modified by Schmertmann *et al.* (1978), the method is now generally preferred to De Beer and Martens' approach.

The method is based on two main assumptions:

(i) the greatest vertical strain in the soil beneath the centre of a loaded foundation of width B occurs at depth B/2 below a square foundation and at depth of B below a long foundation;

(ii) significant stresses caused by the foundation loading can be regarded as insignificant at depths greater than z = 2.0B for a square footing and = 4.0B for a strip footing.

The method involves the use of a vertical strain influence factor, I_z, whose value varies with depth. Values of I_z, for a net foundation pressure increase, Δp, equal to the effective overburden pressure at depth B/2, are shown in Fig. 9.4.

The procedure consists of dividing the sand below the footing into n layers, of thicknesses $\Delta_{z_1}, \Delta_{z_2}, \Delta_{z_3} \ldots \Delta_{z_n}$. If soil conditions permit it is simpler if the layers can be made of equal thickness, Δz. The vertical strain of a layer is taken as equal to the increase in vertical stress at the centre of the layer, i.e. Δp multiplied by I_z, which is then divided by the product of C_r and a factor x. Hence:

$$\rho = C_1 C_2 \Delta p \sum_1^n \frac{I_z}{xC_r} \Delta z_1$$

where

x = 2.5 for a square footing and 3.5 for a long footing

I_z = the strain influence factor, valued for each layer at its centre, and obtained from a diagram similar to Fig. 9.4 but redrawn to correspond to the foundation loading

C_1 = a correction factor for the depth of the foundation

$= 1.0 - 0.5 \dfrac{\sigma'_v}{\Delta p}$ (= 1.0 for a surface footing)

C_2 = a correction factor for creep

$= 1 + 0.2 \log_{10} 10t$ (t = time in years after the application of foundation loading for which the settlement value is required).

As mentioned above, Fig. 9.4 must be redrawn. This is achieved by obtaining a new peak value for I_z from the expression

$$I_z = 0.5 + 0.1 \left(\frac{\Delta p}{\sigma'_{vp}} \right)^{0.5}$$

where σ'_{vp} = the effective vertical overburden pressure at a depth of 0.5B for a square foundation and at a depth of 1.0B for a long foundation.

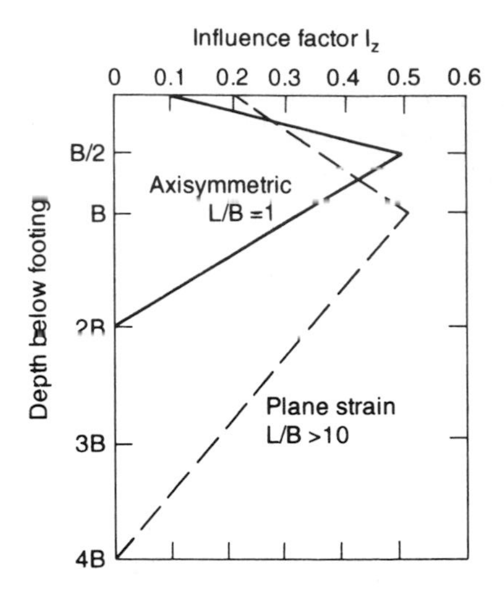

Fig. 9.4 Variation of I_z with depth (after Schmertmann, 1970).

Example 9.3

A foundation, 1.5 m square, will carry a load of 300 kPa and will be founded at a depth of 0.75 m in a deep deposit of granular soil. The soil may be regarded as saturated throughout with a unit weight of 20 kN/m^3, and the approximate N to z relationship is shown in Fig. 9.5a.

If the groundwater level occurs at a depth of 1.5 m below the surface of the soil determine a value for the settlement at the centre of the foundation, (a) by De Beer and Martens' method, (b) by Schmertmann's method.

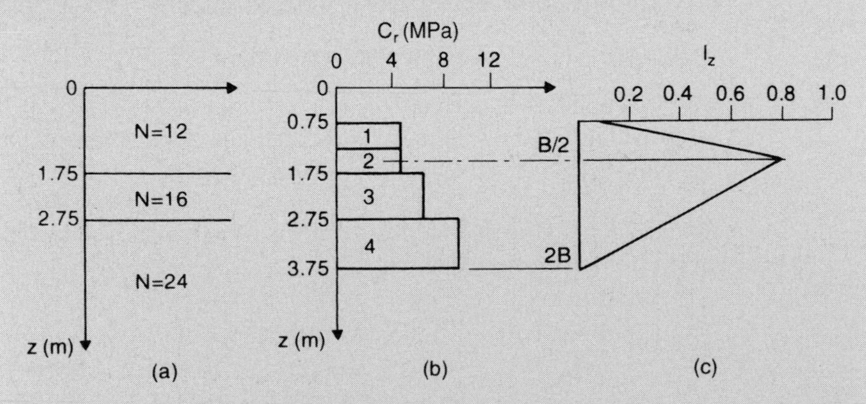

Fig. 9.5 Example 9.3. (a) N to z relationship; (b) C_r to z relationship (part (b) of example); (c) variation of I_z (part (b) of example).

Solution

(a) De Beer and Martens' method

The soil deposit is deep, therefore investigate to a depth of about 3B to 4B below foundation. In conjunction with Fig. 9.5a it is seen that a depth of 5 m below the foundation can be conveniently divided into four layers of soil, two of 1 m and two of 1.5 m thickness, as shown in the tabulated workings, but not in Fig. 9.5.

Net pressure, p = 300 – (0.75 × 20) = 285 kPa

Layer	Thickness, Δz (m)	C_r (kPa)	P_{o1} at layer centre (kPa)	$C_s = \frac{1.5C_r}{p_{o1}}$
1	1.0	400 × 12 = 4800	20 × 1.25 = 25	288
2	1.0	400 × 16 = 6400	(20 × 2.25) – (9.81 × 0.75) = 37.6	255
3	1.5	400 × 24 = 9600	(20 × 3.5) – (9.81 × 2.0) = 50.4	286
4	1.5	9600	(20 × 5.0) – (9.81 × 3.5) = 65.7	219

Use Steinbrenner's method to determine vertical pressure increments (Chapter 4).

Then B = L = 1.5/2 = 0.75 m, and $p_{o1} = p_{o2}$ for each layer.

Layer	B/z = L/z	I_σ	$4I_\sigma$	$\Delta\sigma_z = 4pI_\sigma$	$\ln \frac{p_o + \Delta\sigma_z}{p_o}$ (A)	$\frac{\Delta z}{C_s} \times$ (A)
1	0.75/0.5 = 1.5	0.213	0.852	243	2.3721	0.00824
2	0.75/1.5 = 0.5	0.088	0.352	100	1.2973	0.00509
3	0.75/2.75 = 0.27	0.03	0.12	34	0.5156	0.00270
4	0.75/4.25 = 0.18	0.015	0.06	17	0.2301	0.00158
						Σ0.01761

Total settlement = 0.01761 × 100 = 17.6 mm, say 18 mm

(b) Schmertmann's method

For a square footing significant depths extend to 2.0B and σ'_{vp} is taken as the effective vertical overburden pressure at a depth of 0.5B below the foundation, i.e. in this example, 0.75 m, so that $\sigma'_{vp} = 20(0.75 + 0.75) = 30$ kPa. Net foundation pressure increase $\Delta p = 300 - 20 \times 0.75 = 285$ kPa. Hence:

$$I_z = 0.5 + 0.1\left(\frac{\Delta p}{\sigma'_{vp}}\right)^{0.5}$$

$$= 0.5 + 0.1(285/30)^{0.5} = 0.81$$

The variation of I_z for depths up to 2B below the foundation is shown in Fig. 9.5c. The C_r values shown in Fig. 9.5b are obtained from the N values using the relationship $C_r = 0.4N$ MPa. With these C_r values, it is possible to decide upon the number and thicknesses of the layers that the soil can be divided into. For this example, for the purpose of illustration, only four layers have been chosen and these are shown in Fig. 9.5(b). (For greater accuracy the number of layers should be about 8 for a square footing and up to 16 for a long footing.) For a square foundation Schmertmann recommends that the value of the factor x = 2.5. The calculations are set out below.

Layer	Δz_i (m)	Depth below foundation to centre of layer (m)	C_r (MPa)	I_z	$\frac{I_z \Delta z_i}{xC_r}$
1	0.5	0.25	4.8	0.27	0.011
2	0.5	0.75	4.8	0.81	0.034
3	1.0	1.5	6.4	0.53	0.033
4	1.0	2.5	9.6	0.18	0.008
					Σ0.09

$$C_1 = 1.0 - 0.5\frac{\sigma'_v}{\Delta p} = 1 - \left(0.5 \times \frac{15}{285}\right) = 0.97$$

Assume that $C_2 = 1.0$, then:

$$\rho = 0.97 \times 285 \times 0.09 = 24.9 \text{ mm}$$

Total settlement of centre of foundation = 25 mm

Note The Schmertmann approach is preferred by most engineers to the De Beer and Martens' method.

The plate loading test

The results from a plate loading test can be used to predict the average settlement of a proposed foundation on granular soil. The test should be carried out at the proposed foundation level and the soil tested must be relatively homogeneous for some depth (not with conditions such as illustrated in Fig. 4.9).

If ρ_1 is the settlement of the test footing under a certain value of bearing pressure, then the average settlement of the foundation, ρ, under the same value of bearing pressure can be obtained by the empirical relationship proposed by Terzaghi and Peck (1948):

$$\rho = \rho_1 \left(\frac{2B}{B + B_1}\right)^2$$

where

B_1 = width or diameter of test footing
B = width or diameter of proposed foundation.

One aspect of using the results from a plate loading test for settlement predictions is that it is important to know the position of the groundwater level.

It may be that the bulb of pressure from the test footing is partly or completely above the groundwater level whereas, when the foundation is constructed, the groundwater level will be significantly within the bulb of pressure. Such a situation could lead to actual settlement values as much as twice the values predicted by the formula.

9.3 Consolidation settlement

This effect occurs in clays where the value of permeability prevents the initial excess pore water pressures from draining away immediately. The design loading used to calculate consolidation settlement must be consistent with this effect.

A large wheel load rolling along a roadway resting on a clay will cause an immediate settlement that is in theory completely recoverable once the wheel has passed, but if the same load is applied permanently there will in addition be consolidation. Judgement is necessary in deciding what portion of the superimposed loading carried by a structure will be sustained long enough to cause consolidation, and this involves a quite different procedure from that used in a bearing capacity analysis, which must allow for total dead and superimposed loadings.

9.3.1 *One-dimensional consolidation*

The pore water in a saturated clay will commence to drain away soon after immediate settlement has taken place: the removal of this water leading to the volume change is known as consolidation (Fig. 9.1b). The element contracts both horizontally and vertically under the actions of $\Delta\sigma'_3$ and $\Delta\sigma_1$, which gradually increase in magnitude as the excess pore water pressure, Δu, decreases. Eventually, when $\Delta u = 0$, then $\Delta\sigma'_3 = \Delta\sigma_3$ and $\Delta\sigma'_1 = \Delta\sigma_1$, and at this stage consolidation ceases, although secondary consolidation may still be apparent.

If it can be arranged for the lateral expansion due to the change in shape to equal the lateral compression consequent upon the change in volume, and for these changes to occur together, then there will be no immediate settlement and the resulting compression will be one-dimensional with all the strain occurring in the vertical direction. Settlement by one-dimensional strain is by no means uncommon in practice, and most natural soil deposits have experienced one-dimensional settlement during the process of deposition and consolidation.

The consolidation of a clay layer supporting a foundation whose dimensions are much greater than the layer's thickness is essentially one-dimensional as lateral strain effects are negligible save at the edges.

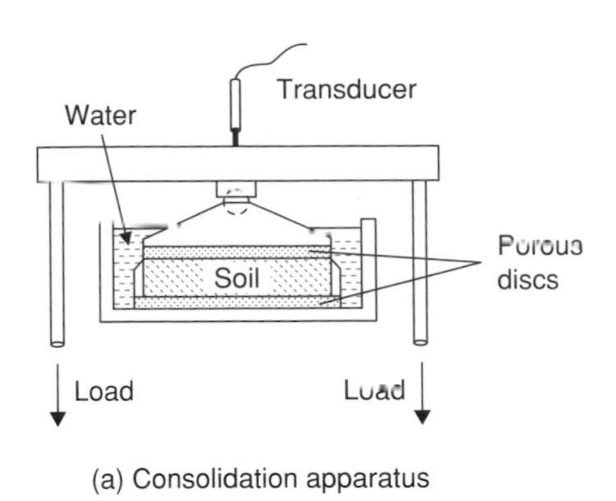

(a) Consolidation apparatus

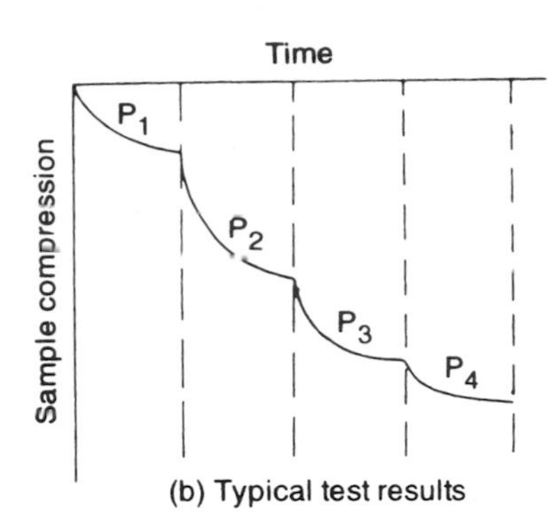

(b) Typical test results

Fig. 9.6 The consolidation test.

9.3.2 The consolidation test

The apparatus generally used in the laboratory to determine the primary compression characteristics of a soil is known as the consolidation test apparatus (or oedometer) and is illustrated in Fig. 9.6a.

The soil sample (generally 75 mm diameter and 20 mm thick) is encased in a steel cutting ring. Porous discs, saturated with air-free water, are placed on top of and below the sample, which is then inserted in the oedometer.

A vertical load is then applied and the resulting compression measured by means of a transducer at intervals of time, readings being logged until the sample has achieved full consolidation (usually for a period of 24 hours). Further load increments are then applied and the procedure repeated, until the full stress range expected *in situ* has been covered by the test (Fig. 9.6b).

The test sample is generally flooded with water soon after the application of the first load increment in order to prevent pore suction.

After the sample has consolidated under its final load increment the pressure is released in stages at 24 hour intervals and the sample allowed to expand. In this way an expansion to time curve can also be obtained.

After the loading has been completely removed the final thickness of the sample can be obtained, from which it is possible to calculate the void ratio of the soil for each stage of consolidation under the load increments. The graph of void ratio to consolidation pressure can then be drawn, such a curve generally being referred to as an e–p curve (Fig. 9.7a).

It should be noted that the values of p refer to effective stress, for after consolidation the excess pore pressures become zero and the applied stress increment is equal to the effective stress increment.

If the sample is recompressed after the initial cycle of compression and expansion, the e–p curve for the whole operation is similar to the curves shown in Fig. 9.7b; the recompression curve is flatter than the original compression curve, primary

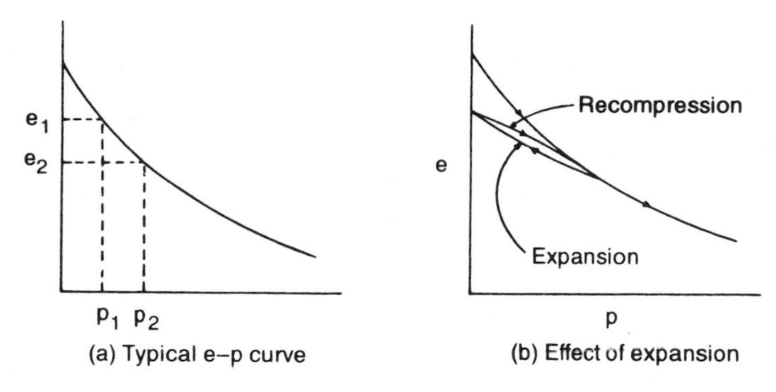

Fig. 9.7 Void ratio to effective pressure curves.

compression being made up of (i) a reversible part and (ii) an irreversible part. Once the consolidation pressure is extended beyond the original consolidation pressure value (the preconsolidation pressure), the e–p curve follows the trend of the original compression curve.

All types of soil, whether sand, silt or clay, have the form of compression curves illustrated in Fig. 9.7. The curves shown can be produced quite quickly in the laboratory for teaching purposes, using a dry sand sample, but consolidation problems are mainly concerned with clays and the oedometer is therefore only used to test these types of soil.

9.3.3 Volumetric change

The volume change per unit of original volume constitutes the volumetric change. If a mass of soil of volume V_1 is compressed to a volume V_2, the assumption is made that the change in volume has been caused by a reduction in the volume of the voids.

$$\text{Volumetric change} = \frac{V_1 - V_2}{V_1}$$

$$= \frac{(1 + e_1) - (1 + e_2)}{1 + e_1}$$

$$= \frac{e_1 - e_2}{1 + e_1}$$

where

e_1 = void ratio at p_1
e_2 = void ratio at p_2.

The slope of the e–p curve is given the symbol 'a', then:

$$a = \frac{e_1 - e_2}{p_1 - p_2} \text{ m}^2/\text{kN}$$

i.e.

$$a = \frac{de}{dp}$$

The slope of the e–p curve is seen to decrease with increase in pressure; in other words, a is not a constant but will vary depending upon the pressure. Settlement problems are usually only concerned with a range of pressure (that between the initial pressure and the final pressure), and over this range a is taken as constant by assuming that the e–p curve between these two pressure values is a straight line.

9.3.4 The Rowe oedometer

An alternative form to the consolidation cell shown in Fig. 9.6 was described by Rowe and Barden (1966) and is listed in BS 1377: Part 6.

The oedometer is hydraulically operated and a various range of cell sizes are available so that test specimens as large as 500 mm diameter and 250 mm thick can be tested. The machine is particularly useful for testing samples from clay deposits where macrofabric effects are significant.

A constant pressure system applies a hydraulic pressure, via a convoluted rubber jack made from rubber some 2 mm thick, on to the top of the test specimen. Vertical settlement is measured at the centre of the sample by means of a hollow brass spindle, 10 mm diameter, attached to the jack and passing out through the centre of the top plate to a suitable dial gauge or transducer.

Drainage of the sample can be made to vary according to the nature of the test and can be either vertical or radial, the latter being arranged to be either inwards or outwards. The expelled water exits via the spindle and it is possible to measure pore water pressures during the test, as well as applying a back pressure to the specimen if required. The apparatus can also be used for permeability tests, as described in BS 1377.

9.3.5 Coefficient of volume compressibility m_v

This value, which is sometimes called the coefficient of volume decrease, represents the compression of a soil, per unit of original thickness, due to a unit increase in pressure, i.e.

$$m_v = \text{Volumetric change/Unit of pressure increase}$$

If H_1 = original thickness and H_2 = final thickness:

$$\text{Volumetric change} = \frac{V_1 - V_2}{V_1} = \frac{H_1 - H_2}{H_1} \qquad \text{(as area is constant)}$$

$$= \frac{e_1 - e_2}{1 + e_1}$$

Now

$$a = \frac{e_1 - e_2}{dp}$$

$$\Rightarrow \quad \text{Volumetric change} = \frac{a\,dp}{1 + e_1}$$

$$\Rightarrow \quad m_v = \frac{a\,dp}{1 + e_1}\,\frac{1}{dp} = \frac{a}{1 + e_1}\ \text{m}^2/\text{MN}$$

For most practical engineering problems m_v values can be calculated for a pressure increment of 100 kPa in excess of the present effective overburden pressure at the sample depth.

Once the coefficient of volume decrease has been obtained we know the compression/unit thickness/unit pressure increase. It is therefore an easy matter to predict the total consolidation settlement of a clay layer of thickness H:

$$\text{Total settlement} = \rho_c = m_v\,dp\,H$$

Typical values of m_v are given in Table 9.3.

In the laboratory consolidation test the compression of the sample is one-dimensional as there is lateral confinement, the initial excess pore water pressure induced in a saturated clay on loading being equal to the magnitude of the applied major principal stress (due to the fact that there is no lateral yield). This applies no matter what type of soil is tested, provided it is saturated.

One-dimensional consolidation can be produced in a triaxial test specimen by means of a special procedure known as the K_0 test (see Bishop and Henkel, 1962).

Table 9.3

Soil	m_v (m^2/MN)
Peat	10.0–2.0
Plastic clay (normally consolidated alluvial clays)	2.0–0.25
Stiff clay	0.25–0.125
Hard clay (boulder clays)	0.125–0.0625

Example 9.4

The following results were obtained from a consolidation test on a sample of saturated clay, each pressure increment having been maintained for 24 hours.

Pressure (kPa)	Thickness of sample after consolidation (mm)
0	20.0
50	19.65
100	19.52
200	19.35
400	19.15
800	18.95
0	19.25

After it had expanded for 24 hours the sample was removed from the apparatus and found to have a moisture content of 25 per cent. The particle specific gravity of the soil was 2.65.

Plot the void-ratio to effective pressure curve and determine the value of the coefficient of volume change for a pressure range of 250–350 kPa.

Solution

$w = 0.25; \qquad G_s = 2.65$

Now $e = wG_s$ (as soil is saturated) $= 0.25 \times 2.65 = 0.662$. This is the void ratio corresponding to a sample thickness of 19.25 mm.

$$\frac{dH}{H_1} = \frac{de}{1 + e_1} \quad \Rightarrow \quad de = \frac{(1 + e_1)}{H_1} dH = \frac{1.662}{19.25} dH = 0.0865\ dH$$

The values of e at the end of each consolidation can be calculated from this expression.

Pressure	H (mm)	dH (mm)	de	e
0	20.0	+0.75	+0.065	0.727
50	19.65	+0.40	+0.035	0.697
100	19.52	+0.27	+0.023	0.685
200	19.35	+0.10	+0.009	0.671
400	19.15	−0.10	−0.009	0.653
800	18.95	−0.30	−0.026	0.636
0	19.25	0	0	0.662

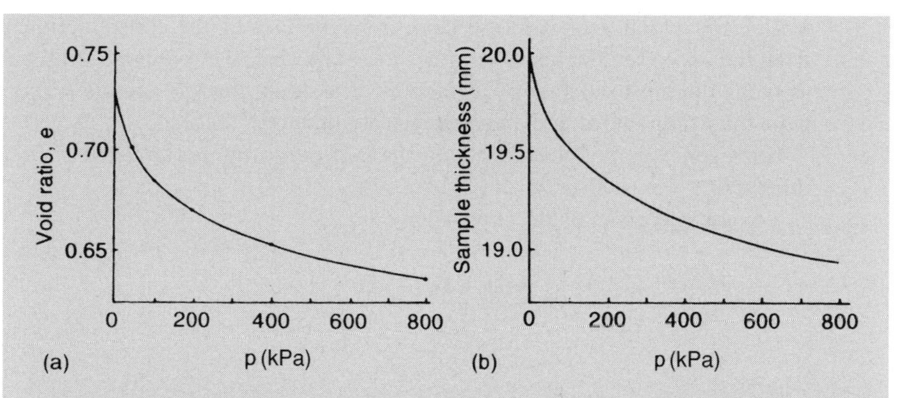

Fig. 9.8 Example 9.4.

From the e–p curve in Fig. 9.8a:

$$e \text{ at } 250 \text{ kPa} = 0.666$$
$$e \text{ at } 350 \text{ kPa} = 0.658$$

$$a = \frac{de}{dp} = \frac{0.666 - 0.658}{100}$$
$$= 0.000\,08 \text{ m}^2/\text{kN}$$

$$\Rightarrow \quad m_v = \frac{a}{1 + e_1} = \frac{0.000\,08}{1.666}$$
$$= 4.8 \times 10^{-5} \text{ m}^2/\text{kN}$$

Alternative method for determining m_v

m_v can be expressed in terms of thicknesses:

$$m_v = \frac{dH}{H_1}\frac{1}{dp} = \frac{1}{H_1}\frac{dH}{dp}$$

dH/dp is the slope of the curve of thickness of sample against pressure. Hence m_v can be obtained by finding the slope of the curve at the required pressure and dividing by the original thickness. The thickness/pressure curve is shown in Fig. 9.8b; from it:

$$H \text{ at } 250 \text{ kPa} = 19.28$$
$$H \text{ at } 350 \text{ kPa} = 19.19$$

$$m_v = \frac{19.28 - 19.19}{19.28 \times 100} = \frac{0.09}{19.28 \times 100}$$
$$= 4.7 \times 10^{-5} \text{ m}^2/\text{kN}$$

If a layer of this clay, 20 m thick, had been subjected to this pressure increase then the consolidation settlement would have been:

$$\rho_c = m_v H\, dp = 0.000\,047 \times 20 \times 100 \times 1000 = 96 \text{ mm}$$

Note The practice of working back from the end of the consolidation test, i.e. from the expanded thickness, in order to obtain an e–p curve is generally accepted as being the most satisfactory as there is little doubt that the sample is more likely to be fully saturated after expansion than at the start of the test.

However, it is possible to obtain the e–p curve by working from the original thickness.

Void ratio is given by the expression

$$e = \frac{V_v}{V_s} = \frac{V - V_s}{V_s} = \frac{A(H - H_s)}{AH_s} = \frac{H - H_s}{H_s}$$

where:

A = area of sample
H = height or thickness
H_s = equivalent height of solids (V_s/A).

Now

$$V_s = \frac{M}{G_s \rho_w}$$

$$\Rightarrow \quad H_s = \frac{M_s}{G_s \rho_w A}$$

By way of illustration let us use the test results of Example 9.4 together with the following information:

Original dimensions of test sample: 75 mm diameter, 20 mm thickness
Mass of sample after removing complete from consolidation apparatus at end of test and drying in oven = 135.6 g.

$$M_s = 135.6 \text{ g}; \qquad A = \frac{\pi}{4} \times 75^2 = 4418 \text{ mm}^2$$

$$\Rightarrow \quad H_s = \frac{135.6 \times 1000}{2.65 \times 1 \times 4418} = 11.58 \text{ mm}$$

Now, as shown above:

$$e = \frac{H - H_s}{H_s}$$

Hence the void ratio to pressure relationship can be found.

Pressure (kPa)	Thickness (H)	$e = \frac{H - H_s}{H_s}$
0	20	$\frac{20 - 11.58}{11.58} = 0.727$
50	19.65	0.697
100	19.52	0.685
200	19.35	0.671
400	19.15	0.653
800	18.95	0.636

Note Such close agreement between the two methods for determining the e–p relationship could only happen in a theoretical example. In practice one often finds large discrepancies between the two methods.

9.3.6 The virgin consolidation curve

Clay is generally formed by the process of sedimentation from a liquid in which the soil particles were gradually deposited and compressed as more material was placed above them. The e–p curve corresponding to this natural process of consolidation is known as the virgin consolidation curve (Fig. 9.9a).

This curve is approximately logarithmic. If the values are plotted to a semi-log scale (e to a natural scale, p to a logarithmic scale), the result is a straight line of equation:

$$e = e_0 - C_C \log_{10} \frac{p_0 + dp}{p_0}$$

Hence e_2 can be expressed in terms of e_1:

$$e_2 = e_1 - C_C \log_{10} \frac{p_2}{p_1}$$

C_C is known as the compression index of the clay.

Compression curve for a normally consolidated clay

A normally consolidated clay is one that has never experienced a consolidation pressure greater than that corresponding to its present overburden. The compression curve of such a soil is shown in Fig. 9.9b.

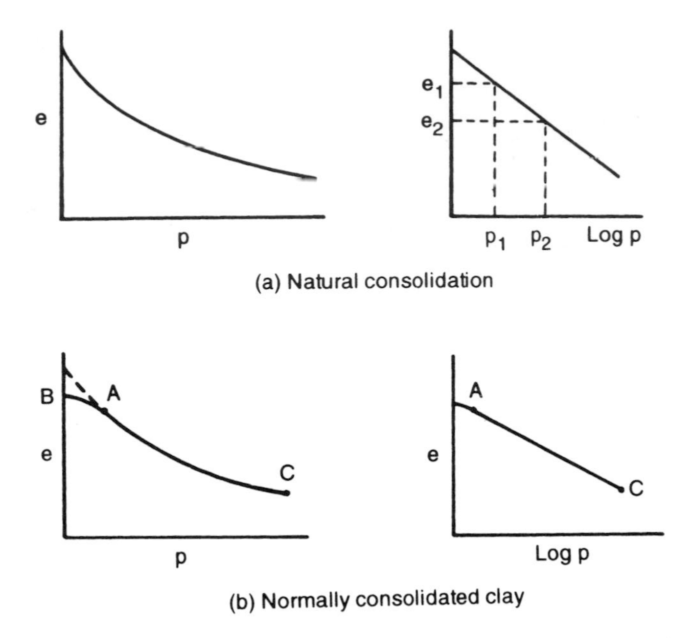

Fig. 9.9 e–p and e–log p curves for natural consolidation and for a normally consolidated clay.

The clay was originally compressed, by the weight of material above, along the virgin consolidation curve to some point A. Owing to the removal of pressure during sampling the soil has expanded to point B. Hence from B to A the soil is being recompressed whereas from A to C the virgin consolidation curve is followed.

The semi-log plot is shown in Fig. 9.9b. As before on the straight line part:

$$e_2 = e_1 - C_C \log_{10} \frac{p_2}{p_1}$$

Compression curve for an overconsolidated clay

An overconsolidated clay is one which has been subjected to a preconsolidation pressure in excess of its existing overburden (Fig. 9.10a), the resulting compression being much less than for a normally consolidated clay. The semi-log plot is no longer a straight line and a compression index value for an overconsolidated clay is no longer a constant.

From the e–p curve it is possible to determine an approximate value for the preconsolidated pressure with the use of a graphical method proposed by Casagrande (1936). First estimate the point of greatest curvature, A, then draw a horizontal line through A (AB) and the tangent to the curve at A (AC). Bisect the angle BAC to give the line AD, and locate the straight part of the compression curve (in Fig. 9.10a the straight part commences at point E). Finally project the straight part of the curve

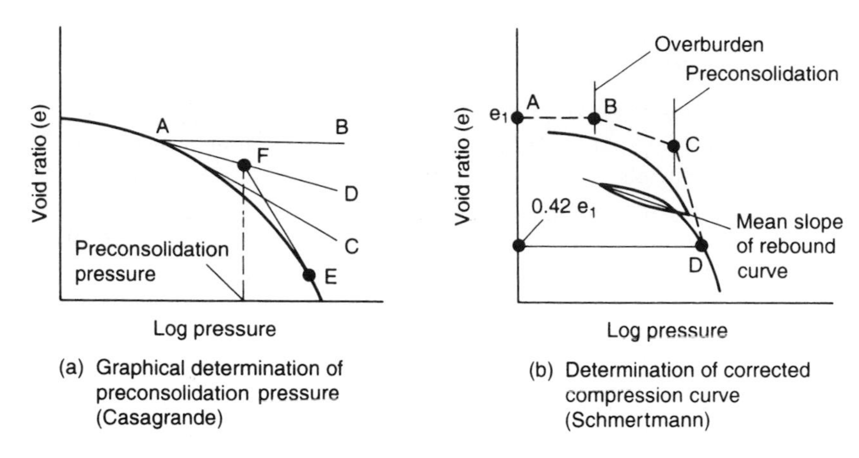

Fig. 9.10 Compression curves for an over-consolidated clay.

upwards to cut AD in F. The point F then gives the value of the preconsolidation pressure.

Evaluation of consolidation settlement from the compression index

$$\frac{dH}{H_1} = \frac{e_1 - e_2}{1 + e_1}$$

$$\Rightarrow \quad dH = \frac{e_1 - e_2}{1 + e_1} H_1$$

$$e_1 - e_2 = C_C \log_{10} \frac{p_2}{p_1}$$

$$\Rightarrow \quad \rho_c = dH = \frac{C_C}{1 + e_1} \log_{10} \frac{p_2}{p_1} H_1$$

This equation is only relevant when a clay is being compressed for the first time and therefore cannot be used for an overconsolidated clay.

Determination of compression index C_C

Terzaghi and Peck (1948) have shown that there is an approximate relationship between the liquid limit of a normally consolidated clay and its compression index. This relationship has been established experimentally and is:

$$C_C \approx 0.009\,(w_L - 10 \text{ per cent})$$

Example 9.5

A soft, normally consolidated clay layer is 15 m thick with a natural moisture content of 45 per cent. The clay has a saturated unit weight of 17.2 kN/m^3, a particle specific gravity of 2.68 and a liquid limit of 65 per cent. A foundation load will subject the centre of the layer to a vertical stress increase of 10.0 kPa.

Determine an approximate value for the settlement of the foundation if groundwater level is at the surface of the clay.

Solution

Initial vertical effective stress at centre of layer

$$= (17.2 - 9.81)\,\frac{15}{2}$$
$$= 55.4 \text{ kPa}$$

Final effective vertical stress = 55.4 + 10 = 65.4 kPa

Initial void ratio, $e_1 = wG_s = 0.45 \times 2.68 = 1.21$

$$C_C = 0.009(65 - 10) = 0.009 \times 55 = 0.495$$

$$\rho_c = \frac{0.495}{2.21} \times \log_{10}\frac{65.4}{55.4} \times 15$$
$$= 0.024 \text{ m} = 240 \text{ mm}$$

This method can be used for a rough settlement analysis of a relatively unimportant small structure on a soft clay layer. For large structures, consolidation tests would be carried out.

9.3.7 Application of consolidation test results

The range of pressure generally considered in a settlement analysis is the increase from p_1 (the existing vertical effective overburden pressure) to p_2 (the vertical effective pressure that will operate once the foundation load has been applied and consolidation has taken place), so that in the previous discussion e_1 represents the void ratio corresponding to the effective overburden pressure and e_2 represents the final void ratio after consolidation. In some text books and papers the initial void ratio, e_1, is given the symbol e_0.

Obtaining a test sample entails removing all of the stresses which are applied to it, this reduction in effective stress causing the sample to either swell or develop negative pore water pressures within itself. Owing to the restraining effect of the sampling tube most soil samples tend to have a negative pore pressure.

In the consolidation test the sample is submerged in water to prevent evaporation losses, with the result that the negative pore pressures will tend to draw in water and the sample consequently swells. To obviate this effect the normal procedure is to start the test by applying the first load increment and then to add the water, but if the sample still tends to swell an increased load increment must be added and the test readings started again. The point e_1 is taken to be the position on the test e–p curve that corresponds to the effective overburden pressure at the depth from which the sample was taken; in the case of a uniform deposit various values of e_1 can be obtained for selected points throughout the layer by reading off the test values of void ratio corresponding to the relevant effective overburden pressures. Generally the test e–p curve lies a little below the actual *in situ* e–p curve, the amount of departure depending upon the degree of disturbance in the test sample. Bearing in mind the inaccuracies involved in any analysis, this departure from the consolidation curve will generally be of small significance unless the sample is severely disturbed and most settlement analyses are based on the actual test results.

An alternative method, mainly applicable to overconsolidated clays, has been proposed by Schmertmann (1953), who points out that e_1 (he uses the symbol e_0) must be equal to wG_s, where w is the *in situ* moisture content at the point considered, and that in a consolidation test on an ideal soil with no disturbance the void ratio of the sample should remain constant at e_1 throughout the pressure range from zero to the effective overburden pressure value. Schmertmann found that the test e–p curve tends to cut the *in situ* virgin consolidation curve at a void ratio value somewhere between 37 and 42 per cent of e_1 and concluded that a reasonable figure for this intersection is $e = 0.42e_1$.

In order to obtain the corrected curve, with disturbance effects removed, the test sample is either loaded through a pressure range that eventually reduces the void ratio of the sample to $0.42e_1$ or else the test is extended far enough for extrapolated values to be obtained, at least one cycle of expansion and recompression being carried out during the test. The approximate value of the preconsolidation pressure is obtained and the test results are put in the form of a semi-log plot of void ratio to log p (Fig. 9.10b). The value of e_1 is obtained from wG_s, w being found from a separate test sample (usually cuttings obtained during the preparation of the consolidation test sample). It is now possible to plot on the test curve (point A) and a horizontal line (AB) is drawn to cut the ordinate of the existing overburden pressure at point B; a line BC is next drawn parallel to the mean slope of the laboratory rebound curve to cut the preconsolidation pressure ordinate at point C, and the value of void ratio equal to $0.42e_1$ is obtained and established on the test curve (point D). Finally points C and D are joined. The corrected curve therefore consists of the three straight lines: AB (parallel to the pressure axis with a constant void ratio value e_1), BC (representing the recompression of the soil up to the preconsolidation pressure), and CD (representing initial compression along the virgin consolidation line).

Apart from the elimination of disturbance effects the method is useful because it permits the use of a formula similar to the compression index of a normally consolidated clay:

$$\rho_c = \frac{C}{1 + e_1} \log_{10} \frac{p_2}{p_1} H$$

where C is the slope of the corrected curve (generally recompression). If the pressure range extends into initial compression the calculation must be carried out in two parts using the two different C values.

9.3.8 General consolidation

In the case of a foundation of finite dimensions, such as a footing sitting on a thick bed of clay, lateral strains will occur and the consolidation is no longer one-dimensional. If two saturated clays of equal compressibility and thickness are subjected to the same size of foundation and loading, the resulting settlements may be quite different even though the consolidation tests on the clays would give identical results. This is because lateral strain effects in the field may induce unequal pore pressures whereas in the consolidation test the induced pore pressure is always equal to the increment of applied stress.

For a saturated soil:

$$\Delta u = \Delta\sigma_3 + A(\Delta\sigma_1 - \Delta\sigma_3) \qquad \text{(see Section 3.11)}$$

Let

p'_1 = initial effective major principal stress
$\Delta\sigma_1$ = increment of total major principal stress due to the foundation loading
Δu = excess pore water pressure induced by the load.

The effective major principal stress on load application will be:

$$p'_1 + \Delta\sigma_1 - \Delta u$$

The effective major principal stress after consolidation will be:

$$p'_1 + \Delta\sigma_1$$

Let

p'_3 = initial effective minor principal stress
$\Delta\sigma_3$ = increment of total minor principal stress due to the foundation loading.

The horizontal effective stress on load application will be:

$$p'_3 + \Delta\sigma_3 - \Delta u$$

If the expression for Δu is examined it will be seen that Δu is greater than $\Delta\sigma_3$. The horizontal effective stress therefore reduces when the load is applied and there will be a lateral expansion of the soil. Hence in the early stages of consolidation the clay will undergo a recompression in the horizontal direction for an effective stress increase of $\Delta u - \Delta\sigma_3$; the strain from this recompression will be small but as consolidation continues the effective stress increases beyond the original value of p'_3 and the strain effects will become larger until consolidation ceases.

Settlement analysis

The method of settlement analysis most commonly in use is that proposed by Skempton and Bjerrum (1957). In this procedure the lateral expansion and compression effects are ignored, since the authors maintain that such a simplification cannot introduce a maximum error of more than 20 per cent and when they compared the actual settlements of several structures with predicted values using their method the greatest difference was in fact only 15 per cent.

Ignoring secondary consolidation, the total settlement of a foundation is given by the expression:

$$\rho = \rho_i + \rho_c$$

where

ρ_i = immediate settlement
ρ_c = consolidation settlement.

In the consolidation test:

$$\rho_{oed} = m_v \Delta\sigma_1 h \qquad (1)$$

where h = sample thickness.

Since there is no lateral strain in the consolidation, $\Delta\sigma_1 = \Delta u$. Hence:

$$\rho_{oed} = m_v \Delta uh$$

or

$$\rho_{c_{oed}} = \int_0^H m_v \Delta u \, dH \qquad (2)$$

where H = thickness of consolidating layer.

In a saturated soil $\Delta u = \Delta\sigma_3 + A(\Delta\sigma_1 - \Delta\sigma_3)$. This may be expressed as:

$$\Delta u = \Delta\sigma_1 \left[A + \frac{\Delta\sigma_3}{\Delta\sigma_1} (1 - A) \right]$$

and, substituting for Δu in Equation (2) we obtain a truer estimation of the consolidation settlement, ρ_c:

$$\rho_c = \int_0^H m_v \Delta\sigma_1 \left[A + \frac{\Delta\sigma_3}{\Delta\sigma_1} (1 - A) \right] dH \qquad (3)$$

Equation (3) can be expressed in terms of Equation (2):

$$\rho_c = \mu\rho = \mu\rho_{c_{oed}}$$
$$- \mu\int_0^{II} m_v \Delta\sigma_1 \, d$$

where

$$\mu = \frac{\int_0^H m_v \Delta\sigma_1 \left[A + \frac{\Delta\sigma_3}{\Delta\sigma_1}(1 - A) \right] dh}{\int_0^H m_v \Delta\sigma_1 \, dH}$$

If m_v and A are assumed constant with depth the equation for μ reduces to:

$$\mu = A + (1 - A)\alpha \tag{4}$$

where

$$\alpha = \frac{\int_0^H \Delta\sigma_3 \, dH}{\int_0^H \Delta\sigma_1 \, dH}$$

Poisson's ratio for a saturated soil is generally taken as 0.5 at the stage when the load is applied so α is a geometrical parameter which can be determined. Various values for α that were obtained by Skempton and Bjerrum are given in Table 9.4.

Table 9.4

H/B	α	
	Circular footing	Strip footing
0	1.00	1.00
0.25	0.67	0.74
0.50	0.50	0.53
1.0	0.38	0.37
2.0	0.30	0.26
4.0	0.28	0.20
10.0	0.26	0.14
∞	0.25	0

The value of the pore pressure coefficient A can now be substituted in Equation (4) and a value for μ obtained, typical results being:

Soft sensitive clays . . .	possibly greater than 1.0
Normally consolidated clays . . .	generally less than 1.0
Average overconsolidated clays . . .	approximately 0.5
Heavily overconsolidated clays . . .	perhaps as little as 0.25

Example 9.6

A sample of the clay of Example 9.4 was subjected to a consolidated undrained triaxial test with the results shown in Fig. 9.11b. The sample was taken from a layer 20 m thick and has a saturated unit weight of 18.5 kN/m^3.

It is proposed to construct a reinforced concrete foundation, length 30 m and width 10 m, on the top of the layer. The uniform bearing pressure will be 200 kPa. Determine the total settlement of the foundation under its centre if the groundwater level occurs at a depth of 5 m below the top of the layer.

Solution

The vertical pressure increment at the centre of the layer can be obtained by splitting the plan area into four rectangles (Fig. 9.11a) and using Fig. 4.6:

$$\Delta\sigma_1 = 110 \text{ kPa}$$

In order to obtain the E value for the soil, $\Delta\sigma_3$ should now be evaluated so that the deviator stress $(\Delta\sigma_1 - \Delta\sigma_3)$ can be obtained.

Alternatively the approximate method can be used:

65 per cent of maximum deviator stress = $0.65 \times 400 = 260$ kPa
Strain at this value = 0.8 per cent (from Fig. 9.11b)

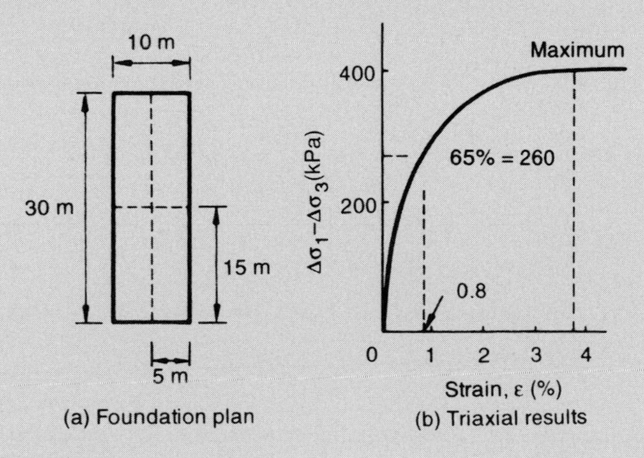

Fig. 9.11 Example 9.6.

Hence:

$$E = \frac{260 \times 100}{0.8} = 32\ 500 \text{ kPa} = 32.5 \text{ MPa}$$

Immediate settlement

Using the rectangles of Fig. 9.11a and Fig. 9.2:

$$\frac{L}{B} = \frac{15}{5} = 3.0 \qquad \frac{H}{B} = \frac{20}{5} = 4.0$$

Hence:

$$I_p = 0.48 \times 4.0 = 1.92$$

$$\rho_i = \frac{pB}{E}(1 - \mu^2)I_p$$

$$= \frac{200}{32\ 500} \times 5 \times 0.75 \times 1.92 \times 0.8 \qquad (0.8 = \text{rigidity factor})$$

$$= 0.036 \text{ m} = 36 \text{ mm}$$

Consolidation settlement

$$\text{Initial effective overburden pressure} = 18.5 \times 10 - 9.81 \times 5 = 136 \text{ kPa}$$

Hence the range of pressure involved is from 136 to 246 kPa.

Using the e–p curve of Fig. 9.8a:

$$e_1 = 0.6800; \qquad e_2 = 0.666$$

$$a = \frac{de}{dp} = \frac{0.680 - 0.666}{110} = \frac{0.014}{110} = 0.000\ 127 \text{ m}^2/\text{kN}$$

$$m_v = \frac{a}{1 + e_1} = \frac{0.000\ 127}{1.680} = 7.6 \times 10^{-5} \text{ m}^2/\text{kN}$$

$$\rho_c = m_v \, dp \, H = 7.6 \times 110 \times 20 \times 10^{-5} = 0.167 \text{ m} = 167 \text{ mm}$$

Total settlement = 36 + 167 = 203 mm

Some reduction could possibly be applied to the value of ρ_c if the value of μ was known.

Alternative method for determining ρ_c

In one-dimensional consolidation the volumetric strain must be equal to the axial strain, i.e.

$$\frac{dH}{H} = \frac{\rho_c}{H} = \frac{de}{1 + e_1}$$

hence:

$$\rho_c = \frac{de}{1 + e_1} H$$

In the example:

$$\rho_c = \frac{0.680 - 0.666}{1.680} \times 20$$
$$= 0.008\ 834 \times 20 = 0.167 \text{ m}$$
$$= 167 \text{ mm}$$

Example 9.7

The plan of a proposed raft foundation is shown in Fig. 9.12a. The uniform bearing pressure from the foundation will be 350 kPa and a site investigation has shown that the upper 7.62 m of the subsoil is a saturated coarse sand of unit weight 19.2 kN/m^3 with groundwater level occurring at a depth of 3.05 m below the top of the sand. The result from a standard penetration test taken at a depth of 4.57 m below the top of the sand gave N = 20. Below the sand there is a 30.5 m thick layer of clay (A = 0.75, E = 16.1 MPa, $E_{swelling}$ = 64.4 MPa). The clay rests on hard sandstone (Fig. 9.12b).

Determine the total settlement under the centre of the foundation.

Solution

Using the De Beer and Martens' approach:

Vertical pressure increments

Gross foundation pressure = 350 kPa
Relief due to excavation of sand = 1.52 × 19.2 = 29 kPa
Net foundation pressure increase, Δp = 350 − 29 = 321 kPa

The foundation is split into four rectangles, as shown in Fig. 9.12a, and Fig. 4.6 is then used to determine values for I_σ.

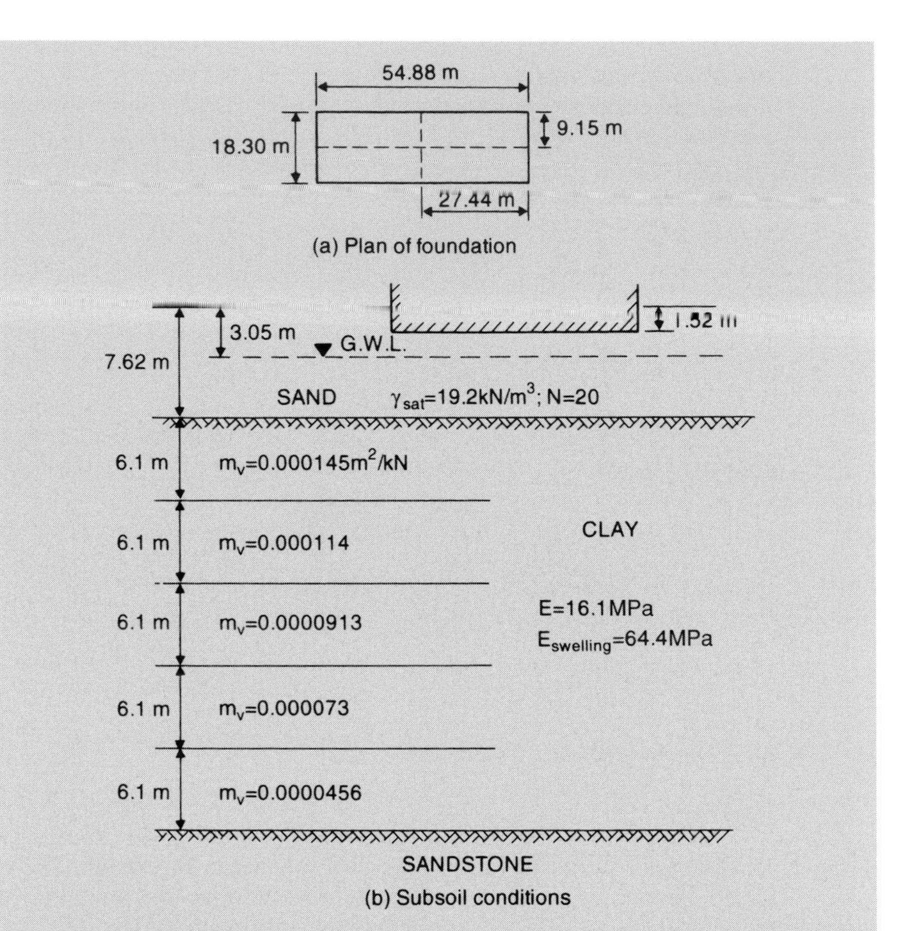

Fig. 9.12 Example 9.7.

Depth below foundation (m)	B/z	L/z	I_σ	$4I_\sigma$	$\Delta\sigma_z$ (kPa)
3.05	3.0	9.0	0.247	0.988	317
9.15	1.0	3.0	0.203	0.812	261
15.25	0.6	1.8	0.152	0.608	195
21.35	0.43	1.29	0.113	0.452	145
27.45	0.33	1.00	0.086	0.344	110
33.55	0.27	0.82	0.067	0.268	86

Immediate settlement

Sand: test value for N = 20:

$$p'_o = 4.57 \times 19.2 - 1.52 \times 9.81 = 73 \text{ kPa}$$

$$C_r = 400 \times 20 = 8000 \text{ kPa}$$

$$C_s = \frac{1.5 \times 8000}{73} = 165$$

$$\rho_i = \frac{6.1}{165} \ln \frac{73 + 317}{73} = 0.062 = 62 \text{ mm}$$

(As the penetration test was carried out on submerged soil there is no need to increase this value to allow for groundwater effects.)

Hence ρ_i in the sand = 62 mm.

Clay: in Fig. 9.2, $H_1 = 6.1$ m and $H_2 = 36.6$ m.

$$\text{For } H_2: \quad \frac{L}{B} = \frac{27.44}{9.15} = 3.0; \qquad \frac{H}{B} = \frac{36.6}{9.15} = 4.0$$

Hence $I_p = 0.475$.

$$\text{For } H_1: \quad \frac{L}{B} = 3.0; \qquad \frac{H}{B} = \frac{6.1}{9.15} = 0.67$$

Hence $I_p = 0.18$.

Settlement under centre of foundation (*note* as heave effects will be allowed for, use gross contact pressure. If heave is not allowed for net foundation pressure should be used).

$$\rho_i = \frac{pB}{E}(1 - \mu^2)4I_p \times \text{Rigidity factor}$$

$$= \frac{350}{16\,100} \times 9.15 \times 0.75 \times 4(0.475 - 0.18) \times 0.8$$

$$= 0.141 \text{ m} = 141 \text{ mm}$$

Heave effects: relief of pressure due to sand excavation = 29 kPa

$$\Rightarrow \quad \text{Heave} = \frac{29}{64\,400} \times 9.15 \times 0.75 \times 4(0.475 - 0.18) \times 0.8$$

$$= 0.0029 \text{ m} = 3 \text{ mm}$$

Hence ρ_i in the clay = 138 mm.

As can be seen from this example the effects of heave are usually only significant when a great depth of material is excavated.

Consolidation settlement

The clay layer has been divided into five layers of thickness, H, equal to 6.1 m.

m_v	$\Delta\sigma_z$	$m_v\Delta\sigma_zH$
0.000145	261	0.231
0.000114	195	0.136
0.0000413	145	0.081
0.000073	110	0.049
0.0000456	86	0.024
		0.521 m = 521 mm

This value of settlement can be reduced by the factor

$$\mu = A + (1 - A)\alpha$$

An approximate value for α can be obtained from Table 9.4.

Hence:

$$\alpha = 0.26$$
$$\mu = 0.75 + 0.25 \times 0.26 = 0.82$$
$$\rho_c = 521 \times 0.82 = 448 \text{ mm}$$
$$\text{Total settlement} = 62 + 138 + 448 = 648 \text{ mm}$$

9.4 Two-dimensional stress paths

As discussed in Chapter 3, the state of stress in a soil sample can be shown graphically by a Mohr circle diagram. In a triaxial compressive test the axial strain of the test specimen increases up to failure and the various states of stress that the sample experiences from the start of the test until failure can obviously be represented by a series of Mohr circles. The same stress states can be represented in a much simpler form by expressing each successive stress state as a point. The line joining these successive points is known as a *stress path*.

Stress paths can be of many forms and we have already used some: the stress–strain relationships plotted in τ–σ space in Chapter 3 to show triaxial test results and the plots in e–log p space used in Fig. 9.9 to illustrate compression curves, etc. In his analysis of foundation settlement problems, Lambe (1964, 1967) used stress paths of maximum shear.

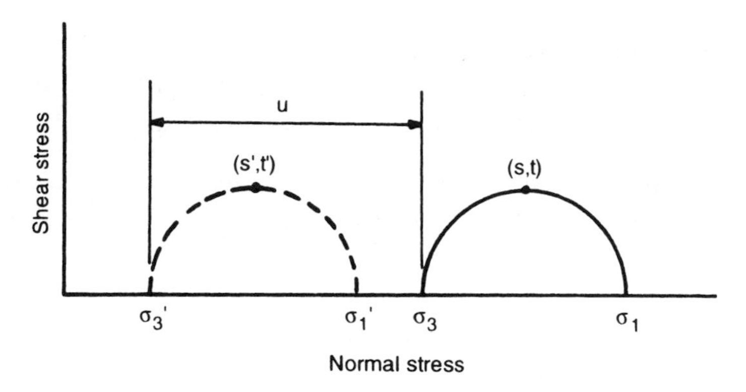

Fig. 9.13 Points of maximum shear stress.

If a Mohr circle diagram of stress is examined (Fig. 9.13) the point of maximum shear has the co-ordinates s and t where:

$$s = \frac{\sigma_1 + \sigma_3}{2} \quad \text{and} \quad t = \frac{\sigma_1 - \sigma_3}{2}$$

σ_1 and σ_3 being the total principal stresses.

In terms of effective stresses, σ'_1 and σ'_3, the point of maximum shear has the co-ordinates s′ and t′ where:

$$s' = \frac{\sigma'_1 + \sigma'_3}{2}$$

If a soil is subjected to a range of values of σ'_1 and σ'_3 the point of maximum shear stress can be obtained for each stress circle; the line joining these points, in the order that they occurred, is termed the stress path or stress vector of maximum shear. Any other point instead of maximum shear can be used to determine a stress path, e.g. the point of maximum obliquity, but Lambe maintains that the stress paths of maximum shear are not only simple to use but also more applicable to consolidation work.

Typical effective stress paths obtained from a series of consolidated undrained triaxial tests on samples of normally consolidated clay together with the effective stress circles at failure are shown in Fig. 9.14.

9.4.1 *Ratios of σ'_3/σ'_1*

If the results of a drained shear test on a soil are considered, the Mohr circle diagram is as shown in Fig. 9.14. The line tangential to the stress circles is the strength envelope, inclined at ϕ' to the normal stress axis. If each Mohr circle is considered it is seen that the ratio σ'_3/σ'_1 is a constant, to which the symbol K_f is applied.

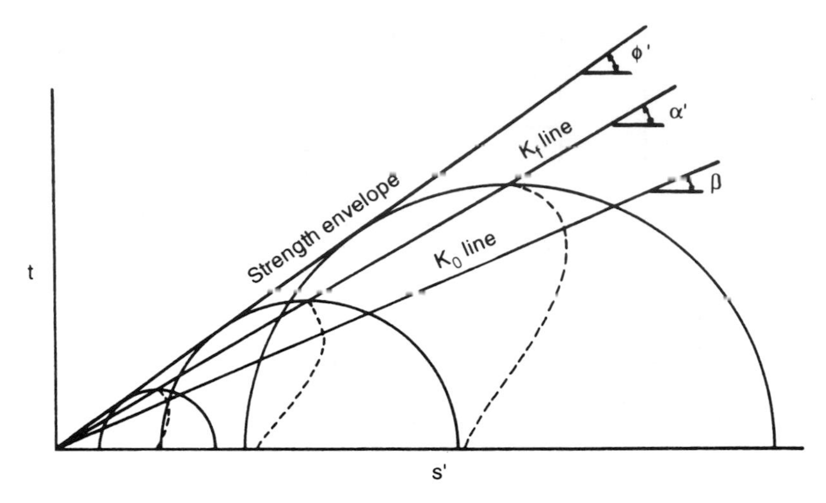

Fig. 9.14 Typical undrained effective stress paths obtained from consolidated undrained triaxial tests on a normally consolidated clay.

The K_f line

If the points of maximum shear for each effective stress circle p'_f and q_f are joined together the stress path of maximum shear stress at failure is obtained. This line is called the K_f line and is inclined at angle α' to the normal stress axis; obviously $\tan \alpha' = \sin \phi'$.

The K_o line

For a soil undergoing one-dimensional consolidation the ratio σ'_3/σ'_1 is again constant and its value is given the symbol K_0. Plotting the maximum shear stress points of these stress circles enables the stress path for one-dimensional consolidation, the K_0 line, to be determined; this line is inclined at angle β to the normal stress axis.

K_0 is the coefficient of earth pressure at rest. For consolidation work K_0 may be defined for a soil with a history of one-dimensional strain as the ratio:

$$K_0 = \frac{\text{Lateral effective stress}}{\text{Vertical effective stress}}$$

9.4.2 Stress paths in the consolidation test

Figure 9.15 shows the stress conditions that arise during and after the application of a pressure increment in the consolidation test. Initially the sample has been consolidated under a previous load and the pore pressure is zero; the Mohr circle is represented by (p, q) the point X, circle I. As soon as the vertical pressure increase, $\Delta\sigma_1$, is applied, the total stresses move from X to Y (circle 1). As the soil is saturated

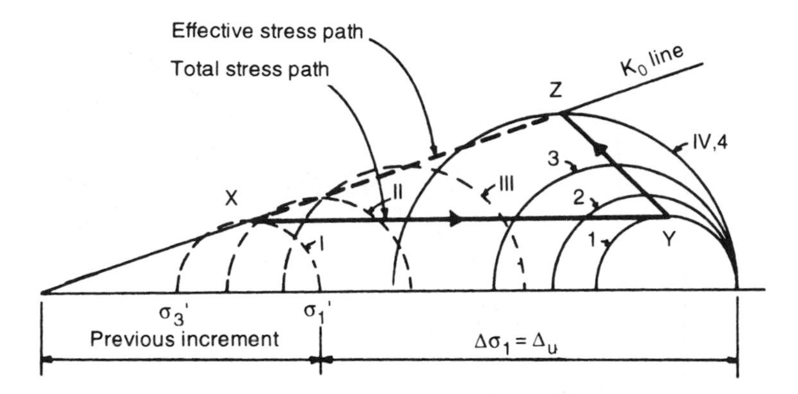

Fig. 9.15 Stress paths in the consolidation test.

$\Delta u = \Delta\sigma_1$ and the effective stress circle is still represented by point X. As consolidation commences the pore water pressure, Δu, begins to decrease and $\Delta\sigma_1'$ begins to increase. The consolidation is one-dimensional and therefore an increase in the major principal effective stress, $\Delta\sigma_1'$, will induce an increase in the minor principal effective stress $\Delta\sigma_3' = K_0\Delta\sigma_1'$. Hence the effective stress circles move steadily towards point Z (circles II, III and IV), where Z represents full consolidation.

The total stress circles can be determined from a study of the effective stress circles. For example the difference between $\Delta\sigma_1$ and $\Delta\sigma_1'$ for circle III represents the pore water pressure within the sample at that time: hence $\Delta\sigma_3$ at this stage in the consolidation is $\Delta\sigma_3'$ for circle III plus the value of the pore water pressure. It can be seen therefore that Δu decreases with consolidation and the size of the Mohr circle for total stress increases until the point Z is reached (circles 2, 3 and 4). Obviously circles 4 and IV are coincident.

9.4.3 *Stress path for general consolidation*

The effective stress plot of Fig. 9.16 represents a typical case of general consolidation. The soil is normally consolidated and point A represents the initial K_0 consolidation; AB is the effective stress path on the application of the foundation load and BC is the effective stress path during consolidation.

Skempton and Bjerrum's assumption that lateral strain effects during consolidation can be ignored presupposes that the strain due to the stress path BC is the same as that produced by the stress path DE. The fact that the method proposed by Skempton and Bjerrum gives reasonable results indicates that the effective stress path during the consolidation of soil in a typical foundation problem is indeed fairly close to the effective stress path DE of Fig. 9.16. There are occasions when this will not be so, however, and the stress path method of analysis can give a more reasonable prediction of settlement values (see Lambe, 1964, 1967). The calculation of settlement in a soft soil layer under an embankment by this procedure has been discussed by Smith (1968a), and the method is also applicable to spoil heaps.

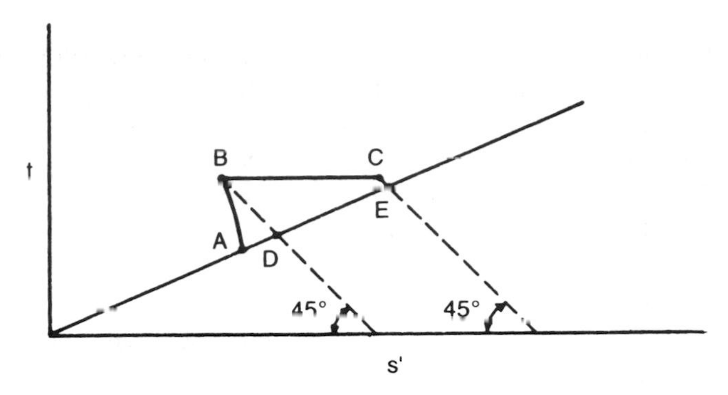

Fig. 9.16 Effective stress path for the general consolidation of a normally consolidated clay.

Example 9.8

A layer of soft, normally consolidated clay is 9.25 m thick and has an existing effective overburden pressure at its centre of 85 kPa.

It is proposed to construct a flexible foundation on the surface of the clay, and the increases in stresses at the centre of the clay, beneath the centre of the foundation, are estimated to be $\Delta\sigma_1 = 28.8$ kPa and $\Delta\sigma_3 = 19.2$ kPa.

Consolidated undrained triaxial tests carried out on representative undisturbed samples of the clay gave the following results:

Cell pressure = 35 kPa

Strain (%)	Deviator stress (kPa)	Pore water pressure (kPa)
0	0	0
1	10.4	0.4
2	20.7	4.8
3	29.0	9.7
4	33.2	13.2
5	35.8	16.6
6	37.3	17.9
6.8	37.8	19.3 (failure)

Cell pressure = 70 kPa

Strain (%)	Deviator stress (kPa)	Pore water pressure (kPa)
0	0	0
1	20.7	4.1
2	42.7	12.8
3	54.4	22.1
4	63.4	30.4
5	66.1	34.8
6	71.7	37.9
7	75.8	40.7 (failure)

By considering a point at the centre of the clay and below the centre of the foundation, draw the effective stress paths for undrained shear obtained from the tests and indicate the effective stress paths for the immediate and consolidation settlements that the foundation will experience.

Assume that $K_0 = 1 - \sin \phi$ and determine an approximate value for the immediate settlement of the foundation.

Solution

The first step is to plot out the two effective stress paths. The calculations are best set out in tabular form:

Cell pressure = 35 kPa

Strain	$\sigma_1 - \sigma_3$	u	$t = \frac{\sigma_1 - \sigma_3}{2}$	$s' = \frac{\sigma'_1 + \sigma'_3}{2}$
0	0	0	0	35
1	10.4	0.4	5.2	39.8
2	20.7	4.8	10.3	40.5
3	29.0	9.7	14.5	39.8
4	33.2	13.2	16.6	38.4
5	35.8	16.6	17.9	36.3
6	37.3	17.9	18.6	35.7
6.8	37.8	19.3	18.9	34.6

Cell pressure = 70 kPa

Strain	$\sigma_1 - \sigma_3$	u	$t = \frac{\sigma_1 - \sigma_3}{2}$	$s' = \frac{\sigma'_1 + \sigma'_3}{2}$
0	0	0	0	70
1	20.7	4.1	10.3	76.2
2	42.7	12.8	21.3	78.5
3	54.4	22.1	27.2	75.1
4	63.4	30.4	31.7	71.3
5	66.1	34.8	33.0	68.2
6	71.7	37.9	35.8	67.9
7	75.8	40.7	37.9	67.2

The stress paths are shown in Fig. 9.17. From the K_f line $\tan \alpha (= \sin \phi) = \tan 28.5° = 0.543$.

$$\Rightarrow \quad K_0 = 1 - 0.543 = 0.457$$

Effective stresses at centre of layer before application of foundation load (initial)

$$\sigma'_{1I} = 85 \text{ kPa}$$

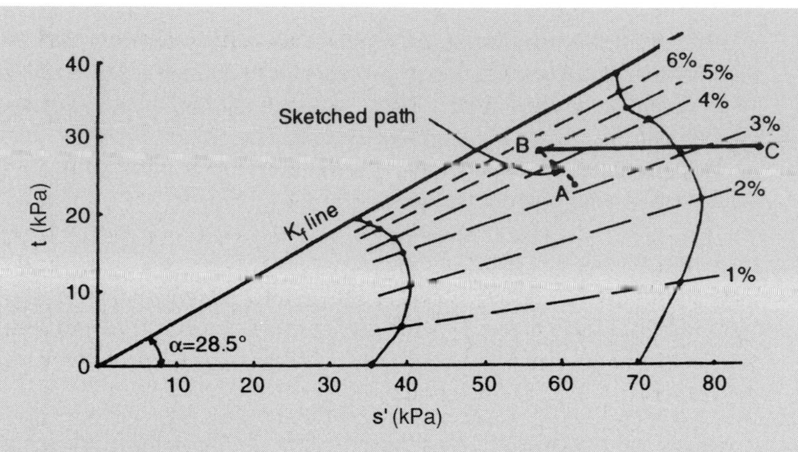

Fig. 9.17 Example 9.8.

Clay is normally consolidated, therefore:

$$\sigma'_{3I} = 0.457 \times 85 = 38.8 \text{ kPa}$$

$$\Rightarrow \text{ s}' = \frac{85 + 38.8}{2} = 61.9; \qquad \text{t} = \frac{85 - 38.8}{2} = 23.1$$

The coordinates s′ and t are plotted on Fig. 9.17 to give the point A, the initial state of stress in the soil.

Effective stress at centre of clay after application and consolidation of foundation load (final)

$$\sigma'_{1F} = \sigma'_{1I} + \Delta\sigma_1 = 85 + 28.8 = 113.8 \text{ kPa}$$
$$\sigma'_{3F} = \sigma'_{3I} + \Delta\sigma_3 = 38.8 + 19.2 = 58.0 \text{ kPa}$$

$$\Rightarrow \text{ s}' = \frac{113.8 + 58}{2} = 85.9; \qquad \text{t} = \frac{113.8 - 58}{2} = 27.9$$

The coordinates s′ and t are plotted in Fig. 9.17 to give the point C, the state of the effective stresses in the soil after consolidation. As illustrated in Fig. 9.16 the stress path from A to B represents the effect of the immediate settlement, whereas the stress path from B to C represents the effects of the consolidation settlement. The problem is to establish the point B, the point that represents the effective stress state in the soil immediately after the application of the foundation load.

During consolidation, at all times,

$$\text{t} = \tfrac{1}{2}(\sigma_1 - \sigma_3) = \tfrac{1}{2}(\sigma'_1 - \sigma'_3).$$

Hence, no matter how the individual values of effective stress vary during consolidation, the value of t remains constant. The line BC must be parallel to the horizontal axis. Hence the point B must lie somewhere along the horizontal line through C.

From A to B the effective undrained stress path is unknown but it is possible to sketch in an approximate, but sufficiently accurate path, by comparing the two test stress paths on either side of it. This has been done in the figure. The immediate settlement can now be found. On the diagram the strain contours (lines joining equal strain values on the two test paths) are drawn. It is seen that the point A lies a little above the 3 per cent strain contour (3.2 per cent). Point B lies on the 5 per cent strain contour. Hence the strain suffered with immediate settlement = 5 − 3.2 = 1.8 per cent.

$$\Rightarrow \quad \rho_i = \frac{1.8}{100} \times 9.25 = 0.167 \text{ m}$$

Exercises

Exercise 9.1

Using the test results from Example 3.9, determine an approximate value for E of the soil and calculate the average settlement of a foundation, 5 m × 1 m, founded on a thick layer of the same soil with a uniform pressure of 600 kPa.

Answer 58 mm

Exercise 9.2

A rectangular, flexible foundation has dimensions L = 4 m and B = 2 m and is loaded with a uniform pressure of 400 kPa. The foundation sits on a layer of deep clay, E = 10 MPa. Determine the immediate settlement values at its centre and at the central points of its edges.

Answer At centre = 92 mm
At centre of long edge = 67 mm
At centre of short edge = 58 mm

Exercise 9.3

A rectangular foundation, 10×2 m^2, is to carry a total uniform pressure of 400 kPa and is to be founded at a depth of 1 m below the surface of a saturated sand of considerable thickness. The bulk unit weight of the sand is 18 kN/m^3 and standard penetration tests carried out below the water table indicate that the deposit has an average N value of 15.

If the water table occurs at the proposed foundation depth, determine a value for the settlement of the centre of the foundation. (Use De Beer and Martens' method.)

Answer 40 mm

Exercise 9.4

A saturated sample of a normally consolidated clay gave the following results when tested in a consolidation apparatus (each loading increment was applied for 24 hours).

Consolidation pressure (kPa)	Thickness of sample (mm)
0	17.32
53.65	16.84
107.3	16.48
214.6	16.18
429.2	15.85
0	16.51

After the sample had been allowed to expand for 24 hours it was found to have a moisture content of 30.2 per cent. The particle specific gravity of the soil was 2.65.

(i) Plot the void ratio to effective pressure.
(ii) Plot the void ratio to log effective pressure and hence determine a value for the compression index of the soil.
(iii) A 6.1 m layer of the soil is subjected to an existing effective overburden pressure at its centre of 107.3 kPa, and a foundation load will increase the pressure at the centre of the layer by 80.5 kPa.

Determine the probable total consolidation settlement of the layer (a) by the coefficient of volume compressibility and (b) by the compression index. Explain why the two methods give slightly different answers.

Answer (a) Settlement by coefficient of volume compressibility = 85 mm
(b) Settlement by compression index = 98 mm

The compression index method is not so accurate as it represents the average of conditions throughout the entire pressure range whereas the coefficient of volume compressibility applies to the actual pressure range considered.

Chapter 10
Rate of Foundation Settlement

The settlement of a foundation in cohesionless soil and the elastic settlement of a foundation in clay can be assumed to occur as soon as the load is applied. The consolidation settlement of a foundation on clay will only take place as water seeps from the soil at a rate depending upon the permeability of the clay.

10.1 Analogy of consolidation settlement

The model shown in Fig. 10.1 helps to give an understanding of the consolidation process. When load is applied to the piston it will be carried initially by the water pressure created, but due to the weep hole there will be a slow bleeding of water from the cylinder accompanied by a progressive settlement of the piston until the spring is compressed to its corresponding load. In the analogy, the spring represents the compressible soil skeleton and the water represents the water in the voids of the soil; the size of the weep hole is analogous to the permeability of the soil.

$$\text{The degree of consolidation, U,} = \frac{\text{Consolidation attained at time t}}{\text{Total consolidation}}$$

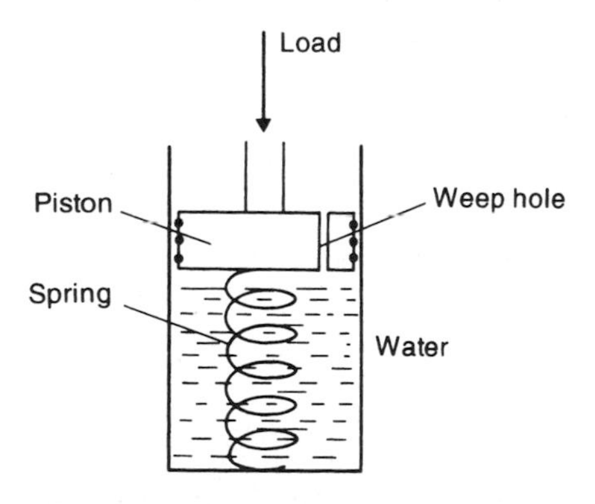

Fig. 10.1 Analogy of consolidation settlement.

10.2 Distribution of the initial excess pore pressure, u_i

If we consider points below the centre of a foundation it is seen that there are three main forms of possible u_i distribution.

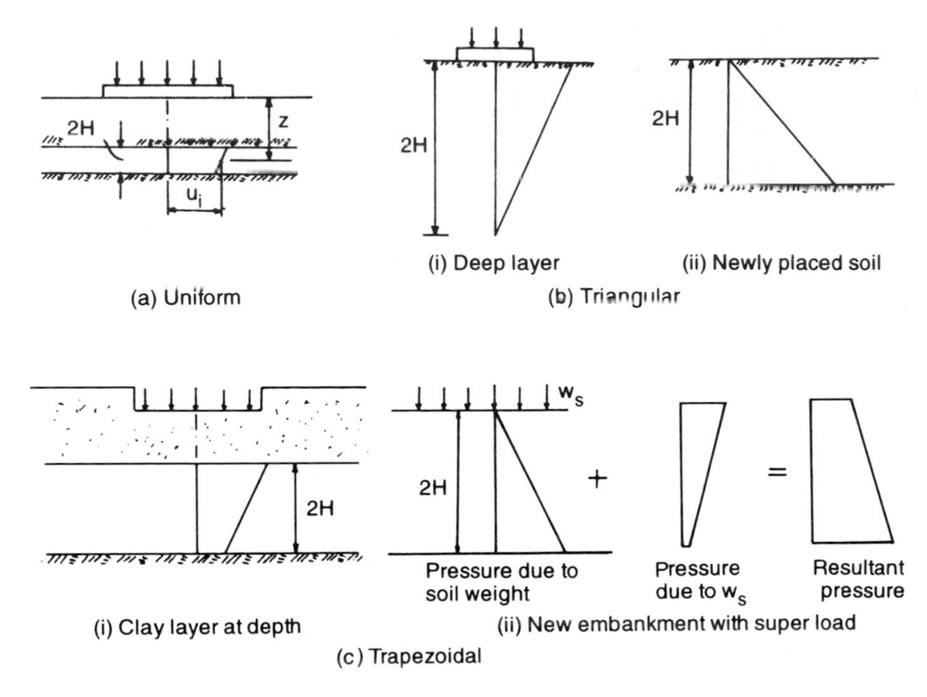

Fig. 10.2 Forms of initial excess pore pressure.

Uniform distribution can occur in thin layers (Fig. 10.2a), so that for all practical purposes u_i is constant and equals $\Delta\sigma_1$ at the centre of the layer.

Triangular distribution is found in a deep layer under a foundation, where u_i varies from a maximum value at the top to a negligible value (taken as zero) at some depth below the foundation (Fig. 10.2b(i)). The depth of this variation depends upon the dimensions of the footing. Figure 10.2b(ii) shows how a triangular distribution may vary from $u_i = 0$ at the top of a layer to $u_i =$ a maximum value at the bottom; this condition can arise with a newly placed layer of soil, the applied pressure being the soil's weight.

Trapezoidal distribution results from the quite common situation of a clay layer located at some depth below the foundation (Fig. 10.2c(i)). In the case of a new embankment carrying a superimposed load, a reversed form of trapezoidal distribution is possible (Fig. 10.2c(ii)).

10.3 Terzaghi's theory of consolidation

Terzaghi's first presented this theory in 1925 and most practical work on the prediction of settlement rates is based upon the differential equation he evolved. The main assumptions in the theory are as follows.

(i) Soil is saturated and homogeneous.
(ii) The coefficient of permeability is constant.

(iii) Darcy's law of saturated flow applies.
(iv) The resulting compression is one dimensional.
(v) Water flows in one direction.
(vi) Volume changes are due solely to changes in void ratio, which are caused by corresponding changes in effective stress.

The expression for flow in a saturated soil has been established in Chapter 2. The rate of volume change in a cube of volume dx.dy.dz is:

$$\left(k_x \frac{\partial^2 h}{\partial x^2} + k_y \frac{\partial^2 h}{\partial y^2} + k_z \frac{\partial^2 h}{\partial z^2} \right) dx.dy.dz$$

For one-dimensional flow (assumption v) there is no component of hydraulic gradient in the x and y directions, and putting $k_z = k$ the expression becomes:

$$\text{Rate of change of volume} = k \frac{\partial^2 h}{\partial z^2} dx.dy.dz$$

The volume changes during consolidation are assumed to be caused by changes in void ratio.

Porosity

$$n = \frac{V_v}{V} = \frac{e}{1 + e}$$

hence

$$V_v = dx.dy.dz \frac{e}{1 + e}$$

Another expression for the rate of change of volume is therefore:

$$\frac{\partial}{\partial t} \left(dx.dy.dz \frac{e}{1 + e} \right)$$

Equating these two expressions:

$$k \frac{\partial^2 h}{\partial z^2} = \frac{1}{1 + e} \frac{\partial e}{\partial t}$$

The head, h, causing flow is the excess hydrostatic head caused by the excess pore water pressure, u.

$$h = \frac{u}{\gamma_w}$$

$$\Rightarrow \quad \frac{k}{\gamma_w}\frac{\partial^2 u}{\partial z^2} = \frac{1}{1+e}\frac{\partial e}{\partial t}$$

With one-dimensional consolidation there are no lateral strain effects and the increment of applied pressure is therefore numerically equal (but of opposite sign) to the increment of induced pore pressure. Hence an increment of applied pressure, dp, will cause an excess pore water pressure of du (= −dp). Now:

$$a = -\frac{de}{dp}$$

hence

$$a = \frac{de}{du} \qquad \text{(see Section 9.3.3)}$$

or

$$de = a\,du$$

Substituting for de:

$$\frac{k}{\gamma_w}(1+e)\frac{\partial^2 u}{\partial z^2} = a\frac{\partial u}{\partial t}$$

$$\Rightarrow \quad c_v\frac{\partial^2 u}{\partial z^2} = \frac{\partial u}{\partial t}$$

where c_v = the coefficient of consolidation and equals

$$\frac{k}{\gamma_w a}(1+e) = \frac{k}{\gamma_w m_v}$$

In the foregoing theory z is measured from the top of the clay and complete drainage is assumed at both the upper and lower surfaces, the thickness of the layer being taken as 2H. The initial excess pore pressure, u_i, = −dp.

The boundary conditions can be expressed mathematically:

when z = 0, u = 0
when z = 2H, u = 0
when t = 0, $u = u_i$

A solution for

$$c_v \frac{\partial^2 u}{\partial z^2} = \frac{\partial u}{\partial t}$$

that satisfies these conditions can be obtained and gives the value of the excess pore pressure at depth z at time t, u_z:

$$u_z = \sum_{m=0}^{m=\infty} \frac{2u_i}{M} \left(\sin \frac{Mz}{H} \right) e^{-M^2T}$$

where

u_i = the initial excess pore pressure, uniform over the whole depth
$M = \frac{1}{2}\pi(2m + 1)$ where m is a positive integer varying from 0 to ∞
$T = \frac{c_v t}{H^2}$, known as the time factor.

Owing to the drainage at the top and bottom of the layer the value of u_i will immediately fall to zero at these points. With the mathematical solution it is possible to determine u at time t for any point within the layer. If these values of pore pressures are plotted, a curve (known as an isochrone) can be drawn through the points (Fig. 10.3b). The maximum excess pore pressure is seen to be at the centre of the layer and, for any point, the applied pressure increment, $\Delta\sigma_1 = u + \Delta\sigma'_1$. After a considerable time u will become equal to zero and $\Delta\sigma_1$ will equal $\Delta\sigma'_1$.

The plot of isochrones for different time intervals is shown in Fig. 10.3c. For a particular point the degree of consolidation, U_z, will be equal to

$$\frac{u_i - u_z}{u_i}$$

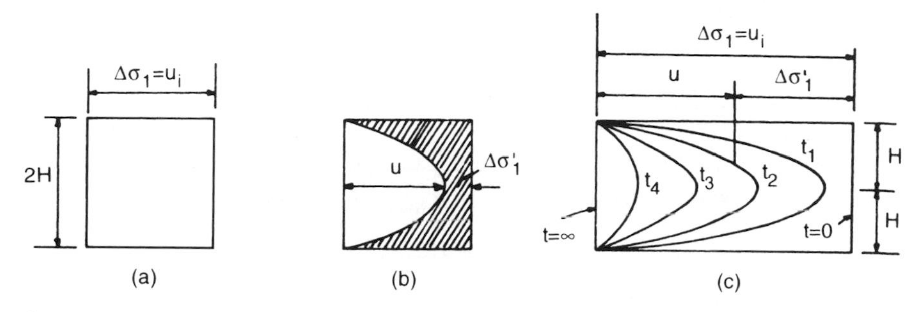

Fig. 10.3 Variation of excess pore pressure with depth and time.

The mathematical expression for U_z is:

$$U_z = 1 - \sum_{m=0}^{m=\infty} \frac{2}{M}\left(\sin\frac{Mz}{H}\right) e^{-m^2T}$$

10.4 Average degree of consolidation

Instead of thinking in terms of U_z, the degree of consolidation of a particular point at depth z, we think in terms of U, the average state of consolidation throughout the whole layer. The amount of consolidation still to be undergone at a certain time is represented by the area enclosed under the particular isochrone, and the total consolidation is represented by the area of the initial excess pore pressure distribution diagram (Fig. 10.3a). The consolidation achieved at this isochrone is therefore the total consolidation less the area under the curve (shown hatched in Fig. 10.3b).

Average degree of consolidation,

$$U = \frac{2Hu_i - \text{Area under isochrone}}{2Hu_i}$$

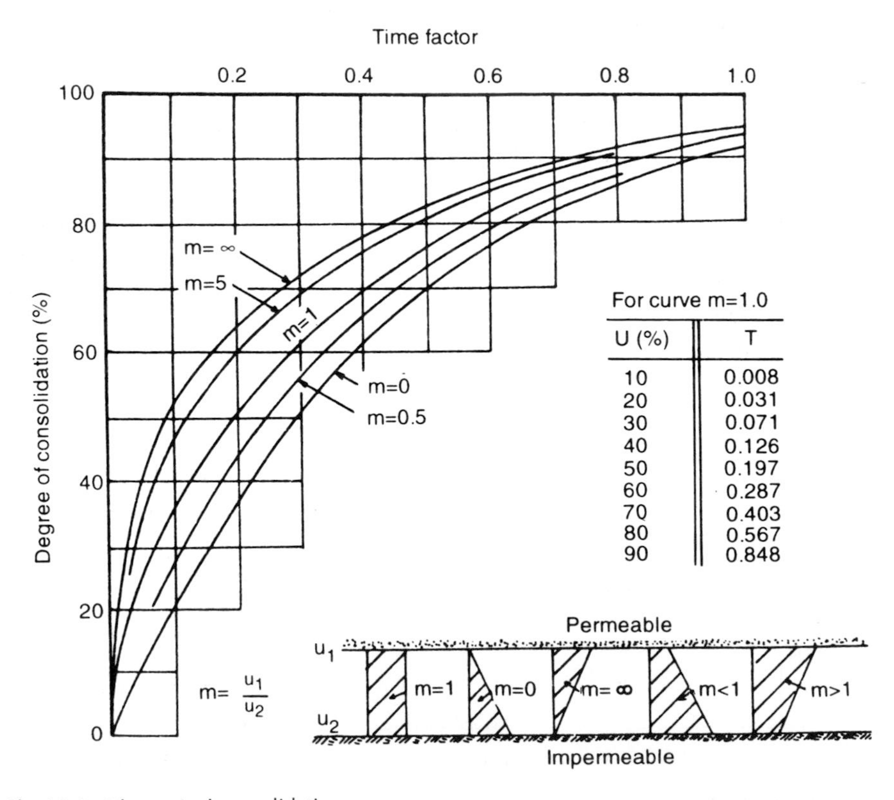

Fig. 10.4 Theoretical consolidation curves.

The mathematical expression for U is:

$$U = 1 - \sum_{m=0}^{m=\infty} \frac{2}{M^2} e^{-M^2T}$$

A theoretical relationship between U and T can therefore be established and is shown in Fig. 10.4, which also gives the relationship for u_i distributions that are not uniform, $m = u_1/u_2$.

10.5 Drainage path length

A consolidating soil layer is usually enclosed, having at its top either the foundation or another layer of soil and beneath it either another soil layer or rock. If the materials above and below the layer are pervious, the water under pressure in the layer will travel either upwards or downwards (a concrete foundation is taken as being pervious compared with a clay layer). This case is known as two-way drainage and the drainage path length, i.e. the maximum length that a water particle can travel (Fig. 10.5a)

$$= \frac{\text{Thickness of layer}}{2} = H$$

If one of the materials is impermeable, water will only travel in one direction – the one-way drainage case – and the length of the drainage path = thickness of layer = 2H (Fig. 10.5b).

The curves of Fig. 10.4 refer to cases of one-way drainage (drainage path length = 2H). Owing to the approximations involved, the curve for m = 1 is often taken for the other cases with the assumption that u_i is the initial excess pore pressure at the centre of the layer. For cases of two-way drainage the curve for m = 1 should be used and the drainage path length, for the determination of T, is taken as H.

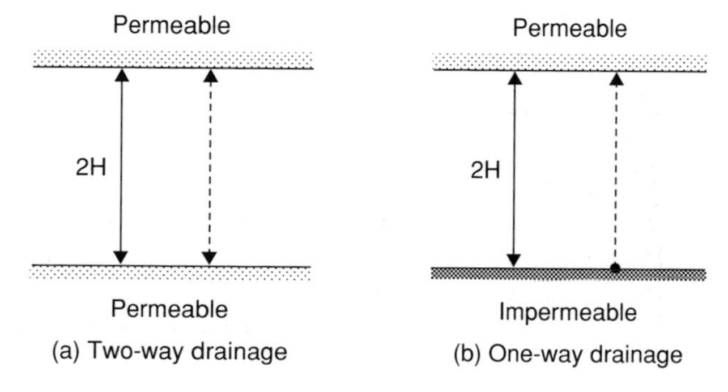

Fig. 10.5 Drainage path length.

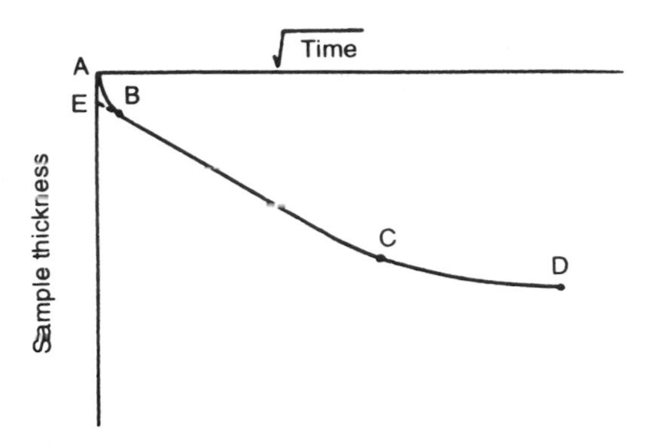

Fig. 10.6 Typical consolidation test results.

10.6 Determination of the coefficient of consolidation, c_v, from the consolidation test

If, for a particular pressure increment applied during a consolidation test, the compression of the test sample is plotted against the square root of time, the result shown in Fig. 10.6 will be obtained.

The curve is seen to consist of three distinct parts: AB, BC and CD.

- *AB (initial compression or frictional lag)*
  A small but rapid compression sometimes occurs at the commencement of the increment and is probably due to the compression of any air present or to the reorientation of some of the larger particles in the sample. In the majority of tests this effect is absent and points A and B are coincident. Initial compression is not considered to be due to any loss of water from the soil and should be treated as a zero error for which a correction is made.
- *BC (primary compression)*
  All the compression in this part of the curve is taken as being due to the expulsion of water from the sample, although some secondary compression will also occur. When the pore pressure has been reduced to a negligible amount it is assumed that 100 per cent consolidation has been attained.
- *CD (secondary compression)*
  The amount by which this effect is evident is a function of the test conditions and can hardly be related to an *in situ* value.

The square root of time 'fitting' method

It will be appreciated that the curve described above is an actual consolidation curve and would not be obtainable from one of the theoretical curves of Fig. 10.4, which can only be used to plot the primary compression range. To evaluate the coefficient of consolidation it is necessary to establish the point C, representing 100 per cent primary consolidation, but it is difficult from a study of the test curve to fix C with

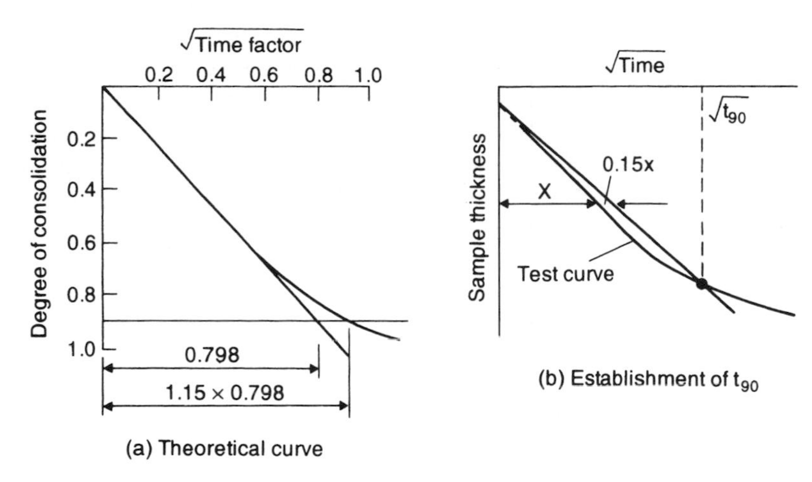

Fig. 10.7 The square root of time 'fitting' method.

accuracy and a procedure in which the test curve is 'fitted' to the theoretical curve becomes necessary.

A method was described by Taylor (1948). If the theoretical curve U against $\sqrt{T}$ is plotted for the case of a uniform initial excess pore pressure distribution, the curve will be like that shown in Fig. 10.7a. Up to values of U equal to about 60 per cent, the curve is a straight line of equation $U = 1.13\sqrt{T}$, but if this straight line is extended to cut the ordinate U = 90 per cent the abscissa of the curve is seen to be 1.15 times the abscissa of the straight line. This fact is used to fit the test and theoretical curves.

With the test curve a corrected zero must first be established by projecting the straight line part of the primary compression back to cut the vertical axis at E (Fig. 10.6). A second line, starting through E, is now drawn such that all abscissas on it are 1.15 times the corresponding values on the laboratory curve, and the point at which this second line cuts the laboratory curve is taken to be the point representing 90 per cent primary consolidation (Fig. 10.7b).

To establish c_v, T_{90} is first found from the theoretical curve that fits the drainage conditions (the curve m = 1); t_{90} is determined from the test curve:

$$T_{90} = \frac{c_v t_{90}}{H^2}$$

i.e.

$$c_v = \frac{T_{90} H^2}{t_{90}}$$

It is seen that the point of 90 per cent consolidation rather than the point for 100 per cent consolidation is used to establish c_v. This is simply a matter of suitability. A consolidation test sample is always drained on both surfaces and in the formula H is taken as half the mean thickness of the sample for the pressure range considered. At first glance it would seem that c_v could not possibly be constant, even for a fairly small pressure range, because as the effective stress is increased the void ratio decreases and both k and m_v decrease rapidly. However, the ratio of k/m_v remains sensibly constant over a large range of pressure so it is justifiable to assume that c_v is in fact constant.

One drawback of the consolidation theory is the assumption that both Poisson's ratio and the elastic modulus of the soil remain constant whereas in reality they both vary as consolidation proceeds. Owing to this continuous variation there is a continuous change in the stress distribution within the soil which, in turn, causes a continuous change in the values of excess pore water pressures. Theories that allow for this effect of the change in applied stress with time have been prepared by Biot (1941) and extended by others, but the approximations involved (together with the sophistication of the mathematics) usually force the user back to the original Terzaghi equation.

10.7 Determination of the permeability coefficient from the consolidation test

Having established c_v, k can be obtained from the formula $k = c_v m_v \gamma_w$. It should be noted that since the mean thickness of the sample is used to determine c_v, m_v should be taken as $a/(1 + \bar{e})$ where $\bar{e}$ is the mean void ratio over the appropriate pressure range.

10.8 Determination of the consolidation coefficient from the triaxial test

It is possible to determine the c_v value of a soil from the consolidation part of the consolidated undrained triaxial test. In this case the consolidation is three-dimensional and the value of c_v obtained is greater than would be the case if the soil were tested in the oedometer. Filter paper drains are usually placed around the sample to create radial drainage so that the time for consolidation is reduced. The effect of three-dimensional drainage is allowed for in the calculation for c_v, but the value obtained is not usually dependable as it is related to the relative permeabilities of the soil and the filter paper (Rowe, 1959).

The time taken for consolidation to occur in the triaxial test generally gives a good indication of the necessary rate of strain for the undrained shear part of the test, but it is not advisable to use this time to determine c_v unless there are no filter drains.

The consolidation characteristics of a partially saturated soil are best obtained from the triaxial test, which can give the initial pore water pressures and the volume change under undrained conditions. Having applied the cell pressure and noted these readings, the pore pressures within the sample are allowed to dissipate while further pore pressure measurements are taken; the accuracy of the results obtained is much greater than with the consolidation test as the difficulty of fitting the theoretical and test curves when air is present is largely removed. The dissipation test is described by Bishop and Henkel (1962).

Example 10.1

Results obtained from a consolidation test on a clay sample for a pressure increment of 100–200 kPa were:

Thickness of sample (mm)	Time (min)
12.200	0
12.141	$\frac{1}{4}$
12.108	1
12.075	$2\frac{1}{4}$
12.046	4
11.985	9
11.922	16
11.865	25
11.827	36
11.809	49
11.800	64

(i) Determine the coefficient of consolidation of the soil.
(ii) How long would a layer of this clay, 10 m thick and drained on its top surface only, take to reach 75 per cent primary consolidation?
(iii) If the void ratios at the beginning and end of the increment were 0.94 and 0.82 respectively, determine the value of the coefficient of permeability.

Solution

(i) The first step is to determine t_{90}. The thickness of the sample is plotted against the square root of time (Fig. 10.8) and if necessary the curve is corrected for zero error to establish the point E. The 1.15 line is next drawn from E and where it cuts the test curve (point F) it gives $\sqrt{t_{90}} = 6.54$. Hence $t_{90} = 42.7$ min.

From the curve for m = 1 (Fig. 10.4), $T_{90} = 0.85$:

$$T = \frac{c_v t}{H^2}$$

Mean thickness of sample during increment (corrected initial thickness 12.168)

$$= \frac{12.168 + 11.800}{2} = 11.984 \text{ mm}$$

$$\Rightarrow \quad H = \frac{11.984}{2} = 5.992 \text{ mm}$$

$$c_v = \frac{0.85 \times 5.992^2}{42.7} = 0.715 \text{ mm}^2/\text{min}$$

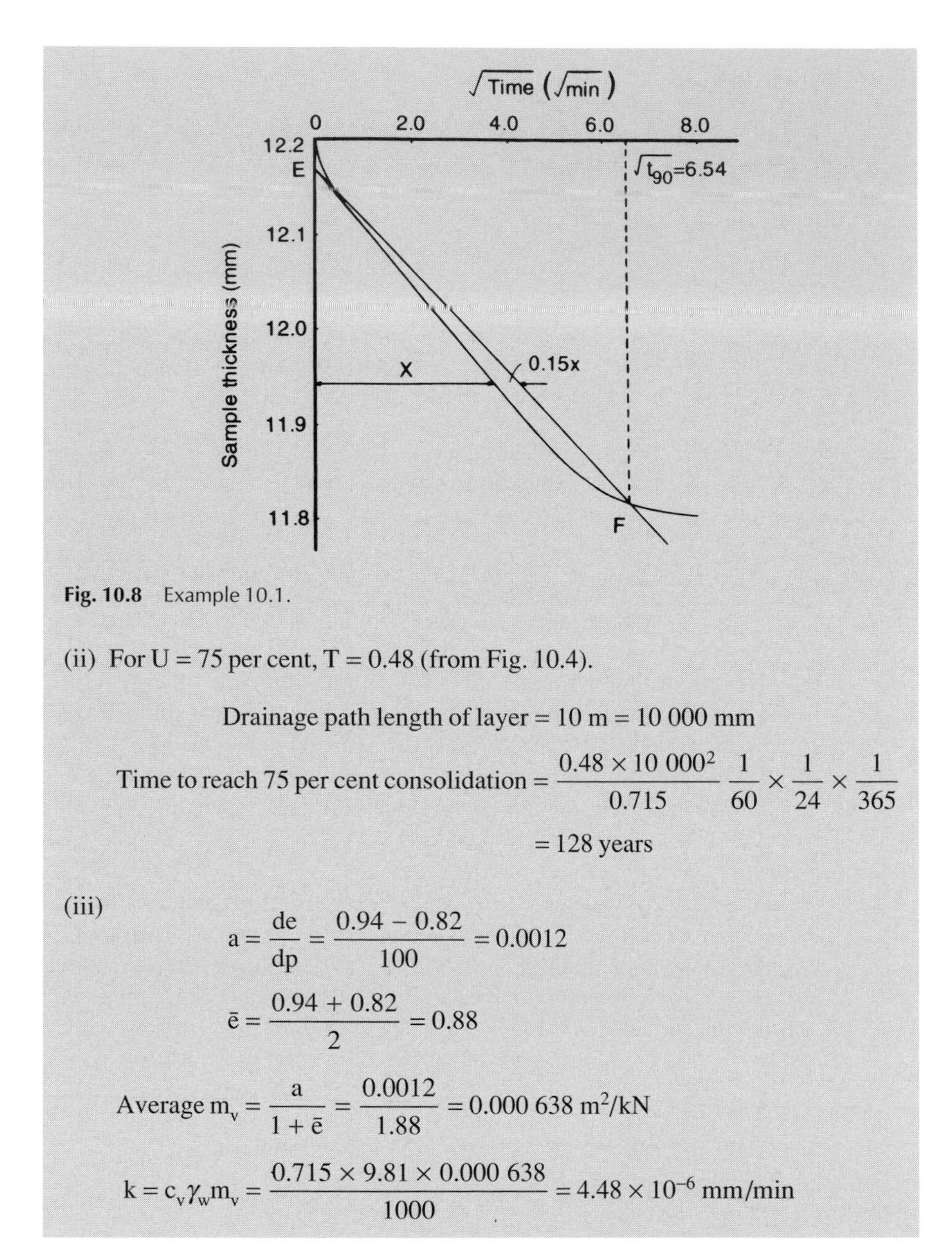

Fig. 10.8 Example 10.1.

(ii) For U = 75 per cent, T = 0.48 (from Fig. 10.4).

$$\text{Drainage path length of layer} = 10 \text{ m} = 10\,000 \text{ mm}$$

$$\text{Time to reach 75 per cent consolidation} = \frac{0.48 \times 10\,000^2}{0.715}\frac{1}{60} \times \frac{1}{24} \times \frac{1}{365}$$

$$= 128 \text{ years}$$

(iii)

$$a = \frac{de}{dp} = \frac{0.94 - 0.82}{100} = 0.0012$$

$$\bar{e} = \frac{0.94 + 0.82}{2} = 0.88$$

$$\text{Average } m_v = \frac{a}{1 + \bar{e}} = \frac{0.0012}{1.88} = 0.000\,638 \text{ m}^2/\text{kN}$$

$$k = c_v \gamma_w m_v = \frac{0.715 \times 9.81 \times 0.000\,638}{1000} = 4.48 \times 10^{-6} \text{ mm/min}$$

10.9 The model law of consolidation

If two layers of the same clay with different drainage path lengths H_1 and H_2 are acted upon by the same pressure increase and reach the same degree of consolidation in times t_1 and t_2 respectively, then theoretically their coefficients of consolidation must be equal as must their time factors, T_1 and T_2:

$$T_1 = \frac{c_{v1}t_1}{H_1^2}; \qquad T_2 = \frac{c_{v2}t_2}{H_2^2}$$

Equating:

$$\frac{t_1}{H_1^2} = \frac{t_2}{H_2^2}$$

This gives a simple method for determining the degree of consolidation in a layer if the simplifying assumption is made that the compression recorded in the consolidation test is solely due to primary compression.

Example 10.2

During a pressure increment a consolidation test sample attained 25 per cent primary consolidation in 5 minutes with a mean thickness of 18 mm. How long would it take a 20 m thick layer of the same soil to reach the same degree of consolidation if (i) the layer was drained on both surfaces and (ii) it was drained on the top surface only?

Solution

In the consolidation test the sample is drained top and bottom

$$\Rightarrow \quad H_1 = \frac{18}{2} = 9.0 \text{ mm}$$

(i) With layer drained on both surfaces $H_2 = 10$ m $= 10\,000$ mm.

$$t_2 = \frac{t_1}{H_1^2} H_2^2 = \frac{5 \times 10\,000^2}{9^2} \times \frac{1}{60} \times \frac{1}{24} \times \frac{1}{365} = 11.7 \text{ years}$$

(ii) With layer drained on top surface only $H_2 = 20$ m.

$$\Rightarrow \quad t_2 = 4 \times 11.7 = 47 \text{ years}$$

Example 10.3

A 19.1 mm thick clay sample, drained top and bottom, reached 30 per cent consolidation in 10 minutes. How long would it take the same sample to reach 50 per cent consolidation?

Solution

As U is known (30 per cent) we can obtain T, either from Fig. 10.4 or by using the relationship that $U = 1.13\sqrt{T}$ (up to U = 60 per cent).

$$T_{30} = \left(\frac{0.3}{1.13}\right)^2 = 0.07$$

$$T = \frac{c_v t}{H^2}, \quad \text{so} \quad c_v = \frac{0.07 \times 9.55^2}{10} = 0.6384\ \text{mm}^2/\text{min}$$

$$T_{50} = \left(\frac{0.5}{1.13}\right)^2 = 0.197 \qquad \text{(or obtain from Fig. 10.4)}$$

$$t_{50} = \frac{T_{50}H^2}{c_v} = \frac{0.197 \times 9.55^2}{0.6384} = 28.1\ \text{min}$$

10.10 Consolidation during construction

A sufficiently accurate solution is generally achieved by assuming that the entire foundation load is applied halfway through the construction period. For large constructions, spread over some years, it is sometimes useful to know the amount of consolidation that will have taken place by the end of construction, the problem being that whilst consolidating the clay is subjected to an increasing load.

Figure 10.9 illustrates the loading diagram during and after construction. While excavation is proceeding swelling may occur (see Example 9.7) such as that which took place in the course of excavation for the piers of Chelsea Bridge, which involved the removal of about 9 m of London Clay and resulted in a heave of 6 mm (Skempton, Peck and MacDonald, 1955). If the coefficient of swelling, c_{vs}, is known it would be fairly straightforward to obtain a solution, first as the pore pressures increase (swelling) and then as they decrease (consolidation), but the assumption is usually made that once the construction weight equals the weight of soil excavated (time t_1 in Fig. 10.9) heave is eliminated and consolidation commences. The treatment of the problem has been discussed by Taylor (1948), who gave a graphical solution, and Lumb (1963), who prepared a theoretical solution for the case of a thin consolidating layer.

By plotting the load–time relationship the time t_1 can be found (Fig. 10.9), the time t_2 being taken as the time in which the net foundation load is applied. The settlement curve, assuming instantaneous application of the load at time t_1, is now plotted and a correction is made to the curve by assuming that the actual consolidation settlement at the end of time t_2 has the same value as the settlement on the instantaneous curve at time $t_2/2$. Point A, corresponding to $t_2/2$, is obtained on the instantaneous curve, and point B is established on the corrected curve by drawing a horizontal from A to meet the ordinate of time t_2 at point B. To establish other points on the corrected curve the procedure is to:

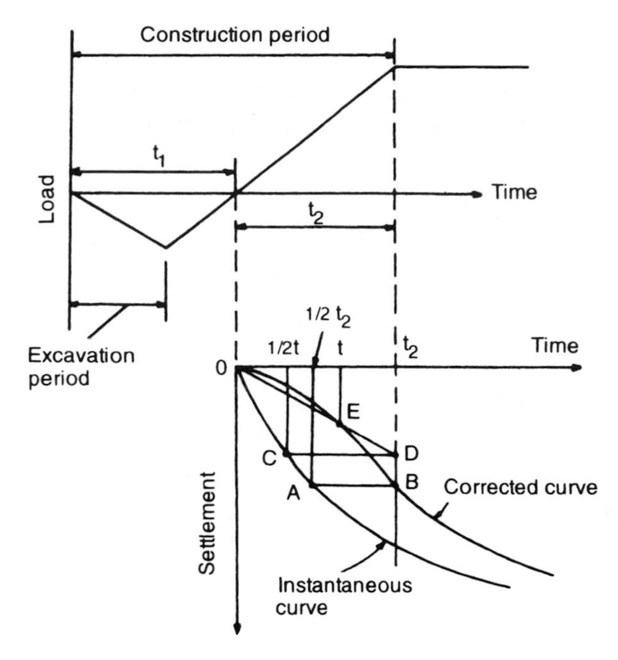

Fig. 10.9 Consolidation during construction.

(i) select a time, t;
(ii) determine the settlement on the instantaneous curve for t/2 (point C);
(iii) draw a horizontal from C to meet the ordinate for t_2 at D, and
(iv) join OD.

Where OD cuts the ordinate for time t gives the point E on the corrected curve, the procedure being repeated with different values of t until sufficient points are established for the curve to be drawn. Points beyond B on the corrected curve are displaced horizontally by the distance AB from the corresponding points on the instantaneous curve.

Example 10.4

If in Example 9.7 the excavation will take 6 months and the structure will be completed in a further 18 months, determine the settlement to time relationship for the central point of the raft during the first 5 years. The clay has a c_v value of 1.86 m^2/year and the sandstone may be considered permeable.

Solution

The initial excess pore water pressure distribution will be roughly trapezoidal. The first step is to determine the values of excess pore pressures at the top and bottom of the clay layer (use Fig. 4.6).

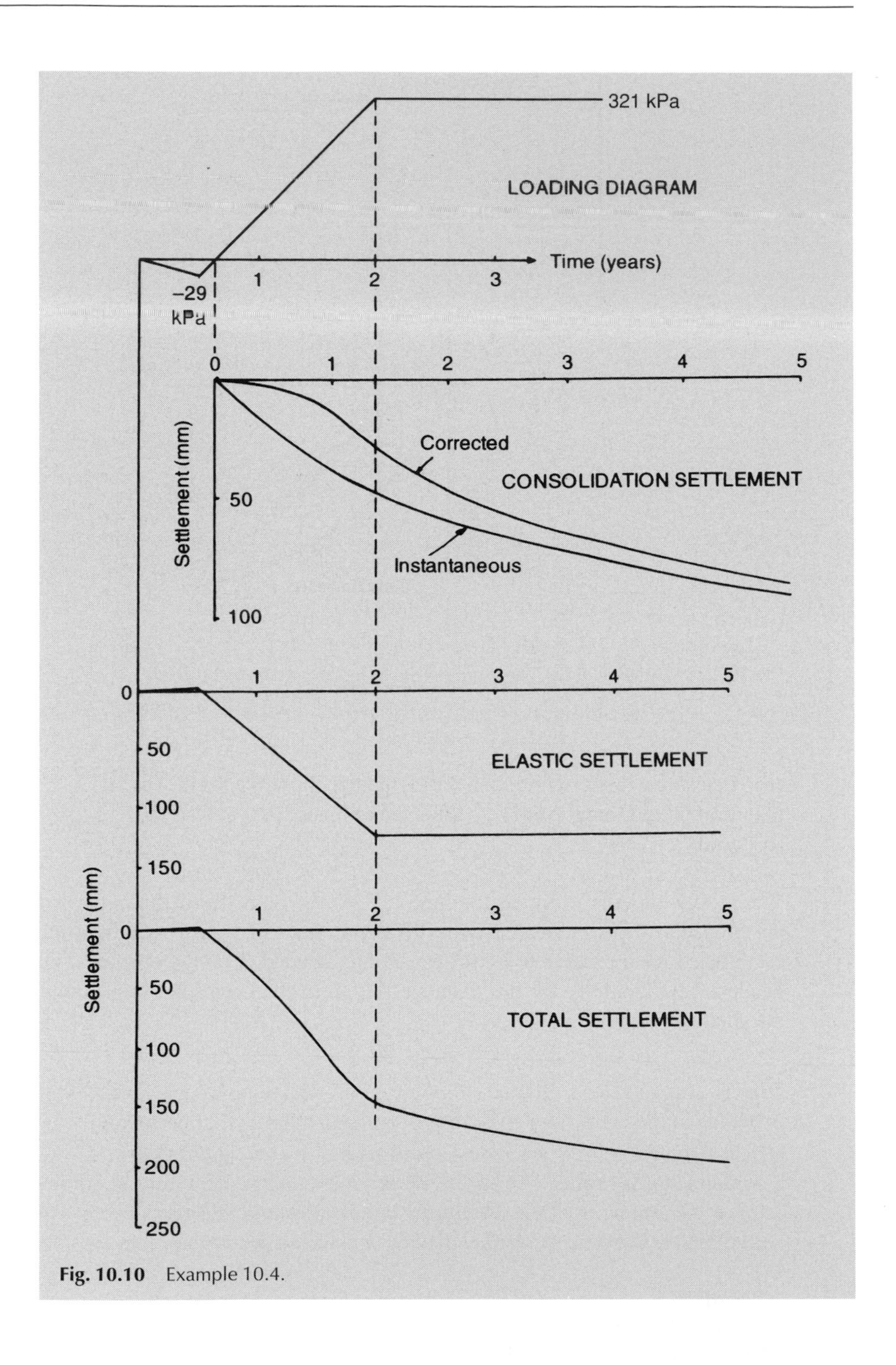

Fig. 10.10 Example 10.4.

	Depth below foundation (m)	$\frac{B}{Z}$	$\frac{L}{Z}$	$I\sigma$	$4I\sigma$	$\Delta\sigma_1$ (kPa)
Top of clay	6.1	1.5	4.5	0.229	0.916	295
Bottom of clay	36.6	0.25	0.75	0.06	0.24	77.3

$$\text{Drainage path length} = \frac{36.6 - 6.1}{2} = 15.25 \text{ m}$$

$$m = \frac{295}{77.3} = 3.82; \text{ values of U are obtained from Fig. 10.4.}$$

t (years)	$T = \frac{c_v t}{H^2}$	U (%)	ρ_c (mm)
1	0.008	10	= 0.1 × 448 = 44.8
2	0.016	15	67.2
3	0.024	18	80.6
4	0.032	22	98.6
5	0.040	24	107.5

Plotting the values of consolidation against time gives the settlement curve for instantaneous loading, which can be corrected to allow for the construction period (Fig. 10.10, which also shows the immediate settlement to time plot). The summation of these two plots gives the total settlement to time relationship.

10.11 Consolidation by drainage in two and three dimensions

The majority of settlement analyses are based on the frequently incorrect assumption that the flow of water in the soil is one dimensional, partly for ease of calculation and partly because in most cases knowledge of soil compression values in three dimensions is limited. There are occasions when this assumption can lead to significant errors (as in the case of an anisotropic soil with a horizontal permeability so much greater than its vertical value that the time–settlement relationship is considerably altered) and when dealing with a foundation which is relatively small compared with the thickness of the consolidating layer some form of analysis allowing for lateral drainage becomes necessary. For an isotropic, homogeneous soil the differential equation for three-dimensional consolidation is:

$$c_v\left(\frac{\partial^2 u}{\partial x^2} + \frac{\partial^2 u}{\partial y^2} + \frac{\partial^2 u}{\partial z^2}\right) = \frac{\partial u}{\partial t}$$

For two dimensions one of the terms in the bracket is dropped.

10.12 Numerical determination of consolidation rates

When a consolidating layer of clay is subjected to an irregular distribution of initial excess pore water pressure, the theoretical solutions are not usually applicable unless the distribution can be approximated to one of the cases considered. In such circumstances the use of a numerical method is fairly common. A spreadsheet can be used for such a purpose and Example 10.5 illustrates the use of a spreadsheet to find the solution.

A brief revision of the relevant mathematics is set out below.

Maclaurin's series

Assuming that f(x) can be expanded as a power series:

$$y = f(x) = a_0 + a_1x + a_2x^2 + a_3x^3 + \cdots a_nx^n$$

$$\frac{dy}{dx} = f'(x) = a_1 + 2a_2x + 3a_3x^2 + 4a_4x^3 + \cdots na_nx^{n-1}$$

$$\frac{d^2y}{dx^2} = f''(x) = 2a_2 + 2.3a_3x + 3.4a_4x^2 + \cdots n(n-1)a_nx^{n-2}$$

$$\frac{d^3y}{dx^3} = f'''(x) = 2.3a_3 + 2.3.4a_4x + \cdots n(n-1)(n-2)a_nx^{n-3}$$

If we put x = 0 in each of the above:

$$a_0 = f(0); \qquad a_1 = f'(0); \qquad a_2 = \frac{f''(0)}{2!}; \qquad a_3 = \frac{f'''(0)}{3!}; \qquad \text{etc.}$$

Generally

$$a_n = \frac{f^n(0)}{n!}$$

Substituting these values:

$$f(x) = f(0) + xf'(0) + \frac{x^2f''(0)}{2!} + \frac{x^3f'''(0)}{3!} + \cdots \frac{x^nf^n(0)}{n!}$$

This is the Maclaurin's series for the expansion of f(x).

Taylor's series

If a curve y = f(x) cuts the y-axis above the origin O at a point A (Fig. 10.11) we can interpret Maclaurin's expression as follows:

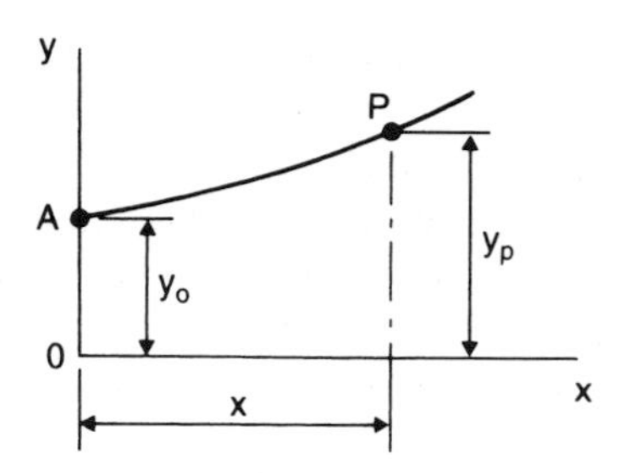

Fig. 10.11 Taylor's series.

Let P be a point on the curve with abscissa x.
Let the values of f(x), f′(x), f″(x), etc., at A be:

$$y_0, y_0', y_0'', \text{etc.}$$

Let the value of f(x) at P be y_p. Then

$$f(x) \text{ at } P = y_p = y_0 + xy_0' + \frac{x^2 y_0''}{2!} + \frac{x^3 y_0'''}{3!} + \cdots$$

This is a Taylor's series and gives us the value of the co-ordinate of P in terms of the ordinate gradient, etc., at A and the distance x between A and P.

Gibson and Lumb (1953) illustrated how the numerical solution of consolidation problems can be obtained by using the explicit finite difference equation. The differential equation for one-dimensional consolidation has been established:

$$c_v \frac{\partial^2 u}{\partial z^2} = \frac{\partial u}{\partial t}$$

Consider part of a grid drawn on to a consolidating layer (Fig. 10.12a). The variation of the excess pore pressure, u, with the depth, z, at a certain time, k, is shown in Fig. 10.12b, and the variation of u at the point O during a time increment from k to k + 1 is illustrated by Fig. 10.12c.

In Fig. 10.12b: from Taylor's theorem:

$$u_{2,k} = u_{0,k} - \Delta z u_{0,k}' + \frac{\Delta z^2}{2!} u_{0,k}'' - \frac{\Delta z^3}{3!} u_{0,k}''' + \cdots$$

$$u_{4,k} = u_{0,k} + \Delta z u_{0,k}' + \frac{\Delta z^2}{2!} u_{0,k}'' + \frac{\Delta z^3}{3!} u_{0,k}''' + \cdots$$

Adding and ignoring terms greater than second order:

$$u_{2,k} + u_{4,k} = 2u_{0,k} = \Delta z^2 u_{0,k}''$$

$$\Rightarrow \quad \frac{\partial^2 u}{\partial z^2} = u_{0,k}'' = \frac{u_{2,k} + u_{4,k} - 2u_{0,k}}{\Delta z^2}$$

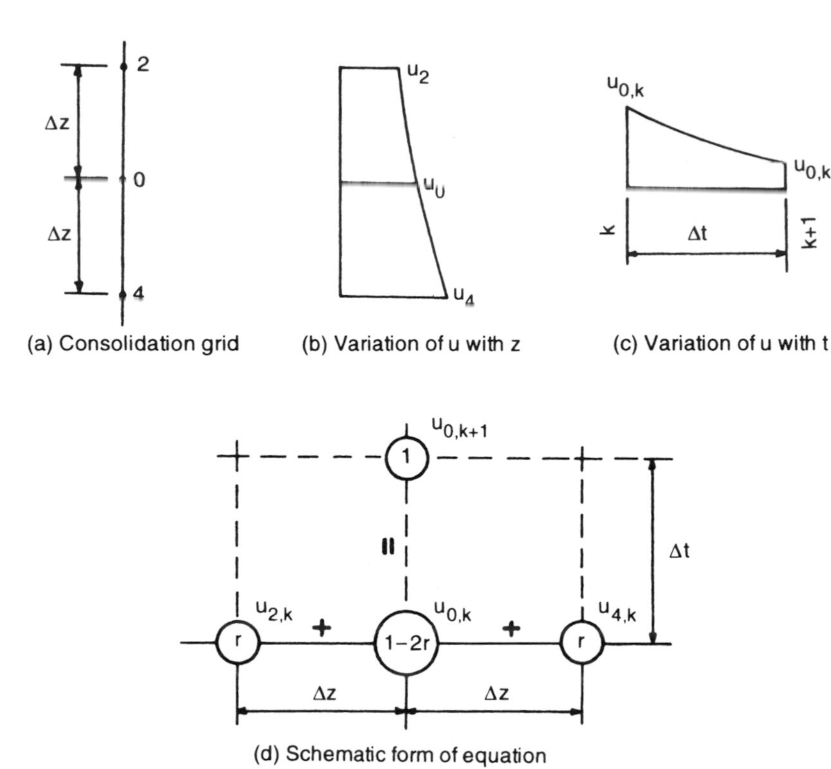

Fig. 10.12 Explicit recurrence formula (general).

In Fig. 10.12c:

$$\frac{\partial u}{\partial t} \text{ is a function } u = f(t)$$

By Taylor's theorem:

$$u_{0,k+1} = u_{0,k} + \Delta t u'_{0,k} + \frac{\Delta t^2}{2!} u''_{0,k} + \cdots$$

Ignoring second derivatives and above:

$$\frac{\partial u}{\partial t} = u'_{0,k} = \frac{u_{0,k+1} - u_{0,k}}{\Delta t}$$

$$\Rightarrow \quad c_v\left(\frac{u_{2,k} + u_{4,k} - 2u_{0,k}}{\Delta z^2}\right) = \frac{u_{0,k+1} - u_{0,k}}{\Delta t}$$

$$\Rightarrow \quad u_{0,k+1} = r(u_{2,k} + u_{4,k} - 2u_{0,k}) + u_{0,k}$$

where

$$r = \frac{c_v \Delta t}{\Delta z^2}$$

The schematic form of this expression is shown in Fig. 10.12d. Hence if a series of points in a consolidating layer are established, Δz apart, it is possible by numerical iteration to work out the values of u at any time interval after consolidation has commenced if the initial excess values, u_i, are known.

Impermeable boundary conditions

Figure 10.13a illustrates this case in which conditions at the boundary are represented by

$$\frac{\partial u}{\partial z} = 0$$

Hence between the points 2_k and 4_k:

$$\frac{\partial u}{\partial z} = \frac{u_{2,k} - u_{4,k}}{2\Delta z} = 0$$

i.e.

$$u_{2,k} = u_{4,k}$$

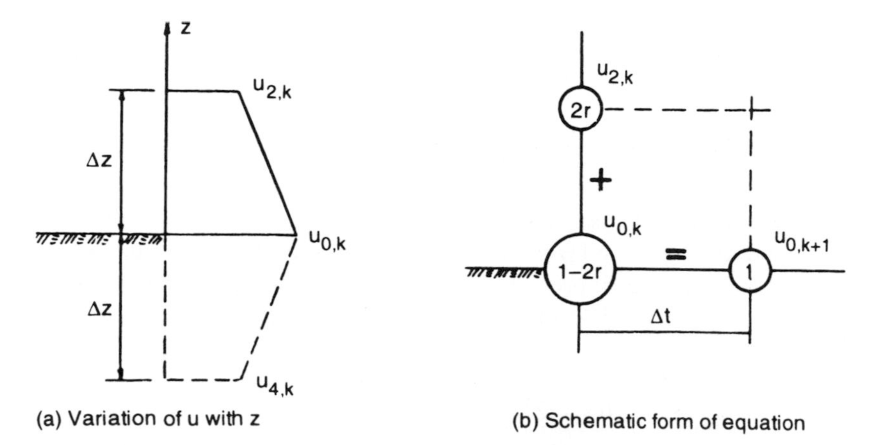

Fig. 10.13 Explicit recurrence formula: treatment for an impermeable boundary.

The equation therefore becomes:

$$u_{0,k+1} = 2r(u_{2,k} - u_{0,k}) + u_{0,k}$$

and is shown in schematic form in Fig. 10.13b.

The boundary equation can also be used at the centre of a double drained layer with a symmetrical initial pore pressure distribution, values for only half the layer needing to be evaluated.

Errors associated with the explicit equation

A full discussion of this subject was given by Crandall (1956), but briefly errors fall into two main groups: truncation errors (due to ignoring the higher derivatives) and rounding-off errors (due to working to only a certain number of decimal places). The size of the space increment, Δz, affects both these errors but in different ways: the smaller Δz is, the less the truncation error that arises but the greater the round-off error tends to become.

The value of r is also important. For stability r must not be greater than 0.5 and, for minimum truncation errors, should be 1/6; the usual practice is to take r as near as possible to 0.5. This restriction means that the time interval must be short and a considerable number of iterations become necessary to obtain the solution for a large time interval. With present software this is not a problem, but if necessary use can be made of either the implicit finite difference equation (Crank and Nicolson, 1947) or the relaxation method (Leibmann, 1955).

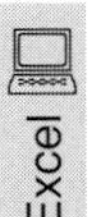

Example 10.5

A layer of clay 4 m thick is drained on its top surface and has a uniform initial excess pore pressure distribution. The consolidation coefficient of the clay is 0.1 m^2/month. Using a numerical method, determine the degree of consolidation that the layer will have undergone 24 months after the commencement of consolidation. Check your answer by the theoretical curves of Fig. 10.4.

Solution

In a numerical solution the grid must first be established: for this example the layer has been split into four layers each of $\Delta z = 1.0$ m (it is important to remember that since Simpson's rule is being applied to determine the degree of consolidation, the layer should be divided into an even number of strips). The initial excess pore pressure values have been taken everywhere throughout the layer as equal to 100 units.

In 24 months:

$$r = \frac{c_v t}{\Delta z^2} = \frac{0.1 \times 24}{1.0} = 2.4$$

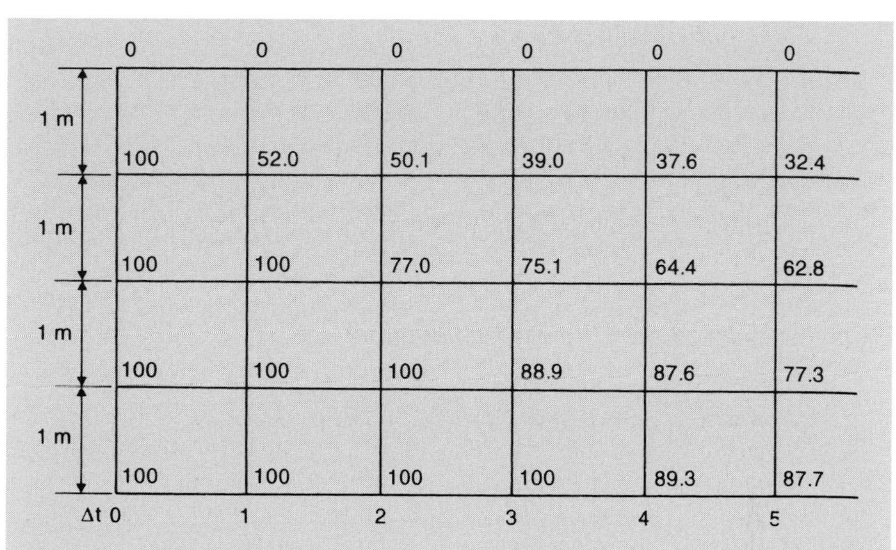

Fig. 10.14 Example 10.5.

For the finite difference equation r must not be greater than 0.5, so use five time increments, i.e. $\Delta t = 4.8$ months and

$$r = \frac{0.1 \times 4.8}{1.0} = 0.48$$

Owing to the instantaneous dissipation at the drained surface the excess pore pressure distribution at time = 0 can be taken as that shown in Fig. 10.14 (the values obtained during the iteration process are also given). The finite difference formula is applied to each point of the grid, except at the drained surface:

$$u_{0,k+1} = r(u_{2,k} + u_{4,k} - 2u_{0,k}) + u_{0,k}$$

For example, with the first time increment the point next to the drained surface has $u = 0.48(0 + 100 - 2 \times 100) + 100 = 52.0$. Note that at the undrained surface the finite difference equation alters.

Degree of consolidation

Area of initial excess pore pressure distribution diagram = 4 × 100 = 400.

Area under final isochrone is obtained by Simpson's rule:

$$\frac{1.0}{3}(87.7 + 4(32.4 + 77.3) + 2 \times 62.8) = 217$$

hence:

$$U = \frac{400 - 217}{400} = 45.7 \text{ per cent}$$

Checking by the theoretical curve:
Total time = 24 months, H = 4 m:

$$T = \frac{c_v t}{H^2} = \frac{0.1 \times 24}{16} = 0.15$$

From Fig. 10.4: U = 45 per cent.

10.13 Construction pore pressures in an earth dam

A knowledge of the induced pore pressures occurring during the construction of an earth dam or embankment is necessary so that stability analyses can be carried out and a suitable construction rate determined. Such a problem is best solved by numerical methods. During the construction of an earth dam (or an embankment) the placing of material above that already in position increases the pore water pressure whilst consolidation has the effect of decreasing it: the problem is one of a layer of soil that is consolidating as it is increasing in thickness. Gibson (1958) examined this condition. If it is assumed that the water in the soil will experience vertical drainage only, the finite difference equation becomes:

$$u_{0,k+1} = r(u_{2,k} + u_{4,k} - 2u_{0,k}) + u_{0,k} + \bar{B}\gamma\Delta z$$

where

Δz = the grid spacing, and also the increment of dam thickness placed in time Δt
γ = unit weight of dam material
$\bar{B}$ = pore pressure coefficient

$$r = \frac{c_v \Delta t}{\Delta z^2}$$

In order that Δz is constant throughout the full height of the dam, all construction periods must be approximated to the same linear relationship and then transformed into a series of steps. The formula can only be applied to a layer that has some finite thickness, and as the layer does not exist initially it is necessary to obtain a solution by some other method for the early stages of construction when the dam is insufficiently thick for the formula to be applicable. Smith (1968b) has shown how a relaxation procedure can be used for this initial stage.

Example 10.6

At a stage in its construction an earth embankment has attained a height of 9.12 m and has the excess pore water pressure distribution shown in Fig. 10.15a. A proposal has been made that further construction will be at the rate of 1.52 m thickness of material placed in one month, the unit weight of the placed material to be 19.2 kN/m^3 and its $\bar{B}$ value about 0.85. Determine approximate values for the excess pore pressures that will exist within the embankment 3 months after further construction is commenced. c_v for the soil = 0.558 m^2/month.

Solution

Check the r value with Δz taken as equal to 1.52 m.

For $\Delta z = 1.52$ m, t = 1.0 month:

$$r = \frac{0.558 \times 1}{(1.52)^2} = 0.241$$

This value of r is satisfactory and has been used in the solution (if r had been greater than 0.5 then Δt and Δz would have had to be varied until r was less than 0.5).

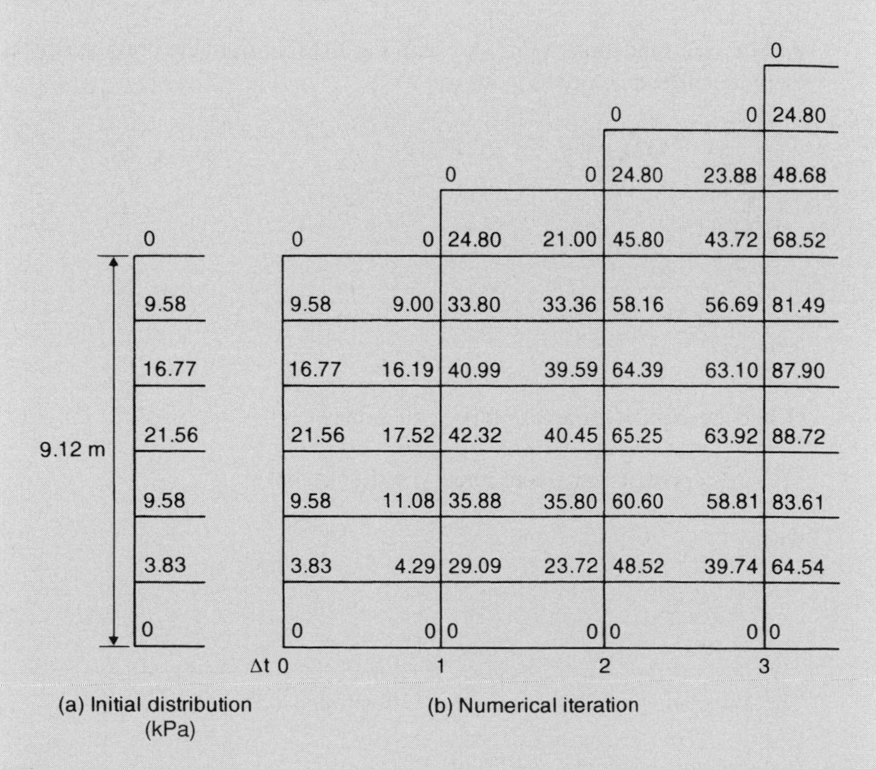

Fig. 10.15 Example 10.6.

A 1.52 m deposit of the soil will induce an excess pressure, throughout the whole embankment, of $1.52 \times 19.2 \times \bar{B} = 24.8$ kPa. This pressure value must be added to the value at each grid point for each time increment. The pore pressure increase is in fact applied gradually over a month, but for a numerical solution we must assume that it is applied in a series of steps, i.e. 24.8 kPa at t = 1 month, at t = 2 months, and at t = 3 months. From t = 0 to t = 1 no increment is assumed to be added and the initial pore pressures will have dissipated further before they are increased.

The numerical iteration is shown in Fig. 10.15b.

10.14 Numerical solutions for two- and three-dimensional consolidation

10.14.1 Two-dimensional consolidation

The differential equation for two-dimensional consolidation has already been given:

$$c_v\left(\frac{\partial^2 u}{\partial x^2} + \frac{\partial^2 u}{\partial y^2}\right) = \frac{\partial u}{\partial t}$$

Part of a consolidation grid is shown in Fig. 10.16a; from the previous discussion of the finite difference equation we can write:

$$\frac{\partial u}{\delta t} = \frac{u_{0,k+1} - u_{0,k}}{\Delta t}$$

$$\frac{\partial^2 u}{\partial y^2} = \frac{c_v}{h^2}(u_{2,k} + u_{4,k} - 2u_{0,k})$$

$$\frac{\partial^2 u}{\partial x^2} = \frac{c_v}{h^2}(u_{1,k} + u_{3,k} - 2u_{0,k})$$

Hence the explicit finite difference equation is:

$$u_{0,k+1} = r(u_{1,k} + u_{2,k} + u_{3,k} + u_{4,k}) + u_{0,k}(1 - 4r)$$

where

$$r = \frac{c_v \Delta t}{h^2}$$

The schematic form of this equation is illustrated in Fig. 10.16b.

Impermeable boundary condition

Impermeable boundaries are treated as for the one-dimensional case.

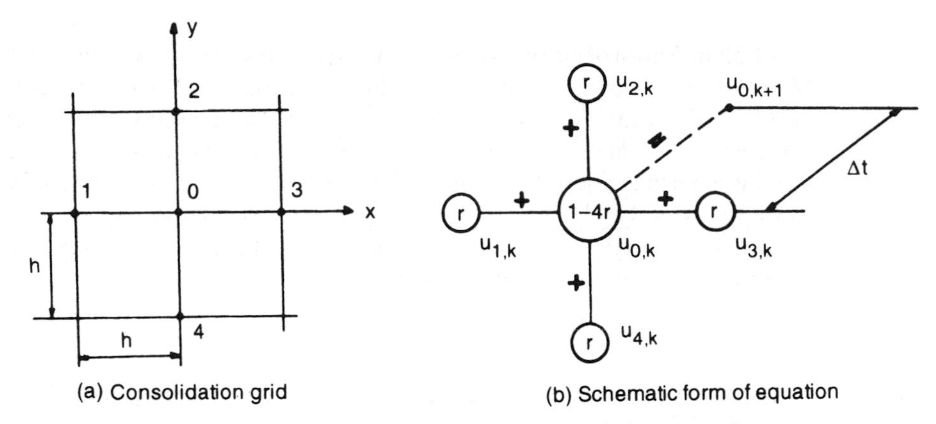

Fig. 10.16 Schematic form of the finite difference equation (two dimensional).

10.14.2 Three-dimensional consolidation

For instances of radial symmetry the differential equation can be expressed in polar co-ordinates:

$$c_v\left(\frac{\partial^2 u}{\partial R^2} + \frac{1}{R}\frac{\partial u}{\partial R} + \frac{\partial^2 u}{\partial z^2}\right) = \frac{\partial u}{\partial t}$$

then

$$\frac{\partial u}{\partial t} = \frac{u_{0,k+1} - u_{0,k}}{\Delta t}$$

$$\frac{\partial^2 u}{\partial z^2} = \frac{u_{2,k} + u_{4,k} - 2u_{0,k}}{\Delta z^2}$$

$$\frac{\partial^2 u}{\partial R^2} = \frac{u_{1,k} + u_{3,k} - 2u_{0,k}}{\Delta R^2}$$

$$\frac{1}{R}\frac{\partial u}{\partial R} = \frac{1}{R}\left(\frac{u_{3,k} - u_{1,k}}{2\Delta R}\right)$$

If we put $\Delta z = \Delta R = h$ the finite difference equation becomes:

$$u_{0,k+1} = r(u_{2,k} + u_{4,k}) + u_{0,k}(1 - 4r) + ru_{1,k}\left(1 - \frac{h}{2R}\right) + ru_{3,k}\left(1 + \frac{h}{2R}\right)$$

where

$$r = \frac{c_v \Delta t}{h^2}$$

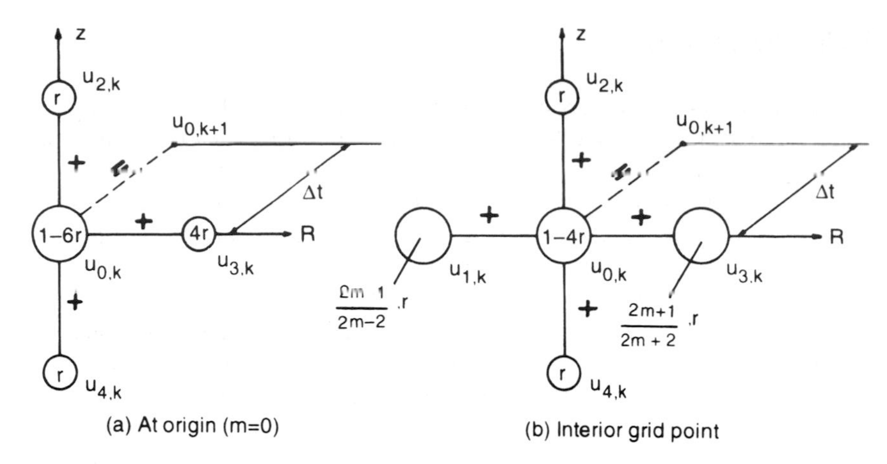

Fig. 10.17 Schematic form of the finite difference equation (three dimensional).

At the origin, where R = 0:

$$\frac{1}{R}\frac{\partial u}{\partial R} \to \frac{\partial^2 u}{\partial R^2}$$

and the equation becomes:

$$u_{0,k+1} = ru_{2,k} + 4ru_{3,k} + ru_{4,k} + u_{0,k}(1 - 6r)$$

Using the convention R = mh, the schematic form for the explicit equation is shown in Fig. 10.17a (for a point at the origin) and Fig. 10.17b (for other interior points).

For drainage in the vertical direction the procedure is the same, but for radial drainage the expression for $u_{0,k+1}$ at a boundary point, where $\partial u/\partial R = 0$, is given by:

$$u_{0,k+1} = r(u_{2,k} + u_{4,k}) + 2ru_{1,k} + u_{0,k}(1 - 4r)$$

Value of r

Scott (1963) pointed out that in three-dimensional work the explicit recurrence formula is stable if r is either equal to or less than 1/6. This is not so severe a restriction as it would at first appear, since with three-dimensional drainage the time required to reach a high degree of consolidation is much less than for one-dimensional drainage. For two-dimensional work r should not exceed 0.5.

10.14.3 Determination of initial excess pore water pressure values

For one-dimensional consolidation problems, u_i can at any point be taken as equal to the increment of the total major principal stress at that point. For two- and three-dimensional problems u_i must be obtained from the formula:

$$u_i = B[\Delta\sigma_3 + A(\Delta\sigma_1 - \Delta\sigma_3)]$$

As the clay is assumed saturated, B = 1.0.

10.15 Sand drains

Sometimes the natural rate of consolidation of a particular soil is too slow, particularly when the layer overlies an impermeable material and, in order that the structure may carry out its intended purpose, the rate of consolidation must be increased. An example of where this type of problem can occur is an embankment designed to carry road traffic. It is essential that most of the settlement has taken place before the pavement is constructed if excessive cracking is to be avoided.

From the Model Law of Consolidation it is known that the rate of consolidation is proportional to the square of the drainage path length. Obviously the consolidation rate is increased if horizontal, as well as vertical, drainage paths are made available to the pore water. This can be achieved by the installation of a system of sand drains, which is essentially a set of vertical boreholes put down through the layer, ideally to a firmer material, and then backfilled with porous material, such as a suitably graded sand. The method was first used across a marsh in California and is described by Porter (1936).

A typical arrangement is shown in Fig. 10.18a. There are occasions when the sand drains are made to puncture through an impermeable layer when there is a pervious layer beneath it. This creates two-way vertical drainage, as well as lateral, and results in a considerable speeding up of construction.

Diameter of drains: vary from 300 to 600 mm. Diameters less than 300 mm are generally difficult to install unless the surrounding soil is considerably remoulded.
Spacing of drains: depends upon the type of soil in which they are placed. Spacings vary between 1.5 and 4.5 m. Sand drains are effective if the spacing, a, is less than the thickness of the consolidating layer, 2H.
Arrangement of grid: sand drains are laid out in either square (Fig. 10.19a) or triangular (Fig. 10.19b) patterns. For triangular arrangements the grid forms a series of equilateral triangles the sides of which are equal to the drain spacing. Barron (1948) maintains that triangular spacing is more economical. In his paper he solved the consolidation theory for sand drains.
Depth of sand drains: dictated by subsoil conditions. Sand drains have been installed to depths of up to 45 m.
Type of sand used: should be clean and able to carry away water yet not permit the fine particles of soil to be washed in.

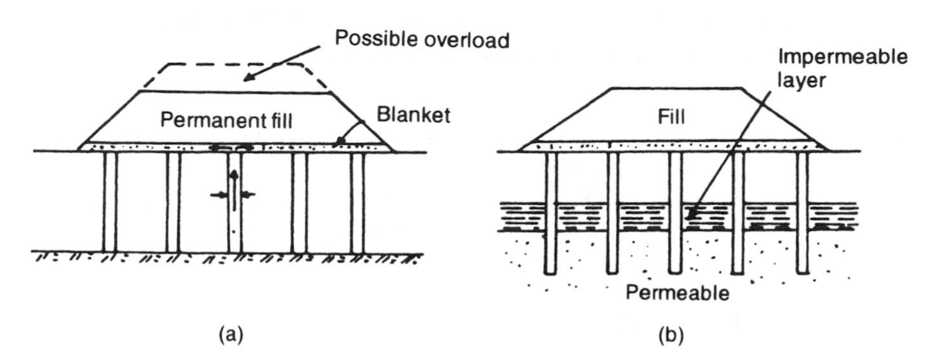

Fig. 10.18 Typical sand drain arrangements.

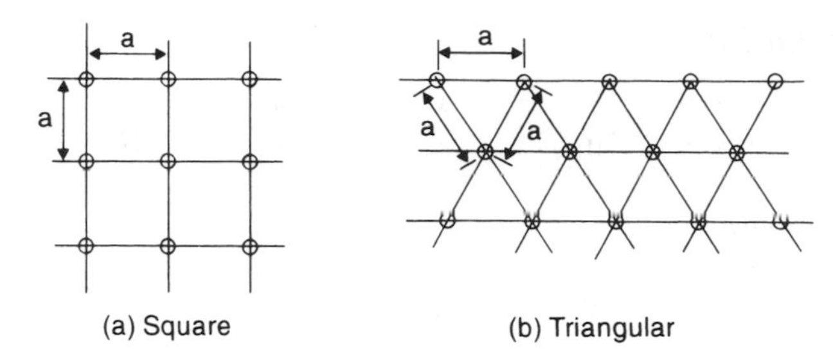

Fig. 10.19 Popular arrangements of sand drains.

Drainage blanket: after drains are installed a blanket of gravel and sand from 0.33 to 1.0 m thick, is spread over the entire area to provide lateral drainage at the base of the fill.

Overfill or surcharge: often used in conjunction with sand drains. It consists of extra fill material placed above the permanent fill to accelerate consolidation. Once piezometer measurements indicate that consolidation has become slow this surcharge is removed.

Strain effects: although there is lateral drainage, lateral strain effects are assumed to be negligible. Hence the consolidation of a soil layer in which sand drains are placed is still obtained from the expression:

$$\rho_c = m_v dp\, 2H$$

Consolidation theory

The three-dimensional consolidation equation is:

$$\frac{\partial u}{\partial t} = c_h \left[\frac{\partial_u^2}{\partial r^2} + \frac{1}{r}\frac{\partial u}{\partial_r} \right] + c_v \frac{\partial_u^2}{\partial z^2}$$

where c_h = coefficient of consolidation for horizontal drainage (when it can be measured: otherwise use c_v).

The various co-ordinate directions of the equation are shown in Fig. 10.20. The equation can be solved by finite differences.

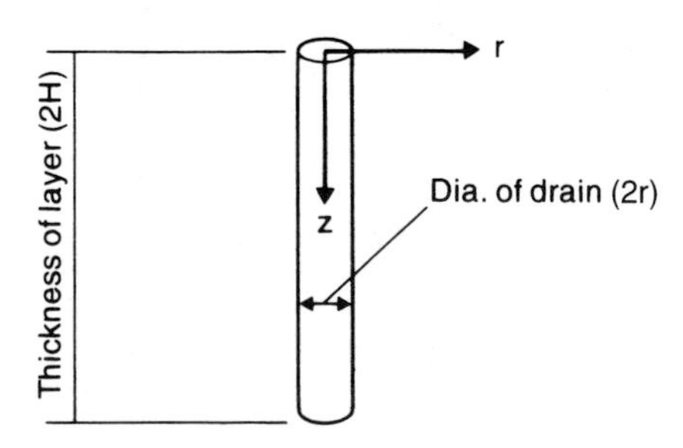

Fig. 10.20 Coordinate directions.

Equivalent radius

The effect of each sand drain extends to the end of its equivalent radius, which differs for square and triangular arrangements (see Fig. 10.19).

For a square system:

Area of square enclosed by grid = a^2

Area of equivalent circle of radius R = a^2

i.e. $\pi R^2 = a^2$ or $R = 0.564a$.

For a triangular system:

A hexagon is formed by bisecting the various grid lines joining adjacent drains (Fig. 10.21). A typical hexagon is shown in the figure from which it is seen that the base of triangle ABC, i.e. the line AB, = a/2.

Now

$$AC = AB \tan \angle CBA = \frac{a}{2} \tan 30° = \frac{a}{2\sqrt{3}}$$

hence:

$$\text{Area of triangle ABC} = \frac{1}{2} \times \frac{a}{2} \times \frac{a}{2\sqrt{3}} = \frac{a^2}{8\sqrt{3}}$$

So that:

$$\text{Total area of the hexagon} = 12 \times \frac{a^2}{8\sqrt{3}} = 0.865a^2$$

Radius of the equivalent circle, $R = 0.525a$

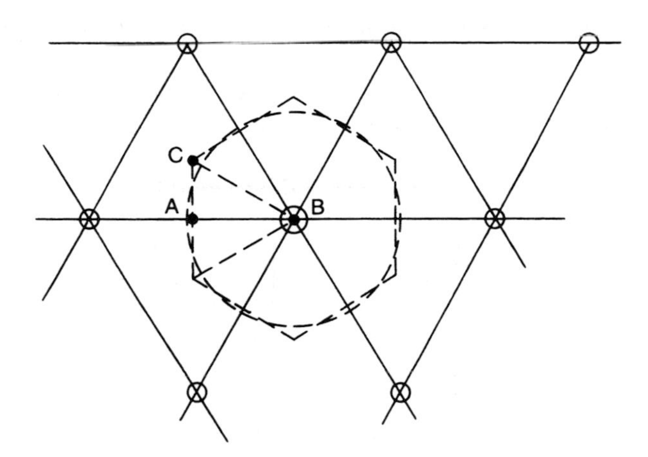

Fig. 10.21 Equivalent radius: triangular system.

Determination of consolidation rates from curves

Barron has produced curves which give the relationship between the degree of consolidation due to radial flow only, U_r, and the corresponding radial time factor, T_r.

$$T_r = \frac{c_h t}{4R^2}$$

where t = time considered.

These curves are reproduced in Fig. 10.22 and it can be seen that they involve the use of factor n. This factor is simply the ratio of the equivalent radius to the sand drain radius.

$$n = \frac{R}{r} \quad \text{and should lie between 5 to 100}$$

To determine U (for both radial and vertical drainage) for a particular time, t, the procedure becomes:

(i) Determine U_z from the normal consolidation curves of U_z against T_z (Fig. 10.4):

$$T_z = \frac{c_v t}{H^2} \quad \text{where} \quad H = \text{vertical drainage path}$$

(ii) Determine U_r from Barron's curves of U_r against T_r.

(iii) Determine resultant percentage consolidation, U, from:

$$U = 100 - \frac{1}{100}(100 - U_z)(100 - U_r)$$

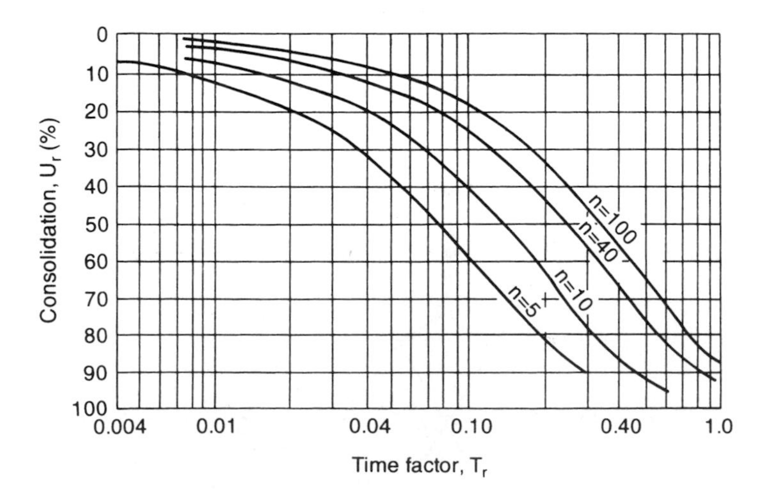

Fig. 10.22 Radial consolidation rates (after Barron, 1948).

Smear effects

The curves in Fig. 10.22 are for idealised drains, perfectly installed, clean and working correctly. Wells are often installed by driving cased holes and then back-filling as the casing is withdrawn, a procedure that causes distortion and remoulding in the adjacent soil. In varved clays (clays with sandwich type layers of silt and sand within them) the finer and more impervious layers are dragged down and smear over the more pervious layers to create a zone of reduced permeability around the perimeter of the drain. This smeared zone reduces the rate of consolidation, and *in situ* measurements to check on the estimated settlement rate are necessary on all but the smallest of jobs.

Effectiveness of sand drains

Sand drains are particularly suitable for soft clays but have little effect on soils with small primary but large secondary effects, such as peat, see Lake (1963).

Example 10.7

A soft clay layer, $m_v = 2.5 \times 10^{-4}$ m²/kN; $c_v = 0.187$ m²/month, is 9.2 m thick and overlies impervious shale. An embankment, to be constructed in six months, will subject the centre of the layer to a pressure increase of 100 kPa. It is expected that a roadway will be placed on top of the embankment one year after the start of construction and maximum allowable settlement after this is to be 25 mm.

Determine a suitable sand drain system to achieve the requirements.

Solution

$$\rho_c = m_v\,dp\,2H = \frac{2.5}{10\,000} \times 100 \times 9.2 \times 1000 = 230 \text{ mm}$$

therefore, minimum settlement that must have occurred by the time the roadway is constructed = 230 − 25 = 205 mm. i.e.

$$U = \frac{205}{230} = 90 \text{ per cent}$$

Assume that settlement commences at half the construction time for the embankment. Then time to reach U = 90 per cent = $12 - \frac{6}{2} = 9$ months.

$$T_z = \frac{c_v t}{H^2} = \frac{0.187 \times 9}{9.2^2} = 0.020$$

From Fig. 10.4 $U_z = 16$ per cent.

Try 450 mm (0.45 m) diameter drains in a triangular pattern.
Select n = 10. Then

$$R/r = 10 \quad \text{and} \quad R = 2.25\ \text{m}$$

hence

$$a = \frac{2.25}{0.525} = 4.3\ \text{m}$$

Select a grid spacing of 3 m.

$$R = 0.525 \times 3 = 1.575\ \text{m}$$

$$n = \frac{1.575}{0.225} = 7$$

$$T_r = \frac{c_v t}{4R^2} = \frac{0.187 \times 9}{4 \times 1.575^2} = 0.169$$

(Note that no value for c_h was given so c_v must be used)

From Fig. 10.22, $U_r = 66$ per cent.

$$U = 100 - \frac{1}{100}(100 - 16)(100 - 66)$$

$$= 71.4 \text{ per cent, which is not sufficient}$$

Try a = 2.25 m; R = 1.18 m; n = 5.25.

$$T_r = \frac{0.187 \times 9}{4 \times 1.18^2} = 0.302$$

From graph, $U_r = 90$ per cent.

$$\text{Total consolidation percentage} = 100 - \frac{1}{100}(100 - 16)(100 - 90)$$

$$= 91.6\%$$

The arrangement is satisfactory.

In practice no sand drain system could be designed as quickly as this. The object of the example is simply to illustrate the method. The question of installation costs must be considered and several schemes would have to be closely examined before a final arrangement could be decided upon.

Exercises

Exercise 10.1

A soil sample in an oedometer test experienced 30 per cent primary consolidation after 10 minutes. How long would it take the sample to reach 80 per cent consolidation?

Answer 80 min

Exercise 10.2

A 5 m thick clay layer has an average c_v value of 5.0×10^{-2} mm²/min. If the layer is subjected to a uniform initial excess pore pressure distribution, determine the time it will take to reach 90 per cent consolidation (i) if drained on both surfaces and (ii) if drained on its upper surface only.

Answer (i) 200 years, (ii) 800 years

Exercise 10.3

In a consolidation test the following readings were obtained for a pressure increment:

Sample thickness (mm)	Time (min)
16.97	0
16.84	$\frac{1}{4}$
16.76	1
16.61	4
16.46	9
16.31	16
16.15	25
16.08	36
16.03	49
15.98	64
15.95	81

(i) Determine the coefficient of consolidation of the sample.
(ii) From the point for U = 90 per cent on the test curve, establish the point for U = 50 per cent and hence obtain the test value for t_{50}. Check your value from the formula

$$t_{50} = \frac{T_{50}H^2}{c_v}$$

Answer $c_v = 1.28$ mm²/min, $t_{50} = 10.2$ min

Exercise 10.4

A sample in a consolidation test had a mean thickness of 18.1 mm during a pressure increment of 150 to 290 kPa. The sample achieved 50 per cent consolidation in 12.5 min. If the initial and final void ratios for the increment were 1.03 and 0.97 respectively, determine a value for the coefficient of permeability of the soil.

Answer $k = 2.78 \times 10^{-6}$ mm/min

Exercise 10.5

A 2 m thick layer of clay, drained at its upper surface only, is subjected to a triangular distribution of initial excess pore water pressure varying from 1000 kPa at the upper surface to 0.0 at the base. The c_v value of the clay is 1.8×10^{-3} m^2/month. By dividing the layer into 4 equal slices, determine, numerically, the degree of consolidation after 4 years.

Note: If the total time is split into seven increments, r = 0.494.

Answer U = 15 per cent

Chapter 11

Compaction and Soil Mechanics Aspects of Highway Design

The process of mechanically pressing together the particles of a soil to increase the density (compaction) is extensively employed in the construction of embankments and in strengthening the subgrades of roads and runways.

Many workers in the field talk about consolidating a soil when they really mean compacting it. Strictly speaking, consolidation is the gradual expulsion of water from the voids of a saturated cohesive soil with consequent reduction in volume, whereas compaction is the packing together of soil particles by the expulsion of air.

The densities achieved by compaction are invariably expressed as dry densities, generally in Mg/m^3 although, occasionally, the units kg/m^3 are used. The moisture content at which maximum dry density is obtained for a given amount of compaction is known as the optimum moisture content.

11.1 Laboratory compaction of soils

11.1.1 British Standard compaction tests

Three different compaction tests are specified in BS 1377 and these are briefly described below.

The 2.5 kg rammer method

An air dried representative sample of the soil under test is passed through a 20 mm sieve and 5 kg is collected. This soil is then thoroughly mixed with enough water to give a fairly low value of water content. For sands and gravelly soils the commencing value of w should be about 5 per cent but for cohesive soils it should be about 8 to 10 per cent less than the plastic limit of the soil tested.

The soil is then compacted in a metal mould of internal diameter 105 mm using a 2.5 kg rammer, of 50 mm diameter, free falling from 300 mm above the top of the soil: see Fig. 11.1. Compaction is effected in three layers, of approximately equal depth. Each layer is given 27 blows, which are spread evenly over the surface of the soil.

The compaction can be considered as satisfactory when the compacted soil is not more than 6 mm above the top of the mould. (Otherwise the test results become inaccurate and should be discarded.) The top of the compacted soil is trimmed level with the top of the mould. The base of the mould is removed and the mould and the test sample it encloses are weighed.

Samples for water content determination are then taken from the top and the base of the soil sample, the rest of the soil being removed from the mould and broken down

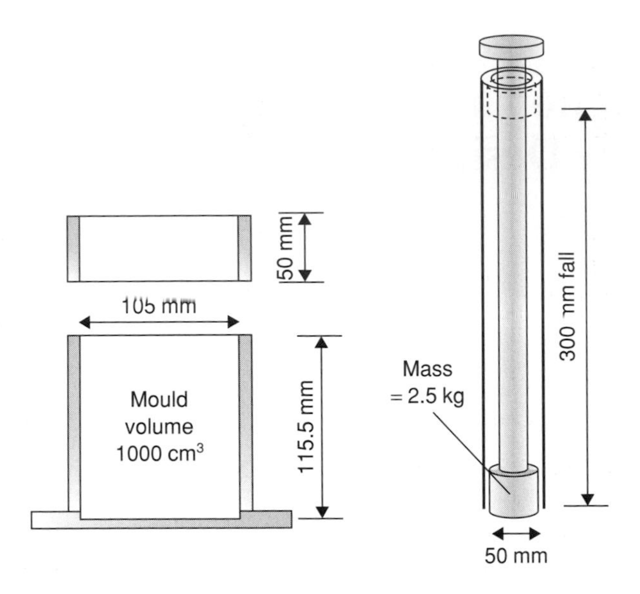

Fig. 11.1 Equipment for the 2.5 kg rammer compaction test.

and mixed with the remainder of the original sample that passed the 20 mm sieve. A suitable increment of water (to give some 2 per cent increase in water content) is thoroughly mixed into the soil and the compaction is repeated. The test should involve not less then five sets of compaction but it is usually continued until the weight of the wet soil in the mould passes some maximum value and begins to decrease.

Eventually, when the test has been completed, the values of water content corresponding to each volume of compacted soil are determined and it becomes possible to plot the dry density to moisture content relationship.

The 4.5 kg rammer method

In this compaction test the mould and the amount of dry soil used are the same as for the 2.5 kg rammer method but a heavier compactive effort is applied to the test sample. The rammer has a mass of 4.5 kg with a free fall of 450 mm above the surface of the soil. The number of blows per layer remains the same, 27, but the number of layers compacted is increased to five.

The vibrating hammer method

It is possible to obtain the dry density/water content relationship for a granular soil with the use of a heavy electric vibrating hammer, such as the Kango. A suitable hammer, according to BS 1377, would have a frequency between 25 and 45 Hz, and a power consumption of 600 to 750 W: it should be in good condition and have been correctly maintained. The hammer is fitted with a special tamper (see Fig. 11.2) and for gravels and sands is considered to give more reliable results than the dynamic compaction techniques just described.

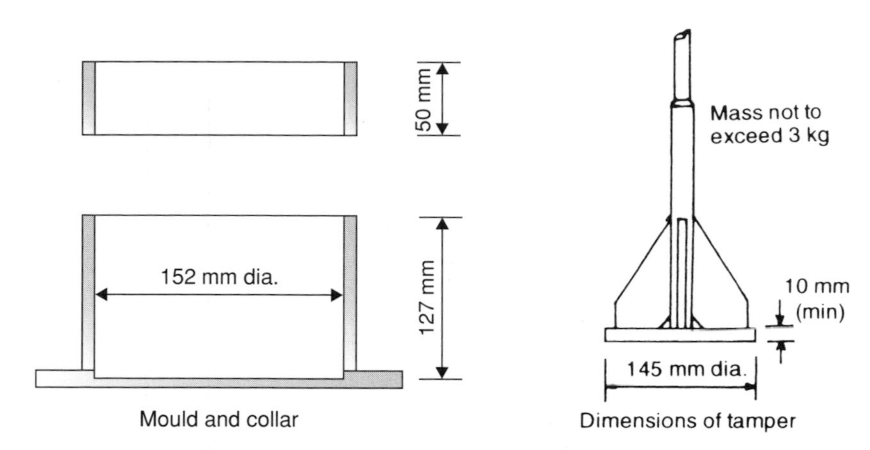

Fig. 11.2 The vibrating hammer compaction test.

The British Standard vibrating hammer test is carried out on soil in the 152 mm diameter mould, with a mould volume of 2305 cm^3, the soil having passed the 37.5 mm sieve.

The soil is mixed with water, as for any compaction test, and is compacted in the mould in three approximately equal layers by pushing the tamper firmly down on to the soil and operating the hammer for 60 seconds, per layer.

The vibrating hammer method should only be used for fine-grained granular soils and for the fraction of medium and coarse grained granular soils passing the 37.5 mm sieve. For highly permeable soils, such as clean gravels and uniformly graded coarse sands, compaction by the vibrating hammer usually gives more dependable results than compaction by either the 2.5 kg rammer or the 4.5 kg rammer.

11.1.2 Soils susceptible to crushing during compaction

The procedure for each of the three compaction tests just described is based on the assumption that the soils tested are not susceptible to crushing during compaction. Because of this, each newly compacted specimen can be broken out of the mould and mixed with the remaining soil for the next compaction.

This technique cannot be used for soils that crush when compacted. With these soils, at least five separate 2.5 kg air dried samples of soil are prepared at different water contents and each sample is compacted, once only, and then discarded.

Preparation of the soil at different water contents

It is important, when water is added to a soil sample, that it is mixed thoroughly to give a uniform dispersion. Inadequate mixing can lead to varying test results and some form of mechanical mixer should be used. Adequate mixing is particularly important with cohesive soils and with highly plastic soils it may be necessary to place the mixed sample in an air tight container for at least 16 hours, in order to allow the moisture to migrate throughout the soil.

11.1.3 Determination of the dry density–moisture content relationship

For each of the three compaction tests described the following readings must be obtained for each compaction:

M_1 = Mass of mould
M_2 = Mass of mould + soil
w = moisture content (as a decimal)

The bulk density and the dry density values for each compaction can now be obtained:

$$\rho_b = \frac{M_2 - M_1}{1000}\ \text{Mg/m}^3 \qquad \text{(for the 2.5 and 4.5 kg rammers)}$$

$$= \frac{M_2 - M_1}{2305}\ \text{Mg/m}^3 \qquad \text{(for the vibrating hammer test)}$$

$$\rho_d = \frac{\rho_b}{1 + w}\ \text{Mg/m}^3$$

When the values of dry density and moisture content are plotted the resulting curve has a peak value of dry density. The corresponding moisture content is known as the optimum moisture content (omc). The reason for this is that at low w values the soil is stiff and difficult to compact, resulting in a low dry density with a high void ratio; as w is increased the water lubricates the soil, increasing the workability and producing high dry density and low void ratio, but beyond omc pore water pressures begin to develop and the water tends to keep the soil particles apart resulting in low dry densities and high void ratios.

With all soils an increase in the compactive effort results in an increase in the maximum dry density and a decrease in the optimum moisture content (Fig. 11.3).

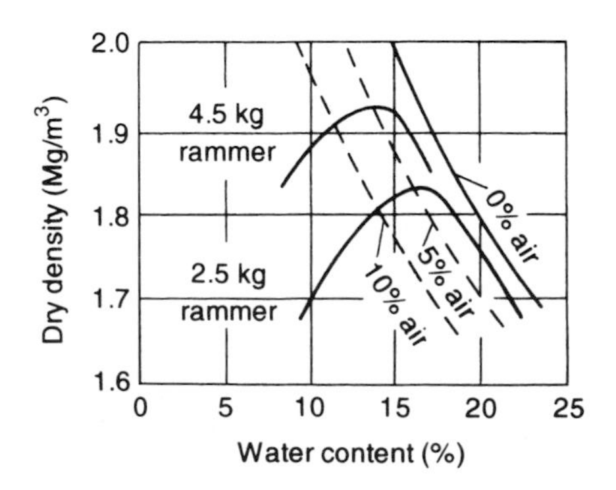

Fig. 11.3 Typical compaction test results.

11.1.4 Percentage air voids, V_a

Saturation line

Figure 11.3 illustrates the saturation line, or zero air voids line as it is often called. It represents the dry densities that would be obtained if all the air in the soil could be expelled, so that after compaction the sample became fully saturated; this state is impossible to achieve by compaction either in the laboratory or in the field, but with the compactive efforts now available it is quite common for a soil to have as little as 5 per cent air voids after compaction.

BS 1377 expresses the percentage air voids as the volume of air in the soil expressed as a percentage of the total volume, rather than as a percentage of the void volume. Hence 5 per cent air voids does not mean the same as 95 per cent degree of saturation.

Air voids line

Just as for the saturation line, it is possible to draw a line showing the dry density to water content relationship for a particular air voids percentage. (See Fig. 11.3.)

$$V_a = \frac{\text{Air volume}}{\text{Total volume}} = \frac{V_v - V_w}{V_v + V_s} = \frac{1 - S_r}{1 + 1/e} \qquad \text{(dividing by } V_v\text{)}$$

hence:

$$V_a = n(1 - S_r) \quad \text{or} \quad S_r = 1 - \frac{V_a}{n}$$

In a partially saturated soil:

$$e = \frac{wG_s}{S_r}$$

and, substituting in the above expression for S_r:

$$e = \frac{wG_s + V_a}{1 - V_a}$$

and substituting in the expression

$$\rho_d = \frac{\rho_w G_s}{1 + e}$$

$$\rho_d = \frac{\rho_w G_s}{1 + \dfrac{wG_s + V_a}{1 - V_a}}$$

which eventually leads to:

$$\rho_d = \rho_w \frac{1 - V_a}{\frac{1}{G_s} + w}$$

It is seen that by putting $V_a = 0$ in the above expression we obtain the relationship between dry density and water content for zero air voids, i.e. the saturation line.

11.1.5 Correction for gravel content

As noted, the standard compaction test is carried out on soil that has passed the 20 mm sieve. Any coarse gravel or cobbles in the soil is removed.

Provided that the percentage of excluded material, X, does not exceed 5 per cent there will be little effect on the derived maximum dry density and optimum moisture content.

If X is much greater than 5 per cent then the test values will be affected and may be quite different to those pertaining to the natural soil. BS 1377 admits that there is no generally accepted method to allow for this problem and suggests that for X up to 25 per cent some form of correction to allow for the removal of the gravel can be applied. A suggested method is set out below.

The percentage of material retained on the 20 mm sieve, X, can be obtained either by weighing the amount on the sieve (if the soil is oven dried) or else from the particle size distribution curve.

In the compacted soil this percentage of coarse gravel and cobbles has been replaced by an equal mass of soil of a smaller size. Generally the particle specific gravity of the gravel will be greater than that of the soil of smaller size so the volume of the gravel excluded would occupy a smaller volume than that of the soil which replaced it.

If ρ_d = maximum dry density obtained from the test then:

$$\text{Mass of gravel excluded} = \frac{X}{100}\rho_d$$

Considering unit volume, the volume of soil that replaced the gravel = X, so:

$$\text{Volume of gravel omitted} = \frac{X}{100}\frac{\rho_d}{\rho_w G_g}$$

where G_g = particle specific gravity of excluded gravel.

$$\Rightarrow \quad \text{Difference in volume} = \frac{X}{100} - \frac{X}{100}\frac{\rho_d}{\rho_w G_g} = \frac{X}{100}\left(1 - \frac{\rho_d}{\rho_w G_g}\right)$$

$$\text{Corrected maximum dry density} = \frac{\rho_d}{1 - \frac{X}{100}\left(1 - \frac{\rho_d}{\rho_w G_g}\right)}$$

$$= \frac{\rho_d}{1 + \frac{X}{100}\left(\frac{\rho_d}{\rho_w} G_g - 1\right)}$$

Similarly:

$$\text{Corrected optimum moisture content} = \frac{100 - X}{100} w$$

where w = the optimum moisture content obtained from the test.

Even when a gravel correction is applied compaction test results are not representative for a soil with X much greater than 25 per cent. BS 1377 mentions that in such cases the CBR mould (a mould 152 mm in diameter and 127 mm high) can be used but gives little guidance as to procedure. General practice has been to pass the soil through a 37.5 mm sieve before test. The compaction procedure is similar to that for the standard test except that the soil is compacted in the bigger mould in three equal layers with the same 2.5 kg rammer falling 300 mm but the number of blows per layer is increased to 62.

Correction for the excluded gravel (i.e. the particles greater than 37.5 mm) can be carried out in the manner proposed for the 20 mm size.

Example 11.1

The following results were obtained from a compaction test using the 2.5 kg rammer.

Mass of mould + wet soil (g)	2783	3057	3224	3281	3250	3196
Moisture content (per cent)	8.1	9.9	12.0	14.3	16.1	18.2

The weight of the compaction mould, less its collar and base, was 1130 g and the soil had a particle specific gravity of 2.70.

Plot the curve of dry density against moisture content and determine the optimum moisture content. On your diagram plot the lines for 5 per cent and 0 per cent air voids.

Solution

The calculations are best tabulated:

w (per cent)	Mass of mould + wet soil (g)	Mass of wet soil $= M_1$ (g)	Mass of dry soil, $M_2 = M_1/(1 + w)$ (g)	Dry density $\rho_d = M_2/1000$ (Mg/m^3)
8.1	2783	1653	1529	1.53
9.9	3057	1927	1753	1.75
12.0	3224	2094	1870	1.87
14.3	3281	2151	1882	1.88
16.1	3250	2120	1826	1.83
18.2	3196	2066	1748	1.75

The relationship between ρ_d and V_a is given by the expression:

$$\rho_d = \rho_w \frac{1 - V_a}{\dfrac{1}{G_s} + w}$$

Values for the 0 per cent and 5 per cent air voids lines can be obtained by substituting $V_a = 0$ and $V_a = 0.05$ in the above formula along with different values for w.

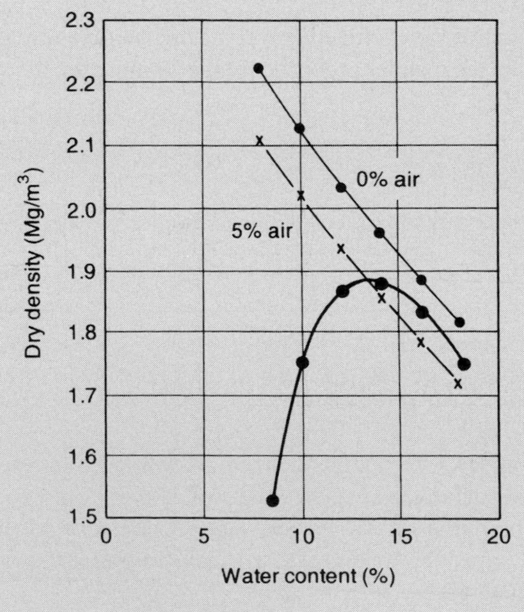

Fig. 11.4 Example 11.1.

w (per cent)	8	10	12	14	16	18
			ρ_d (Mg/m^3)			
$V_a = 0$	2.22	2.13	2.04	1.96	1.89	1.81
$V_a = 0.05$	2.11	2.02	1.94	1.86	1.79	1.73

The complete plot is shown in Fig. 11.4.

Example 11.2

A 2.5 kg rammer compaction test on a soil sample that had been passed through a 20 mm sieve gave a maximum dry density of 1.91 Mg/m^3 and an optimum moisture content of 13.7 per cent. If the percentage mass of the soil retained on the sieve was 20, determine more correct values for $\rho_{d_{max}}$ and the omc. The particle specific gravity of the retained particles was 2.78.

Solution

$$\rho_{d_{max}} = \frac{\rho_d}{1 + \dfrac{X}{100}\left(\dfrac{\rho_d}{\rho_w G_s} - 1\right)}$$

$$= \frac{1.91}{1 + 0.2\left(\dfrac{1.91}{1.0 \times 2.78} - 1\right)}$$

$$= 2.04 \text{ Mg/m}^3$$

$$\text{omc} = \frac{100 - X}{100} w = \frac{80}{100} \times 0.137 = 11 \text{ per cent}$$

11.2 Main types of compaction plant

The three main types of compaction equipment are:

(i) rollers;
(ii) rammers, and
(iii) vibrators.

Rollers are by far the most common.

The compaction plant described below is categorised in terms of total mass. The static mass per metre width of roll is the total mass on the roll divided by the width of the roll. Where a roller has more than one axle the machine's category is determined using the roller which gives the greatest value of mass per unit width.

Smooth wheeled roller

Probably the most commonly used roller in the world is the smooth wheeled roller. It consists of hollow steel drums so that its weight distribution can be altered by the addition of ballast (sand or water) to the rolls. These rollers vary in mass/m width from some 2100 kg to over 54 000 kg and are suitable for the compaction of most soils except uniform and silty sands. They are usually self-propelled by diesel engine although towable units are also available.

The successful operation of smooth wheeled rollers is difficult (and often impossible) when site conditions are wet, and in these circumstances rollers that can be towed by either track-laying or wheeled tractors are used. Both dead weight and vibratory units are available commercially.

Vibratory roller

The smooth wheeled vibratory roller is either self propelled or towed and has a mass/m width value from 270 kg to over 5000 kg. The compactive effort is raised by vibration, generally in the form of a rotating shaft (powered by the propulsion unit) that carries out-of-balance weights. Tests have shown that the best results on both heavy clays and granular soils are obtained when the frequency of vibration is in the range 2200–2400 cycles per minute. Vibration is obviously more effective in granular soil but tests carried out by the Road Research Laboratory (1952) showed that, in a cohesive silty clay, the effect of vibration on a 200 mm layer doubled the compactive effort.

A vibrating roller operating without vibration is regarded as being a smooth wheeled roller.

Pneumatic-tyred roller

In its usual form the pneumatic-tyred roller is a container or platform mounted between two axles, the rear axle generally having three wheels and the forward axle two (so arranged that they track in with the rear wheels), although some models have five wheels at the back and four at the front. The dead load is supplied by weights placed in the container to give a mass per wheel range from 1000 kg to over 12 000 kg. The mass per wheel is simply the total mass of the unit divided by the number of wheels. A certain amount of vertical movement of the wheels is provided for so that the roller can exert a steady pressure on uneven ground – a useful facility in the initial stages of a fill.

This type of roller originated as a towed unit but is now also produced in a self-propelled form; it is suitable for most types of soil and has particular advantages on wet cohesive materials.

Sheepsfoot roller

This roller consists of a hollow steel drum from which the feet project, dead weight being provided by placing water or wet sand inside the drum. It is generally used as a towed assembly (although self-propelled units are available), with the drums mounted either singly or in pairs.

The feet are usually either club-shaped (100 × 75 mm) or tapered (57 × 57 mm), the number on a 5000 kg roller varying between 64 and 88. Variations in the shape of the feet have been tried in America with a view to increasing the operating speed.

The sheepsfoot roller is only satisfactory on cohesive soils, but at low moisture contents the resulting compaction of such soils is probably better than can be obtained with other forms of plant. Their use in the UK is rather infrequent because of the generally wet conditions.

Wedge foot compactors

These four driven roller vehicles look somewhat like a tractor without its tracks. They are excellent for compacting soft clays and are often used on dam cores. The feet of these rollers have a minimum end area of 0.1 m^2 and the total area of the feet exceeds 15 per cent of the area of the cylinder swept by the feet.

The grid roller

This is a towed unit consisting of rolls made up from 38 mm diameter steel bars at 130 mm centres, giving spaces of 90 mm square. The usual mass of the roller is about 5500 kg which can be increased to around 11 000 kg by the addition of dead weights; there are generally two rolls, but a third can be added to give greater coverage. The grid roller is suitable for many soil types, but wet clays tend to adhere to the grid and convert it into a form of smooth roller.

Rammers and vibrators

Manually controlled power rammers can be used for all soil types and are useful when rolling is impractical due to restricted site conditions. Vibrating plates produce high dry densities at low moisture content in sand and gravels and are particularly useful when other plant cannot be used.

11.3 Moisture content value for *in situ* compaction

With the main types of compaction plant the optimum moisture contents of many soils are fairly close to their natural moisture contents and in such cases compaction can be carried out without variation of water content.

In Britain, the *Design manual for roads and bridges* (Highways Agency, 1995) (DMRB) is used as the basis for highway design and is generally referred to in parallel with the *Specification for highway works* (Highways Agency, 2004a) (SHW). The DMRB emphasises the importance of placing and compacting soils whose natural

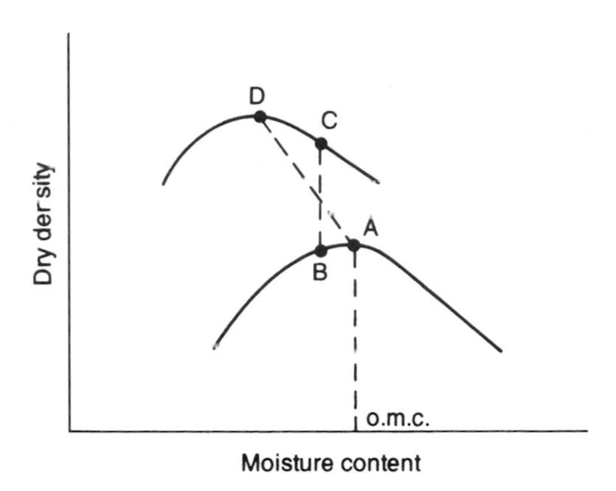

Fig. 11.5 Overcompaction.

moisture contents render them in an *acceptable* state. Soils in an acceptable state will generally achieve the degree of compaction required and will be able to be excavated, transported and placed satisfactorily. The DMRB gives guidance on the limits of acceptability for different types of soil, and the methods of test by which acceptability should be assessed for each. In both the SHW and DMRB different types of soil (e.g. cohesive, granular) together with their anticipated use (e.g. general fill, landscape fill) are identified apart by a *Class* type. The DMRB recommends different types of test (e.g. moisture content, compaction, MCV (see Section 11.6.8)) and appropriate limits of acceptability for each class, and if the soil is placed at a moisture content within the limits, a satisfactory degree of compaction should be achieved.

Overcompaction

Care should be taken to ensure that the compactive effort in the field does not place the soil into the range beyond optimum moisture content. In Fig. 11.5 the point A represents the maximum dry density corresponding to the optimum moisture content. If the soil being compacted has a moisture content just below the optimum value the dry density attained will be point B, but if compaction is continued after this stage the optimum moisture content decreases (to point D) and the soil will reach the density shown by point C. Although the dry density is higher, point C is well past the optimum value and the soil will therefore be much softer than if compaction had been stopped once point B had been reached.

11.4 Specification of the field compacted density

11.4.1 Compactive effort in the field

The amount of compactive effort delivered to a point in a soil during compaction depends upon both the mass of the compacting unit and the number of times that it runs over the point (i.e. the number of passes). Obviously the greater the number of passes the greater the compactive effort but, as discussed in the previous section, this greater compactive effort will not necessarily achieve a higher dry density. The number of passes must correspond to the compactive effort required for maximum dry density when the actual water content of the soil is the optimum moisture content and is usually somewhere between 3 and 10. Specification by *method compaction* is when the contractor is instructed to use a particular compaction machine and a fixed number of passes.

11.4.2 Relative compaction

Laboratory compaction tests use different compactive efforts from those of many of the big machines so results from these tests cannot always be used directly to predict the maximum dry density values that will be achieved in the field. What can be used is the *relative compaction*, which is the percentage ratio of the *in situ* maximum dry density of the compacted fill material to the maximum dry density obtained with the relevant laboratory compaction test. Specification by *end product compaction* is when the contractor is directed to achieve a certain minimum value relative compaction and is allowed to select his own plant.

11.4.3 Method compaction

The Highways Agency Specification (2004a), gives two tables (6.1 and 6.4) from which, knowing the classification characteristics of the soil, it is possible to decide upon the most suitable compaction machine together with the number of passes that it must use. With this information a specification by method compaction can be prepared. The tables are too large to reproduce here.

11.4.4 End-product compaction

The values of the maximum dry density and the optimum moisture content are obtained using the 2.5 kg rammer or the vibrating hammer test, depending upon which is more relevant to the expected field compaction. The value of the required relative compaction should be equal to or greater than the value quoted in Table 6.1 of the Highways Agency Specification (2004a), usually 90 to 95 per cent.

11.4.5 Air voids percentage

Another form of end-product specification is to instruct that a certain minimum value of air voids percentage is to be obtained in the compacted soil. This value of V_a was for many years taken as between 5 and 10 per cent, but more recent research has suggested that a value of less than 5 per cent is required to remove the risk of collapse on inundation sometime after construction (Trenter and Charles, 1996, Charles *et al.*, 1998).

11.5 *In situ* tests carried out during earthwork construction

Tests for bulk density and moisture content must be carried out at regular intervals if proper control of the compaction is to be achieved. American practice, now widely used in the UK, is to take at least four density tests per 8 hour shift with a minimum of one test for each 400 m^3 of earthwork compacted.

11.5.1 Bulk density determination

Core-cutter method

Details of the core-cutter apparatus, which is suitable for cohesive soils, are given in Fig. 11.6. After the cutter has been first pressed into the soil and then dug out, the soil is trimmed to the size of the cutter and both cutter and soil are weighed; given the weight and dimensions of the cutter, the bulk density of the soil can be obtained.

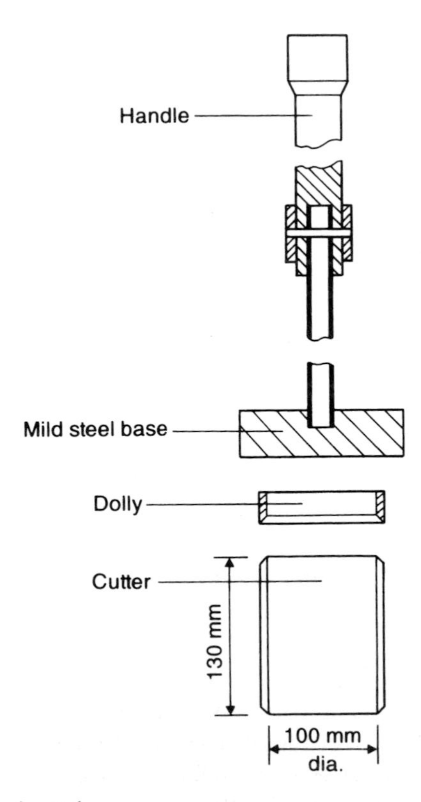

Fig. 11.6 Core cutter for clay soil.

Sand replacement method

For granular soils the apparatus shown in Fig. 11.7 is used. A small round hole (about 100 mm diameter and 150 mm deep) is dug and the mass of the excavated material is carefully determined. The volume of the hole thus formed is obtained by pouring into it sand of known density from a special graduated container; given the weight of sand in the container before and after the test, the weight of sand in the hole and hence the volume of the hole can be determined.

The apparatus shown in Fig. 11.7 is suitable for fine to medium grained soils and is known as the small pouring cylinder method.

For coarse grained soils a larger pouring cylinder is used. This cylinder has an internal diameter of 215 mm and a height of 170 mm to the valve or shutter. The excavated hole in this case should be about 200 mm in diameter and some 250 mm deep. This larger pouring cylinder can also be used for fine to medium grained soils.

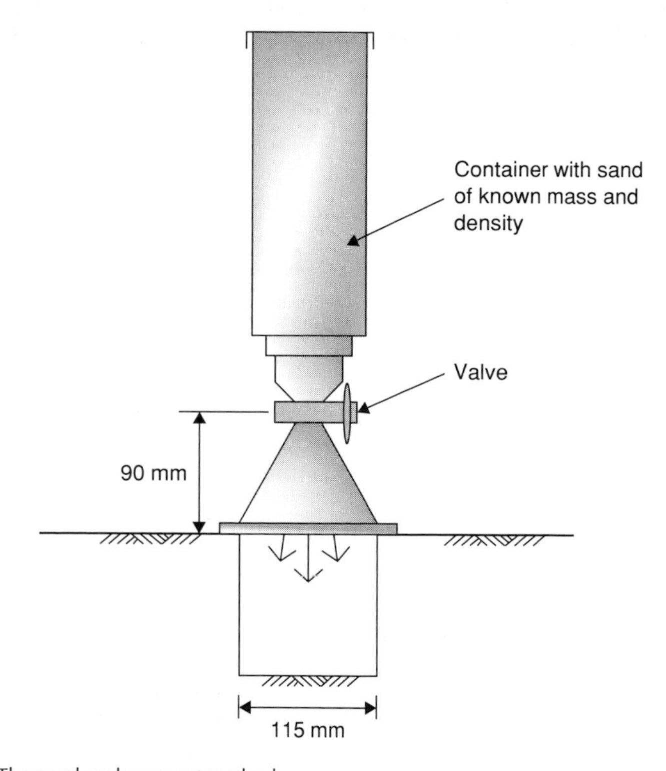

Fig. 11.7 The sand replacement method.

The penetration needle

This apparatus can be used for spot checks on the bulk density of cohesive soils and consists basically of a needle attached to a spring-loaded plunger, an array of interchangeable needle tips being available ranging from 6.45 to 645 mm^2 according to the type of soil tested. A calibration of penetration against density is obtained by pushing the needle into specially prepared samples at different densities and noting the penetration.

Nuclear radiation

An instrument that can measure both the density of a soil and its water content was described by Meigh and Skipp in 1960. It consists of an aluminium probe which is pushed into the soil. The neutrons emitted from the source lose their energy by collision and, since the number of slow neutrons thus produced depends upon the amount of water present, measurements of the water content, the bulk density and the dry density of the soil can be obtained. Since 1960 the apparatus has been considerably improved and, with the degree of sophistication now available, the apparatus gives much more rapid and dependable results than those obtained from the density tests. However, it should be noted that some local authorities will not permit the use of apparatus that uses a nuclear source.

11.5.2 Moisture content determination

Quick moisture content determinations are essential if compaction work is to proceed smoothly, and the Highways Agency *Manual of contract documents for highway works, MCHW 2* (2004b) permits the use of quicker drying techniques than are specified in BS 1377. The most common of these tests are described below. Nevertheless it is standard practice to calibrate these results occasionally by collecting suitable soil samples and determining their water content values accurately by oven drying.

Time domain reflectometry (TDR)

The TDR method for the measurement of soil water content is relatively new and an early review of the method was given by Topp and Davis (1985). The technique involves the determination of the propagation velocity of an electromagnetic pulse sent down a fork-like probe installed in the soil (Fig. 11.8). The velocity is determined by measuring the time taken for the pulse to travel down the probe and be

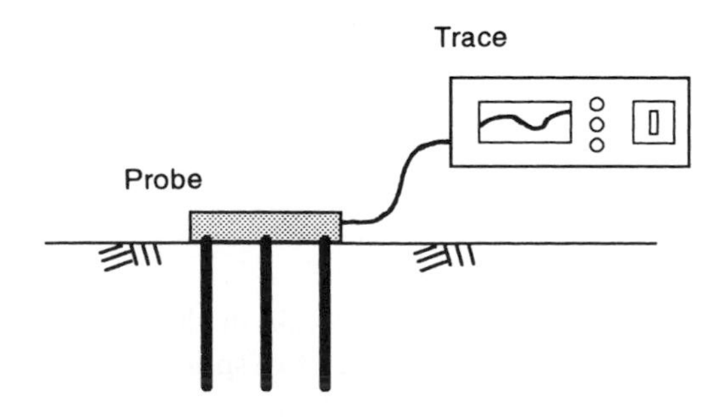

Fig. 11.8 Time domain reflectometry probe.

reflected back from its end. The propagation velocity depends on the dielectric constant of the material in contact with the probe (i.e. the soil). The dielectric constant of free water is 80 and that of a soil matrix is typically 3 to 6. Hence as the moisture content of the soil changes, there is a measurable change in the dielectric constant of the system, which affects the velocity of the pulse. Therefore, by measuring the time taken by the pulse we can establish the moisture content of the soil around the probe.

Soil moisture capacitance probe

Descriptions of the capacitance technique for determining soil moisture content are given by Dean *et al.* (1987) and Bell *et al.* (1987). The probe utilises the same principle as TDR, i.e. that the capacitance of a material depends on its dielectric constant. The probe is used inside a PVC access tube installed vertically into the soil. Indirect measurements of the capacitance of the surrounding soil are determined at the required depth and these are translated into soil moisture content using a simple formula. A description of the use of a capacitance probe for measuring soil moisture content is given by Smith *et al.* (1997).

Neutron probe

Nuclear moisture gauges have been discussed in the previous section. With the introduction of the TDR and capacitance techniques just described, they are becoming less common as the alternatives offer much greater freedom with respect to health and safety legislation aspects.

Speedy moisture tester

A known mass of the wet soil is placed in a pressurised container and a quantity of calcium carbide is added, the chamber then being sealed and the two materials brought into contact by shaking. The reaction of the carbide on the water in the soil produces acetylene gas, the amount (and hence the pressure) of which depends upon the quantity of water in the soil. A pressure gauge, calibrated for moisture content, is fitted at the base of the cylinder and from this the moisture content can be read off as soon as the needle records a steady level. A characteristic of the apparatus that must be corrected for is that the value of moisture content obtained is in terms of the wet weight of soil.

11.6 Highway design

Highway design in the UK is based on the *Design manual for roads and bridges* (Highways Agency, 1995, 1996).

The basic components of a road

Both a road and a runway consist of two basic parts, the pavement and the subgrade.

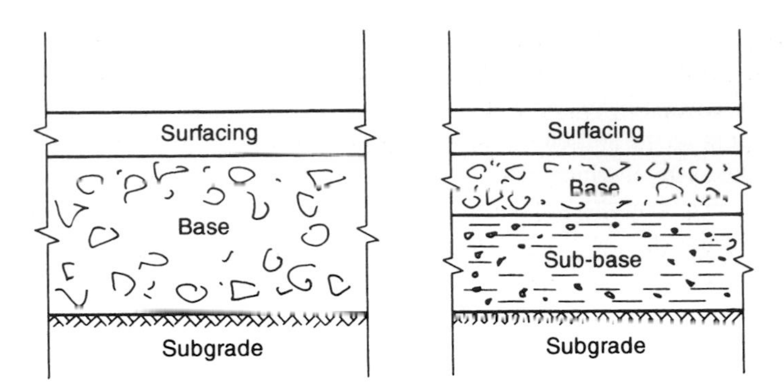

Fig. 11.9 Pavement construction.

- *Pavement*: distributes wheel loads over an area so that the bearing capacity of the subgrade is not exceeded. It usually consists of two or more layers of material: a top layer or wearing surface which is durable and waterproof, and a base material. For economical reasons the base material is sometimes split into two layers, a base and a sub-base (Fig. 11.9).
- *Subgrade*: the natural soil upon which the pavement is laid. The subgrade is seldom strong enough to carry a wheel load directly. There are two possibilities:

  (a) improve the strength of the subgrade and thereby reduce the required pavement thickness;
  (b) design and construct a sufficiently thick pavement to suit the subgrade.

Types of pavement

- *Flexible*: lean concrete bases, cement-bound granular bases, tar or bitumen-bound macadam, all overlaid with bituminous surfaces.
- *Rigid*: reinforced concrete base and surface.

Choice of pavement depends largely upon the local economic considerations.

11.6.1 Assessment of subgrade strength and stiffness

The strength of the subgrade is the main factor in determining the thickness of the pavement although its susceptibility to frost must also be considered.

The value of the stiffness modulus of the subgrade is required if the stresses and strains in the pavement and the subgrade are to be calculated.

The California Bearing Ratio test

Subgrade strength is expressed in terms of its California Bearing Ratio (CBR) value. The CBR value is measured by an empirical test devised by the California State

Highway Association and is simply the resistance to a penetration of 2.54 mm of a standard cylindrical plunger of 49.6 mm diameter, expressed as a percentage of the known resistance of the plunger to various penetrations in crushed aggregate, notably 13.24 kN at 2.5 mm penetration and 19.96 kN at 5.0 mm penetration.

The reason for the odd dimension of 49.6 mm for the diameter of the plunger is that the test was originally devised in the USA and used a plunger with a cross-sectional area of 3.0 square inches. This area translates as 1935 mm^2 and, as the test is international, it is impossible to vary this area and therefore the plunger has a diameter of 49.6 mm.

Laboratory CBR test

The laboratory CBR test is generally carried out on remoulded samples of the subgrade, and is described in BS 1377. The usual form of the apparatus is illustrated in Fig. 11.10. The sample must be compacted to the expected field dry density at the appropriate water content. The appropriate water content is the *in situ* value used for the field compaction. However if the final strength of the subgrade is required a further CBR test must be carried out with the soil at the same dry density but with the water content adjusted to the value that will eventually be reached in the subgrade after construction. This value of water content is called the *equilibrium moisture content.*

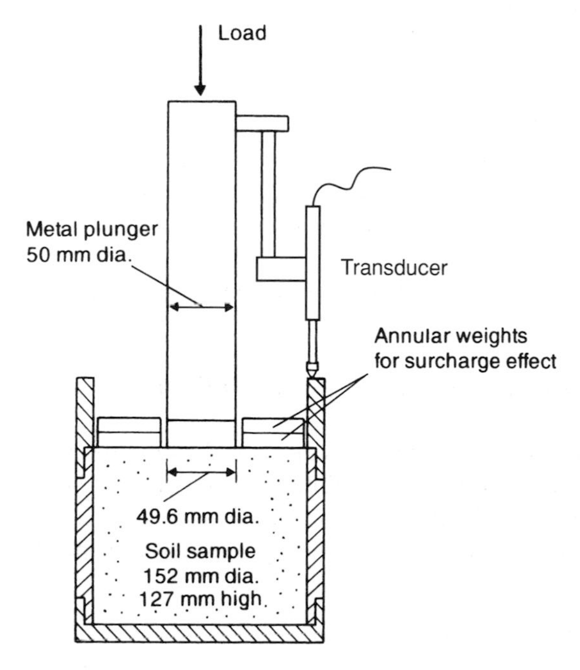

Fig. 11.10 The California bearing ratio test. The dial gauge may be replaced by a displacement transducer.

The dry density value can only be truly determined from full-sized field tests using the compaction equipment that will eventually be used for the road construction. Where this is impracticable, the dry density can be taken as that corresponding to 5 per cent air voids at a moisture content corresponding to the omc of the standard compaction test. In some soils this will not be satisfactory – it is impossible to give a general rule; for example, in silts, a spongy condition may well be achieved if a compaction to 5 per cent air voids is attempted. This state of the soil would not be allowed to happen *in situ* and the laboratory tester must therefore increase the air voids percentage until the condition disappears.

With its collar, the mould into which the soil is placed has a diameter of 152 mm and a depth of 177 mm. The soil is broken down, passed through a 20 mm sieve and adjusted to have the appropriate moisture content. The final compacted sample has dimensions 152 mm diameter and 127 mm height.

(i) *Static compaction*: sufficient wet soil to fill the mould when compacted is weighed out and placed in the mould. The soil is now compressed into the mould in a compression machine to the required height dimension. This is the most satisfactory method.
(ii) *Dynamic compaction*: the weighed wet soil is compacted into the mould in five layers using either the BS 2.5 kg rammer or the 4.5 kg rammer. The number of blows for each layer is determined by experience, several trial runs being necessary before the amount of compactive effort required to leave the soil less than 6 mm proud of the mould top is determined. The soil is now trimmed and the mould weighed so that the density can be determined.

After compaction the plunger is first seated into the top of the sample under a specific load: 50 N for CBR values up to 30 per cent and 250 N for soils with CBR values above 30 per cent.

The plunger is made to penetrate into the soil at the rate of 1.0 mm/minute and the plunger load is recorded for each 0.25 mm penetration up to a maximum of 7.5 mm.

Test results are plotted in the form of a load–penetration diagram by drawing a curve through the experimental points. Usually the curve will be convex upwards (Curve Test 1 in Fig. 11.11), but sometimes the initial part of the curve is concave upwards and, over this section, a correction becomes necessary. The correction consists of drawing a tangent to the curve at its steepest slope and producing it back to cut the penetration axis. This point is regarded as the origin of the penetration scale for the corrected curve.

The plunger resistance at 2.5 mm is expressed as a percentage of 13.24 kN and the plunger resistance at 5.00 mm is expressed as a percentage of 19.96 kN. The higher of these two percentages is taken as the CBR value of the soil tested.

Surcharge effect

When a subgrade or a sub-base material is to be tested the increase in strength due to the road construction material placed above can be allowed for. Surcharge weights, in the form of annular discs with a mass of 2 kg can be placed on top of the soil test

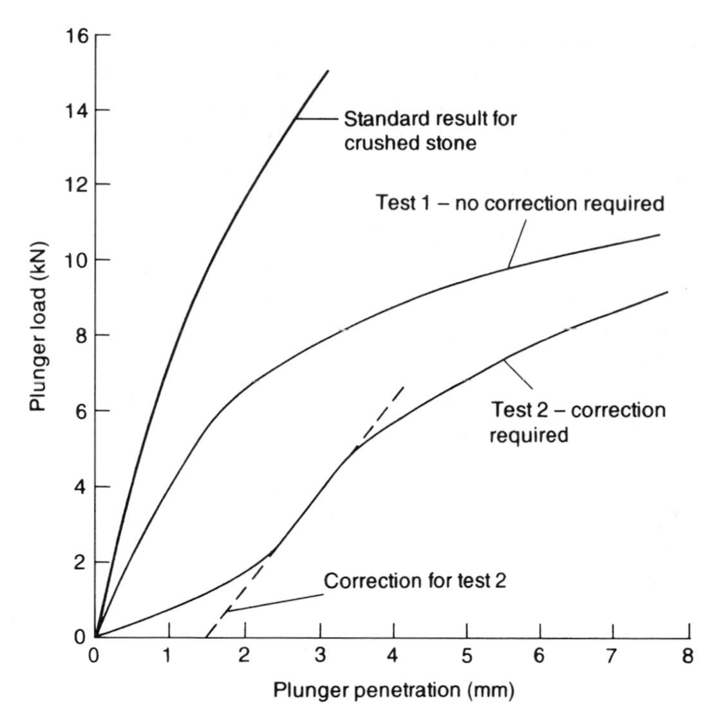

Fig. 11.11 Typical CBR results.

sample. The plunger penetrates through a hole in the disc to reach the soil. Each 2 kg disc is roughly equivalent to 75 mm of surcharge material.

In situ *CBR test*

When an existing road is to be reconstructed the value of the subgrade strength must be determined. For such a situation the CBR test can be carried out *in situ* but it must be remembered that the water content of the soil should be the equilibrium value so that the test should be carried out in a newly excavated pit, not less than 1 m deep, on the freshly exposed soil.

There are alternatives to the CBR test. Black (1979) has shown that *in situ* subgrade strengths can be measured by cone penetrometer tests, carried out in boreholes. Also the equilibrium subgrade strength of the most common cohesive soils in Britain can be obtained by a soil suction method described by Black and Lister (1979).

Estimation of CBR values

Table C1 in the Transport and Road Research Laboratory Report LR 1132 (Powell *et al.*, 1984) gives estimates of equilibrium CBR values for most common soils in Britain for various conditions of construction, groundwater and pavement thickness. The TRRL table is reproduced here as Table 11.1.

Table 11.1 Equilibrium suction-index CBR values. Reproduced from TRRL Report LR1132 (1984).

Type of soil	Plasticity index	High water table						Low water table					
		Construction conditions:						Construction conditions:					
		Poor		Average		Good		Poor		Average		Good	
		Thin	Thick	Thin	Thick	Thin	Thick	Thin	Thick	Thin	Thick	Thin	Thick
Heavy clay	70	1.5	2	2	2	2	2	1.5	2	2	2	2	2.5
	60	1.5	2	2	2	2	2.5	1.5	2	2	2	2	2.5
	50	1.5	2	2	2.5	2	2.5	2	2	2	2.5	2	2.5
	40	2	2.5	2.5	3	2.5	3	2.5	2.5	3	3	3	3.5
Silty clay	30	2.5	3.5	3	4	3.5	5	3	3.5	4	4	4	6
Sandy clay	20	2.5	4	4	5	4.5	7	3	4	5	6	6	8
	10	1.5	3.5	3	6	3.5	7	2.5	4	4.5	7	6	>8
Silt*	–	1	1	1	1	2	2	1	1	2	2	2	2
Sand (poorly graded)	–	← 20 →											
Sand (well graded)	–	← 40 →											
Sandy gravel (well graded)	–	← 60 →											

* estimated assuming some probability of material saturating.

A high water table is taken as one that is 300 mm below foundation level whilst a low water table is one that is 1000 mm below foundation level. For water tables at depths between these limits the CBR value may be found by interpolation.

In Table 11.1 a thick pavement has a total thickness of 1200 mm including any capping layer (used for motorways) and a thin pavement has a total thickness of 300 mm. For pavement thicknesses between these limits the value of the CBR may be interpolated.

11.6.2 Drainage and weather protection

Whenever practical the water table should be prevented from rising above a depth of 300 mm below the foundation level. This is usually achieved by the installation of French drains at a depth of 600 mm.

11.6.3 Capping layer

If the subgrade is weak, a *capping layer* of relatively cheap material can be placed between the subgrade and the sub-base to improve the strength of the structure. The *Design manual for roads and bridges* (Highways Agency, 1995) gives the required thickness of capping and sub-base layers: the thickness of each depends on the CBR of the subgrade. The method given in the manual permits alternative combinations of capping and sub-base layer thickness. However, for subgrades of CBR less than 15 per cent, the manual states that the minimum thickness of aggregate layer (capping or sub-base) shall be 150 mm, and for a subgrade of CBR less than 3 per cent, the first layer of aggregate placed (as capping or sub-base) shall be at least 200 mm thick. Possible measures to be taken for cases where the subgrade is particularly soft (CBR less than 2 per cent) are also given in the manual.

11.6.4 Subgrade stiffness

Once the CBR value of a subgrade has been determined it can be converted to a value of stiffness or elastic modulus. The TRRL Laboratory Report LR1132 (Powell *et al.*, 1984) suggests the following lower bound relationship:

$$E = 17.6(\text{CBR})^{0.64}\text{MPa}$$

where CBR is in per cent.

Example 11.3

A CBR test on a sample of subgrade gave the following results:

Plunger penetration (mm)	0.25	0.5	0.75	1.0	1.25	1.5	1.75	2.0
Plunger load (kN)	1.0	1.6	2.4	3.6	4.5	5.3	6.0	6.8
Plunger penetration (mm)	2.25	2.5	2.75	3.0	3.25	3.5	3.75	4.0
Plunger load (kN)	7.5	8.3	9.0	9.4	10.1	10.7	11.2	11.7
Plunger penetration (mm)	4.25	4.5	4.75	5.0	5.25	5.5	5.75	6.0
Plunger load (kN)	12.2	12.7	13.0	13.5	14.1	14.4	14.6	14.9

The standard force penetration curve, corresponding to 100 per cent CBR has the following values:

Plunger penetration (mm)	2	4	6	8
Plunger load (kN)	11.5	17.6	22.2	26.3

Determine the CBR value of the subgrade.

Solution

The standard penetration curve is shown drawn in Fig. 11.12.

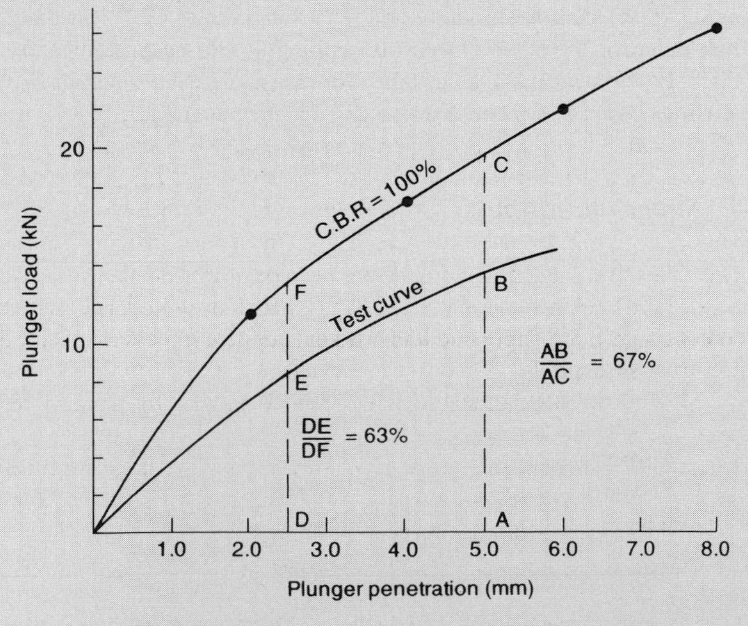

Fig. 11.12 Example 11.3.

The test points are plotted and a smooth curve drawn through them. In this case there is no need for the correction procedure as the curve is concave upwards in its initial stages.

From the test curve it is seen that at 2.5 mm the plunger load is 8.3 kN and at 5.0 mm the penetration is 13.5 kN.

The CBR value is therefore either

$$\frac{8.3}{13.24} \times 100 \quad \text{or} \quad \frac{13.5}{19.96} \times 100$$

whichever is the greater, i.e. 63 per cent or 67 per cent.

The CBR value of the subgrade is 67 per cent, say 65 per cent.

Note CBR values are rounded off as follows:

to nearest 1 per cent for CBR values up to 30 per cent;
to nearest 5 per cent for CBR values between 30 and 100 per cent;
to nearest 10 per cent for CBR values greater than 100 per cent.

11.6.5 Frost susceptibility of subgrades and base materials

Cohesive soils: can be regarded as non-frost-susceptible when I_p is greater than 15 per cent for well drained soils and 20 per cent for poorly drained soils (i.e. water table within 600 mm of formation level).

Non-cohesive soils: except for limestone gravels, may be regarded as non-frost-susceptible if with less than 10 per cent fines.

Limestone gravels are likely to be frost-susceptible if the average saturation moisture content of the limestone aggregate exceeds 2 per cent.

Chalks: all crushed chalks are frost-susceptible. Magnitude of heave increases linearly with the saturation moisture content of the chalk aggregate.

Limestone: all oolitic and magnesian limestones with an average saturation moisture content of 3 per cent or more must be regarded as frost-susceptible.

All hard limestones with less than 2 per cent of average saturation moisture content within the aggregate and with 10 per cent or less fines can be regarded as non-frost-susceptible.

Granites: crushed granites with less than 10 per cent fines can be regarded as non-frost susceptible.

Burnt colliery shales: very liable to frost heave. No relationship is known, so that tests on representative samples are regarded as essential before the material is used in the top 450 mm of the road structure.

Slags: crushed, graded slags are not liable to frost heave if they have less than 10 per cent fines.

Pulverised fuel ash: coarse fuel ashes with less than 40 per cent fines are unlikely to be frost-susceptible.

Fine ashes may be frost-susceptible and tests should be carried out before such materials are used in the top 450 mm.

11.6.6 Traffic assessment

The *Design manual for roads and bridges* (Highways Agency, 1996) describes two methods for the estimation of design traffic for road design: the *standard method* and the *full traffic assessment method*. The standard method is used for the design of new roads and this section considers only this method. In this method, the flow of commercial vehicles (i.e. those greater than 15 kN unladen vehicle weight) is established in order to determine the cumulative design traffic. Commercial vehicles are placed into categories depending on the type of the vehicle (e.g. coach or lorry), the number of axles and the vehicle rigidity. Buses and coaches are placed in the *public service vehicle* (PSV) category, and lorries are placed into one of two *other goods vehicles* categories (OGV1 and OGV2) depending on their size. Traffic flow data should be determined from traffic studies and, from these, the percentage of OGV2 vehicles in the total flow is determined. To establish the cumulative design traffic, use is made of design charts. The charts give the cumulative design traffic (in millions of standard axles) depending on the forecast traffic flow at opening together with the percentage of OGV2 vehicles in the flow.

11.6.7 Design life of a road

An important part in the design of a road is the decision as to the number of years of life for which the road can be economically built. The DMRB suggests a design life of 20 years for bituminous roads, because their life may be extended by a strengthening overlay. With a concrete road a major extension of life involves considerable problems and the DMRB therefore recommends a design life of 40 years.

11.6.8 The moisture condition value, MCV

The selection of a satisfactory soil for earthworks in road construction involves the visual identification of unsuitable soils, the classification tests described in Chapter 1 together with the use of at least one of the three compaction tests described in this chapter. The compaction test chosen is the one that uses a compactive effort nearest to the expected construction compactive effort and is used to determine the optimum moisture content value, i.e. the upper value of water content beyond which the soil becomes unworkable. The system can give good results, particularly with experienced engineers, but there are occasions when the assessment of a particular soil's suitability for earthworks is still difficult.

The moisture condition test is an attempt to remove some of the selection difficulties and was proposed by Parsons in 1976. It is essentially a strength test in which the compactive effort necessary to achieve near full compaction of the test sample is determined.

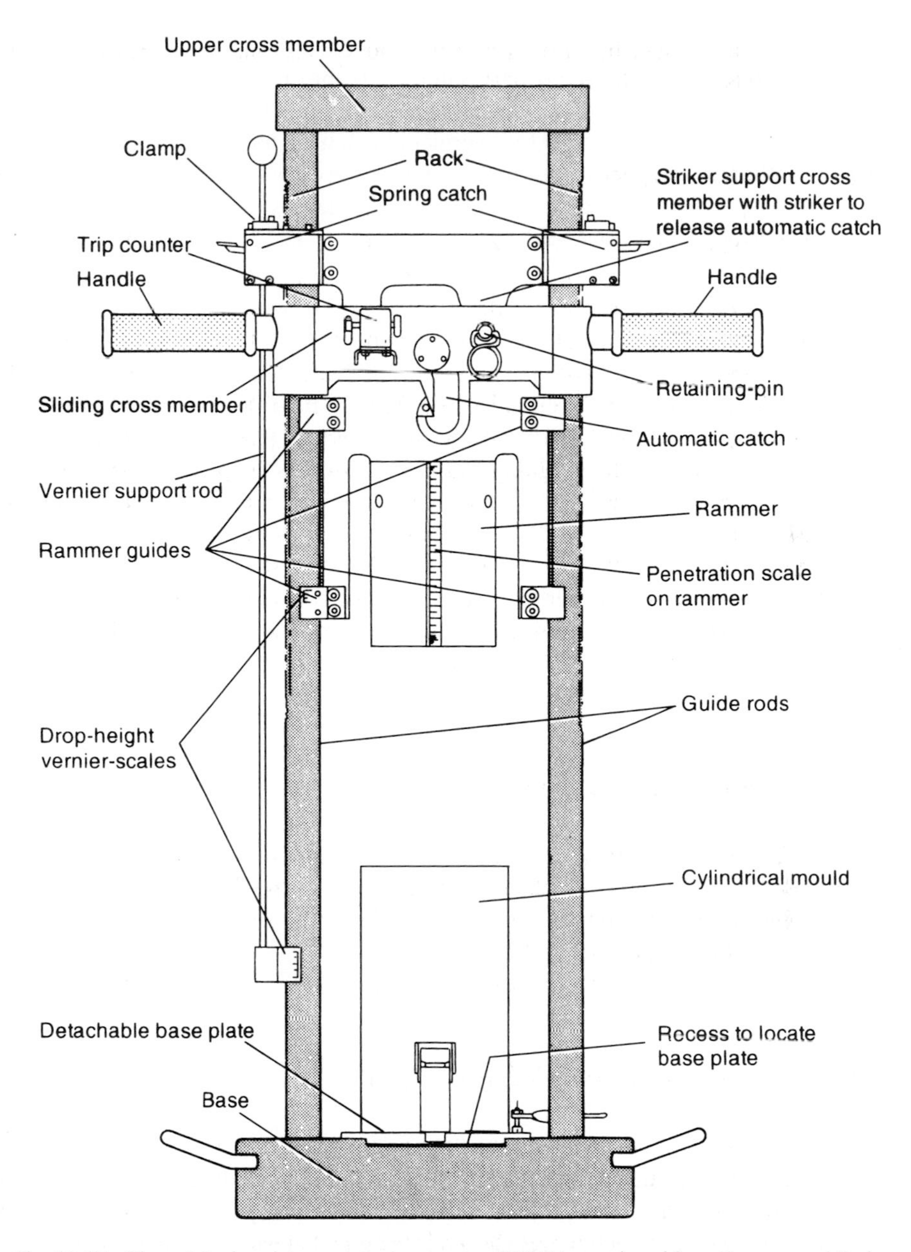

Fig. 11.13 The original moisture content apparatus (MCA) (reproduced from Parsons and Boden, 1979). A modified version is in use nowadays.

The moisture condition value, MCV, is a measure of this compactive effort and is correlated with the undrained shear strength, c_u, or the CBR value, that the soil will attain when subjected to the same level of compaction (Parsons and Boden, 1979).

Details of the apparatus are shown in Fig. 11.13. Basically the test consists of placing a 1.5 kg sample of soil that has passed through a 20 mm sieve into a cylindrical mould of internal diameter 100 mm. The sample is then compacted to maximum bulk density with blows from a 7 kg rammer, 97 mm in diameter and falling 250 mm. After a selected number of blows (see Example 11.4), the penetration of the rammer into the mould is measured by a vernier attachment and noted. The test is terminated when no further significant penetration is noted or as soon as water is seen to extrude from the base of the mould. The latter requirement is essential if the water content of the sample is not to change.

Required calculations

The difference in penetration for a given number of blows, B, and a further three times as many blows (i.e. 4B blows) is calculated and is plotted against the logarithm of B. The calculation is repeated for all relevant B values so that a plot similar to Fig. 11.14 of Example 11.4 is obtained. The MCV is taken to be $10 \log_{10} B$ where B = the number of blows at which the change of penetration = 5 mm. As is seen from Fig. 11.14 the chart can be prepared so that the value of $10 \log_{10} B$, to the nearest 0.1, can be read directly from the plot. The value of 5 mm for the change in penetration value was arbitrarily selected as the point at which no further significant increase in density can occur. This avoids having to extrapolate the point at which zero penetration change occurs.

The MCV for a soil at its natural moisture content can be obtained with one test, the wet soil being first passed through a 20 mm sieve and a 1.5 kg sample collected.

Example 11.4

A sample of silty clay was subjected to a moisture condition test at its natural water content. The results obtained are set out below.

No. of blows, B	Penetration (mm)
1	45.6
2	55.3
3	62.7
4	67.1
6	74.5
8	79.5
12	86.1
16	90.7
24	96.6
32	99.4
48	101.9
64	102.9
96	103.5
128	103.8

Determine the MCV for the soil.

Solution

The calculations are best tabulated:

No. of blows, B	Penetration (mm)	Change in penetration with additional 3B blows (mm)
1	45.6	21.5
2	55.3	24.2
3	62.7	23.4
4	67.1	23.6
6	74.5	22.1
8	79.5	19.9
12	86.1	15.8
16	90.7	12.2
24	96.6	6.9
32	99.4	4.4
48	101.9	
64	102.9	
96	103.5	
128	103.8	

The plot of change in penetration to number of blows is shown in Fig. 11.14. The plot crosses the line at B = 30.2. Hence the MCV of the soil is $10 \times \log_{10} 30.2 = 14.8$. The value of MCV can be read directly on the bottom scale.

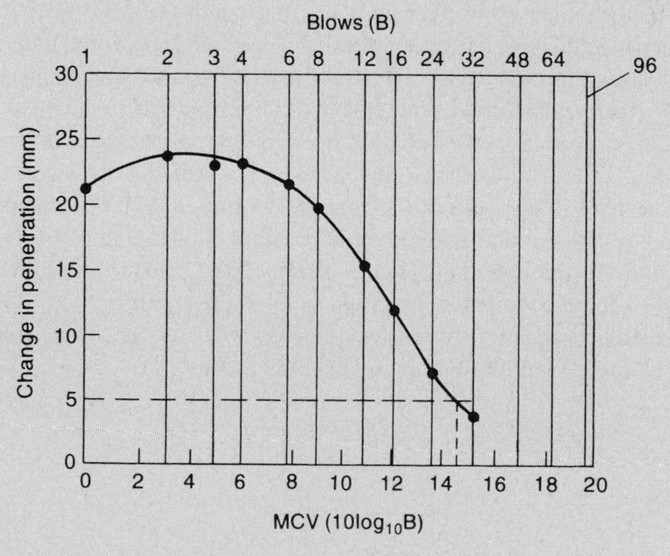

Fig. 11.14 Example 11.4.

The calibration line of a soil

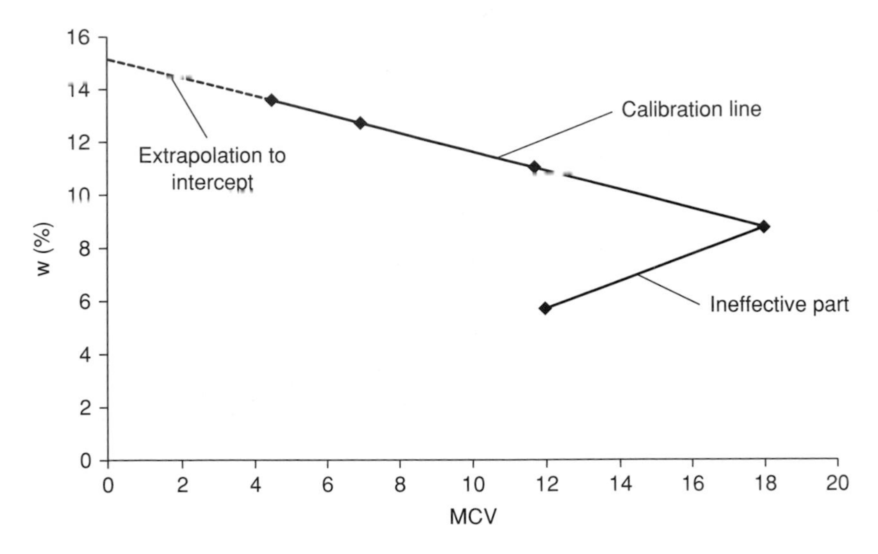

Fig. 11.15 Typical MCV calibration line.

If the relationship of MCV to water content is required, then a sample (approximately 25 kg) of the soil is air dried and passed through the 20 mm BS sieve. At least four sub-samples are taken and mixed thoroughly with different quantities of water to give a suitable range of moisture contents. Ideally, the range of moisture contents should be such as to yield MCVs between 3 and 15.

Test samples of mass 1.5 kg are prepared from each sub-sample and tested in the same manner as for the natural test and a series of MCVs are thus obtained. The moisture contents of the test samples are determined and a plot of the series of moisture contents against MCVs is produced. The best-fit plot is linear and is referred to as the *MCV calibration line*. An example of a calibration line is shown in Fig. 11.15.

The purpose of the best straight line plot is to gain a value of the *sensitivity* of the soil and a range of moisture contents within which the MCV test is applicable. The sensitivity is a measure of the change in value of MCV for a specified change in moisture content and is equal to the reciprocal of the slope of the calibration line. If required, the calibration line is extrapolated back to the moisture content axis (y-axis) to give the *intercept*. Often an ineffective part of the calibration line can be detected if samples are tested at sufficiently low moisture contents (Fig. 11.15).

The equation of the calibration line is:

$$w = a - b(\text{MCV})$$

where

w = moisture content at MCV (per cent)
a = intercept (per cent)
b = slope of line.

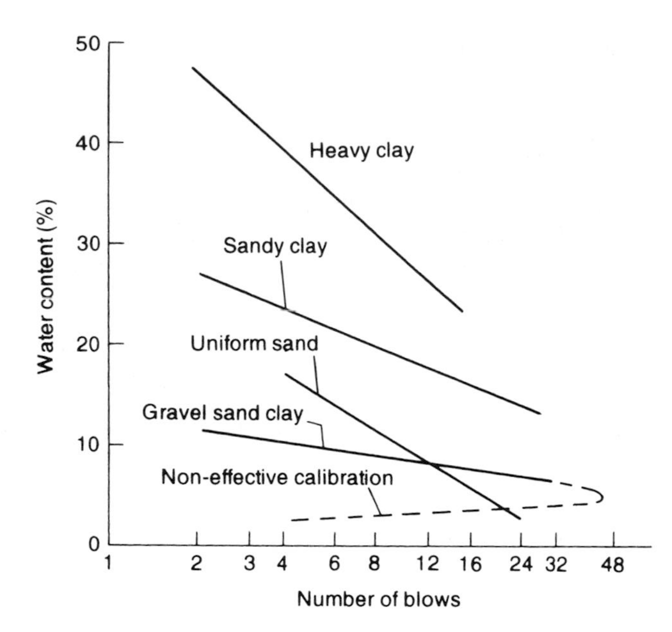

Fig. 11.16 Typical calibration line for different soils (after Parsons, 1976).

The use of calibration lines in site investigation

To determine the equation of a calibration line it is only necessary to know its slope, b, and its intercept, a. This information can be extremely useful in deciding whether or not a soil is suitable for earthworks. Typical calibration lines for different soils are shown in Fig. 11.16. It is important to realise, however, that MCV calibration test results determined in the laboratory do not give an accurate picture of the conditions likely to be experienced in the field. As a result, a degree of care and judgement is required when using values of sensitivity and intercept to judge a soil's *in situ* suitability for earthworking.

The value of the intercept gives a measure of the potential of the soil to retain moisture when in a state of low compaction. This potential increases with the value of the intercept.

The slope of the line gives a measure of the soil's change in MCV value, i.e. in strength, with variation in water content. The steeper the slope the more acceptable the soil. A soil with a very low slope will suffer significant changes in its MCV value over a small range of water content and could well be an ideal soil for compaction on a fine day yet utterly unsuitable in rainy or misty conditions. Such a soil has a very high sensitivity to moisture content changes and would therefore be considered unacceptable for earthworks. Sensitivity to water content change is simply the reciprocal of the slope of the calibration line.

Suitability of the moisture condition test

Although no difficulty will be experienced with the majority of soils, particularly those of a cohesive nature, problems can arise with granular soils with a low fines content, such as glacial tills.

A guide to determine whether the MCV test can be satisfactorily applied to a particular soil is given in SHW *Manual of contract documents for highway works, MCHW 2* (Highways Agency, 2004b) and this has since been revised by Oliphant and Winter (1997). Briefly the suitability of the moisture condition test can be assessed by considering the soil's proportion of fines, sand and gravel, obtained from a particle size distribution test.

Tests using the MCA are intended to replace the previous techniques of defining an upper limit of moisture content for soil acceptability. Generally speaking a soil with an MCV of not less than 8.5, which is comparable to the *in situ* compactive effort of present day plant, will be acceptable for earthworks.

The procedures discussed above have been used successfully on Scottish trunk road projects for a number of years. However it must be stated that in England, opinions vary from considering the approach to being relatively new and still under development to considering it to be unsuitable. There are plenty of detractors, such as Baird (1988).

Exercises

Exercise 11.1

The results of a compaction test on a soil are set out below.

Moisture content (per cent)	9.0	10.2	12.5	13.4	14.8	16.0
Bulk density (Mg/m^3)	1.923	2.051	2.220	2.220	2.179	2.096

Plot the dry density to moisture content curve and determine the maximum dry density and the optimum moisture content.

If the particle specific gravity of the soil was 2.68, determine the air void percentage at maximum dry density.

Answer $\rho_{d_{max}} = 1.97$ Mg/m^3
omc = 13 per cent
Percentage air voids = 2.0

Exercise 11.2

A 2.5 kg rammer compaction test was carried out in a 105 mm diameter mould of volume 1000 cm^3 and a mass of 1125 g.

Test results were:

Moisture content (per cent)	10.0	11.0	12.0	13.0	14.0
Mass of wet soil and mould (g)	3168	3300	3334	3350	3320

Plot the curve of dry density against moisture content and determine the test value for $\rho_{d_{max}}$ and omc.

On your diagram plot the zero and 5 per cent air voids lines (take G_s as 2.65). If the percentage of gravel omitted from the test (particle specific gravity = 2.73) was 10 per cent, determine more correct values for $\rho_{d_{max}}$ and the omc.

Answer From test: $\rho_{d_{max}} = 1.97\ \text{Mg/m}^3$; omc = 12 per cent
Corrected values: $\rho_{d_{max}} = 2.03\ \text{Mg/m}^3$; omc = 11 per cent

Exercise 11.3

An MCV test carried out on a soil sample gave the following results:

No. of blows, B	Penetration (mm)
1	38.0
2	45.1
3	64.0
4	72.3
6	78.1
8	80.1
12	81.0
16	81.5
24	81.7
32	81.7
48	81.7

Determine the MCV of the soil.

Answer MCV = 9.5

Exercise 11.4

An MCV test carried out on a glacial till gave the results set out below:

w (per cent)	5.6	6.8	7.8	8.0	8.5	9.0	9.5
MCV	13.3	13.5	13.7	13.1	11.1	8.5	6.0

Plot the MCV/w relationship and hence determine the equation of the calibration line and determine its sensitivity.

Show that the MCV values obtained for the two lowest water content values form the ineffective part of the calibration line.

Answer w = 10.7 – 0.21MCV
Sensitivity = 1/0.21 = 4.8

Chapter 12
Unsaturated Soils

12.1 Unsaturated soils

When the voids of a soil contain both air and water the soil is said to be partially saturated, or unsaturated.

The pressure of the air in the soil is given the symbol u_a and the pressure of the water is given the symbol u_w.

The degree of saturation is defined by the ratio:

$$S_r = \frac{\text{Volume of water}}{\text{Volume of voids}}$$

Research into understanding the properties and behaviour of unsaturated soils is a rapidly developing area of soil mechanics. In the past ten years much attention has been paid to the topic, which has led to four major international conferences and numerous regional meetings being held on all of the aspects involved in this important subject area. In addition, leading journals have published specialist supplements on the topic as the subject gains wider interest. As the subject becomes increasingly important to researchers and practising engineers, many specialist texts have become available; however, this chapter summarises many of the more significant aspects of the subject.

12.1.1 Soil suction

As discussed in Chapter 2, many of the deposits of gravels, sands, and silts encountered above the water table are unsaturated. Light and heavy clays lying within this region can generally be assumed to be saturated although, in their upper regions, they can sometimes be partly saturated. The chapter also mentions that rain water, or indeed any form of free water, percolates downwards from the surface towards the water table and that most of the soils, whose permeability allows this water to pass through, have the capacity to trap and hold on to some of it.

This capacity of a soil mass to retain water within its structure is related to the prevailing suction and to the soil properties within the whole matrix of the soil, e.g. void and soil particle sizes, amount of held water, etc. For this reason it is often referred to as 'matrix suction', although many authors refer to it as 'matric suction', the term that is used in this book.

It is generally accepted that the amount of matric suction, s, present in an unsaturated soil is the difference between the values of the air pressure, u_a, and the water pressure, u_w.

$$s = u_a - u_w$$

If u_a is constant, then the variation in the suction value of an unsaturated soil depends upon the value of the pore water pressure within it. This value is itself related to the degree of saturation of the soil.

12.1.2 The water retention curve

If a slight suction is applied to a saturated soil, no net outflow of water from the pores is caused. However, as the suction is increased, water starts to flow out of the larger pores within the soil matrix. As the suction is increased further, more water flows from the smaller pores until at some limit, corresponding to a very high suction, only the very narrow pores contain water. Additionally, the thickness of the adsorbed water envelopes around the soil particles reduces as the suction increases. Increasing suction is thus associated with decreasing soil wetness or moisture content. The amount of water remaining in the soil is a function of the pore sizes and volumes and hence a function of the matric suction. This function can be determined experimentally and may be represented graphically as the *water retention curve*, such as the examples shown in Fig. 12.1.

The amount of water in the pores for a particular value of suction will depend on whether the soil is wetting or drying. This gives rise to the phenomenon known as hysteresis, and the shape of the water retention curve for each process is shown in Fig. 12.1. A full descriptive text on water retention curves and hysteresis is given by Fredlund and Rahardjo (1993).

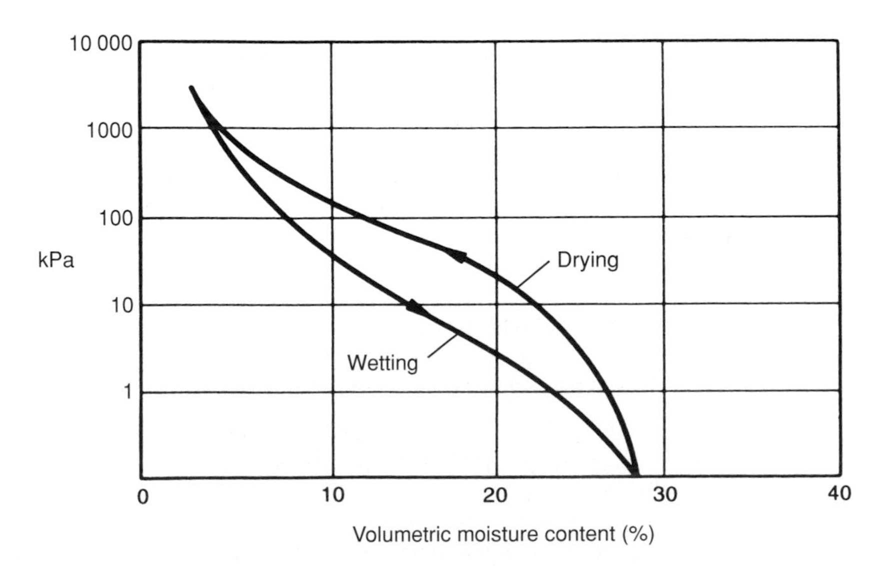

Fig. 12.1 Example wetting and drying water retention curves.

12.1.3 Water vapour

So far this discussion of soil water has only considered water in its liquid phase. It is now necessary to consider the effect that water vapour which, along with air or gas, also occupies the voids of an unsaturated soil, has on the final distribution of water held in the soil.

Water vapour is created by the evaporation of water at an air–water interface. If the space above the interface is a closed system evaporation will continue until the rates of evaporation and recondensation at the interface become equal and the vapour space is saturated with water vapour molecules.

This equilibrium pressure is known as the saturated vapour pressure of water and, if the air–water interface is level, has a constant value for a particular vapour temperature.

If the amount of water is insufficient, so that it all evaporates before the saturated pressure value is reached, then the actual vapour pressure achieved is expressed as a percentage of the saturation pressure and is termed the 'relative humidity'.

A sample of moist soil placed inside a similar sealed container will create water vapour, by evaporation from the soil, until a certain vapour equilibrium pressure value is achieved. The value of this pressure is less than the saturated pressure of water vapour at the same temperature because of the effect of the suction of the soil within which the water is held.

Further reductions in the value of the water vapour pressure, in a closed system, can be caused by the following:

(i) If the air–water interface is not level but is in the form of a meniscus then additional downward forces, caused by the surface tension effects of the curved water surface, can reduce the vapour pressure.

(ii) Salts, dissolved in the soil water can also reduce the pressure of the water vapour in the soil.

It should be noted that dissolved sodium chloride (i.e. common salt), when dissolved in the water, reduces the vapour pressure by an amount directly related to the amount of salt in solution. This property is often used in the calibration of some types of suction measuring apparatus.

It is often convenient to express the value of the vapour pressure of soil water as a relative humidity, i.e. as a percentage of the saturated vapour pressure of water at the same temperature.

12.2 Measurement of soil suction

From a geotechnical point of view there are two components of soil suction as follows:

(1) *Matric suction*: that part of the water retention energy created by the soil matrix.
(2) *Osmotic suction*: that part of the water retention energy created by the presence of dissolved salts in the soil water.

It should be noted that these two forms of soil suction are completely independent and have no effect on each other.

The total suction exhibited by a soil is obviously the summation of the matric and the osmotic suctions.

If a soil is granular and free of salt there is no osmotic suction and the matric and total suctions are equal. However, clays contain salts and these salts cause a reduction in the vapour pressure. This results in an increase in the total suction, and this increase is the energy needed to transfer water into the vapour phase (i.e. the osmotic suction).

Developments in soil suction measurements

Since Croney and Coleman's work (1953, 1958, 1960) a large amount of research has concentrated on both the laboratory and the *in situ* measurement of pore water pressures in soil.

As the water pressure in an unsaturated soil is usually negative it was soon realised that piezometers evolved to measure positive water pressures were generally only able to measure negative water pressures up to values of about 80 kPa when used as tensiometers (Penman, 1995). The problem was cavitation, i.e. the appearance of gas and vapour cavities within the measuring equipment that eventually led to a complete breakdown of the equipment.

Commercial equipment for unsaturated soils has gradually evolved and there are now several types available which can be used to measure soil suction values. Amongst them are psychrometers, porous blocks, filter papers, suction plates, pressure plates and tensiometers, the last being the most popular for *in situ* measurements. A useful survey was prepared by Ridley and Wray (1995).

The psychrometer method

A psychrometer is used to measure humidity and is therefore suitable to measure total soil suction, i.e. the summation of the matric and the osmotic components. The equipment and its operation have been described by Fredlund and Rahardjo (1993).

A sample of the soil to be tested is placed in a plastic container. A hole is then drilled to the centre of the specimen, a calibrated psychrometer inserted and the drilled hole backfilled with extra soil material.

The whole unit is finally sealed with plastic sheeting and placed in an air tight container, where it is left for three days with its temperature maintained at 25°C. After this time the soil sample is deemed to have achieved both thermal and vapour pressure equilibrium and relative humidity measurements can be taken.

The filter paper method

With this technique, described by Campbell and Gee (1986), both total and matric suctions can be measured. In a typical test the soil specimen is prepared in a cylindrical plastic container and a dry filter paper disc is placed over its upper surface. (Fig. 12.2). This filter will measure the matric suction.

A perforated glass disc is placed over the filter paper and a further filter paper is then placed over the glass. As this top filter paper is not in actual contact with the soil sample it can only measure the total suction.

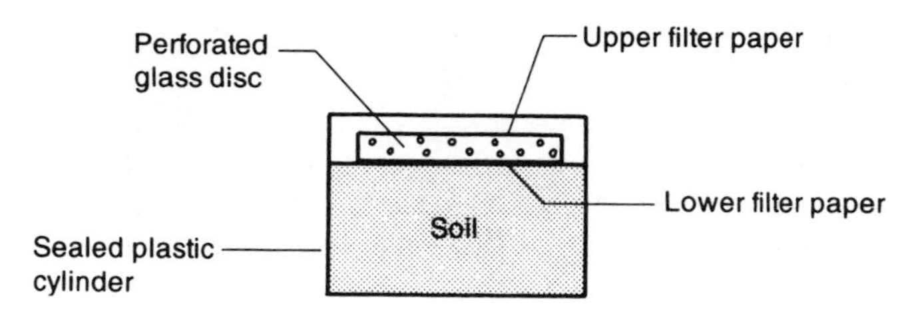

Fig. 12.2 Soil suction measurement – an arrangement for the filter paper method.

The assembled specimen/filter paper is left for at least a month, at a temperature of 25°C, in order to obtain thermal and vapour equilibrium. At the end of this time the assembly is dismantled and the water contents of the specimen and the two filter papers determined. The water contents obtained for the filter papers can be converted into the required suction values by using a suction/water content curve for the filter paper material.

The tensiometer

Stannard (1992) presented a review of the standard tensiometer and covered the relevant theory, its construction and possible uses. The apparatus is mainly used for *in situ* measurements and consists of a porous ceramic cup placed in contact with the soil to be tested.

A borehole is put down to the required depth and the ceramic filter lowered into position. Water is then allowed to exit from a water reservoir within the tensiometer and to enter the soil. The operation continues until the tensile stress holding the water in the tensiometer equals the stress holding the water in the soil (i.e. the total soil suction).

The tensile stress in the water in the tensiometer is measured by some pressure measuring device, e.g. a manometer, a vacuum gauge or a transducer, and is taken to be the value of the total soil suction.

The tensiometer must be fully de-aired during installation, as one would expect if accurate results are to be obtained. The response time of the type of apparatus described is only a few minutes but it has the disadvantage, until recently, that it could only be used to measure suctions up to about 100 kPa.

However, developments are continually taking place, and details of a new tensiometer, developed by Imperial College, of much smaller dimensions and capable of measuring pore water pressures as low as –1500 kPa have been presented by Ridley and Burland (1993).

12.3 Soil structure changes with water content

A simplified picture of a soil skeleton has been given in Fig. 1.9a. Although useful in illustrating the basic structure of soils, the concept that the soil particles are arranged in all directions in more or less symmetrical patterns very rarely applies.

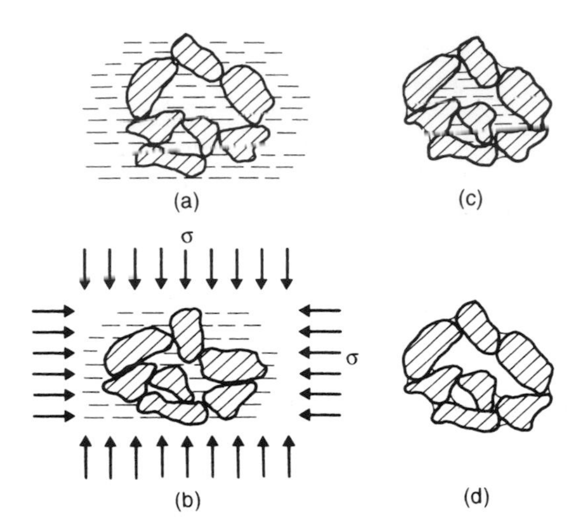

Fig. 12.3 Structural changes in a granular soil resulting from increases in applied stress and pressure deficiency (after Jennings and Burland, 1962).

Figure 12.3 is a reproduction of the diagram prepared by Jennings and Burland (1962) to help them to describe the structural changes that can occur in a non-uniform granular soil when it is subjected to water content and/or applied loading variations.

12.3.1 Granular soils

In a non-uniform saturated granular material the grains tend to form arches, as illustrated in Fig. 12.3a. The effect is, of course, three dimensional so that the soil has a honeycombed structure. The whole soil mass is obviously fairly compressible.

The application of an external stress system causes both shear and normal forces to develop at the grain contact points. If the soil is capable of drainage then there will be a reduction in volume of the soil mass, as the induced internal stress cause individual grains to slide or roll across each other, Fig. 12.3b.

A similar effect is obtained if the compression is caused by surface tension effects instead of by applied loading. This situation can arise if water conditions alter so that the soil is no longer submerged but is still saturated, Fig. 12.3c. In this case the compressive forces are maintained by the action of the menisci that have formed around the edges of the soil.

If further loss of water occurs then air will enter the soil and the menisci will retreat into the inner voids of the soil mass.

The intergranular forces are now derived from the high curvature of the menisci at the particle contact points (Fig. 12.3d). These menisci induce only normal forces between the particles which, as there are now no shear forces present, tend to 'bond' together to form a strong and stable soil structure.

If, when in this state, the soil is subjected to externally applied loading it will offer considerably greater resistance to the shear forces induced at the particle contacts than it would have done had it been submerged.

12.3.2 Clayey soils

The shape of a clay particle hardly resembles that of a granular soil particle. Clay particles consist of combined minute thin flat sheets of silica and aluminium or other minerals. The surfaces of these sheets possess an electro negativity whilst their edges are electrically charged, either positive, neutral or negative, depending upon the mineral involved.

It is these electrical forces, acting on the particles at the time of its formation, which were responsible for the structure and therefore the strength of a clay soil.

12.3.3 Swelling and collapse phenomena of clayey soils

In the dry state the structure of a clay soil is somewhat analogous to that of a granular soil in that the clay tends to form itself into 'grains' or 'packets'. These grains consist of numerous clay particles tightly bonded together but, if the applied loading is increased, they will have little tendency to slip or roll and will only tend to distort, unlike granular soil particles.

If such a loaded soil is wetted the bonding between each 'packet' of clay particles is removed and the packets tend to be displaced relative to one another. At the same time each packet, as it takes in water, tends to expand.

The final behaviour of the soil is governed by the magnitude of both the applied loads and the change in the water content. With low loads the soil can be expected to expand whereas, at higher loads, it is more likely that it will decrease in volume.

In their paper, Jennings and Burland point out that, instead of a dry clay swelling as it becomes wet, as is predicted from Terzaghi's effective stress theory, it may first collapse:

> 'It is quite possible that soaking of the soil under a load will result first in a rapid reduction in volume due to removal of intergranular "bonds" and then a slow increase in volume due to particles taking up water.'

The possible catastrophic effects that can be caused when a dry compacted soil is flooded with water have been discussed by Charles and Burford (1987). They also give details of the complete failure of a block of eight two storey houses built on a stiff clay backfill, which had a maximum depth of 12 m. Soon after construction of the brickwork it was observed that settlement had occurred at the centre of the block after heavy rain. Tests carried out by the Building Research Establishment showed that the compacted fill was susceptible to collapse if flooded.

The block was never occupied and was finally demolished some 9 years later by which time the amount of settlement of the block had reached some 0.3 m.

12.4 Stress states in unsaturated soils

As defined previously, the effective stress is the stress that controls changes in both the volume and the strength of a soil.

The effective stress equation for saturated soils was evolved by Terzaghi in 1923 and is described in Chapter 3:

$$\sigma' = \sigma - u$$

where

σ' = the effective stress
σ = the total applied stress
u = pore water pressure.

As there are two void fluids, liquid and gas, in an unsaturated soil the direct use of Terzaghi's effective stress equation is obviously not possible.

It should also be noted that an unsaturated soil, or a saturated soil that becomes unsaturated, can experience the problems of shrinkage, swelling and structural collapse, described in the previous section, as well as the problems of variation in strength and consolidation characteristics associated with saturated soils.

12.4.1 The χ parameter

It was felt by most soil workers in the 1950s and early 1960s that it must be possible to apply the principle of effective stress to partially saturated soils. In 1955, Bishop suggested an effective stress formula for unsaturated soils which combined the total stress, σ, the pore air pressure, u_a, and the pore water pressure, u_w, into a single effective stress tensor, σ':

$$\sigma' = \sigma - [u_a - \chi(u_a - u_w)]$$

where χ is a parameter related to the degree of saturation and, to some extent, the soil structure, having a value of 0.0 for a dry soil and 1.0 for a fully saturated one.

Only two years later, Jennings and Burland (1962) reported on the results they had obtained from consolidation tests carried out on samples of artificially prepared silty sand, silt and silty clay and questioned the validity of Bishop's equation.

Virgin consolidation curves were prepared for each soil by (i) consolidating from a slurry, and (ii) soaking a set of the prepared oedometer samples under various normal pressures. These virgin consolidation curves were compared with consolidation curves obtained from tests on unsoaked samples of each soil.

It was found that, for each soil type, the effective stress equation only applied over a range of S_r values. For the silty clay the range was a small as 15 per cent.

The oedometer testes also dramatically illustrated how the effective stress principle, in its proposed form, could not be used to predict the possibility of an unsaturated soil collapse. At the end of the dry tests on the unsaturated samples the oedometers were filled with water and the samples allowed to soak.

It is well known that soaking a soil reduces the suction forces within it to zero. This decrease in suction must mean that the effective stress also decreases. A decrease in effective stress should cause expansion yet, in every case, the samples, when soaked under constant loading, suffered further volume reduction, i.e. collapse. The collapse phenomenon is exactly the reverse of what should happen according to the effective stress principle, which clearly did not apply.

Since the early 1960s, research into the determination of satisfactory χ values was carried out, with invariably disappointing results. In some tests χ values greater than 1.0 and, in others, χ values less than 0.0 were obtained.

After some time the reason for the erratic values obtained for χ began to be understood. The problem is caused by the presence of 'meniscus water', and a good description of its effects has been given by Wheeler and Karube in the introduction to their 1995 paper. The following is a summary.

In a saturated soil, no free air exists, every void within the soil being full of water. Pore water, in this state, is often referred to as 'bulk water' although it is possible that it contains a small amount of dissolved air.

If a saturated soil is slowly dried out the outer limits of the bulk water tend to evaporate and the outer soil voids begin to empty of water and take in air. Voids of an unsaturated soil are therefore filled with water, or a water and air mixture, or simply air.

With air and water filled voids small lenses of water form menisci around the particle contacts (see Fig. 12.3d). With clays there will also be adsorded water, so strongly attached to the soil particles that it can be regarded as being part of the soil skeleton.

Although the volume of meniscus water in an unsaturated soil may be very small it can have a dramatic effect on the mechanical behaviour of the soil, an effect that cannot be estimated.

A further problem is the inability of a single effective stress tensor, σ', to replicate both swelling and collapse effects. If a dry soil is inundated with water its suction decreases and the soil will either collapse or swell, depending upon the relative magnitude of the applied loading.

The inability to determine whether a decrease in suction acts like an increase or a decrease in the effective stress of a saturated soil means that, for an unsaturated soil, it is not possible to combine $\sigma - u_a$ and $u_a - u_w$ into a practical single value for σ'. This fact was generally accepted by the end of the 1980s (Alonso *et al.*, 1990).

12.4.2 Recent developments

Two stress state variables

In an undrained soil there are three stress parameters, $\sigma - u_a$, $\sigma - u_w$ and $u_a - u_w$. Fredlund and Morgenstern, in their paper of 1977, agreed with the approach adopted by Coleman (1962) and Bishop and Blight (1963) in that only two of the three stress variables are necessary to define the stress state of an unsaturated soil. The common choice is to use the net stress, $\sigma - u_a$, and the matric soil suction, $u_a - u_w$, as the two independent stress state variables. This is mainly because u_a is the atmospheric pressure and can usually be assumed to have a value of zero. With this approach constitutive models of unsaturated soil behaviour, both for volumetric change and for shear strength, have been proposed and examined since the 1970s.

A possibly useful formula for the shear strength of an unsaturated soil was presented by Fredlund *et al.* in 1978:

$$\tau = c' + (\sigma - u_a) \tan \phi' + (u_a - u_w) \tan \phi^b$$

where c' and ϕ' are the cohesive and angle of friction for an equivalent saturated soil and ϕ^b the angle of friction with respect to changes in suction.

However, many authors subsequently showed that, although ϕ' is more or less constant for most soils, ϕ^b is not. Escario and Juca (1989) suggested that the equation should be rewritten as:

$$\tau = c' + (\sigma' - u_a) \tan \phi' + f(u_a - u_w)$$

where $f(u_a - u_w)$ is a nonlinear function of suction.

Wheeler and Karube (1995) have prepared a review of several recent models, some of which use more complex parameters.

Elasto-plastic critical state models for unsaturated soil

Critical state modelling is described in Chapter 13, and this approach is the most recent line of research in unsaturated soils. It is an attempt to allow for the possible occurrence of irreversible plastic strains in an unsaturated soil and to link them to the volumetric and/or shear behaviour of the soil.

Such an elasto-plastic critical state model for unsaturated soil was presented in a qualitative form by Alonso, Gens and Hight in 1987, and the model was developed into a full mathematical form by Alonso *et al.* (1990). Wheeler and Sivakumar (1995), using experimental data they obtained from controlled suction triaxial tests on compacted kaolin, proposed further modifications to the model.

There is little doubt that the strength and volume change behaviour of unsaturated soils together with the associated problems of swelling and collapse will be important areas of research for some years to come.

Chapter 13
Critical State Theory

Up to this point the material contained in the chapters of this book has used three models of soil behaviour:

- the Mohr–Coulomb model, for the prediction of soil shear strengths;
- the soil modelled as an elastic medium, for the estimation of stresses induced by applied loads and for immediate settlement problems;
- the soil modelled as analogous in behaviour to that of a dashpot and a spring supported piston, for consolidation settlement evaluations.

13.1 Critical state theory

Over the last half century a fourth model of soil behaviour has been established and stems from the work of Roscoe *et al.*, who, in 1958, suggested that, within saturated remoulded clays subjected to loadings that created a constant and low rate of increasing strain, there existed both a critical void ratio line and a yield surface.

Reporting on various triaxial test results the authors showed that, when subjected to this form of loading, clays would reach, and pass through, a failure point without collapse and would then continue to suffer deformation as both the void ratio and the relevant stress paths followed a yield surface until a critical void ratio value was achieved.

At this critical void ratio value the values of the void ratio, the pore water pressure and the stresses within the soil remain constant, even with further deformations, provided that the rate of strain is not changed.

This important concept has led to the theory of critical state, an attempt to create a soil model that brings together the relationships between its shear strength and its void ratio, and which can be applied to any type of soil.

The theory has been established as a research tool for several years and is now accepted for use in limit state design. It is hoped that the material contained in this chapter will provide the reader with a suitable introduction to the subject.

13.2 Symbols

Critical state theory is a three-dimensional approach and therefore uses three parameters: p and q, the equivalent of the s and t parameters used in Chapter 9, and a third parameter, v, the specific volume, which is defined in Chapter 1 and is the total volume of soil that contains a unit volume of solids. From Chapter 1: $v = (1 + e)$.

As explained in Chapter 3, in the triaxial test, where $\sigma_2 = \sigma_3$, we can say that:

$$\sigma_{oct} = \frac{1}{3}(\sigma_a + 2\sigma_r); \quad \tau_{oct} = \frac{\sqrt{2}}{3}(\sigma_1 - \sigma_3)$$

In order to avoid the term $\frac{1}{3}\sqrt{2}$, p and q were defined as:

$$p = \frac{1}{3}(\sigma_1 + 2\sigma_a) = \sigma_{oct} \qquad (1)$$

$$q = (\sigma_1 - \sigma_3) = \frac{3}{\sqrt{2}}\tau_{oct} \qquad (2)$$

Similar expressions apply for effective stress:

$$p' = \frac{1}{3}(\sigma'_1 + 2\sigma'_a) = \sigma'_{oct} \qquad (3)$$

$$q' = (\sigma'_1 - \sigma'_3) = \frac{3}{\sqrt{2}}\tau'_{oct} \qquad (4)$$

As critical state theory uses the results obtained from soil samples subjected to triaxial tests the above formulae for p, p′, q and q′ are used in this chapter.

The advantage of these parameters is their association with the strains that they cause. Changes in p′ are associated with volumetric strains and changes in q with shear strains.

For the general three-dimensional state, Equations (1) to (4) have the form:

$$p = \frac{1}{3}(\sigma_1 + \sigma_2 + \sigma_3)$$

$$q = \frac{1}{\sqrt{2}}[(\sigma_1 - \sigma_2)^2 + (\sigma_2 - \sigma_3)^2 + (\sigma_3 - \sigma_1)^2]$$

It should be noted that, when dealing with consolidation aspects, the v–ln p′ plot is used instead of the e–ln p′ plot used in earlier chapters.

13.3 Critical state

In a drained test the void ratio of a soil changes during shear. If several samples of the same soil are tested at different initial densities it is found that, if p′ is constant, the samples all fail at the same void ratio. If the deformation is allowed to continue the sample will remain at the same void ratio and only deform by shear distortion. This condition is referred to as the *critical state*.

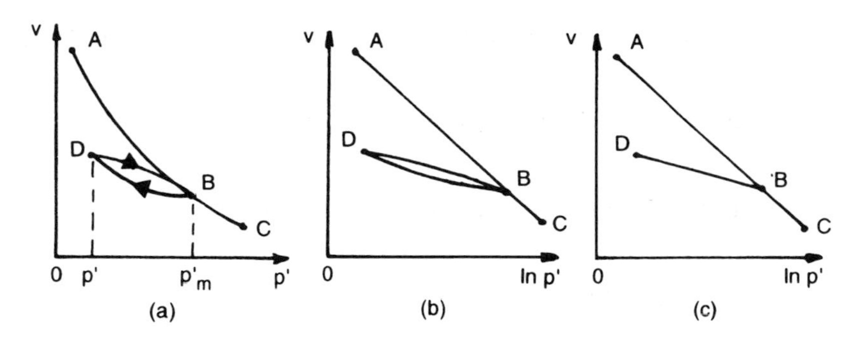

Fig. 13.1 Typical shape of the isotropic normal consolidation of a saturated cohesive soil.

13.4 Isotropic consolidation

Most soil samples tested in the triaxial apparatus are isotropically consolidated, i.e. consolidated under an all-round hydrostatic pressure, before the commencement of the shearing part of the test. It is appreciated that other forms of consolidation are possible, e.g. K_0 consolidation, but these forms will not be considered here.

The form of the compression curve for an isotropically consolidated clay is shown in Fig. 13.1a. It should be noted that the plot is in the form of a v–p′ plot, the vertical axis being 0 : V and the horizontal axis 0 : p′. The v–ln p′ plot is shown in Fig. 13.1b and from this diagram we see that, if we are prepared to ignore the slight differences between the expansion and the recompression curves, the semi-log plot of the isotropic consolidation curve for most clays can be assumed to be made up from a set of straight lines and to have the idealised form of Fig. 13.1c.

Any point on the line ABC represents normal consolidation whereas a point on the line BD, or indeed any point below ABC, represents overconsolidation. As line DB represents the idealised condition that the expansion and recompression curves coincide, it is probably best to give it a new name, and it is therefore usually called the swelling line.

If the maximum previous pressure on a swelling line is p'_m and the pressure at D, a point on the swelling line, is p′ then we can say that the degree of overconsolidation represented by point D is $R_p = p'_m/p'$. (Note the use of the subscript ‘p’ in R_p to indicate isotropic consolidation.)

Figure 13.2 is a close-up of Fig. 13.1c. In the diagram let the slope of AC, the normal isotropic consolidation line, be $-\lambda$, and the slope of the swelling line, DB, be $-\kappa$. N = the specific volume of a soil normally consolidated at ln p′ value of 0.0. This gives ln p′ = 0. Then the equation of line AC is:

$$v = N - \lambda \ln p'$$

A swelling line, such as BD, can lie anywhere beneath the line AC as its position is dependent upon the value of the maximum pressure on the line, p_m, which determines the position of B.

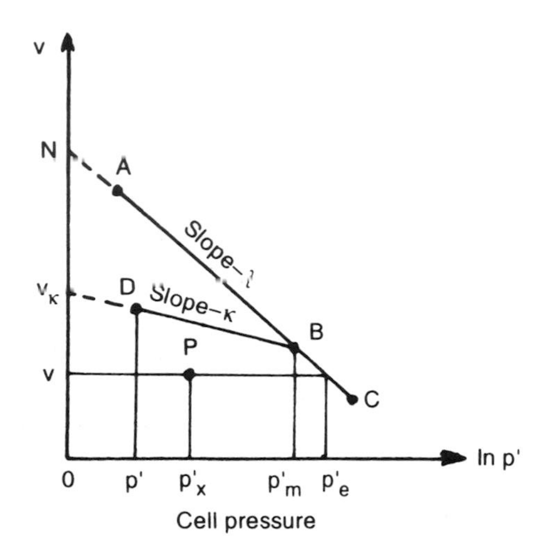

Fig. 13.2

Let v_κ = the specific volume of an overconsolidated soil at p' = unity (i.e. 1.0 kPa). Then the equation of line DB is:

$$v = v_\kappa - \kappa \ln p'$$

λ, N and κ are measured values and must be found from appropriate tests.

Note The normal consolidation line, AC, is often referred to as the λ line, i.e. the lambda line, and the swelling line BD is often called the κ line, i.e. the kappa line.

13.4.1 Equivalent isotropic consolidation pressure, p'_e

Consider a particular specific volume, v. Then the value of consolidation pressure which corresponds to v on the normal isotropic consolidation curve is known as the equivalent consolidation pressure and is given the symbol p'_e. In Fig. 13.2 the point P represents a soil with a specific volume, v, and an existing effective consolidation pressure p'_x. The procedure for determining p'_e is illustrated in the diagram. Note that as P is below AB, it represents a state of overconsolidation.

For a normally consolidated clay, subjected to an undrained triaxial test, $p'_e = \sigma'_r$ but with drained tests p'_e will vary (see Example 13.3).

13.4.2 Comparison between isotropic and one-dimensional consolidation

If a sample of clay is subjected to one-dimensional consolidation in an oedometer and another sample of the clay is subjected to isotropic consolidation in a triaxial cell

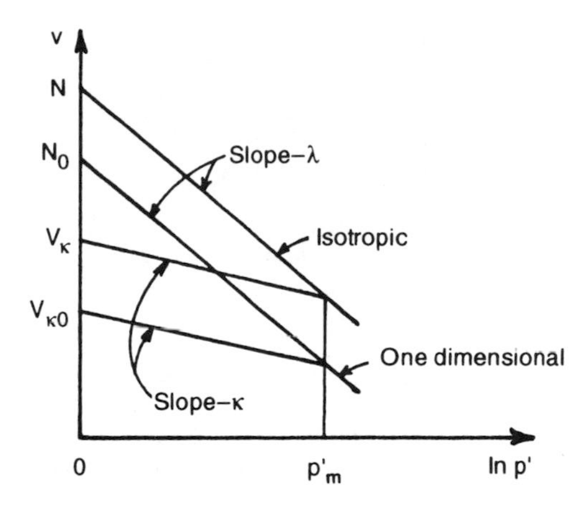

Fig. 13.3 Isotropic and one-dimensional consolidation.

then the idealised forms of the v–ln p′ plots for the tests will be more or less as illustrated in Fig. 13.3.

The values of the slopes of the two normal consolidation lines are very close and, for all practical purposes, can both be assumed to be equal to $-\lambda$. Similarly the slopes of the swelling lines can both be taken as equal to $-\kappa$.

Note that the values of ln p′ for the one-dimensional test are taken as equal to ln σ', where σ' = the normal stress acting on the oedometer sample.

As the compression index C_c, defined in Chapter 9, is expressed in terms of common logarithms we see that:

$$\lambda \approx \frac{C_c}{2.3}$$

13.5 Stress paths in three-dimensional stress space

We have considered two different forms of two-dimensional stress paths in Chapters 3 and 9 and we must now examine the form of these paths if they were plotted in three-dimensional space defined by p′, q and v.

Undrained tests

If we consider the plane q–p′ then we can plot the effective stress paths for undrained shear in a manner similar to the previous two-dimensional stress paths. Remember that $q = \sigma_1 - \sigma_3$ and that

$$p' = \frac{\sigma_1 + 2\sigma_3}{3}$$

The resulting diagram is shown in Fig. 13.4a. The points A_1, A_2 and A_3 lie on the isotropic normal consolidation line and their respective stress paths reach the failure

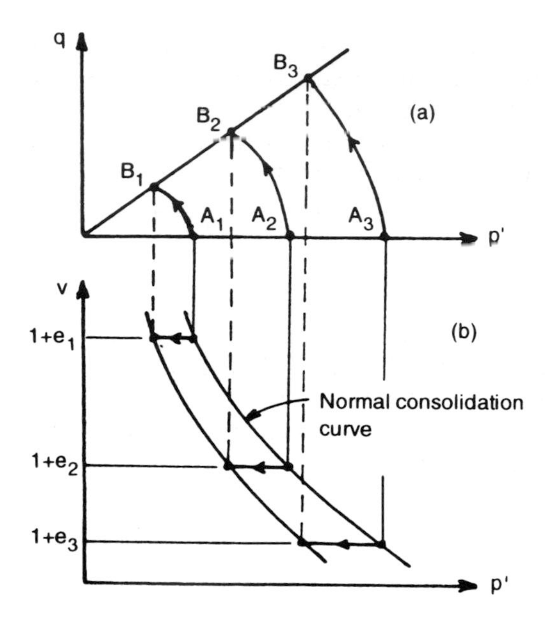

Fig. 13.4 Stress paths for undrained shear.

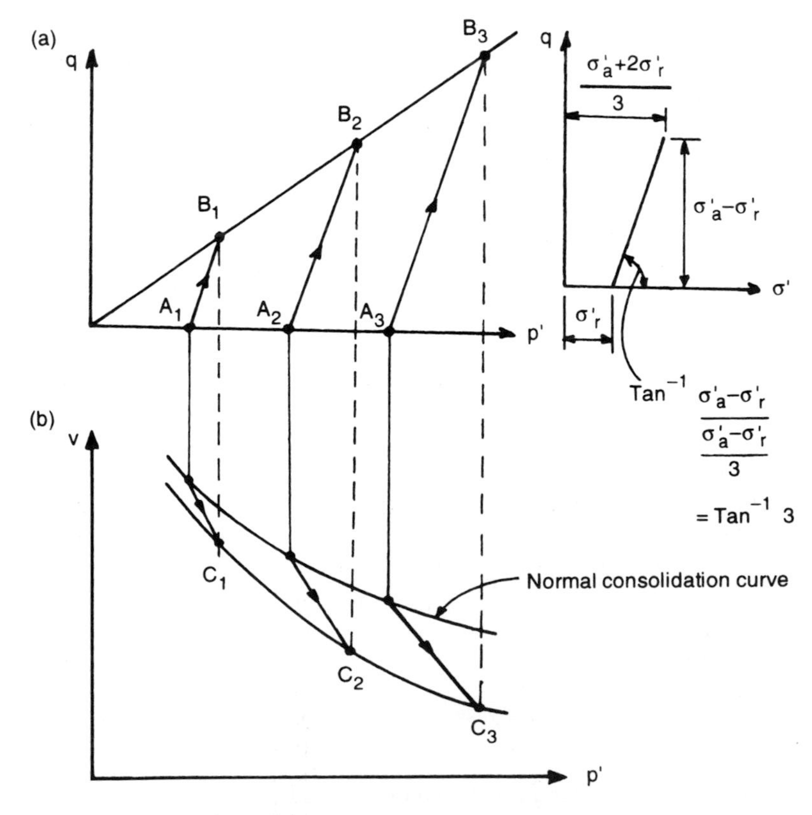

Fig. 13.5 Stress paths for drained shear.

boundary at points B_1, B_2 and B_3. As the tests are undrained the values of void ratio at points B_1, B_2, B_3 are the same as they were when the soil was at the stress states A_1, A_2 and A_3 respectively. Knowing the e values we can determine the values of specific volume and prepare the corresponding plot on the v–p′ plane (Fig. 13.4b).

It is seen that the failure points B_1, B_2 and B_3 lie on a straight line in the q–p′ plane and on a curve, similar to the normal consolidation curve, in the v–p′ plane.

Drained tests

The effective stress paths for drained shear are shown in Fig. 13.5. For the q–p′ plane the plot consists of straight lines which are inclined to the horizontal at $\tan^{-1} 3$. The reason why is illustrated in Fig. 13.5.

The points C_1, C_2 and C_3 represent the failure points after drained shear, so the void ratio values at these points are less than those at the corresponding A points.

The stress paths in the v–p′ plane are illustrated in Fig. 13.5b. As with the undrained case, the failure points C_1, C_2 and C_3 lie on a curved line similar to the normal consolidation line.

13.6 The critical state line

Parry (1960) published a comprehensive set of results obtained from drained and undrained triaxial tests carried out on normally and overconsolidated samples of Weald clay. A few of his results of tests on normally consolidated samples are reproduced in the first four columns of Table 13.1 (converted into SI units). With this

Table 13.1 Results of triaxial compression tests on normally consolidated clay samples (after Parry, 1960).

		Undrained tests			
σ_r (kPa)	$\sigma_{af} - \sigma_r$ (kPa)	u_f (kPa)	w_f (%)	p'_f (kPa)	v
103.4	68.3	50.3	25.1	75.9	1.67
206.9	119.3	113.8	23.0	132.9	1.61
310.3	172.4	171.7	21.5	196.1	1.57
413.7	224.8	227.5	20.3	261.1	1.54
827.4	468.9	458.5	18.5	525.2	1.49

		Drained tests		
σ'_r (kPa)	$\sigma'_{af} - \sigma'_r$ (kPa)	w_f (%)	p'_f (kPa)	v
103.4	114.5	23.0	141.6	1.61
206.9	244.8	20.4	288.5	1.54
310.3	348.2	19.3	426.4	1.51
413.7	481.3	18.5	574.1	1.49
827.4	930.8	16.1	1138.0	1.43

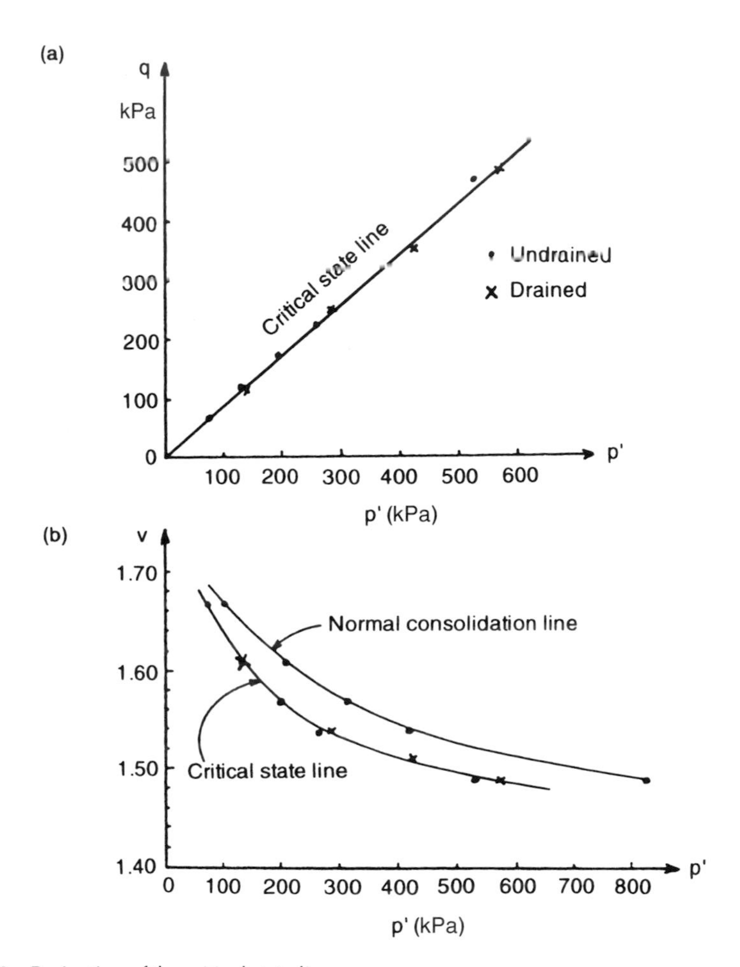

Fig. 13.6 Projection of the critical state line.

information and taking $G_s = 2.75$, the tabulated values of q, p′ and v were calculated. The (p′, q) points obtained from each of the test results are plotted in Fig. 13.6a and the (p′, v) points are plotted in Fig. 13.6b.

We can deduce from these diagrams that there must be a single line of failure points within the p′–q–v space which projects as a straight line on to the q–p′ plane and projects as a curved line, close to the normal consolidation line, on to the v–p′ plane. This line is known as the critical state line and its position is illustrated in Fig. 13.7.

The equation of the critical state line

The line's projection on to the q–p′ plane is a straight line with the equation $q = Mp'$, where M is the slope of the line.

The projection of the critical state line on to the v–p′ plane is unfortunately curved but if we consider the projection on to the v : ln p′ plane we obtain a straight

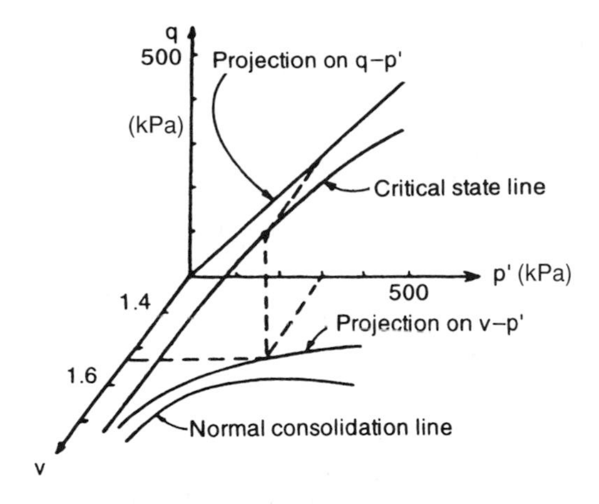

Fig. 13.7 Position of the critical state line.

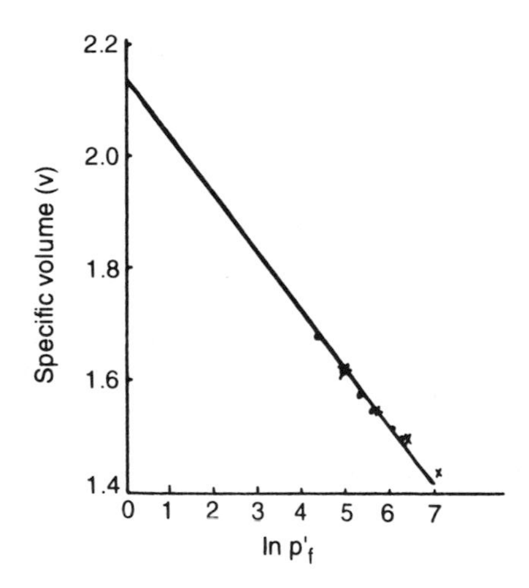

Fig. 13.8 v–ln p'_f of the values tabulated in Table 13.1.

line with a slope that can be assumed to be equal to the slope of the normal consolidation line.

The values for p'_f are tabulated in Table 13.1 and it is a simple matter to obtain a set of ln p'_f values so that a v–ln p'_f plot can be obtained. Figure 13.8 shows the v–ln p'_f plot for Parry's results from Table 13.1.

If we use the symbol Γ (capital gamma) to represent the value of v which corresponds to a ln p′ = 0 (i.e. a p′ value of unity, usually taken as 1.0 kPa) then the equation of the straight line projection is:

$$v = \Gamma - \lambda \ln p'$$

which can be written as:

$$\frac{\Gamma - v}{\lambda} = \ln p'$$

$$\Rightarrow \quad p' = \exp \frac{\Gamma - v}{\lambda}$$

Hence, the critical state line is that line which satisfies the two equations:

$$q = Mp' \quad \text{and} \quad p' = \exp \frac{\Gamma - v}{\lambda}$$

The values of M, N, Γ and λ vary with the type of soil. From Figs 13.6 and 13.8 we see that the values for remoulded Weald clay are approximately $M = 0.85$; $N = 2.13$; $\Gamma = 2.09$ and $\lambda = 0.10$.

13.7 Representation of triaxial tests in p′–q–v space

The results of drained and undrained triaxial compression tests can be represented in the three-dimensional stress space p′–q–v. For both tests the sample is first consolidated to some point A on the normal consolidation curve corresponding to some particular value of specific volume, v_0. From this stage the two tests must be considered separately.

13.7.1 The undrained test

If the sample is now sheared undrained the stress path will move upwards from A until it meets the critical state line at point B where failure occurs. As the test is undrained the value of the specific volume remains constant at v_0 so that the stress path, no matter how it wanders, is restricted to a plane passing through A and parallel to the p′–q plane.

This plane, for a normally consolidated clay, will not be bigger than the area shown in Fig. 13.9 and can be referred to as the undrained plane passing through A or, alternatively, as the undrained plane at distance v_0 from the origin. Hence, knowing the position of A, we can obtain the position of the failure point, B, by drawing the undrained plane through A and noting where it intersects the critical state line.

Expressions for p′ and q can easily be obtained if we remember that the void ratio at failure, e_f, is equal to the void ratio immediately after consolidation, e_0, i.e. $v_0 = v_f$. Now

$$p' = \exp \frac{\Gamma - v_f}{\lambda}$$

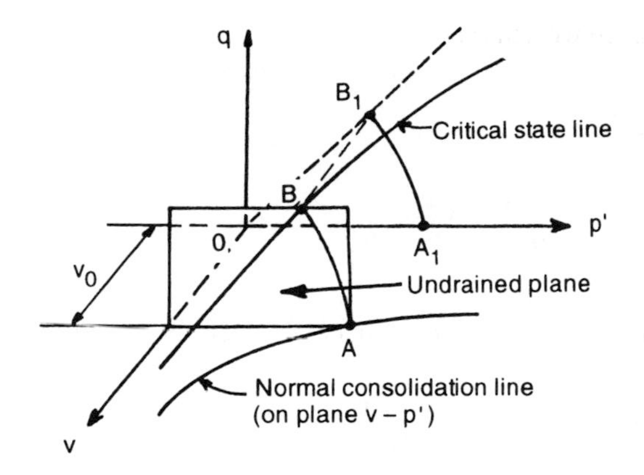

Fig. 13.9 The undrained shear test.

and

$$q = Mp' = M \exp \frac{\Gamma - v_f}{\lambda}$$

Example 13.1

A sample of Weald clay is consolidated in a triaxial cell with a cell pressure of 200 kPa and is then sheared undrained. Using the values for M, N, Γ and λ obtained from Figs 13.6 and 13.8 determine the values of q, p′ and v at failure.

Solution

$$M = 0.85;\ N = 2.13;\ \Gamma = 2.09;\ \lambda = 0.10$$

$$v_0 = N - \lambda \ln p' = 2.13 - 0.1 \ln 200 = 1.60$$

$$\Rightarrow \quad v_f = 1.60$$

$$q = M \exp \frac{\Gamma - v_f}{\lambda} = 0.85 \exp \frac{2.09 - 1.60}{0.1} = 114\ \text{kPa}$$

$$p' = \frac{q}{M} = \frac{114}{0.85} = 134\ \text{kPa}$$

13.7.2 The drained test

From Fig. 13.5 we know that the projection of the drained stress path on to the q–p′ plane is a straight line inclined at angle $\tan^{-1} 3$ to the horizontal. This means that the stress path of drained shear, no matter how it wanders on its journey from A to B, must always lie within the rectangle shown in Fig. 13.10a. In the figure we see that the projection of stress path AB on to the plane q–p′ is the line A_1B_1.

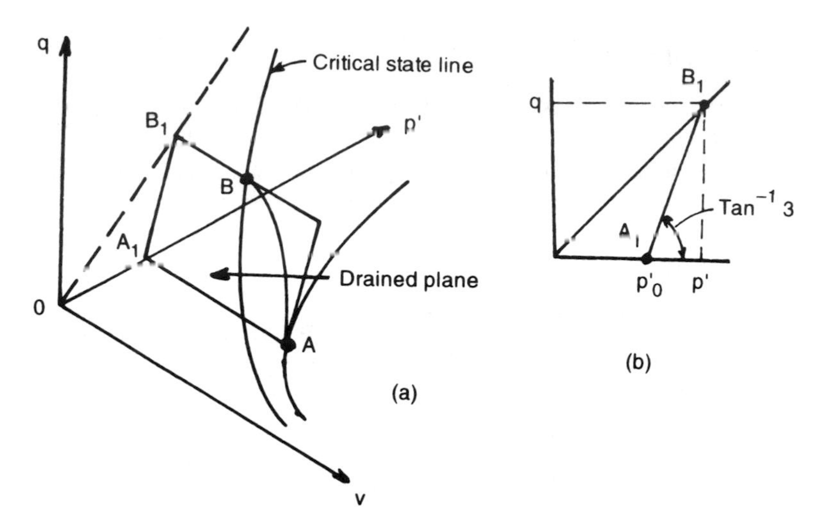

Fig. 13.10 The drained shear test.

Hence, in a manner similar to the undrained test, if we know that a soil has been isotropically consolidated to point A on the normal consolidation curve and is to be subjected to drained shear then we also know that its failure point, B, is at the intersection of the drained plane drawn through A with the critical state line.

From Fig. 13.10b:

$$q = 3(p' - p'_0)$$

and we know that

$$q = Mp'$$

$$\Rightarrow \quad 3(p' - p'_0) = Mp'$$

i.e.

$$p' = \frac{3p'_0}{3 - M}$$

and

$$q = \frac{3Mp'_0}{3 - M}$$

The specific volume at failure, v_f, can therefore be obtained from the formula:

$$v_f = \lambda - \ln \frac{3p'_0}{3 - M}$$

Example 13.2

A sample of clay was subjected to isotropic normal consolidation at a pressure of 350 kPa. The sample was then sheared in a drained state.

Determine the values of q, p′ and v at failure if the properties of the clay were: $M = 0.89$; $N = 2.87$; $\Gamma = 2.76$ and $\lambda = 0.16$.

Solution

$$q = \frac{3Mp'_0}{3-M} = \frac{3 \times 0.89 \times 350}{3 - 0.89} = 443 \text{ kPa}$$

$$p' = \frac{q}{M} = \frac{443}{0.89} = 498 \text{ kPa}$$

$$v_f = \Gamma - \lambda \ln p' = 2.76 - 0.16 \ln 498 = 1.77$$

13.8 The Roscoe surface

For any value of the consolidation pressure p'_0 there will be a corresponding position for A and hence an infinite number of possible planes, drained or undrained, on which stress paths travelling from A to B may lie. A number of planes with their stress paths are shown in Figs 13.11a (undrained) and 13.11b (drained).

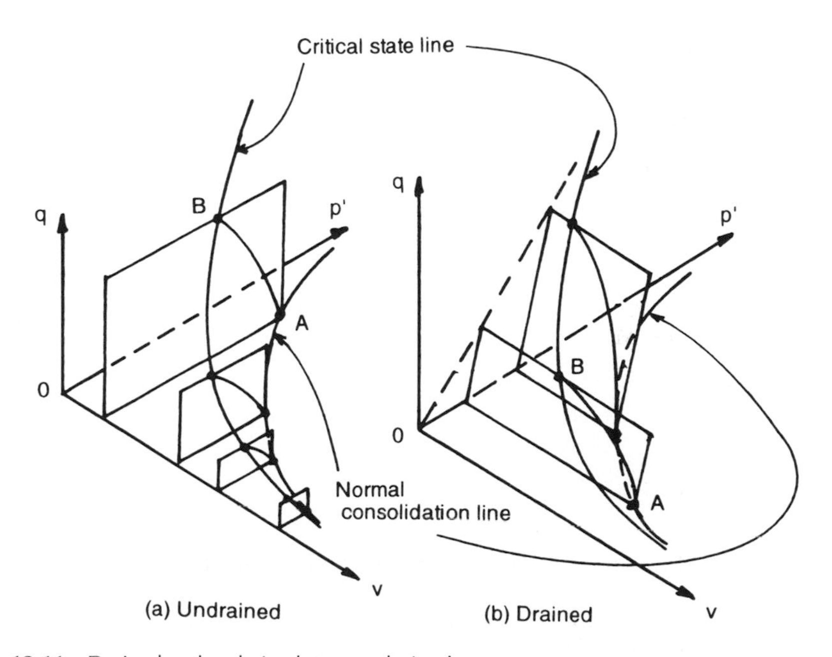

Fig. 13.11 Drained and undrained stress paths in p′–q–v space.

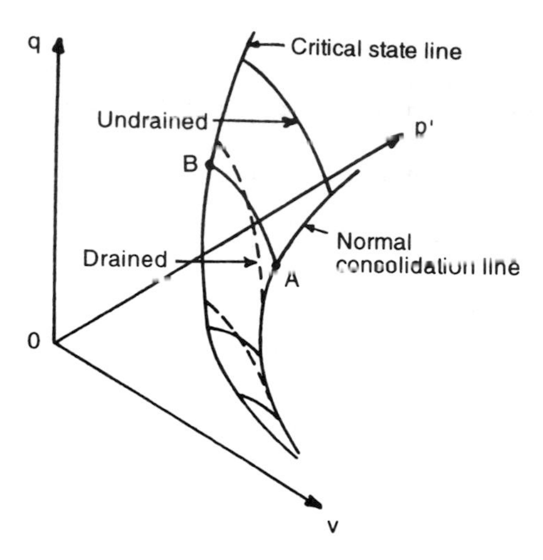

Fig. 13.12 The Roscoe surface.

If we place the two sets of stress paths together we see that they appear to lie on a three-dimensional surface bounded by the critical state line at the top and by the normal consolidation line at the bottom. It can be shown that both sets of stress paths do lie on this surface by the technique of normalisation. If we take the results of a set of undrained and drained compression tests and divide the test q and p′ values by their corresponding p'_e values the resulting plots tend to lie on a single unique line of the form illustrated in Fig. 13.12. Undrained and drained stress paths plotted in p′–q–v space therefore lie on the same three-dimensional surface. This surface is called the Roscoe surface.

Example 13.3

A sample of saturated clay had an initial volume of 86.2 ml and was isotropically consolidated at a cell pressure of 300 kPa, which ensured normal consolidation. During consolidation 6.2 ml of water was expelled into the drainage burette and the void ratio of the sample at this stage was estimated to be 0.893. The sample was then subjected to drained shear, and readings of deviator stress and volume change were taken at increments of axial strain with the following results:

ε_a (%)	$\sigma'_a - \sigma'_r$ (kPa)	ΔV (ml)
5	210	2.47
10	330	5.12
15	415	6.72
20	478	7.76
22 (failure)	507	8.08

If N = 2.92 and $\lambda = 0.18$ for the soil, plot the stress path normalised to p'_e.

Solution

Volume after consolidation, $V_0 = 86.2 - 6.2 = 80.0$ ml

$$\text{Volumetric strain, } \varepsilon_v = \frac{\Delta V}{V_0}$$

Let the specific volume at a particular value of volumetric strain, ε_v, be

$$v = (1 + e)$$

then:

$$\varepsilon_v = \frac{\Delta V}{V_0} = \frac{(1 + e) - (1 + e_0)}{1 + e_0} = \frac{v - v_0}{v_0}$$

$$\Rightarrow \quad v = v_0(1 - \varepsilon_v)$$

Remembering that:

$$p'_e = \exp\frac{N - v}{\lambda}$$

we can now determine the values tabulated below:

ε_a	$\sigma'_a - \sigma'_r$ (kPa)	ΔV (ml)	ε_v	v	p'_e (kPa)	p' (kPa)	p'/p'_e	q/p'_e
0	0	0	0	1.893	300.0	300.0	1.0	0
5	210	2.47	0.031	1.834	417.1	370.0	0.89	0.503
10	330	5.12	0.064	1.772	588.6	410.0	0.70	0.56
15	415	6.72	0.084	1.734	727.0	438.3	0.60	0.57
20	478	7.76	0.097	1.709	835.3	459.3	0.55	0.572
22	507	8.08	0.101	1.702	868.4	469.0	0.54	0.584

The normalised plot is shown in Fig. 13.13.

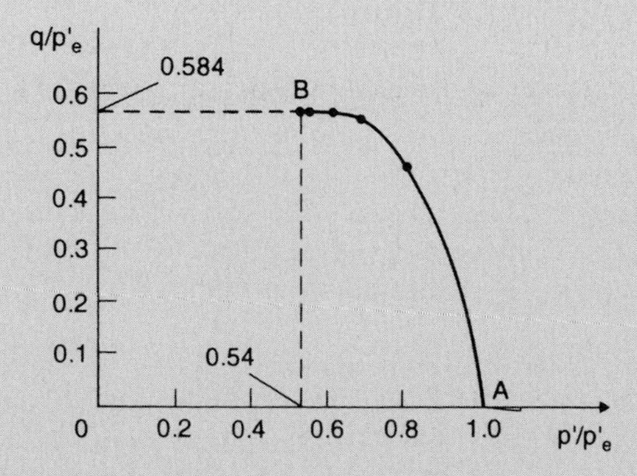

Fig. 13.13 Example 13.3.

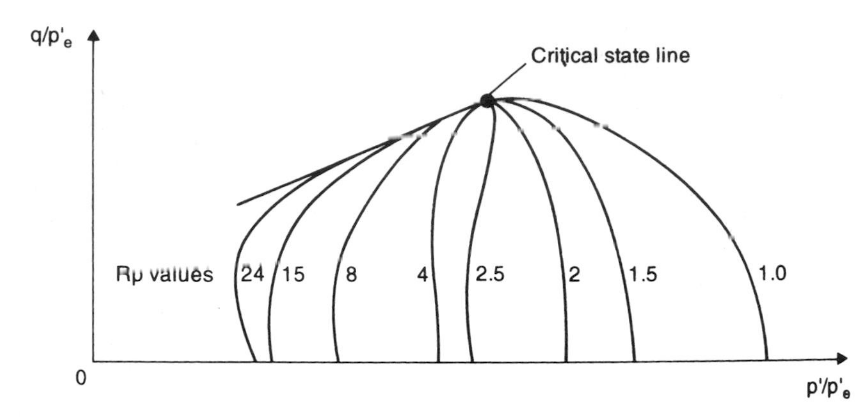

Fig. 13.14 Undrained stress paths of overconsolidated clay.

Overconsolidated clays

We have established that the stress paths of normally consolidated clays lie on the Roscoe surface and, in order to complete the picture, we must determine whether the stress paths of overconsolidated clays also lie on this surface or whether they have a unique surface of their own.

Figure 13.14 shows a series of normalised stress paths of undrained shear obtained from tests carried out on overconsolidated clays. As expected, with $R_p = 1.0$ the stress path lies on the Roscoe surface (as the soil is normally consolidated). For lightly overconsolidated clays, i.e. for R_p values up to about 2.5, the stress paths rise upwards, more or less vertically in the initial loading stage, towards the Roscoe surface but, before reaching it, they bend slightly and gradually make their ways to the critical state line where failure occurs.

The stress paths for the more heavily overconsolidated clays initially rise up more or less vertically and then incline inwards during the final loading stages to become tangential to a common straight line as they make their way towards the critical state line. This straight line is a boundary known as the Hvorslev surface.

13.9 The overall state boundary

The Roscoe surface has the property that any stress state outside it cannot exist in a soil. It is a boundary between possible stress states and impossible stress states and is therefore usually referred to as a state boundary. The Hvorslev surface is a similar state boundary and links up with the Roscoe surface at the critical state line (point B in Fig. 13.15). The Hvorslev surface cannot extend to the q/p'_e axis because of the line of no tension, OC, which rises from the origin at a slope of 1 : 3.

Note If we assume that soil cannot carry tension then tension failure must occur if ever σ'_a is less than σ'_r. Now the lowest possible value of q is 0 which means that the tension failure boundary must pass through the origin. The highest possible value for q will occur when $\sigma'_r = 0$ and q therefore equals σ'_a. At this stage $p' = \sigma'_a/3$ which means that $q/p' = 3$.

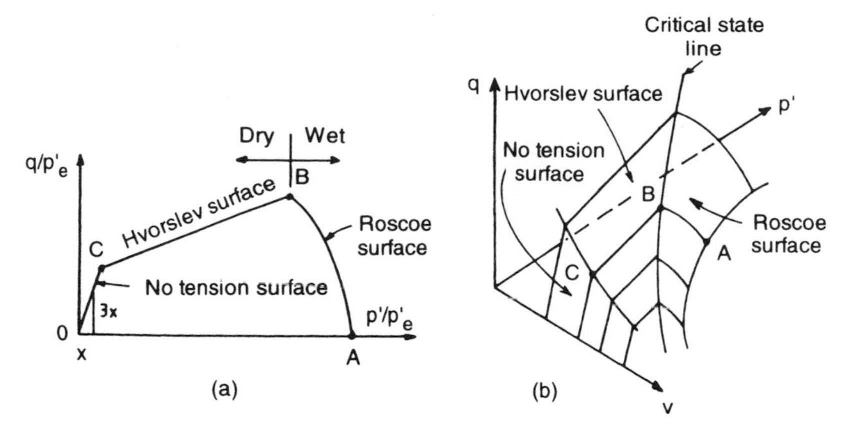

Fig. 13.15 The overall state boundary surface.

Hence, if we select some value for p'/p'_e, say x, then $p'_e = p'/x$ and $q/p'_e = 3x$. The unified plot of the complete state boundary, in q/p'_e–p'/p'_e space is shown in Fig. 13.15a and in p′–q–v space in Fig. 13.15b.

Wet and dry regions

If we examine Fig. 13.15 we see that the state boundary surface for normally consolidated soils is further from the origin than the critical state line. In this region, any soil travelling along a drained stress path from A to B suffers a gradual reduction in specific volume, meaning that the water content at the end of the test must be less than it was immediately before the shearing stage. We can say therefore that soils in this region, when subjected to drained shear, have an initial state that is *wetter* than the critical state.

Heavily overconsolidated clays have initial consolidation points A on the other side of the critical state linc and if these soils are subjected to drained shear they will expand as they approach failure conditions with a corresponding increase in water content. Such soils have an initial state that is *drier* than the critical state.

13.10 Equation of the Hvorslev surface

We have established that the Hvorslev surface is a straight line when projected in q/p'_e–p'/p'_e space, as shown in Fig. 13.16.

We can therefore write the equation for the surface in the form y = mx + c, i.e.

$$\frac{q}{p'_e} = C + \frac{mp'}{p'_e}$$

i.e.

$$q = Cp'_e + mp'$$

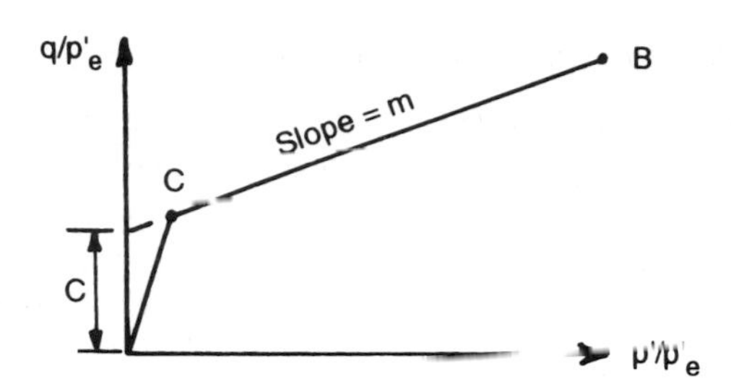

Fig. 13.16 The Horslev surface.

Now

$$p'_e = \exp\left(\frac{N - v}{\lambda}\right)$$

i.e.

$$q = C \exp\left(\frac{N - v}{\lambda}\right) + mp'$$

But, as the Hvorslev surface cuts the critical state line at point B, it must satisfy the critical state equations:

$$q = Mp' \quad \text{and} \quad v = \Gamma - \lambda \ln p'$$

Substituting for q and v gives:

$$Mp' = C \exp\left(\frac{N - \Gamma + \lambda \ln p'}{\lambda}\right) + mp'$$

$$= C\left[\exp\left(\frac{N - \Gamma}{\lambda}\right)\right][\exp(\ln p')] + mp'$$

$$= Cp' \exp\left(\frac{N - \Gamma}{\lambda}\right) + mp'$$

i.e.

$$C = \frac{M - m}{\exp\left(\frac{N - \Gamma}{\lambda}\right)} = (M - m)\exp\left(\frac{\Gamma - N}{\lambda}\right)$$

$$\Rightarrow \quad q = (M - m)\left[\exp\left(\frac{\Gamma - N}{\lambda}\right)\right]\left[\exp\left(\frac{N - v}{\lambda}\right)\right] + mp'$$

$$= (M - m)\exp\left(\frac{\Gamma - V}{\lambda}\right) + mp'$$

Tendency of overconsolidated clays towards critical state

The question must now be asked: Do undrained and drained stress paths for heavily overconsolidated clays reach the critical state line as do normally consolidated and lightly overconsolidated clays?

In an ideal situation the stress paths of heavily overconsolidated clays will reach the Hvorslev surface and then will continue to move up this surface to the critical state line. But ideal situations rarely occur, and a more realistic picture is as follows.

Undrained stress paths

In undrained shear there is no change in specific volume, and the stress paths of heavily overconsolidated clays after reaching the Hvorslev surface will continue upwards to the critical state line, where failure will occur. Failure can always occur before the critical line has been reached if the irregularities in the soil are of significance.

Drained stress paths

With drained shear the stress paths of heavily overconsolidated clays reach the Hvorslev surface on the dry side of critical. At this stage the soil fails, i.e. it has achieved its maximum value of q. However, if the test is allowed to continue the stress path will move up towards the critical state line. The specific volume will increase but, because of this, the value of p'_e will decrease as q also decreases so that the ratio of q/p'_e can increase to allow the stress path to reach the critical state line. That this situation will happen in an actual soil test is speculative but there is sufficient experimental evidence now available to say that there is a tendency for these stress paths to move towards the critical state line after failure.

13.11 Residual and critical strength states

The stress conditions that apply at the critical state line represent the ultimate strength of the soil (i.e. its critical state strength) and this is the lowest strength that the soil will reach provided that the strains within it are reasonably uniform and not excessive in magnitude. The residual strength of a soil operates, in the case of clays, only after the soil has been subjected to considerable strains with layers of soil sliding over other layers.

It is important that the difference between these two strengths is appreciated. Skempton (1964) showed that the residual angle of friction of London clay, ϕ_r, can be less than 10° whereas Schofield and Wroth (1968) reported that the same soil at critical state conditions has an angle of friction ϕ_{cv} of $22\frac{1}{2}$°.

Further study

The material in this chapter is simply an introduction to critical state soil mechanics. Readers interested in pursuing this subject further should refer to Atkinson and Bransby (1978).

Chapter 14
Site Investigation and Ground Improvement

A site investigation, or soil survey, is an essential part of the preliminary design work on any important structure in order to obtain information regarding the sequence of strata and the groundwater level, and also to collect samples for identification and testing. In addition a site investigation is often necessary to assess the safety of an existing structure or to investigate a case where failure has occurred.

British Standard Code of Practice BS 5930, *Site investigations*, lists the following as the main objectives of a site investigation:

(i) to assess the general suitability of the site for the proposed works;
(ii) to enable an adequate and economic design to be prepared;
(iii) to foresee and provide against difficulties that may arise during construction due to ground and other local conditions;
(iv) to predict any adverse effect of the proposed construction on neighbouring structures.

Site investigations are also generally required for the investigation of the safety of an existing structure and for the investigation of a structural failure that has taken place.

14.1 Desk study

The desk study is generally the first stage in a site investigation. It involves collecting and collating published information about the site under investigation and pulling it all together to build a conceptual model of the site. This model can then be used to guide the rest of the investigation, especially the ground investigation. Much of the information gathered at the desk study stage is contained in maps, published reports, aerial photographs and personal recollection.

Sources of information

The sources of information available to the engineer include geological maps, topographical maps (Ordnance Survey maps), soil survey maps, aerial photographs, mining records, groundwater information, existing site investigation reports, local history literature, meteorological records and river and coastal information. Details of a few of these are provided below but a thorough description of the sources of desk study information is given by Clayton *et al.* (1995).

Geological maps

Geological maps provide information on the extent of rock and soil deposits at a particular site. The significance of the geological information must be correctly interpreted by the engineer to assist in the further planning of the site investigation. Geological maps are produced by the British Geological Survey (BGS).

Topographical maps

Ordnance Survey maps provide information on, for example, the relief of the land, site accessibility, and the land forms present. A study of the sequence of maps for the same location produced at different periods in time, can reveal features which are now concealed and identify features which are experiencing change.

Soil survey maps

A pedological soil survey involves the classification, mapping and description of the surface soils in the area and is generally of main interest to agriculturists. The soil studied is the top 1–1.5 m which is that part of the profile which is significantly affected by vegetation and the elements. The maps produced give a good indication of the surface soil type and its drainage properties. The surface soil type can often be related to the parent soil lying beneath, and so soil types below 1.5 m can often be interpreted from the maps.

Aerial photographs

With careful interpretation of aerial photographs it is possible to deduce information on land forms, topography, land use, historical land use, and geotechnical behaviour. The photographs allow a visual inspection of a site when access to the site is restricted. The interpretation of aerial photography is discussed by Dumbleton and West (1970).

Existing site investigation reports

These can often be the most valuable source of geotechnical information. If a site investigation has been performed in the vicinity in the past, then information may already exist on the rock and soil types, drainage, access, etc. The report may also contain details of the properties of the soils and test results.

14.2 Site reconnaissance

A walk over the site can often help to give an idea of the work that will be required. Differences in vegetation often indicate changes in subsoil conditions, and any cutting, quarry or river on or near the site should be examined. Site access, overhead restrictions, signs of slope instability are further examples of aspects which can be observed during the walk over survey.

14.3 Ground investigation

14.3.1 Site exploration methods

Test or trial pits

A test pit is simply a hole dug in the ground that is large enough for a ladder to be inserted, thus permitting a close examination of the sides. With this method groundwater conditions can be established exactly and undisturbed soil samples are obtainable relatively easily. Below a depth of about 4 m, the problems of strutting and the removal of excavated material become increasingly important and the cost of trial pits increases rapidly; in excavations below groundwater level the expense may be prohibitive.

Hand auger or post-hole auger

The hand auger (attached to drill rods and turned by hand) is often used in soft soils for borings to about 6 m and is useful for site exploration work in connection with roads. In cohesive soils the clay auger shown in Fig. 14.1 is used, but for gravels a gravel (or worm) auger can be employed.

Boring rig

In most site investigations the boreholes are taken down by some form of well-boring equipment and can extend to considerable depths, the operation usually being carried out in the dry in the UK whereas in the USA wash boring techniques are more common.

The auger is replaced by a clay cutter – a much heavier unit weighing about 55 kg (Fig. 14.2) – and power is provided to lift and lower it. Boring consists of dropping the cutter from some 1.5 to 3 m above the soil and is largely carried out by hand,

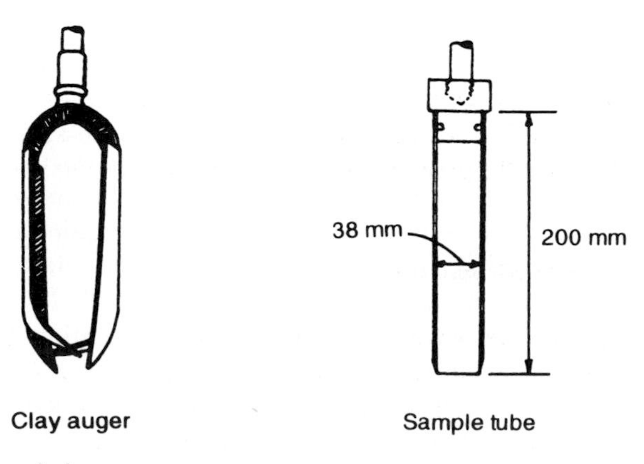

Fig. 14.1 The post-hole auger.

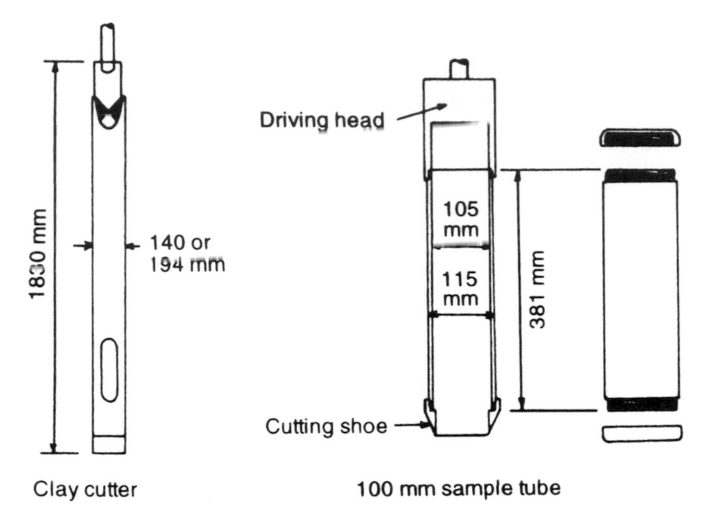

Fig. 14.2 Clay cutter and sample tube.

although this practice is going out of fashion and when site conditions are suitable the operation is often powered. In compact sands and gravels water is generally added if the deposit is not already wet. The material is removed by means of a shell which is dropped in a similar manner to the clay cutter: it is fitted with a clack (a hinged lid) that closes as the shell is withdrawn and retains the loose particles. In extremely hard granular deposits a chisel is sometimes necessary to achieve break up of the soil.

Boreholes in sands and gravels, and most deep boreholes in clay, must be lined with steel tubes to prevent collapse of the sides, the casings (of slightly larger diameter than the cutter) being hammered or surged downwards as boring proceeds.

In order to prove bedrock a minimum penetration of 1.5 to 3 m is generally required. Penetration into soft rocks is sometimes possible with the chisel but for hard rocks a diamond drill becomes necessary, particularly if rock cores are to be obtained.

Setting out of trial pits and/or boreholes

An accurate setting out by theodolite is not necessary, since lining in with structures marked on an Ordnance Survey map or using a compass survey will give all the accuracy needed. For heavy structures, boreholes require to be some 15 to 30 m apart and should be taken down to about 1.5 times the width of the structure unless rock is encountered at lower depths. For roads the boreholes need not be closer than 300 m centres unless vegetation changes indicate variations in soil conditions, and need not go beyond 3 m below formation level.

Guidance on site investigation in the form of a handbook for engineers, is given by Clayton *et al.* (1995).

14.3.2 Sampling

Two types of soil sample can be obtained: disturbed sample and undisturbed sample.

Disturbed samples

The auger parings or the contents of the shell can be collected as disturbed soil samples. Such soil has been remoulded and is of no use for shear strength tests but is useful for identification tests, w_L and w_P, particle size distribution, etc.

Disturbed samples are usually collected in airtight tins or jars or in plastic sampling bags, and are labelled to give the borehole number, the depth, and a description.

Undisturbed samples (cohesive soil)

In a trial pit samples can be cut out by hand if care is taken. Such a sample must be placed in an airtight container and as a further precaution should first be given at least two coats of paraffin wax.

The hand auger can be used to obtain useful samples for unconfined compression tests and employs 38 mm sampling tubes with a length of 200 mm (Fig. 14.1). The auger is first removed from the rods and the tube fitted in its place, after which the tube is driven into the soil at the bottom of the borehole, given a half turn, and withdrawn. Finally, the ends of the tube are sealed with paraffin wax.

With the boring rig, 100 mm diameter undisturbed samples are collected, the sampling tube being 105 mm diameter and usually 381 mm long (Fig. 14.2) but dimensions can vary, e.g. tubes 106 mm diameter and 457 mm long are also used. The tube is first fitted with a special cutting shoe and then driven into the ground by a falling weight in a similar manner to the standard penetration test; during driving any entrapped water, air or slush can escape through a non-return value fitted in the driving head at the top of the tube. After collection the sample is sealed at both ends with paraffin wax and, as a further precaution, sealing caps are screwed on to the tube.

For soils such as soft clays and silts that are sensitive to disturbance a *thin-walled* sample tube can be used. Because of the softness of the soil to be collected the tube is simply machined at its end to form a cutting edge and does not have a separate cutting shoe. The thin-walled sampler is similar in appearance to the sample tube shown in Fig. 14.1 but can have an internal diameter of up to about 200 mm.

These sampling techniques involve the removal of the boring rods from the hole, the replacement of the cutting edge with the sampler, the reinsertion of the rods, the collection of the sample, the removal of the rods, the replacement of the sampler with the cutting edge and, finally, the reinsertion of the rods so that boring may proceed. This is a most time-consuming operation and for deep bores, such as occur in site investigations for off-shore oil rigs, techniques have been developed to enable samplers to be inserted down through the drill rods so that soil samples can be collected much more quickly.

Degree of sample disturbance

No matter how careful the technique employed there will inevitably be some disturbance of the soil during its collection as an 'undisturbed' sample, the least disturbance occurring in samples cut from the floor or sides of a trial pit. With sample tubes, jacking is preferable to hammering although if the blows are applied in a regular pattern there is little difference between the two.

The degree of disturbance has been related (Hvorslev, 1949) to the area ratio of the sample tube:

$$\text{Area ratio} = \frac{D_e^2 - D_i^2}{D_i^2}$$

where D_e and D_i are the external and internal diameters of the tube respectively. It is generally agreed that, for good undisturbed 100 mm diameter samples, the area ratio should not exceed 25 per cent, but in fact most cutting heads have area ratios $\approx$ 28 per cent. For 38 mm samples the area ratio should not exceed 20 per cent. Thin-walled sample tubes, of any diameter, have an area ratio of about 10 per cent.

Undisturbed samples (sands)

If care is taken it may be possible to extract a sand sample by cutting from the bottom or sides of a trial pit. In a borehole, above groundwater level, sand is damp and there is enough temporary cohesion to allow samples to be collected in sampling tubes, but below groundwater level tube sampling is not possible. Various techniques employing chemicals or temporarily freezing the groundwater have been tried, but they are expensive and not very satisfactory; the use of compressed air in conjunction with the sampler evolved by Bishop (1948), however, enables a reasonably undisturbed sample to be obtained.

Owing to the fact that sand is easily disturbed during transportation any tests on the soil in the undisturbed state should be carried out on the site, the usual practice being to use the results of penetration tests instead of sampling.

Frequency of sampling

Samples, both disturbed and undisturbed, should be taken at every change of stratum and at least at every 1.5 m in apparently homogeneous material.

Continuous sampling

In some cases, particularly where the soil consists of layers of clay, separated by thin bands of sand and silt and even peat, it may be necessary to obtain a continuous core of the soil deposits for closer examination in the laboratory. Such sampling techniques are highly specialised and require the elimination of friction between the soil sample and the walls of the sampler. A sampler which reduces side friction by the use of thin strips of metal foil placed between the soil and the tube was developed by Kjellman *et al.* (1950) and is capable of collecting a core 68 mm in diameter and up to 25 m in length.

14.3.3 In situ *tests*

Penetration tests

The standard penetration test that was described in Chapter 8 is normally used for cohesionless soils, although Terzaghi and Peck (1948) give an approximate relationship for clays:

For square footings, $q_a \approx 16\ N$ (kPa)
For continuous footings, $q_a \approx 12\ N$ (kPa)

where

q_a = safe bearing capacity (F = 3)
N = uncorrected test number of blows.

The Dutch cone test, also described in Chapter 8, can be used for both cohesive and cohesionless soils.

Plate loading test

Loading tests are more applicable to cohesionless soils than to cohesive soils due to the time necessary for the latter to reach full consolidation. Generally two tests are carried out as a check on each other, different-sized plates of the same shape being used in granular soils so that the settlement of the proposed foundation can be evolved from the relationship between the two plates. The loading is applied in increments (usually one-fifth of the proposed bearing pressure) and is increased up to two or three times the proposed loading. Additional increments should only be added when there has been no detectable settlement in the preceding 24 hours. Measurements are usually taken to 0.01 mm, and where there is no definite failure point the ultimate bearing capacity is assumed to be the pressure causing a settlement equal to 20 per cent of the plate width.

Vane test

In soft sensitive clays it is difficult to obtain samples that have only a slight degree of disturbance, and *in situ* shear tests are usually carried out by means of the vane test (Fig. 14.3). The apparatus consists of a 75 mm diameter vane, with four small blades 150 mm long. For stiff soils a smaller vane, 50 mm diameter and 100 mm high, may be used. The vanes are pushed into the clay a distance of not less than three times the diameter of the borehole ahead of the boring to eliminate disturbance effects, and the undrained strength of the clay is obtained from the relationship

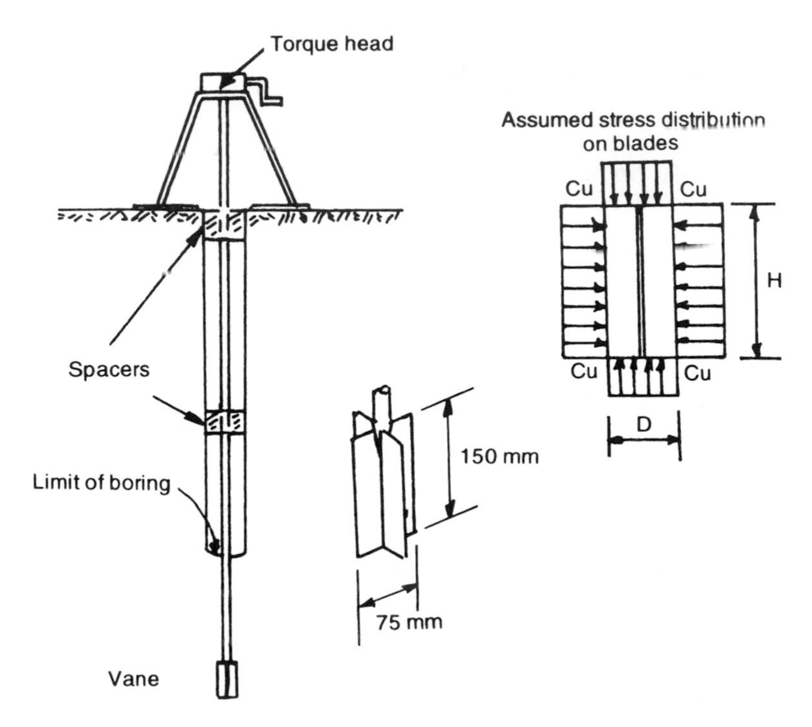

Fig. 14.3 The vane test.

with the torque necessary to turn the vane. The rate of turning the rods, throughout the test, is kept within the range 6–12° per minute. After the soil has sheared, its remoulded strength can be determined by noting the minimum torque when the vane is rotated rapidly.

Figure 14.3 indicates that the torque head is mounted at the top of the rods. This is standard practice for most site investigation work but, for deep bores, it is now possible to use apparatus in which the torque motor is mounted down near to the vane, in order to remove the whip in the rods.

Because of this development the vane has largely superseded the standard penetration test, for deep testing. The latter test has the disadvantage that the load must always be applied at the top of the rods so that some of the energy from a blow must be dissipated in them. This energy loss becomes more significant the deeper the bore, so that the test results become more suspect.

The actual stress distribution generated by a cylinder of soil being rotated by the blades of a vane which has been either jacked or hammered into it is a matter for conjecture. BS 1377 has adopted the simplifying assumption that the soil's resistance to shear is equivalent to a uniform shear stress, equal to the undrained strength of the soil, c_u, acting on both the perimeter and the ends of the cylinder (see Fig. 14.3).

For equilibrium the applied torque, T, = moment of resistance of vane blades. The torque due to the ends can be obtained by considering an elemental annulus and integrating over the whole area:

$$\text{End torque} = 2 \times c_u \int_0^r 2\pi r \, dr \, r = 2c_u \left[2\pi \frac{r^3}{3} \right]_0^{D/2} = c_u \frac{\pi D^3}{6}$$

$$\text{Side torque} = c_u \pi DH \times \frac{D}{2} = c_u \frac{\pi D^2 H}{2}$$

$$T = c_u \frac{\pi D^2 H}{2} \left(1 + \frac{D}{3H} \right)$$

where

D = measured width of vane
H = measured height of vane.

Example 14.1

A vane, used to test a deposit of soft alluvial clay, required a torque of 67.5 Nm.
The dimensions of the vane were: D = 75 mm; H = 150 mm.
Determine a value for the undrained shear strength of the clay.

Solution

$$T = c_u \frac{\pi D^2 H}{2} \left(1 + \frac{D}{3H} \right)$$

i.e.

$$67.5 = c_u \pi \times \frac{0.075^2 \times 0.15}{2} \left(1 + \frac{0.075}{0.45} \right) \times 1000 \text{ kPa}$$

$$\Rightarrow \quad c_u = 44 \text{ kPa}$$

14.3.4 Water level observations

It is not possible to determine accurate groundwater conditions during the boring and sampling operations, except possibly in granular soils.

Standpipes

In clays and silts it takes some time for water to fill in a borehole, and the normal procedure for obtaining the groundwater level is to insert an open-ended tube, usually 50 mm in diameter and perforated at its end, into the borehole (Fig. 14.4). The tube is packed around with gravel and sealed in position with puddle clay, the borehole then being backfilled to prevent access to rainwater. Observations must be taken for several weeks until equilibrium is achieved.

By inserting more than one tube, different strata can be cut off by puddled clay and the various water heads obtained separately. When a general water level is to be

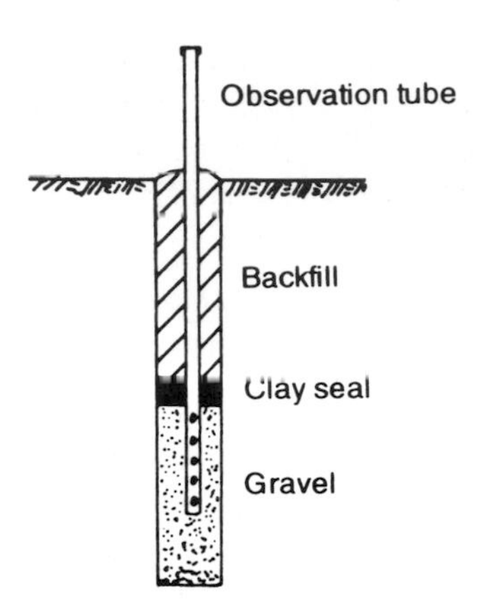

Fig. 14.4 Ground water level observation in a borehole.

obtained, the gravel is usually extended to within a short distance of the top of the borehole and then sealed with the puddle clay.

Pore water pressure measurement

Open-ended tubes have a tendency to silt up, as well as exhibiting a slow response to rapid pore water pressure changes that can be caused by tidal variations or changes in foundation loadings. Casagrande type standpipe piezometers are more commonly used. They have a porous intake filter and are sealed into the soil with either bentonite plugs or with a bentonite/cement grout. When a faster response is necessary a closed piezometer system, such as a pneumatic piezometer or a vibrating wire piezometer (see Chapter 5), is used instead of an open one.

The advantages of the electrical system are that (i) pressure is measured at the tip so that piezometric levels below the gauge house level can be recorded, (ii) the ancillary equipment is compact, and (iii) the time response of these instruments to pore pressure changes is fairly rapid. Disadvantages include the fact that the readings from an electric tip depend upon an initial calibration that cannot be checked once the tip has been installed (unless it is dug out), and the risk of calibration drift (especially if the tip is to be in operation for some time). The general tendency appears to be to use the hydraulic tip whenever possible.

Instrumentation in geotechnical engineering is dealt with in detail by Dunnicliff (1993).

14.3.5 Soil profile

From the results of a site investigation vertical sections (soil profiles) are generally prepared, showing to scale the sequence and thickness of the strata.

Foundation engineers are mainly interested in the materials below the subsoil, and with stratified sedimentary deposits conditions may be more or less homogeneous.

Boulder clay deposits can also be homogeneous although unstratified, but they often have an erratic structure in which pockets of different soils are scattered through the main deposit and make it difficult to obtain an average value for the deposit's characteristics. Furthermore, the boulder clay itself may vary considerably, and at certain levels it can even decrease in strength with increasing depth.

Secondary structure of deposits

Besides the primary structure of stratification, many clays contain a secondary structure of hair cracks, joints and slickensides. The cracks (often referred to as macroscopic fissures) and joints generally occurred with shrinkage when at some stage in its development the deposit was exposed to the atmosphere and dried out; slickensides are smooth, highly polished surfaces probably caused by movement along the joints. If the effect of these fissures is ignored in the testing programme the strength characteristics obtained may bear little relationship to the properties of the clay mass.

With the application of a foundation load there is little chance of the fissures opening up, but in cuttings (due to the expansion caused by stress relief) some fissures may open and allow the ingress of rain water which will eventually soften the upper region of the deposit and lead to local slips. Fissures are more prevalent in overconsolidated clays, where stress relief occurs, than in normally consolidated clays, but any evidence of fissuring should be reported in the boring record.

14.4 Site investigation reports

The site investigation report is the final product of the exploration programme. It consists of a summary of the ground conditions encountered, a list of the tests carried out and recommendations as to possible foundation arrangements.

The recipient of the report is the client, the person or company who pays for the work done. Such people are rarely engineers and therefore appoint an architect or consulting engineer to design any proposed development. The person appointed is naturally concerned with the financial aspects of the work he is supervising and this applies to the site investigation work as well as to the later construction.

Obviously someone must keep a rein on expenditure, but if this attitude is too strictly maintained it can have detrimental effects on the efficacy of a site investigation. It is not unknown for a consultant to employ a soils investigation firm to prepare a report on a development site and to specify, in advance, the number of boreholes, the number of samples to be collected and the number of laboratory tests that will be carried out. If relatively homogeneous subsoil conditions are encountered, such a procedure can lead to unnecessary costs, whereas with variable conditions, the money allocated may be totally inadequate if a meaningful report is to be achieved.

Ideally, the soils engineer should be allowed to modify the site investigation programme as work proceeds. Such a system could obviously be abused but, with reputable soils firms, can prove to be both efficient and economical.

Reports are generally prepared in sections, headed as described in the following section.

RECORD OF BOREHOLE 109

Ref. No. 866 **Dia. of boring** 150mm to 13.70m LCP
Ground level 3.40m 76mm to 17.40m DD

PROGRESS			SAMPLE/TEST		STRATA			
HOLE	CASING	WATER	DEPTH	TYPE	LEGEND	DEPTH	LEVEL	DESCRIPTION
28/8/98			0.30	D1		0.30	3.10	Brown sandy TOPSOIL
			1.88 - 2.13	S(29)				medium dense to dense red
				D2				brown silty SAND and FINE
		Met at	0.30 - 2.18	B1				GRAVEL
		2.00	2.20	W				
			2.50	D3		2.6	0.80	
			3.15 - 3.45	C(6)				Medium dense brown silty
			3.45 - 3.75	(17)				SAND with clayey layers,
				D4				containing occasional gravel
			2.70 - 3.90	B2				
			4.10 - 4.55	U-/120				
4.56	4.56	4.00	4.10	D5		4.56	-1.16	
29/8/98		1.50						Stiff light brown laminated
			5.50 5.95	U1/80				silty CLAY, with layers of
			6.00	D6				sand
			6.90	D7				
						7.10	-3.70	
								Medium dense becoming
			7.45 - 7.75	S(29)				dense brown SAND
				D8				
			8.45 - 8.75	S(22)				
				D9				
			9.45 - 9.75	S(26)				
				D10				
			10.45 - 10.75	S(46)				
10.73	10.73	8.00		D11				
30/8/98		1.50	11.20	D12		11.20	-7.80	
			11.30 - 11.75	U2/120				Compact brown silty SAND
			11.80	D13		11.80	-8.40	with layers of silty clay
								Compact brown SAND and
								GRAVEL
13.40	12.90	9.90	12.90 - 13.35	U3/150		12.80	-9.40	
31/8/98		8.00	13.35	D14				Dense grey-brown clayey
								SAND with occasional gravel
13.70	13.44	13.00	13.70	D15		13.70	-10.30	
4/9/98		1.50	13.70 - 15.10	1.40•				Hard mottled red-brown,
								grey and green coarse
		WATER						grained BASALTIC TUFF
		FLUSH	15.10 - 16.34	1.24				
16.34	13.44	-0.36				16.13	-12.73	
5/9/98		1.04	16.34 - 17.40	1.06				Soft and medium hard
								weathered mottled red-brown,
								grey-green BASALTIC TUFF
						17.19	-13.79	
								Hard mottled grey-green
17.40	18.41	-0.36				17.40	-14.00	BASALTIC TUFF

Remarks: Penetration test continued beyond normal drive from 3.45m. 40mm diameter perforated standpipe 18.00m long inserted, surrounded by gravel filter with bentonite seal and screw cap at surface.

Key:

D	Disturbed sample	**S(30)**	Standard penetration test	**U1/70**	Undisturbed sample 100mm dia
B	Bulk sample	**C(27)**	Cone penetration test	**/70**	Blows to drive sample 450mm
W	Water sample	**(27)**	No blows for 300mm pentn.	**U-/70**	U/d sample - no recovery
•	Core recovery	**V**	*In situ* vane shear test	**LCP**	Light cable percussion
				DD	Rotary diamond drilled

Fig. 14.5 Borehole log.

Preamble

This introductory section consists of a brief summary which gives the location of the site, the date of the investigation and name of the client, the types of bores put down and the equipment used.

Description of site

Here a general description of the site is given: whether it is an open field or a redevelopment of a site where old foundations, cellars and walls, etc., remain. Some mention is made of the general geology of the area, whether there are old mineral workings at depth and, if so, whether the report has considered their possible effects or not. A map, showing the site location and the positions of any boreholes put down, is usually included in the report.

Description of subsoil conditions encountered

This section should consist of a short, and readable, description of the general subsoil conditions over the site with reference to the borehole journals. Generally the significance of any *in situ* testing carried out is mentioned.

Borehole journals

A borehole journal is a list of all the materials encountered during the boring. A journal is best shown in sectional form so that the depths at which the various materials were met can be easily seen. A typical borehole journal is shown in Fig. 14.5. It should include a note of all the information that was found, groundwater conditions, numbers and types of samples taken, list of *in situ* tests, time taken by boring, etc.

Description of laboratory soil tests

This is simply a list of the tests carried out together with a set of laboratory sheets showing particle size distribution curves, liquid limit plots, Mohr circle plots, etc.

Conclusions

It is in this section that firm recommendations as to possible foundation types and modes of construction should be given. Unless specified otherwise, it is the responsibility of the architect or consultant to decide on the actual structure and the construction. For this reason the writer of the report should endeavour to list possible alternatives: whether strip foundations are possible, if piling is a sensible proposition, etc. For each type listed an estimation of its size, working load and settlement should be included.

If the investigation has been limited by specification or finance and the conclusions have been based on scant information, it is important that the fact is mentioned so that any possible allegations of negligence may be refuted.

14.5 Ground improvement

A simple definition of *in situ* improvement of a soil deposit is the increase in its shear strength along with a reduction in its compressibility.

14.5.1 Drainage or consolidation techniques

Surcharge loading

Surcharge loading is probably the simplest method of ground improvement and can be applied to cohesive soils. The technique involves subjecting the surface of the soil to a temporary loading using some method such as the placing of temporary earth fill, water filled tanks or tension piles secured to some form of framework, etc. The soil experiences consolidation under this loading and both its stiffness and shear strength increase. The time taken for full consolidation depends upon the length of the drainage path which can be decreased by the insertion of vertical drainage wells.

Stage construction

This technique is also used for cohesive soils and involves determining the rate of construction that will allow the soil to consolidate and increase in strength sufficiently to maintain an adequate factor of safety against bearing capacity failure for the corresponding increment of construction loading. By proceeding in constructional steps the foundation soil eventually becomes sufficiently strong to support the full construction loading. Because the soil settles during the construction phase the method is usually applied to earth embankments rather than to rigid foundations. The stress path method evolved by Lambe (1964, 1967), which has been described in Chapter 9, can be used for this approach but it is usually also necessary to monitor the actual soil behaviour during construction using some form of instrumentation installed at the start of the work.

Electro-osmosis

This method causes water within a soil to drain away under the action of an electrical potential and can be very effective in fine grained soils, such as silts and clayey silts where well point systems (see Chapters 2 and 10) cannot be used because of the low permeability of the soil. The system was first used by Germany in World War II during the construction of U-boat pens at Trondheim, in Norway, and its application has been described by Casagrande (1947).

Steel or aluminium rods, from 10 to 100 mm diameter are driven into the soil over the area to be treated. These rods act as anodes, their corresponding cathodes being conventional well points. An electrical potential of some 50 volts per metre is created by direct current and the water within the soil is gradually driven towards the well points, which are pumped out at intervals.

The method is only rarely used, possibly because of the high installation and running costs.

14.5.2 Compactive techniques

Possibly the most common method for improving ground is by compaction of the soil in series of layers. Compaction is described in Chapter 11 and is particularly suitable for fill material. However, with existing soil deposits, modern compaction plant can only improve the soil for a depth of 1 or 2 m below its surface so that, for the improvement of a deep soil deposit where deep compaction is required, some other method must be used.

Vibro-compaction

This method cannot be used in cohesive soils and is most effective in granular soils, although soils with up to some 25 per cent silt can also be treated. A large vibrating probe, suspended from a crane, is lowered into the ground. The probe penetrates downwards under its own weight and compacts the surrounding soil, up to a distance of about 2.5 m from the probe, by virtue of the temporary reduction in effective stress caused by the vibration. Probes are normally spaced at 1.5 to 3.0 m and can compact suitable soils to a depth of about 12 m.

Vibro-flotation

In order to assist the penetration of the vibrating probe into the ground, water jets can be fitted at the top and bottom of the probe. In this case the process is referred to as vibro-flotation and the probe is called a *vibrofloat*.

Vibro-replacement

This technique can be used to improve the load-carrying capacity of soft silts and clays. Essentially the soil is reinforced by the insertion of stone columns. This is achieved with the use of a vibrating probe, similar to the technique used in vibro-compaction. The probe is allowed to penetrate the soil and does so by displacing the soil radially. Once the required depth has been reached the probe is withdrawn and the hole created by the probe backfilled with graded aggregate, up to 75 mm in size. The probe is then reintroduced to both compact and radially displace the aggregate. The process is repeated until the required stone column has been created. With soft clays soil is removed, not displaced, by means of water jets fitted to the probe. The method is really only suitable for light foundation loads as heavy loads can cause excessive settlement

Dynamic consolidation

This method involves the dropping of a large weight, 100 to 400 kN, from a height of 5 to 30 m, on to the surface of the soil. It is seen that the energy delivered to the soil per blow can be as high as 12 000 kN m although the energy values normally used lie between 1500 and 5000 kN m. The impact of the weight with the soil creates shock waves that can penetrate to a depth of 10 m. In cohesionless soils these shock waves create liquefaction, immediately followed by compaction of the soil, whereas in cohesive soils they create excessive pore water pressures, which are followed by the consolidation of the soil.

The work is generally carried out by a specialist contractor whose engineering judgement can be used to give a reasonable estimate of the energy requirements for a particular site.

Before the work is commenced the area to be treated is covered with a layer of granular material of thickness between 0.5 and 1.0 m. The layer acts as a working platform for the equipment and helps to prevent excessive penetration of the weight. It also provides a pre-load surcharge of some 10 to 20 kN/m^2 and helps to drain away water as it is driven out of the soil.

The weight, or *pounder*, is usually dropped five to ten times at each selected point, the points being spaced on a square grid, 5 to 15 m in dimension. In the case of cohesive soils not all the blows are delivered at once as it is necessary to have pauses in order that full consolidation for a particular compaction of a treated area is first achieved. These pauses can extend to weeks in some cases.

14.5.3 Grouting techniques

The engineering properties of a soil can be improved by the injection of chemical fluids which solidify and hence strengthen the soil structure. Obviously the system is only effective if the voids of the soil can be penetrated by the grout, and it therefore has little application for cohesive soils, except when fissures require to be sealed. Grouting is mainly restricted to granular soil and weathered rocks. The procedure is expensive and is only used when other methods of soil improvement are not applicable.

Cement grouts

Cement grouts are used to seal fissured rocks and to decrease permeability in sand and gravel deposits.

Bentonite and bitumen slurries

Suspensions of bentonite and emulsions of bitumen can be used to reduce the permeability of sands and gravels provided the grain size is not less than medium sand.

Chemical grouts

For fine sands chemical grouts such as sodium silicate, which comes in the form of a syrupy liquid, are made to react with a compound, such as calcium chloride, to form a stiff silica gel. The two agents can be mixed together along with a retarding agent so that gelification does not happen until the grouting process is completed, or they can be injected separately so that they react together within the soil mass. The latter process has been used successfully for many years but has the disadvantage that a large number of grout holes, spread at not more than 700 mm, are required.

14.5.4 Geotextiles

The use of fabrics in ground improvement techniques is a recent development which has been highly successful and has taken place over the last 30 years.

The first use of fabrics was as a temporary expedient whereby the surfaces of soft soils were covered with fibre grids on to which temporary roads could then be constructed. Nowadays fabrics are used in the permanent construction of most forms of earthworks.

Fabrics range from natural products, such as cotton, jute and wool, to the polymer plastics produced from long chain hydrocarbon molecules which are now being increasingly used and are briefly described in Chapter 7. A generic term, *geotextiles*, is used to cover all the various different fabrics. In this chapter we will only concern ourselves with plastic materials, which now account for at least 75 per cent of the fabrics used in civil engineering.

Functions of geotextiles

Geotextiles are incorporated into a soil structure to satisfy at least one of the following functions:

(i) separation;
(ii) filtration;
(iii) drainage;
(iv) reinforcement.

Separation

The base of a pavement construction may be subjected to separation if it is placed directly on to the surface of a soft subgrade. Separation is the upward migration of particles of the fine subgrade soil accompanied by the downward movement of the denser base particles. Such intermixing of soil particles can create a weak zone at the interface between the two materials resulting in considerable reduction in bearing capacity strength.

The placing of a relatively weak strength geotextile fabric on the surface of a soft subgrade, prior to constructing the base, is all that is necessary to provide a permanent solution to separation between the two materials.

Filtration

Where a cohesive soil is subjected to seepage a suitable geotextile can be used to prevent the migration of the fine soil particles in exactly the same way as the granular filters described in Chapter 2. A geotextile filter, placed at the end of the seepage path, operates in a different manner to a granular filter. Soil particles tend to collect at the boundary between the soil and the geotextile and this appears to induce a self-filtration effect within the soil.

Drainage

Special types of permeable geotextile fabrics can be used to form drainage layers in basements and behind retaining walls in exactly the same manner as the layers of granular material illustrated in Figs 6.21c, d and e.

Reinforcement

The use of plastic reinforcement in reinforced soil retaining walls is now well established and is increasing. The technique is mentioned in Chapter 7.

In the construction of an earth embankment on top of a soft foundation soil a layer of geotextile fabric, placed on the surface of the soft soil, can give enough tensile strength to allow it to support an incremental layer of the embankment without spreading or edge failure during consolidation and thus permit stage construction to be carried out.

The sub-bases of roads supported by soft subgrades can be strengthened by the inclusion of layers of a geotextile fabric.

14.6 Environmental geotechnics

Environmental geotechnics brings together the principles of geotechnical engineering with the concerns for the protection of the environment and the subject is becoming increasingly important to the geotechnical engineer. Applications of environmental geotechnics include contaminated land (both its control and reclamation), containment of toxic wastes, design of landfill sites, and the management of mining wastes. These applications have a number of common features which epitomise environmental geotechnics problems: soil water flow problems, soil chemistry, and local and national government legislation. Many of the environmental geotechnics issues concern the leaching of toxins into the soil and groundwater supplies, and so the soil properties which are of greatest significance are permeability, void ratio and plasticity. The study of environmental geotechnics is a subject in its own right and is beyond the scope of this book. Readers interested in this aspect of geotechnics should refer to the texts of Attewell (1993), Cairney (1993), and Harris (1994) for a description of the subject.

References

Alonso, E.E., Gens, A. and Hight, D.W. (1987) 'Special problem soils'. General Report. *Proc 9th Eur. Conf. Soil Mech.*, **3**, Dublin.

Alonso, E.E., Gens, A. and Josa, A. (1990) 'A constitutive model for partially saturated soils', *Géotechnique*, **40**, (3).

American Society for Testing and Materials (1980) *Natural building stones, soil and rock*. Annual book of ASTM Standards. Philadelphia, Pennsylvania.

Atkinson, J.H. and Bransby, P.L. (1978) *The mechanics of soils – an introduction to critical state soil mechanics*. McGraw-Hill, London.

Atterberg, A. (1911) 'Die Plastizitat der Tone'. *Int. Mitt. für Bodenkunden*, I; (1) 10–43, Berlin.

Attewell, P. (1993) *Ground pollution: Environment, geology, engineering and law*. E & FN Spon, London.

Baird, H.G. (1988) 'Earthworks control – assessment of "suitability"'. *Ground Engineering*, **22**, (4).

Barden, L. (1974) 'Sheet pile wall design based on Rowe's method'. Part III of CIRIA Technical Report No. 54 – *A comparison of quay wall design methods*, London.

Barron, R.A. (1948) 'Consolidation of fine grained soils by drain wells'. *Transactions ASCE*, **113**.

Bazaraa, A.R.S.S. (1967) 'The use of the standard penetration test for estimating settlements of shallow foundations on sand'. PhD Thesis, University of Illinois.

Begemann, H.K.S. (1965) 'The friction jacket cone as an aid in determining the soil profile'. *Proc. 6th Int. Conf. ISSMFE*, **1**, pp. 17–20, Montreal.

Bell, A.L. (1915) 'Lateral pressure and resistance of clay and the supporting power of clay foundations'. *Minutes Proc. Inst. Civil. Engrs*, London.

Bell, J.P., Dean, T.J. and Hodnett, M.G. (1987) 'Soil moisture measurement by an improved capacitance technique, Part II. Field techniques, evaluation and calibration'. *Jour. Hydrology*, **93**, pp. 79–90.

Berezantzev, V.G., Khristorforov, V. and Golubkov, V. (1961) 'Load bearing capacity and deformation of piled foundations'. *Proc. 5th Int. Conf. ISSMFE*, **2**, pp. 11–15, Paris.

Bertam, G.E. (1940) *An experimental investigation of protective filters*. Harvard Graduate School of Engineering. Pub. No. 267.

Biot, M.A. (1941) 'General theory of three dimensional consolidation'. *Jour. Appl. Phys.*

Bishop, A.W. (1948) 'A new sampling tool for use in cohesionless sands below groundwater level'. *Géotechnique*, **1**, pp. 125–31.

Bishop, A.W. (1954) 'The use of pore-pressure coefficients in practice'. *Géotechnique*, **4**, (2).

Bishop, A.W. (1955) 'The use of the slip circle in the stability analysis of slopes'. *Géotechnique*, **5**, (1).

Bishop, A.W. (1960) 'Opening discussion'. *Proc. Conf. on Pore Pressure and Suction in Soils*. Butterworth, London.

Bishop, A.W. (1966) 'The strength of soils as engineering materials'. *Géotechnique*, **16**, (2).

Bishop, A.W. and Blight, G.E. (1963) 'Some aspects of effective stress in saturated and partially saturated soils'. *Géotechnique*, **13**, (3).

Bishop, A.W. and Henkel, D.J. (1962) *The measurement of soil properties in the triaxial test*. Edward Arnold, London.
Bishop, A.W. and Morgenstern, N. (1960) 'Stability coefficients for earth slopes'. *Géotechnique*, **10**, (4).
Bishop, A.W., Alpan, I., Blight, G.E. and Donald, I.B. (1960) 'Factors controlling the strength of partially saturated cohesive soils'. *Proc. of Research Conf. on Shear Strength of Cohesive Soils*, ASCE, Colorado.
Bishop, A.W., Green, G.E., Garga, V.K., Andresen, A. and Brown, J.D. (1971) 'A new ring shear apparatus and its application to the measurement of residual strength'. *Géotechnique*, **21**, (4).
Black, W.P.M. (1979) *The strength of clay subgrades: its measurement by a penetrometer*. Department of Environment, Department of Transport, TRRL Report LR901. Crowthorne, Berks.
Black, W.P.M. and Lister, N.W. (1979) *The strength of clay fill subgrades: its prediction in relation to road performance*. Department of Environment, Department of Transport, TRRL Report LR889, Crowthorne, Berks.
Boussinesq, J. (1885) *Application des potentiels à l'étude de l'équilibre et de mouvement des solides élastiques*. Gauthier–Villard, Paris.
Bowles, J.E. (1982) *Foundation analysis and design*. (3rd edn) McGraw-Hill Book Co., New York.
Bright, N.J. and Roberts, J.J. (2004) *Students' guide to the Eurocodes*. British Standards Institution, London.
British Standards Institution (1986) *Code of practice for foundations*. BS 8004. London.
British Standards Institution (1990) *British Standard methods of test for soils for civil engineering purposes*. BS 1377. London.
British Standards Institution (1994) *Code of practice for earth retaining structures*. BS 8002. London. Reprinted with revisions, 2001.
British Standards Institution (1995) *Code of practice for strengthened/reinforced soils and other fills*. BS 8006. London.
British Standards Institution (1999) *Code of practice for site investigations*. BS 5930. London.
British Standards Institution (2004) *Eurocode 7: Geotechnical design – Part 1: General rules*. BS EN 1997-1, London.
British Steel (1997) *British Steel piling handbook*. 7th edn. British Steel, Scunthorpe.
Broms, B.B. (1966) 'Methods of calculating the ultimate bearing capacity of piles, a summary'. *Sols–Soils*, **5**, (18 and 19), pp. 21–31.
Broms, B.B. (1971) Lateral earth pressures due to compaction of cohesionless soils. *Proc. 4th Int. Conf. Soil Mech.*, Budapest, pp. 373–384.
Brown, J.D. and Meyerhof, G.C. (1969) 'Experimental study of bearing capacity in layered clays'. *Proc. 7th Int. Conf. ISSMFE*, **2**, pp. 45–51, Mexico City.
Burland, J.B., Potts, D.M. and Walsh, N.M. (1981) 'The overall stability of free and propped embedded cantilever retaining walls', *Ground Engineering*, **14**, (5), London, pp. 28–38.
Cairney, T. (1993) *Contaminated land: Problems and solutions*. Blackie Academic, London.
Campbell, G.S. and Gee, G.W. (1986) 'Water potential: miscellaneous methods'. In *Methods of Soil Analysis. Part 1. Physical and Mineralogical Methods – Agronomy Monograph No. 9*. 2nd edn. American Society of Agronomy–Soil Science Society of America.
Casagrande, A. (1936) 'The determination of the preconsolidation loads and its practical significance'. *Proc. 1st Int. Conf. ISSMFE*, **3**, pp. 60–64, Harvard University, Cambridge, Mass.
Casagrande, A. (1937) 'Seepage through earth dams'. *Jour. New England Water Works Association*, **51**, (2). Reprinted in Harvard University Publ. No. 209 (1937) and in 1940 in *Contributions to soil mechanics 1925–1940*, Boston Society of Civil Engineers, Boston.

Casagrande, A. (1947a) 'Classification and identification of soils'. *Proc. Am. Soc. Civ. Engrs*, 73.

Casagrande, L. (1947b) *The application of electro-osmosis to practical problems in foundations and earthworks*. Department Scientific and Industrial Res., Building Res. Tech. Paper No. 30, HMSO, London.

Charles, J.A. and Burford, D. (1987) 'Settlement and ground water in opencast mining backfills'. *Ground water effects in geotechnical engineering. Proc. 9th Int. Conf. ISSMFE*, Dublin, Vol. 1.

Charles, J.A., Skinner, H.D. and Watts, K.S. (1998) 'The specification of fills to support buildings on shallow foundations: the "95% fixation"'. *Ground Engineering*, January, London.

Clayton, C.R.I. (1993) *Retaining structures*. Proceedings of the conference held by ICE, at Cambridge, July 1992, Thomas Telford, London.

Clayton, C.R.I. (1995) *The standard penetration test (SPT): methods and use*. CIRIA Report 143, London.

Clayton, C.R.I. and Jukes, A.W. (1978) 'A one-point cone penetrometer liquid limit test'. *Géotechnique*, **28**, (4), p. 469.

Clayton, C.R.I. and Symons, I.F. (1992) The pressure of compacted fill on retaining walls (Technical Note). *Géotechnique*, **42**, (1), pp. 127–130.

Clayton, C.R.I., Matthews, M.C. and Simons, N.E. (1995) *Site investigation: A handbook for engineers*. 2nd edn. Blackwell Science, Oxford.

Coffman, B.S. (1960) *Estimating the relative density of sands*. Civ. Engng., New York.

Coleman, J.D. (1962) 'Stress strain relations for partly saturated soil'. Correspondence in *Géotechnique*, **12**, (4).

Coulomb, C.A. (1766) 'Essais sur une application des règles des maxims et minimis à quelques problèmes de statique relatifs à l'architecture'. *Mem. Acad. Roy. Pres. Divers*, Sav. 5,7, Paris.

Crandall, S.H. (1956) *Engineering analysis*. Addison Wesley, Reading, Mass.

Crank, J. and Nicolson, P. (1947) 'A practical method for numerical evaluation of partial differential equations of the heat conduction type'. *Proc. Camb. Phil. Soc. Math. Phys. Sci*.

Croney, D. and Coleman, J.D. (1953) 'Soil moisture suction properties and their bearing on the moisture distribution in soils'. *Proc 3rd Int. Conf. ISSMFE*, Zurich.

Croney, D. and Coleman, J.D. (1958) *Field studies of the movement of soil moisture*. Tech. Paper No. 41, DSIR, London.

Croney, D. and Coleman, J.D. (1960) 'Pore pressure and suction in soil'. *Proc. Conf. on Pore Pressure and Suction in Soils*. Butterworth, London.

Culmann, K. (1866) *Die Graphische Statik, Section 8, Theorie der Stütz und Futtermauern*, Meyer and Zeller, Zurich.

Darcy, H. (1856) *Les fontaines publiques de la ville de Dijon*. Paris.

Dean, T.J., Bell, J.P. and Baty, A.J.B. (1987) 'Soil moisture measurement by an improved capacitance technique, Part 1. Sensor design and performance.' *Jour. Hydrology*, **93**, pp. 67–78.

De Beer, E.E. (1963) 'The scale effect in the transposition of the results of deep sounding tests on the ultimate bearing capacity of piles and caisson foundations'. *Géotechnique*, **13**, (1).

De Beer, E.E. (1970) 'Experimental determination of the shape factor and the bearing capacity factors for sand'. *Géotechnique*, **20**, (4), pp. 387–411.

De Beer, E.E. and Martens, A. (1957) Method of computation of an upper limit for the influence of the homogenity of sand layers in the settlement of bridges. *Proc. 4th Int. Conf. ISSMFE*, **1**, London.

De Cock, F., Legrand, C. and Huybrechts, N. (2003) Axial static pile load test in compression or in tension – Recommendations from ISSMGE subcommittee ERTC 3 – Piles. *Proc. 13th Eur. Conf. Soil Mech. and Geotech. Enging*, **3**, pp. 717–741, Prague.

Department of Transport (1978) *Reinforced earth retaining walls and bridge abutments for embankments*. Technical Memorandum (Bridges) BE 3/78, revised 1987.

Dumbleton, M.J. and West, G. (1970) *Air-photograph interpretation for road engineers in Britain*. Transport and Road Research Laboratory Report LR369, Crowthorne, Berks.

Dunnicliff, J. (1993) *Geotechnical instrumentation for monitoring field performance*. John Wiley & Sons, Chichester.

Escario, V. and Juca, A. (1989) 'Strength and deformation of partly saturated soils'. *Proc. 12th Int. Conf. ISSMFE*, Rio de Janeiro, Vol. 1.

Fadum, R.E. (1948) 'Influence values for estimating stresses in elastic foundations'. *Proc. 2nd Int. Conf. ISSMFE*, Rotterdam, **3**.

Fellenius, W. (1927) *Erdstatische Berechnungen*. Ernst, Berlin.

Fellenius, W. (1936) 'Calculation of stability of earth dams'. *Trans. 2nd Congress on Large Dams*.

Fox, E.N. (1948) 'The mean elastic settlement of a uniformly loaded area and its practical significance'. *Proc. 2nd Int. Conf. ISSMFE*, Rotterdam.

Frank, R., Bauduin, C., Driscoll, R., Kavvadas, M., Krebs Ovesen, N., Orr, T. and Schuppener, B. (2004) *Designer's Guide to EN 1997-1, Eurocode 7: Geotechnical design – general rules*. Thomas Telford, London.

Fredlund, D.G. and Morgenstern, N.R. (1977) 'Stress state variables for unsaturated soils.' *Jour. Geotech. Eng. Div. ASCE*, **103**, (GT5).

Fredlund, D.G. and Rahardjo, H. (1993) *Soil mechanics for unsaturated soils*. John Wiley & Sons, New York.

Fredlund, D.G., Morgenstern, N.R. and Widger, R.A. (1978) 'The shear strength of unsaturated soils'. *Canadian Geotechnical Jour.*, **15**, (3).

Gassler, G. (1990) 'In-situ techniques of reinforced soil'. *Performance of reinforced soil structures*. (Edited by A. McGowan, K.C. Yeo and K.Z. Andrewes.) Held at University of Strathclyde, Thomas Telford, London, pp. 185–196.

Gibbs, H.J. and Holtz, W.G. (1957) 'Research on determining the density of sands by spoon penetration test'. *Proc. 4th Int. Conf. ISSMFE*, **1**, London.

Gibson, R.E. (1958) 'The progress of consolidation in a clay layer increasing in thickness with time'. *Géotechnique*, **8**.

Gibson, R.E. and Lumb, P. (1953) Numerical solution for some problems in the consolidation of clay. *Proc. Inst. Civ. Engrs*, Part 1, London.

Grim, R.E. (1968) *Clay mineralogy*. 2nd Ed. McGraw-Hill Book Co., New York.

Hansen, J.B. (1957) 'Foundation of structures – General report'. *Proc. 4th Int. Conf. ISSMFE*, London.

Hansen, J.B. (1970) 'A revised and extended formula for bearing capacity'. *Danish Geotech. Inst.*, Bulletin 28, Copenhagen.

Harr, M.E. (1962) *Groundwater and seepage*. McGraw-Hill Book Co., New York.

Harris, M. (1994) *Contaminated land: investigation, assessment and remediation*. Thomas Telford, London.

Hazen, A. (1892) 'Some physical properties of sands and gravels with special reference to their use in filtration'. *24th Annual Report*, Mass. State Board of Health, Massachusetts.

Head, K.H. (1992) *Manual of soil laboratory testing*. Vols 1, 2 and 3 (1,238 pps). Pentech Press, London.

Henkel, D.J. (1965) 'The shear strength of saturated remoulded clays'. *Proc. Res. Conf. on Shear Strength of Cohesive Soils*. ASCE, Boulder, Colorado.

Her Majesty's Stationery Office (1952) *Soil mechanics for road engineers*. London.

Highways Agency, Scottish Office Development Department, The Welsh Office and The Department of the Environment for Northern Ireland (1995) *Earthworks: Design and preparation of contract documents, HA 44/91; Design manual for roads and bridges*, Volume 4, Geotechnics and drainage. London.

Highways Agency, Scottish Office Development Department, The Welsh Office and The Department of the Environment for Northern Ireland (1996) *Pavement design and maintenance, HD 24/96; Design manual for roads and bridges*, Volume 7, Pavement design and maintenance. London.

Highways Agency, Scottish Office Development Department, The Welsh Office and The Department of the Environment for Northern Ireland (2004a) *Manual of contract documents for highway works, MCHW 1; Specification for highway works*: Series 600. Earthworks, London.

Highways Agency, Scottish Office Development Department, The Welsh Office and The Department of the Environment for Northern Ireland (2004b) *Manual of contract documents for highway works, MCHW 2; Notes for guidance on the Specification for highway works*: Series NG 600. Earthworks, London.

Hvorslev, M.J. (1937) 'On the strength properties of remoulded cohesive soils'. (In Danish) *Danmarks Naturvidenskabelige Samfund, Ingeniörridenskabelige Skrifter*, Series A, No. 45, Copenhagen.

Hvorslev, M.J. (1949) *Subsurface exploration and sampling of soils for civil engineering purposes*. Waterways Expt. Sta., US Corps of Engineers, Vicksburg.

Hvorslev, M.J. (1965) 'Physical components of the shear strength of saturated clays'. *Proc. Res. Conf. on Shear Strength of Cohesive Soils*. ASCE, Boulder, Colorado.

Institution of Structure Engineers (1951) *Earth retaining structures*. CP2. Institution of Structure Engineers, London.

ISSMFE (1985) *Lexicon in eight languages*. Pub. Int. Soc. for Soil Mechs. and Found. Engrg.

Jaky, J. (1944) 'The coefficient of earth pressure at rest'. *Jour. Soc. Hungarian Architects and Engineers*, **78**, (22).

Janbu, N. (1957) 'Earth pressure and bearing capacity calculations by generalised procedure of slices'. *Proc. 4th Int. Conf. ISSMFE*, London.

Jennings, J.E.B. and Burland, J.B. (1962) Limitations to the use of effective stresses in partially saturated soils. *Géotechnique*, **13**, (2).

Jones, C.J.F.P. (1996) *Earth reinforcement and soil structures*. Revised edn. ASCE Press, New York: Thomas Telford, London.

Jumikis, A.R. (1962) *Soil mechanics*. Van Nostrand, New York, London.

Jürgenson, L. (1934) 'The application of theories of elasticity and plasticity to foundation problems'. *Proc. Boston Soc. Civil Engrs*, Boston.

Kerisel, J. and Absi, E. (1990) *Active and passive earth pressure tables*. 3rd edn. Balkema, Rotterdam.

Kjellman, W., Kallstenius, T. and Wager, O. (1950) 'Soil sampler with metal foils'. *Proc. Royal Swed. Geot. Inst.*, No. 1.

Lake, J.R. (1963) 'A full-scale experiment to determine the effectiveness of vertical sand drains in peat under a road embankment in Dunbartonshire, Scotland'. *Euro. Conf. Soil Mechs and Found. Engng.*, Weisbaden, Germany.

Lambe, T.W. (1964) 'Methods of estimating settlement'. Conf. on the Design of Foundations for Control of Settlements. *Jour. ASCE*, **90**, (SM5).

Lambe, T.W. (1967) 'Stress path method'. *Jour. ASCE*, **93**, (SM6).

Leibmann, G. (1955) 'The solution of transient heat flow and heat transfer problems by relaxation'. *Br. Jour. Appld. Physics*.

Logan, J. (1964) 'Estimating transmissibility from routine production tests of water wells'. *Ground Water*, **2**, 35–37.

Lousberg, M., Calembert, L. *et al.* (1974) 'Penetration testing in Belgium'. State of the art report, Euro. Symp. on Penetration Testing, Swedish Geotech. Soc., *National Swedish Building Research Publication*, **1**, pp. 7–17.

Lumb, P. (1963) 'Rate of settlement of a clay layer due to a gradually applied load'. *Civ. Engine. Pub. Works Review*. London.

Lunne, T., Robertson, P.K. and Powell, J.J.M. (1997) *Cone penetration testing in geotechnical practice*. E & FN Spon, London.

Marshall, T.J. (1958) 'A relation between permeability and size distribution of pores', *Soil Science*, **9**, (8), pp. 1–8.

Meigh, A.C. (1987) *Cone penetration testing methods and interpretation*. CIRIA, Butterworth, London.

Meigh, A.C. and Nixon, I.K. (1961) 'Comparison of *in situ* tests for granular soils'. *Proc. 5th Int. Conf. ISSMFE*, **1**, pp. 449–507, Paris.

Meigh, A.C. and Skipp, B.O. (1960) 'Gamma-ray and neutron methods of measuring soil density and moisture'. *Géotechnique*, **10**, (2).

Meyerhof, G.G. (1951) 'The ultimate bearing capacity of foundations'. *Géotechnique*, **1**, (4), pp. 301–332.

Meyerhof, G.G. (1953) 'The bearing capacity of foundations under eccentric and inclined loads'. *Proc. 3rd Int. Conf. ISSMFE*, Zurich.

Meyerhof, G.G. (1956) 'Penetration tests and bearing capacity of cohesionless soils'. *Proc. ASCE, Jour. Soil. Mech. Found. Div.*, **85**, (SM6), pp. 1–19.

Meyerhof, G.G. (1959) 'Compaction of sands and bearing capacity of piles'. *Proc. ASCE, Jour. Soil Mech. Found. Div.*, **85**, (SM6), pp. 1–30.

Meyerhof, G.G. (1963) 'Some recent research on the bearing capacity of foundations' *Canadian Geotech. Jour.*, **1**, (1), pp. 16–23.

Meyerhof, G.G. (1974) 'State-of-the-art of penetration testing in countries outside Europe'. *Proc. 1st Euro. Symp. on Penetration Testing*, **2**, pp. 40–48, Stockholm.

Meyerhof, G.G. (1976) 'Bearing capacity and settlement of piled foundations'. *Jour. Geotech. Engng. Div.*, ASCE. **102**, (GT3), pp. 195–228.

Mikkelsen, P.E. and Green, G.E. (2003) 'Piezometers in fully grouted boreholes'. *Proc. Symp Field Measurements in Geomech.*, Oslo, Norway.

Muir Wood, D. (1991) *Soil behaviour and critical state soil mechanics*. Cambridge University Press.

Myles, B. and Bridle, R.J. (1991) 'Fired soil nails – the machine'. *Ground Engineering*, July/August, London.

Newmark, N.M. (1942) *Influence charts for computation of stresses in elastic foundations*. University of Illinois Engng Exp. Stn., Bull. No. 338.

O'Connor, M.J. and Mitchell, R.J. (1977) 'An extension to the Bishop and Morgenstern slope stability charts'. *Canadian Geotechnical Jour.*, **14**, 144–151.

Oliphant, J. and Winter, M.G. (1997) Limits of use of the moisture condition apparatus. *Proceedings Institution of Civil Engineers, Transp.*, **123**, 17–29.

Padfield, C.J. and Mair, R.J. (1984) *Design of retaining walls embedded in stiff clay*. CIRIA Report 104, London.

Palmer, D.J. and Stuart, J.G. (1957) 'Some observations on the standard penetration test and a correlation of the test with a new penetrometer'. *Proc. 5th Int. Conf. ISSMFE.*

Parry, R.H.G. (1960) 'Triaxial compression and extension tests on remoulded saturated clay'. *Géotechnique*, **10**.

Parsons, A.W. (1976) *The rapid determination of the moisture condition of earthwork material*. Department of Environment, Department of Transport, TRRL Report LR750, Crowthorne, Berks.

Parsons, A.W. and Boden, J.B. (1979) *The moisture condition test and its potential applications in earthworks*. Department of Environment, Department of Transport, TRRL Report SR522, Crowthorne, Berks.

Penman, A.D.M. (1961) 'A study of the response time of various types of piezometer'. *Proc. Conf. on Pore Pressure and Suction in Soils*. Butterworth, London.

Penman, A.D.M. (1983) 'Latest geotechnical developments relating to embankment dams'. *Ground Engineering*, **16**, (4).

Penman, A.D.M. (1985) 'Tailings dams'. *Ground Engineering*, **18**, (2).

Penman, A.D.M. (1995) 'The effect of gas on measured pore pressures'. *Proc. 1st Int. Conf. on Unsaturated Soils*., Paris, Vol. 1.

Pokharel, G. and Ochiai, T. (1997) 'Design and construction of a new soil nailing (PAN Wall®) method', *Proc 3rd Int. Conf. Ground Improvement Geosytems*. (Edited by M.C.R. Davies and F. Schlosser), London.

Porter, O.J. (1936) 'Studies of fill construction over mud flats including a description of experimental construction using vertical sand drains to hasten stabilisation'. *Proc. Int. Conf. ISSMFE*., Harvard, USA.

Potts, D.M. and Fourie, A.B. (1984) 'The behaviour of a propped retaining wall: Results of a numerical experiment'. *Géotechnique*, **34**, pp. 383–404.

Poulos, H.G. and Davis, E.H. (1974) *Elastic solutions for soil and rock mechanics*. John Wiley & Sons Inc., New York.

Powell, W.D., Potter, J.F., Mayhew, H.C. and Nunn, M.E. (1984) *The structural design of bituminous roads*. Department of Environment, Department of Transport, TRRL Report LR1132, Crowthorne, Berks.

Prandtl, L. (1921) 'Uber die Eindringungsfestigkeit plastischer Baustoffe und die Festigkeit von Schneiden'. *Zeitschrift fur Angewandte Mathematik*, **1**, (1), pp. 15–20.

Rankine, W.J.M. (1857) 'On the stability of loose earth'. *Philosophical Trans. Royal Soc.*, Part 1, **147**, pp. 9–27. London.

RDGC (1991) *Renforcement des sols par clouage: Programme Clouterre*. Projets Nationaux de Recherche Developpement en Génie Civil, Paris.

Ridley, A.M. and Burland, J.B. (1993) 'A new instrument for the measurement of soil moisture suction'. *Géotechnique*, **43**, (2).

Ridley, A.M. and Wray, W.K. (1995) 'Suction measurement: A review of current theory and practices'. *Proc. 1st. Int. Conf. on Unsaturated Soils*. Paris, Vol. 3.

Road Research Laboratory (1952) *Soil mechanics for road engineers*. Department of Scientific and Industrial Research. HMSO. London.

Roscoe, K.H., Schofeld, A.N. and Wroth, C.P. (1958) 'On the yielding of soils', *Géotechnique*, **8**, (1).

Rowe, P.W. (1952) Anchored sheet-pile walls. *Proc. Inst. Civ. Engrs*., Part 1, **1**, pp. 27–70, London.

Rowe, P.W. (1957) Sheet-pile walls in clay. *Proc. Inst. Civ. Engrs*., **7**, pp. 629–654, London.

Rowe, P.W. (1958) 'Measurements in sheet-pile walls driven into clay'. *Proc. Brussels Conf. 58 on Earth Pressure Problems*. Belgian group of ISSMFE, Brussels.

Rowe, P.W. and Barden, L. (1966) 'A new consolidation cell'. *Géotechnique*, **16**, (2).

Schlosser, F. and Long, N.T. (1974) Recent results in French research on reinforced earth. *Jour. Const. Div. Am. Soc. Civ. Engrs*, **100**.

Schlosser F. (1982) 'Behaviour and design of soil nailing'. *Proceedings International Symposium on Recent Developments in Ground Improvement Techniques*, Bangkok, Balkema, pp. 399–413.

Schlosser, F. and de Buhan, P. (1990) 'Theory and design related to the performance of reinforced soil structures, State-of-the-art review'. *Performance of reinforced soil structures.* (Edited by A. McGowan, K.C. Yeo and K.Z. Andrewes.) held at University of Strathclyde, Thomas Telford, London, pp. 1–14.

Schmertmann. J.H. (1953) 'Estimating the true consolidation behaviour of clay from laboratory test results'. *Proc. ASCE*, 120.

Schmertmann, J.H. (1970) 'Static cone to compute settlement over sand'. *Proc. ASCE, Jour. Soil Mech. Found. Div.*, **96** (SM3), pp. 1001–1043.

Schmertmann, J.H., Hartman, I.P. and Brown, P.R. (1978) 'Improved strain influence factor diagrams'. *Proc. ASCE, Jour. Geotech. Engng. Div.*, **104** (GT8), pp. 1131–1135.

Schofield, A.N. and Wroth, C.P. (1968) *Critical state soil mechanics.* McGraw-Hill, London.

Scott, R.F. (1963) *Principles of soil mechanics.* Addison Wesley, Reading, Mass.

Skempton, A.W. (1951) 'The bearing capacity of clays'. *Proc. Building Research Congress*, pp. 180–189, London.

Skempton, A.W. (1953) 'The colloidal activity of clays'. *Proc. 3rd Int. Conf. ISSMFE* **1**, Zurich.

Skempton, A.W. (1954) 'The pore pressure coefficient A and B'. *Géotechnique*, **4**, (4).

Skempton, A.W. (1960) 'Effective stress in soils, concrete and rocks'. *Proc. Conf. on Pore Pressure and Suction in Soils.* Butterworth, London.

Skempton, A.W. (1964) 'Long term stability of slopes'. *Géotechnique*, **14**, (2).

Skempton, A.W. and Bjerrum, L. (1957) 'A contribution to settlement analysis of foundations on clay'. *Géotechnique*, **7**, (4), pp. 168–178.

Skempton, A.W. and Northley, A. (1952) 'The sensitivity of clays'. *Géotechnique*, **2**, (2).

Skempton, A.W., Peck, R.V. and MacDonald, D.H. (1955) 'Settlement analysis of six structures in Chicago and London'. *Proc. Inst. Civil Engrs.*

Smith, G.N. (1968a) 'Determining the settlements of embankments on soft clay'. *Highways and Pub. Works*, London.

Smith, G.N. (1968b) 'Construction pore pressures in an earth dam'. *Civ. Engng. Publ. Wks. Review.* London.

Smith, G.N. (1971) *An introduction to matrix and finite element methods in civil engineering.* Applied Science Publishers, London.

Smith, I.G.N., Oliphant, J. and Wallis, S.G. (1997) 'Field validation of a computer model for forecasting mean weekly *in situ* moisture condition value.' *The Electronic Jour. Geotechnical Engineering*, Issue 2, http://geotech.civen.okstate.edu/ejge/ppr9701/index.htm.

Smoltczyk, U. (1985) Axial pile loading test – Part 1: static loading. *Geotechnical Testing Journal*, **8**, pp. 79–90.

Sokolovski, V.V. (1960) *Statics of soils media.* (Trans. by D.H. Jones and A.N. Schofield.) Butterworth & Co. Ltd., London.

Stannard, D.I. (1992) 'Tensiometers – theory, construction and use'. *Geotechnical Testing Jour.*, **15**, (1).

Steinbrenner, W. (1934) *Tafeln zur Setzungsberechnung.* Die Strasse, **1**, pp. 121–124.

Sultan, H.A. and Seed, H.B. (1967) 'Stability of sloping core earth dams'. *Proc. Am. Soc. Civ. Engrs*, **93**, (SM4).

Sutherland, H.B. (1963) 'The use of *in situ* tests to estimate the allowable bearing pressures of cohesionless soils'. *Structural Engineer*, **41**, (3), pp. 85–92.

Swartzendruber, D. (1961) 'Modification of Darcy's law for the flow of water in soils'. *Soil Science*, **93**. London.

Taylor, D.W. (1948) *Fundamentals of soil mechanics.* John Wiley & Sons Inc., New York; Chapman & Hall, London.

Terzaghi, K. (1925) *Erdbeaumechanik auf bodenphysikalischer grundlage.* Deuticke, Vienna.

Terzaghi, K. (1943) *Theoretical soil mechanics.* John Wiley, London and New York.

Terzaghi, K. and Peck, R.B. (1948) *Soil mechanics in engineering practice.* Chapman and Hall, London; John Wiley & Sons Inc., New York.

Terzaghi, K., Peck, R.B. and Mesri, G. (1996) *Soil mechanics in engineering practice.* 3rd edn., John Wiley, New York.

Thorburn, S. (1963) 'Tentative correction chart for the standard penetration test in non-cohesive soils'. *Civ. Engng. Public Works Rev.*

Timoshenko, S.P. and Goodier, J.N. (1951) *Theory of plasticity.* 2nd edn. McGraw-Hill, New York.

Topp, G.C. and Davis, J.L. (1985) 'Measurement of soil water content using time-domain reflectometry (TDR): a field evaluation'. *Soil Science Society of America Jour.*, **49**, pp. 19–24.

Trenter, N.A. and Charles, J.A. (1996) 'A model specification for engineered fills for building purposes'. *Proceedings Institution of Civil Engineers, Geotechnical Engineering*, **119**, (4), pp. 219–230.

Tsagareli, Z.V. (1967) *New methods of light weight wall construction.* (In Russian.) Stroiizdat, Moscow.

Vesic, A.S. (1963) *Bearing capacity of deep foundations in sand.* Soil Mechanics Lab. Report, Georgia Inst. of Technology.

Vesic, A.S. (1970) 'Tests on instrumented piles, Ogeechee River site'. *Jour. Soil Mechs, and Found Div.*, ASCE, **96**, (SM2), pp. 561–584.

Vesic, A.S. (1973) 'Analysis of ultimate loads of shallow foundations'. *Jour. Soil Mechs, and Found. Div.*, ASCE, **99**, (SMI), pp. 45–73.

Vesic, A.S. (1975) 'Bearing capacity of shallow foundations'. Chapter 3 in *Foundation Engng. Handbook*, H.F. Winterkorn & H.Y. Fang, Van Nostrand, Reinhold Co., New York.

Vidal, H. (1966) 'La terre armée'. *Annales de l'Institut Technique du Bâtiment et des Travaux Publics*, **19**, (223 and 224), France.

Wheeler, S.J. and Karube, D. (1995) 'State of the art report: constitutive modelling.' *Proc. 1st Int. Conf. Unsaturated Soils.* Paris, Vol. 3.

Wheeler, S.J. and Sivakumar, V. (1995) 'An elasto-plastic critical state framework for unsaturated soil'. *Géotechnique*, **45**, (1), pp. 35–53.

Whitaker, T. (1957) 'Experiments with model piles in groups'. *Géotechnique*, 7, pp. 147–167.

Wilson, G. (1941) 'The calculation of the bearing capacity of footings on clay'. *Jour. Inst. Civ. Engrs*, London.

Wise, W.R. (1992) 'A new insight on pore structure and permeability'. *Water Resources Research*, **28**, (2), pp. 189–198.

Index